Subsystems of Second Order Arithmetic
Second Edition

Foundations of mathematics is the study of the most basic concepts and logical structure of mathematics, with an eye to the unity of human knowledge. Almost all of the problems studied in this book are motivated by an overriding foundational question: What are the appropriate axioms for mathematics? Through a series of case studies, these axioms are examined to prove particular theorems in core mathematical areas such as algebra, analysis, and topology, focusing on the language of second order arithmetic, the weakest language rich enough to express and develop the bulk of mathematics.

In many cases, if a mathematical theorem is proved from appropriately weak set existence axioms, then the axioms will be logically equivalent to the theorem. Furthermore, only a few specific set existence axioms arise repeatedly in this context, which in turn correspond to classical foundational programs. This is the theme of reverse mathematics, which dominates the first half of the book. The second part focuses on models of these and other subsystems of second order arithmetic. Additional results are presented in an appendix.

Stephen G. Simpson is a mathematician and professor at Pennsylvania State University. The winner of the Grove Award for Interdisciplinary Research Initiation, Simpson specializes in research involving mathematical logic, foundations of mathematics, and combinatorics.

PERSPECTIVES IN LOGIC

The *Perspectives in Logic* series publishes substantial, high-quality books whose central theme lies in any area or aspect of logic. Works that present new material not available in book form are particularly welcome. The series ranges from introductory texts suitable for beginning graduate courses to specialized monographs at the frontiers of research. Each book offers an illuminating perspective for its intended audience.

The series has its origins in the old *Perspectives in Mathematical Logic* series edited by the Ω-Group for "Mathematische Logik" of the Heidelberger Akademie der Wissenchaften, whose beginnings date back to the 1960s. The Association for Symbolic Logic has assumed editorial responsibility for the series and changed its name to reflect its interest in books that span the full range of disciplines in which logic plays an important role.

For more information, see http://www.aslonline.org/books_perspectives.html

PERSPECTIVES IN LOGIC

Subsystems of Second Order Arithmetic

Second Edition

STEPHEN G. SIMPSON

Pennsylvania State University

ASSOCIATION FOR SYMBOLIC LOGIC

CAMBRIDGE UNIVERSITY PRESS
Cambridge, New York, Melbourne, Madrid, Cape Town, Singapore,
São Paulo, Delhi, Dubai, Tokyo

Cambridge University Press
32 Avenue of the Americas, New York NY 10013-2473, USA

Published in the United States of America by Cambridge University Press, New York

www.cambridge.org
Information on this title: www.cambridge.org/9780521150149

Association for Symbolic Logic
Richard A. Shore, Publisher
Department of Mathematics, Cornell University, Ithaca NY 14853
www.aslonline.org

First published 2009
This digitally printed version 2010

A catalogue record for this publication is available from the British Library

Library of Congress Cataloging in Publication Data

Simpson, Stephen G. (Stephen George), 1945–
Subsystems of second order arithmetic / Stephen G. Simpson. –2nd ed.
    p. cm. – (Perspectives in logic ; 1)
Includes bibliographical references and index.
ISBN 978-0-521-88439-6 (hardback)
1. Predicate calculus. I. Title. II. Series.
        QA9.7.S537   2009
    511.3–dc22   2008052364

ISBN 978-0-521-88439-6 Hardback
ISBN 978-0-521-15014-9 Paperback

CONTENTS

Part B. Models of Subsystems of Z_2

LIST OF TABLES

PREFACE

Foundations of mathematics is the study of the most basic concepts and logical structure of mathematics, with an eye to the unity of human knowledge. Among the most basic mathematical concepts are: number, shape, set, function, algorithm, mathematical axiom, mathematical definition, and mathematical proof. Typical questions in foundations of mathematics include: What is a number? What is a shape? What is a set? What is a function? What is an algorithm? What is a mathematical axiom? What is a mathematical definition? What is a mathematical proof? What are the most basic concepts of mathematics? What is the logical structure of mathematics? What are the appropriate axioms for numbers? What are the appropriate axioms for shapes? What are the appropriate axioms for sets? What are the appropriate axioms for functions?

Obviously, foundations of mathematics is a subject of the greatest mathematical and philosophical importance. Beyond this, foundations of mathematics is a rich subject with a long history, going back to Aristotle and Euclid and continuing in the hands of outstanding modern figures such as Descartes, Cauchy, Weierstraß, Dedekind, Peano, Frege, Russell, Cantor, Hilbert, Brouwer, Weyl, von Neumann, Skolem, Tarski, Heyting, and Gödel. An excellent reference for the modern era in foundations of mathematics is van Heijenoort [272].

In the late 19th and early 20th centuries, virtually all leading mathematicians were intensely interested in foundations of mathematics and spoke and wrote extensively on this subject. Today that is no longer the case. Regrettably, foundations of mathematics is now out of fashion. Today, most of the leading mathematicians are ignorant of foundations and focus mostly on structural questions. Today, foundations of mathematics is out of favor even among mathematical logicians, the majority of whom prefer to concentrate on methodological or other non-foundational issues.

This book is a contribution to foundations of mathematics. Almost all of the problems studied in this book are motivated by an overriding foundational question: *What are the appropriate axioms for mathematics?* We undertake a series of case studies to discover which are the appropriate axioms for proving particular theorems in core mathematical areas such as algebra, analysis, and topology. We focus on the language of second order arithmetic, because that language is the weakest one that is rich enough to express and develop the bulk of core mathematics. It turns out that, in many particular cases, if a mathematical theorem is proved from appropriately weak set existence axioms, then the axioms will be logically equivalent to the theorem. Furthermore, only a few specific set existence axioms arise repeatedly in this context: recursive comprehension, weak König's lemma, arithmetical comprehension, arithmetical transfinite recursion, Π_1^1 comprehension; corresponding to the formal systems RCA_0, WKL_0, ACA_0, ATR_0, Π_1^1-CA_0; which in turn correspond to classical foundational programs: constructivism, finitistic reductionism, predicativism, and predicative reductionism. This is the theme of Reverse Mathematics, which dominates Part A of this book. Part B focuses on models of these and other subsystems of second order arithmetic. Additional results are presented in an appendix.

The formalization of mathematics within second order arithmetic goes back to Dedekind and was developed by Hilbert and Bernays [115, supplement IV]. The present book may be viewed as a continuation of Hilbert/Bernays [115]. I hope that the present book will help to revive the study of foundations of mathematics and thereby earn for itself a permanent place in the history of the subject.

The first edition of this book [249] was published in January 1999. The second edition differs from the first only in that I have corrected some typographical errors and updated some bibliographical entries. Recent advances are in research papers by numerous authors, published in *Reverse Mathematics* 2001 [228] and in scholarly journals. The Web page for this book is http://www.math.psu.edu/simpson/sosoa/. I would like to develop this Web page into a forum for research and scholarship, not only in subsystems of second order arithmetic, but in foundations of mathematics generally.

Stephen G. Simpson
November 2008

ACKNOWLEDGMENTS

Much of my work on subsystems of second order arithmetic has been carried on in collaboration with my doctoral and postdoctoral advisees at Berkeley and Penn State, including: Stephen Binns, Stephen Brackin, Douglas Brown, Natasha Dobrinen, Qi Feng, Fernando Ferreira, Mariagnese Giusto, Kostas Hatzikiriakou, Jeffry Hirst, James Humphreys, Michael Jamieson, Alberto Marcone, Carl Mummert, Ju Rao, Rick Smith, John Steel, Kazuyuki Tanaka, Robert Van Wesep, Galen Weitkamp, Takeshi Yamazaki, and Xiaokang Yu. I also acknowledge the collaboration and encouragement of numerous colleagues including: Peter Aczel, Jeremy Avigad, Jon Barwise, Michael Beeson, Errett Bishop, Andreas Blass, Lenore Blum, Douglas Bridges, Wilfried Buchholz, John Burgess, Samuel Buss, Douglas Cenzer, Peter Cholak, Chi-Tat Chong, Rolando Chuaqui, John Clemens, Peter Clote, Carlos Di Prisco, Rod Downey, Ali Enayat, Herbert Enderton, Harvey Friedman, Robin Gandy, William Gasarch, Noam Greenberg, Petr Hájek, Valentina Harizanov, Victor Harnik, Leo Harrington, Christoph Heinatsch, Ward Henson, Peter Hinman, Denis Hirschfeldt, William Howard, Martin Hyland, Gerhard Jäger, Haim Judah, Irving Kaplansky, Alexander Kechris, Jerome Keisler, Jeffrey Ketland, Bjørn Kjos-Hanssen, Stephen Kleene, Julia Knight, Ulrich Kohlenbach, Roman Kossak, Georg Kreisel, Antonín Kučera, Masahiro Kumabe, Richard Laver, Steffen Lempp, Manuel Lerman, Azriel Lévy, Alain Louveau, Angus Macintyre, Michael Makkai, Richard Mansfield, David Marker, Donald Martin, Adrian Mathias, Alex McAllister, Kenneth McAloon, Timothy McNicholl, George Metakides, Joseph Mileti, Joseph Miller, Grigori Mints, Michael Möllerfeld, Antonio Montalban, Yiannis Moschovakis, Gert Müller, Roman Murawski, Jan Mycielski, Michael Mytilinaios, Anil Nerode, Andre Nies, Charles Parsons, Marian Pour-El, Michael Rathjen, Jeffrey Remmel, Jean-Pierre Ressayre, Ian Richards, Hartley Rogers, Gerald Sacks, Ramez Sami, Andre Scedrov, James Schmerl, Kurt Schütte, Helmut Schwichtenberg, Dana Scott, Saharon Shelah, John Shepherdson, Naoki Shioji, Joseph Shoenfield, Richard Shore, Wilfried Sieg, Jack Silver, Ksenija Simic, Theodore Slaman, Craig Smoryński, Robert Soare, Reed Solomon, Robert Solovay,

Rick Sommer, Andrea Sorbi, Gaisi Takeuti, Dirk van Dalen, Lou van den Dries, Daniel Velleman, Stanley Wainer, Dongping Yang, Yue Yang, and especially Solomon Feferman, Carl Jockusch, and Wolfram Pohlers.

I acknowledge the help of various institutions including: the Alfred P. Sloan Foundation, the American Mathematical Society, the Association for Symbolic Logic, the Centre National de Recherche Scientifique, the Deutsche Forschungsgemeinschaft, the National Science Foundation, the Omega Group, Oxford University, the Pennsylvania State University, the Raymond N. Shibley Foundation, the Science Research Council, Springer-Verlag, Stanford University, the University of California at Berkeley, the University of Illinois at Urbana–Champaign, the University of Munich, the University of Paris, the University of Tennessee, and the Volkswagen Foundation.

A preliminary version of this book was written with a software package called MathText. I acknowledge important help from Robert Huff, the author of MathText, and Janet Huff. Padma Raghavan wrote additional software to help me convert the manuscript from MathText to LaTeX. The first edition [249] was published by Springer-Verlag with editorial assistance from the Association for Symbolic Logic. The second edition is being published by the Association for Symbolic Logic and Cambridge University Press. I acknowledge help from Samuel Buss, Ward Henson, Reinhard Kahle, Steffen Lempp, Manuel Lerman, and Thanh-Ha LeThi.

I thank my darling wife, Padma Raghavan, for her encouragement and emotional support while I was bringing this project to a conclusion, both in 1997–1998 and again in 2008.

Stephen G. Simpson
November 2008

Chapter I

INTRODUCTION

I.1. The Main Question

The purpose of this book is to use the tools of mathematical logic to study certain problems in foundations of mathematics. We are especially interested in the question of which set existence axioms are needed to prove the known theorems of mathematics.

The scope of this initial question is very broad, but we can narrow it down somewhat by dividing mathematics into two parts. On the one hand there is set-theoretic mathematics, and on the other hand there is what we call "non-set-theoretic" or "ordinary" mathematics. By *set-theoretic mathematics* we mean those branches of mathematics that were created by the set-theoretic revolution which took place approximately a century ago. We have in mind such branches as general topology, abstract functional analysis, the study of uncountable discrete algebraic structures, and of course abstract set theory itself.

We identify as *ordinary* or *non-set-theoretic* that body of mathematics which is prior to or independent of the introduction of abstract set-theoretic concepts. We have in mind such branches as geometry, number theory, calculus, differential equations, real and complex analysis, countable algebra, the topology of complete separable metric spaces, mathematical logic, and computability theory.

The distinction between set-theoretic and ordinary mathematics corresponds roughly to the distinction between "uncountable mathematics" and "countable mathematics". This formulation is valid if we stipulate that "countable mathematics" includes the study of possibly uncountable complete separable metric spaces. (A metric space is said to be separable if it has a countable dense subset.) Thus for instance the study of continuous functions of a real variable is certainly part of ordinary mathematics, even though it involves an uncountable algebraic structure, namely the real number system. The point is that in ordinary mathematics, the real line partakes of countability since it is always viewed as a separable metric space, never as being endowed with the discrete topology.

1

In this book we want to restrict our attention to ordinary, non-set-theoretic mathematics. The reason for this restriction is that the set existence axioms which are needed for set-theoretic mathematics are likely to be much stronger than those which are needed for ordinary mathematics. Thus our broad set existence question really consists of two subquestions which have little to do with each other. Furthermore, while nobody doubts the importance of strong set existence axioms in set theory itself and in set-theoretic mathematics generally, the role of set existence axioms in ordinary mathematics is much more problematical and interesting.

We therefore formulate our *Main Question* as follows: *Which set existence axioms are needed to prove the theorems of ordinary, non-set-theoretic mathematics?*

In any investigation of the Main Question, there arises the problem of choosing an appropriate language and appropriate set existence axioms. Since in ordinary mathematics the objects studied are almost always countable or separable, it would seem appropriate to consider a language in which countable objects occupy center stage. For this reason, we study the Main Question in the context of the language of second order arithmetic. This language is denoted L_2 and will be described in the next section. All of the set existence axioms which we consider in this book will be expressed as formulas of the language L_2.

I.2. Subsystems of Z_2

In this section we define Z_2, the formal system of second order arithmetic. We also introduce the concept of a subsystem of Z_2.

The *language of second order arithmetic* is a two-sorted language. This means that there are two distinct sorts of variables which are intended to range over two different kinds of object. Variables of the first sort are known as *number variables*, are denoted by $i, j, k, m, n, \ldots$, and are intended to range over the set $\omega = \{0, 1, 2, \ldots\}$ of all natural numbers. Variables of the second sort are known as *set variables*, are denoted by $X, Y, Z, \ldots$, and are intended to range over all subsets of ω.

The terms and formulas of the language of second order arithmetic are as follows. *Numerical terms* are number variables, the constant symbols 0 and 1, and $t_1 + t_2$ and $t_1 \cdot t_2$ whenever t_1 and t_2 are numerical terms. Here $+$ and $\cdot$ are binary operation symbols intended to denote addition and multiplication of natural numbers. (Numerical terms are intended to denote natural numbers.) *Atomic formulas* are $t_1 = t_2$, $t_1 < t_2$, and $t_1 \in X$ where t_1 and t_2 are numerical terms and X is any set variable. (The intended meanings of these respective atomic formulas are that t_1 equals t_2, t_1 is less than t_2, and t_1 is an element of X.) *Formulas* are built up from atomic formulas by means of propositional connectives $\land, \lor, \neg, \rightarrow,$

$\leftrightarrow$ (and, or, not, implies, if and only if), *number quantifiers* $\forall n, \exists n$ (for all n, there exists n), and *set quantifiers* $\forall X, \exists X$ (for all X, there exists X). A *sentence* is a formula with no free variables.

DEFINITION I.2.1 (language of second order arithmetic). L_2 is defined to be the language of second order arithmetic as described above.

In writing terms and formulas of L_2, we shall use parentheses and brackets to indicate grouping, as is customary in mathematical logic textbooks. We shall also use some obvious abbreviations. For instance, $2 + 2 = 4$ stands for $(1 + 1) + (1 + 1) = ((1 + 1) + 1) + 1$, $(m + n)^2 \notin X$ stands for $\neg((m + n) \cdot (m + n) \in X)$, $s \leq t$ stands for $s < t \vee s = t$, and $\varphi \wedge \psi \wedge \theta$ stands for $(\varphi \wedge \psi) \wedge \theta$.

The semantics of the language L_2 are given by the following definition.

DEFINITION I.2.2 (L_2-structures). A *model for* L_2, also called a *structure for* L_2 or an L_2-*structure*, is an ordered 7-tuple

$$M = (|M|, S_M, +_M, \cdot_M, 0_M, 1_M, <_M),$$

where $|M|$ is a set which serves as the range of the number variables, S_M is a set of subsets of $|M|$ serving as the range of the set variables, $+_M$ and $\cdot_M$ are binary operations on $|M|$, 0_M and 1_M are distinguished elements of $|M|$, and $<_M$ is a binary relation on $|M|$. We always assume that the sets $|M|$ and S_M are disjoint and nonempty. Formulas of L_2 are interpreted in M in the obvious way.

In discussing a particular model M as above, it is useful to consider formulas with parameters from $|M| \cup S_M$. We make the following slightly more general definition.

DEFINITION I.2.3 (parameters). Let B be any subset of $|M| \cup S_M$. By a *formula with parameters from* B we mean a formula of the extended language $L_2(B)$. Here $L_2(B)$ consists of L_2 augmented by new constant symbols corresponding to the elements of B. By a *sentence with parameters from* B we mean a sentence of $L_2(B)$, i.e., a formula of $L_2(B)$ which has no free variables.

In the language $L_2(|M| \cup S_M)$, constant symbols corresponding to elements of S_M (respectively $|M|$) are treated syntactically as unquantified set variables (respectively unquantified number variables). Sentences and formulas with parameters from $|M| \cup S_M$ are interpreted in M in the obvious way. A set $A \subseteq |M|$ is said to be *definable over M allowing parameters from* B if there exists a formula $\varphi(n)$ with parameters from B and no free variables other than n such that

$$A = \{a \in |M| : M \models \varphi(a)\}.$$

Here $M \models \varphi(a)$ means that M *satisfies* $\varphi(a)$, i.e., $\varphi(a)$ is true in M.

We now discuss some specific L_2-structures. The *intended model* for L_2 is of course the model

$$(\omega, P(\omega), +, \cdot, 0, 1, <)$$

where ω is the set of natural numbers, $P(\omega)$ is the set of all subsets of ω, and $+, \cdot, 0, 1, <$ are as usual. By an ω-*model* we mean an L_2-structure of the form

$$(\omega, S, +, \cdot, 0, 1, <)$$

where $\emptyset \neq S \subseteq P(\omega)$. Thus an ω-model differs from the intended model only by having a possibly smaller collection S of sets to serve as the range of the set variables. We sometimes speak of the ω-*model* S when we really mean the ω-model $(\omega, S, +, \cdot, 0, 1, <)$. In some parts of this book we shall be concerned with a special class of ω-models known as β-models. This class will be defined in §I.5.

We now present the formal system of second order arithmetic.

DEFINITION I.2.4 (second order arithmetic). The *axioms of second order arithmetic* consist of the universal closures of the following L_2-formulas:

(i) basic axioms:
$$n + 1 \neq 0$$
$$m + 1 = n + 1 \rightarrow m = n$$
$$m + 0 = m$$
$$m + (n + 1) = (m + n) + 1$$
$$m \cdot 0 = 0$$
$$m \cdot (n + 1) = (m \cdot n) + m$$
$$\neg m < 0$$
$$m < n + 1 \leftrightarrow (m < n \vee m = n)$$

(ii) induction axiom:
$$(0 \in X \wedge \forall n \, (n \in X \rightarrow n + 1 \in X)) \rightarrow \forall n \, (n \in X)$$

(iii) comprehension scheme:
$$\exists X \, \forall n \, (n \in X \leftrightarrow \varphi(n))$$

where $\varphi(n)$ is any formula of L_2 in which X does not occur freely.

Intuitively, the given instance of the comprehension scheme says that there exists a set $X = \{n : \varphi(n)\}$ = the set of all n such that $\varphi(n)$ holds. This set is said to be *defined by* the given formula $\varphi(n)$. For example, if $\varphi(n)$ is the formula $\exists m \, (m + m = n)$, then this instance of the comprehension scheme asserts the existence of the set of even numbers.

In the comprehension scheme, $\varphi(n)$ may contain free variables in addition to n. These free variables may be referred to as *parameters* of this instance of the comprehension scheme. Such terminology is in harmony with definition I.2.3 and the discussion following it. For example, taking $\varphi(n)$ to be the formula $n \notin Y$, we have an instance of comprehension,

$$\forall Y \, \exists X \, \forall n \, (n \in X \leftrightarrow n \notin Y),$$

asserting that for any given set Y there exists a set $X =$ the complement of Y. Here the variable Y plays the role of a parameter.

Note that an L_2-structure M satisfies I.2.4(iii), the comprehension scheme, if and only if S_M contains all subsets of $|M|$ which are definable over M allowing parameters from $|M| \cup S_M$. In particular, the comprehension scheme is valid in the intended model. Note also that the basic axioms I.2.4(i) and the induction axiom I.2.4(ii) are valid in any ω-model. In fact, any ω-model satisfies the full *second order induction scheme*, i.e., the universal closure of

$$(\varphi(0) \wedge \forall n \, (\varphi(n) \rightarrow \varphi(n+1))) \rightarrow \forall n \, \varphi(n),$$

where $\varphi(n)$ is any formula of L_2. In addition, the second order induction scheme is valid in any model of I.2.4(ii) plus I.2.4(iii).

By *second order arithmetic* we mean the formal system in the language L_2 consisting of the axioms of second order arithmetic, together with all formulas of L_2 which are deducible from those axioms by means of the usual logical axioms and rules of inference. The formal system of second order arithmetic is also known as Z_2, for obvious reasons, or Π^1_∞-CA_0, for reasons which will become clear in §I.5.

In general, a *formal system* is defined by specifying a language and some axioms. Any formula of the given language which is logically deducible from the given axioms is said to be a *theorem* of the given formal system. At all times we assume the usual logical rules and axioms, including equality axioms and the law of the excluded middle.

This book will be largely concerned with certain specific subsystems of second order arithmetic and the formalization of ordinary mathematics within those systems. By a *subsystem of* Z_2 we mean of course a formal system in the language L_2 each of whose axioms is a theorem of Z_2. When introducing a new subsystem of Z_2, we shall specify the axioms of the system by writing down some formulas of L_2. The axioms are then taken to be the universal closures of those formulas.

If T is any subsystem of Z_2, a *model of* T is any L_2-structure satisfying the axioms of T. By Gödel's completeness theorem applied to the two-sorted language L_2, we have the following important principle: A given L_2-sentence σ is a theorem of T if and only if all models of T satisfy σ. An *ω-model of* T is of course any ω-model which satisfies the axioms of T, and similarly a *β-model of* T is any β-model satisfying the axioms of T. Chapters VII, VIII, and IX of this book constitute a thorough study of models of subsystems of Z_2. Chapter VII is concerned with β-models, chapter VIII is concerned with ω-models other than β-models, and chapter IX is concerned with models other than ω-models.

All of the subsystems of Z_2 which we shall consider consist of the basic axioms I.2.4(i), the induction axiom I.2.4(ii), and some set existence axioms. The various subsystems will differ from each other only with respect to their set existence axioms. Recall from §I.1 that our Main Question concerns the role of set existence axioms in ordinary mathematics. Thus, a principal theme of this book will be the formal development of specific portions of ordinary mathematics within specific subsystems of Z_2. We shall see that subsystems of Z_2 provide a setting in which the Main Question can be investigated in a precise and fruitful way. Although Z_2 has infinitely many subsystems, it will turn out that only a handful of them are useful in our study of the Main Question.

Notes for §I.2. The formal system Z_2 of second order arithmetic was introduced in Hilbert/Bernays [115] (in an equivalent form, using a somewhat different language and axioms). The development of a portion of ordinary mathematics within Z_2 is outlined in Supplement IV of Hilbert/Bernays [115]. The present book may be regarded as a continuation of the research begun by Hilbert and Bernays.

I.3. The System ACA_0

The previous section contained generalities about subsystems of Z_2. The purpose of this section is to introduce a particular subsystem of Z_2 which is of central importance, namely ACA_0.

In our designation ACA_0, the acronym ACA stands for arithmetical comprehension axiom. This is because ACA_0 contains axioms asserting the existence of any set which is arithmetically definable from given sets (in a sense to be made precise below). The subscript 0 denotes restricted induction. This means that ACA_0 does not include the full second order induction scheme (as defined in §I.2). We assume only the induction axiom I.2.4(ii).

We now proceed to the definition of ACA_0.

DEFINITION I.3.1 (arithmetical formulas). A formula of L_2, or more generally a formula of $L_2(|M| \cup S_M)$ where M is any L_2-structure, is said to be *arithmetical* if it contains no set quantifiers, i.e., all of the quantifiers appearing in the formula are number quantifiers.

Note that arithmetical formulas of L_2 may contain free set variables, as well as free and bound number variables and number quantifiers. Arithmetical formulas of $L_2(|M| \cup S_M)$ may additionally contain set parameters and number parameters, i.e., constant symbols denoting fixed elements of $\mathcal{S}_M$ and $|M|$ respectively.

Examples of arithmetical formulas of L$_2$ are

$$\forall n\, (n \in X \rightarrow \exists m\, (m + m = n)),$$

asserting that all elements of the set X are even, and

$$\forall m\, \forall k\, (n = m \cdot k \rightarrow (m = 1 \vee k = 1)) \wedge n > 1 \wedge n \in X,$$

asserting that n is a prime number and is an element of X. An example of a non-arithmetical formula is

$$\exists Y\, \forall n\, (n \in X \leftrightarrow \exists i\, \exists j\, (i \in Y \wedge j \in Y \wedge i + n = j))$$

asserting that X is the set of differences of elements of some set Y.

DEFINITION I.3.2 (arithmetical comprehension). The *arithmetical comprehension scheme* is the restriction of the comprehension scheme I.2.4(iii) to arithmetical formulas $\varphi(n)$. Thus we have the universal closure of

$$\exists X\, \forall n\, (n \in X \leftrightarrow \varphi(n))$$

whenever $\varphi(n)$ is a formula of L$_2$ which is arithmetical and in which X does not occur freely. ACA$_0$ is the subsystem of Z$_2$ whose axioms are the arithmetical comprehension scheme, the induction axiom I.2.4(ii), and the basic axioms I.2.4(i).

Note that an L$_2$-structure

$$M = (|M|, \mathcal{S}_M, +_M, \cdot_M, 0_M, 1_M, <_M)$$

satisfies the arithmetical comprehension scheme if and only if $\mathcal{S}_M$ contains all subsets of $|M|$ which are definable over M by arithmetical formulas with parameters from $|M| \cup \mathcal{S}_M$. Thus, a model of ACA$_0$ is any such L$_2$-structure which in addition satisfies the induction axiom and the basic axioms.

An easy consequence of the arithmetical comprehension scheme and the induction axiom is the *arithmetical induction scheme*:

$$(\varphi(0) \wedge \forall n\, (\varphi(n) \rightarrow \varphi(n + 1))) \rightarrow \forall n\, \varphi(n)$$

for all L$_2$-formulas $\varphi(n)$ which are arithmetical. Thus any model of ACA$_0$ is also a model of the arithmetical induction scheme. (Note however that ACA$_0$ does not include the second order induction scheme, as defined in §I.2.)

REMARK I.3.3 (first order arithmetic). We wish to remark that there is a close relationship between ACA$_0$ and first order arithmetic. Let L$_1$ be the *language of first order arithmetic*, i.e., L$_1$ is just L$_2$ with the set variables omitted. *First order arithmetic* is the formal system Z$_1$ whose language is L_1 and whose axioms are the basic axioms I.2.4(i) plus the *first order induction scheme*:

$$(\varphi(0) \wedge \forall n \, (\varphi(n) \rightarrow \varphi(n+1))) \rightarrow \forall n \, \varphi(n)$$

for all L_1-formulas $\varphi(n)$. In the literature of mathematical logic, first order arithmetic is sometimes known as *Peano arithmetic*, PA. By the previous paragraph, every theorem of Z_1 is a theorem of ACA_0. In model-theoretic terms, this means that for any model $(|M|, S_M, +_M, \cdot_M, 0_M, 1_M, <_M)$ of ACA_0, its first order part $(|M|, +_M, \cdot_M, 0_M, 1_M, <_M)$ is a model of Z_1. In §IX.1 we shall prove a converse to this result: Given a model

$$(|M|, +_M, \cdot_M, 0_M, 1_M, <_M) \tag{1}$$

of first order arithmetic, we can find $S_M \subseteq P(|M|)$ such that

$$(|M|, S_M, +_M, \cdot_M, 0_M, 1_M, <_M)$$

is a model of ACA_0. (Namely, we can take $S_M = \mathrm{Def}(M) =$ the set of all $A \subseteq |M|$ such that A is definable over (1) allowing parameters from $|M|$.) It follows that, for any L_1-sentence σ, σ is a theorem of ACA_0 if and only if σ is a theorem of Z_1. In other words, ACA_0 is a *conservative extension* of first order arithmetic. This may also be expressed by saying that Z_1, or equivalently PA, is the *first order part* of ACA_0. For details, see §IX.1.

REMARK I.3.4 (ω-models of ACA_0). Assuming familiarity with some basic concepts of recursive function theory, we can characterize the ω-models of ACA_0 as follows. $S \subseteq P(\omega)$ is an ω-model of ACA_0 if and only if

(i) $S \neq \emptyset$;
(ii) $A \in S$ and $B \in S$ imply $A \oplus B \in S$;
(iii) $A \in S$ and $B \leq_T A$ imply $B \in S$;
(iv) $A \in S$ implies $\mathrm{TJ}(A) \in S$.

(This result is proved in §VIII.1.)

Here $A \oplus B$ is the *recursive join* of A and B, defined by

$$A \oplus B = \{2n : n \in A\} \cup \{2n+1 : n \in B\}.$$

$B \leq_T A$ means that B is *Turing reducible* to A, i.e., B is *recursive in* A, i.e., the characteristic function of B is computable assuming an oracle for the characteristic function of A. $\mathrm{TJ}(A)$ denotes the *Turing jump* of A, i.e., the complete recursively enumerable set relative to A.

In particular, ACA_0 has a minimum (i.e., unique smallest) ω-model, namely

$$\mathrm{ARITH} = \{A \in P(\omega) : \exists n \in \omega \, (A \leq_T \mathrm{TJ}(n, \emptyset))\},$$

where $\mathrm{TJ}(n, X)$ is defined inductively by $\mathrm{TJ}(0, X) = X$, $\mathrm{TJ}(n+1, X) = \mathrm{TJ}(\mathrm{TJ}(n, X))$. More generally, given a set $B \in P(\omega)$, there is a unique smallest ω-model of ACA_0 containing B, consisting of all sets which are arithmetical in B. (For $A, B \in P(\omega)$, we say that A is *arithmetical in B* if $A \leq_T \mathrm{TJ}(n, B)$ for some $n \in \omega$. This is equivalent to saying that A is definable in some or any ω-model $(\omega, \mathcal{S}, +, \cdot, 0, 1, <)$, $B \in \mathcal{S} \subseteq P(\omega)$, by an arithmetical formula with B as a parameter.)

Models of ACA$_0$ are discussed further in §§VIII.1, IX.1, and IX.4. The development of ordinary mathematics within ACA$_0$ is discussed in §I.4 and in chapters II, III, and IV.

Notes for §I.3. By remark I.3.3, the system ACA$_0$ is closely related to first order arithmetic. First order arithmetic is one of the best known and most studied formal systems in the literature of mathematical logic. See for instance Hilbert/Bernays [115], Mendelson [185, chapter 3], Takeuti [261, chapter 2], Shoenfield [222, chapter 8], Hájek/Pudlák [100], and Kaye [137]. By remark I.3.4, ω-models of ACA$_0$ are closely related to basic concepts of recursion theory such as relative recursiveness, the Turing jump operator, and the arithmetical hierarchy. For an introduction to these concepts, see for instance Rogers [208, chapters 13–15], Shoenfield [222, chapter 7], Cutland [43], or Lerman [161, chapters I–III].

I.4. Mathematics within ACA$_0$

The formal system ACA$_0$ was introduced in the previous section. We now outline the development of certain portions of ordinary mathematics within ACA$_0$. The material presented in this section will be restated and greatly refined and extended in chapters II, III, and IV. The present discussion is intended as a partial preview of those chapters.

If X and Y are set variables, we use $X = Y$ and $X \subseteq Y$ as abbreviations for the formulas $\forall n (n \in X \leftrightarrow n \in Y)$ and $\forall n (n \in X \to n \in Y)$ respectively.

Within ACA$_0$, we define $\mathbb{N}$ to be the unique set X such that $\forall n (n \in X)$. (The existence of this set follows from arithmetical comprehension applied to the formula $\varphi(n) \equiv n = n$.) Thus, in any model

$$M = (|M|, \mathcal{S}_M, +_M, \cdot_M, 0_M, 1_M, <_M)$$

of ACA$_0$, $\mathbb{N}$ denotes $|M|$, the set of natural numbers in the sense of M, and we have $|M| \in \mathcal{S}_M$. We shall distinguish between $\mathbb{N}$ and ω, reserving ω to denote the set of natural numbers in the sense of "the real world," i.e., the metatheory in which we are working, whatever that metatheory might be.

Within ACA$_0$, we define a *numerical pairing function* by

$$(m, n) = (m + n)^2 + m.$$

Within ACA$_0$ we can prove that, for all $m, n, i, j \in \mathbb{N}$, $(m, n) = (i, j)$ if and only if $m = i$ and $n = j$. Moreover, using arithmetical comprehension, we can prove that for all sets $X, Y \subseteq \mathbb{N}$, there exists a set $X \times Y \subseteq \mathbb{N}$ consisting of all (m, n) such that $m \in X$ and $n \in Y$. In particular we have $\mathbb{N} \times \mathbb{N} \subseteq \mathbb{N}$.

For $X, Y \subseteq \mathbb{N}$, a *function* $f : X \to Y$ is defined to be a set $f \subseteq X \times Y$ such that for all $m \in X$ there is exactly one $n \in Y$ such that $(m, n) \in f$. For $m \in X$, $f(m)$ is defined to be the unique n such that $(m, n) \in f$. The usual properties of such functions can be proved in ACA_0. In particular, we have *primitive recursion*. This means that, given $f : X \to Y$ and $g : \mathbb{N} \times X \times Y \to Y$, there is a unique $h : \mathbb{N} \times X \to Y$ defined by $h(0, m) = f(m)$, $h(n+1, m) = g(n, m, h(n, m))$ for all $n \in \mathbb{N}$ and $m \in X$. The existence of h is proved by arithmetical comprehension, and the uniqueness of h is proved by arithmetical induction. (For details, see §II.3.) In particular, we have the *exponential function* $\exp(m, n) = m^n$, defined by $m^0 = 1$, $m^{n+1} = m^n \cdot m$ for all $m, n \in \mathbb{N}$. The usual properties of the exponential function can be proved in ACA_0.

In developing ordinary mathematics within ACA_0, our first major task is to set up the *number systems*, i.e., the natural numbers, the integers, the rational number system, and the real number system.

The natural number system is essentially already given to us by the language and axioms of ACA_0. Thus, within ACA_0, a *natural number* is defined to be an element of $\mathbb{N}$, and the *natural number system* is defined to be the structure $\mathbb{N}, +_\mathbb{N}, \cdot_\mathbb{N}, 0_\mathbb{N}, 1_\mathbb{N}, <_\mathbb{N}, =_\mathbb{N}$, where $+_\mathbb{N} : \mathbb{N} \times \mathbb{N} \to \mathbb{N}$ is defined by $m +_\mathbb{N} n = m + n$, etc. (Thus for instance $+_\mathbb{N}$ is the set of triples $((m, n), k) \in (\mathbb{N} \times \mathbb{N}) \times \mathbb{N}$ such that $m + n = k$. The existence of this set follows from arithmetical comprehension.) This means that, when we are working within any particular model $M = (|M|, \mathcal{S}_M, +_M, \cdot_M, 0_M, 1_M, <_M)$ of ACA_0, a natural number is any element of $|M|$, and the role of the natural number system is played by $|M|, +_M, \cdot_M, 0_M, 1_M, <_M, =_M$. (Here $=_M$ is the identity relation on $|M|$.)

Basic properties of the natural number system, such as uniqueness of prime power decomposition, can be proved in ACA_0 using arithmetical induction. (Here one can follow the usual development within first order arithmetic, as presented in textbooks of mathematical logic. Alternatively, see chapter II.)

In order to define the set $\mathbb{Z}$ of *integers* within (any model of) ACA_0, we first use arithmetical comprehension to prove the existence of an equivalence relation $\equiv_\mathbb{Z} \subseteq (\mathbb{N} \times \mathbb{N}) \times (\mathbb{N} \times \mathbb{N})$ defined by $(m, n) \equiv_\mathbb{Z} (i, j)$ if and only if $m + j = n + i$. We then use arithmetical comprehension again, this time with $\equiv_\mathbb{Z}$ as a parameter, to prove the existence of the set $\mathbb{Z}$ consisting of all $(m, n) \in \mathbb{N} \times \mathbb{N}$ such that that (m, n) is the minimum element of its equivalence class with respect to $\equiv_\mathbb{Z}$. (Here minimality is taken with respect to $<_\mathbb{N}$, using the fact that $\mathbb{N} \times \mathbb{N}$ is a subset of $\mathbb{N}$. Thus $\mathbb{Z}$ consists of one element of each $\equiv_\mathbb{Z}$-equivalence class.) Define $+_\mathbb{Z} : \mathbb{Z} \times \mathbb{Z} \to \mathbb{Z}$ by letting $(m, n) +_\mathbb{Z} (i, j)$ be the unique element

of $\mathbb{Z}$ such that $(m, n) +_{\mathbb{Z}} (i, j) \equiv_{\mathbb{Z}} (m + i, n + j)$. Here again arithmetical comprehension is used to prove the existence of $+_{\mathbb{Z}}$. Similarly,
define $-_{\mathbb{Z}} \colon \mathbb{Z} \to \mathbb{Z}$ by $-_{\mathbb{Z}}(m, n) \equiv_{\mathbb{Z}} (n, m)$, and define $\cdot_{\mathbb{Z}} \colon \mathbb{Z} \times \mathbb{Z} \to \mathbb{Z}$ by
$(m, n) \cdot_{\mathbb{Z}} (i, j) \equiv_{\mathbb{Z}} (mi + nj, mj + ni)$. Let $0_{\mathbb{Z}} = (0, 0)$ and $1_{\mathbb{Z}} = (1, 0)$.
Define a relation $<_{\mathbb{Z}} \subseteq \mathbb{Z} \times \mathbb{Z}$ by letting $(m, n) <_{\mathbb{Z}} (i, j)$ if and only if
$m + j < n + i$. Finally, let $=_{\mathbb{Z}}$ be the identity relation on $\mathbb{Z}$. This com-
pletes our definition of the system of integers within ACA$_0$. We can prove
within ACA$_0$ that the system $\mathbb{Z}, +_{\mathbb{Z}}, -_{\mathbb{Z}}, \cdot_{\mathbb{Z}}, 0_{\mathbb{Z}}, 1_{\mathbb{Z}}, <_{\mathbb{Z}}, =_{\mathbb{Z}}$ has the usual
properties of an ordered integral domain, the Euclidean property, etc.

In a similar manner, we can define within ACA$_0$ the set of *rational
numbers*, $\mathbb{Q}$. Let $\mathbb{Z}^+ = \{a \in \mathbb{Z} \colon 0 <_{\mathbb{Z}} a\}$ be the set of positive integers,
and let $\equiv_{\mathbb{Q}}$ be the equivalence relation on $\mathbb{Z} \times \mathbb{Z}^+$ defined by $(a, b) \equiv_{\mathbb{Q}}$
(c, d) if and only if $a \cdot_{\mathbb{Z}} d = b \cdot_{\mathbb{Z}} c$. Then $\mathbb{Q}$ is defined to be the set
of all $(a, b) \in \mathbb{Z} \times \mathbb{Z}^+$ such that (a, b) is the $<_{\mathbb{N}}$-minimum element of
its $\equiv_{\mathbb{Q}}$-equivalence class. Operations $+_{\mathbb{Q}}, -_{\mathbb{Q}}, \cdot_{\mathbb{Q}}$ on $\mathbb{Q}$ are defined by
$(a, b) +_{\mathbb{Q}} (c, d) \equiv_{\mathbb{Q}} (a \cdot_{\mathbb{Z}} d +_{\mathbb{Z}} b \cdot_{\mathbb{Z}} c, b \cdot_{\mathbb{Z}} d)$, $-_{\mathbb{Q}}(a, b) \equiv_{\mathbb{Q}} (-_{\mathbb{Z}}a, b)$, and
$(a, b) \cdot_{\mathbb{Q}} (c, d) \equiv_{\mathbb{Q}} (a \cdot_{\mathbb{Z}} c, b \cdot_{\mathbb{Z}} d)$. We let $0_{\mathbb{Q}} \equiv_{\mathbb{Q}} (0_{\mathbb{Z}}, 1_{\mathbb{Z}})$ and $1_{\mathbb{Q}} \equiv_{\mathbb{Q}} (1_{\mathbb{Z}}, 1_{\mathbb{Z}})$,
and we define a binary relation $<_{\mathbb{Q}}$ on $\mathbb{Q}$ by letting $(a, b) <_{\mathbb{Q}} (c, d)$ if and
only if $a \cdot_{\mathbb{Z}} d <_{\mathbb{Z}} b \cdot_{\mathbb{Z}} c$. Finally $=_{\mathbb{Q}}$ is the identity relation on $\mathbb{Q}$. We can
then prove within ACA$_0$ that the rational number system $\mathbb{Q}, +_{\mathbb{Q}}, -_{\mathbb{Q}}, \cdot_{\mathbb{Q}},$
$0_{\mathbb{Q}}, 1_{\mathbb{Q}}, <_{\mathbb{Q}}, =_{\mathbb{Q}}$ has the usual properties of an ordered field, etc.

We make the usual identifications whereby $\mathbb{N}$ is regarded as a subset
of $\mathbb{Z}$ and $\mathbb{Z}$ is regarded as a subset of $\mathbb{Q}$. (Namely $m \in \mathbb{N}$ is identified
with $(m, 0) \in \mathbb{Z}$, and $a \in \mathbb{Z}$ is identified with $(a, 1_{\mathbb{Z}}) \in \mathbb{Q}$.) We use $+$
ambiguously to denote $+_{\mathbb{N}}, +_{\mathbb{Z}}$, or $+_{\mathbb{Q}}$ and similarly for $-, \cdot, 0, 1, <$. For
$q, r \in \mathbb{Q}$ we write $q - r = q + (-r)$, and if $r \neq 0$, $q/r =$ the unique $q' \in \mathbb{Q}$
such that $q = q' \cdot r$. The function $\exp(q, a) = q^a$ for $q \in \mathbb{Q} \setminus \{0\}$ and
$a \in \mathbb{Z}$ is obtained by primitive recursion in the obvious way. The *absolute
value* function $|\,| \colon \mathbb{Q} \to \mathbb{Q}$ is defined by $|q| = q$ if $q \geq 0$, $-q$ otherwise.

REMARK I.4.1. The idea behind our definitions of $\mathbb{Z}$ and $\mathbb{Q}$ within ACA$_0$
is that $(m, n) \in \mathbb{N} \times \mathbb{N}$ corresponds to the integer $m - n$, while $(a, b) \in$
$\mathbb{Z} \times \mathbb{Z}^+$ corresponds to the rational number a/b. Our treatment of $\mathbb{Z}$ and
$\mathbb{Q}$ is similar to the classical Dedekind construction. The major difference
is that we define $\mathbb{Z}$ and $\mathbb{Q}$ to be sets of representatives of the equivalence
classes of $\equiv_{\mathbb{Z}}$ and $\equiv_{\mathbb{Q}}$ respectively, while Dedekind uses the equivalence
classes themselves. Our reason for using representatives is that we are
limited to the language of second order arithmetic, while Dedekind was
working in a richer set-theoretic context.

A *sequence of rational numbers* is defined to be a function $f \colon \mathbb{N} \to \mathbb{Q}$.
We denote such a sequence as $\langle q_n \colon n \in \mathbb{N} \rangle$, or simply $\langle q_n \rangle$, where $q_n =$
$f(n)$. Similarly, a *double sequence of rational numbers* is a function $f \colon \mathbb{N} \times$
$\mathbb{N} \to \mathbb{Q}$, denoted $\langle q_{mn} \colon m, n \in \mathbb{N} \rangle$ or simply $\langle q_{mn} \rangle$, where $q_{mn} = f(m, n)$.

DEFINITION I.4.2 (real numbers). Within ACA_0, a *real number* is defined to be a Cauchy sequence of rational numbers, i.e., a sequence of rational numbers $x = \langle q_n : n \in \mathbb{N} \rangle$ such that

$$\forall \epsilon \, (\epsilon > 0 \rightarrow \exists m \, \forall n \, (m < n \rightarrow |q_m - q_n| < \epsilon)).$$

(But see remark I.4.4 below.) Here ϵ ranges over $\mathbb{Q}$. If $x = \langle q_n \rangle$ and $y = \langle q_n' \rangle$ are real numbers, we write $x =_{\mathbb{R}} y$ to mean that $\lim_n |q_n - q_n'| = 0$, i.e.,

$$\forall \epsilon \, (\epsilon > 0 \rightarrow \exists m \, \forall n \, (m < n \rightarrow |q_n - q_n'| < \epsilon)),$$

and we write $x <_{\mathbb{R}} y$ to mean that

$$\exists \epsilon \, (\epsilon > 0 \wedge \exists m \, \forall n \, (m < n \rightarrow q_n + \epsilon < q_n')).$$

Also $x +_{\mathbb{R}} y = \langle q_n + q_n' \rangle$, $x \cdot_{\mathbb{R}} y = \langle q_n \cdot q_n' \rangle$, $-_{\mathbb{R}} x = \langle -q_n \rangle$, $0_{\mathbb{R}} = \langle 0 \rangle$, $1_{\mathbb{R}} = \langle 1 \rangle$.

Informally, we use $\mathbb{R}$ to denote the set of all real numbers. Thus $x \in \mathbb{R}$ means that x is a real number. (Formally, we cannot speak of the set $\mathbb{R}$ within the language of second order arithmetic, since it is a set of sets.) We shall usually omit the subscript $\mathbb{R}$ in $+_{\mathbb{R}}, -_{\mathbb{R}}, \cdot_{\mathbb{R}}, 0_{\mathbb{R}}, 1_{\mathbb{R}}, <_{\mathbb{R}}, =_{\mathbb{R}}$. Thus the *real number system* consists of $\mathbb{R}, +, -, \cdot, 0, 1, <, =$. We shall sometimes identify a rational number $q \in \mathbb{Q}$ with the corresponding real number $x_q = \langle q \rangle$.

REMARK I.4.3. Note that we have not attempted to select elements of the $=_{\mathbb{R}}$-equivalence classes. The reason is that there is no convenient way to do so in ACA_0. Instead, we must accustom ourselves to the fact that $=$ on $\mathbb{R}$ (i.e., $=_{\mathbb{R}}$) is an equivalence relation other than the identity relation. This will not cause any serious difficulties.

REMARK I.4.4. The above definition of the real number system is similar but not identical to the one which we shall actually use in our detailed discussion of ordinary mathematics within ACA_0, chapters II through IV. The reason for the discrepancy is that the above definition, while suitable for use in ACA_0 and intuitively appealing, is not suitable for use in weaker systems such as RCA_0. (RCA_0 will be introduced in §§I.7 and I.8 below.) The definition used for the detailed development is slightly less natural, but it has the advantage of working smoothly in weaker systems. In any case, the two definitions are equivalent over ACA_0, equivalent in the sense that the two versions of the real number system which they define can be proved in ACA_0 to be isomorphic.

Within ACA_0 one can prove that the real number system has the usual properties of an Archimedean ordered field, etc. The *complex numbers* can be introduced as usual as pairs of real numbers. Within ACA_0, it is straightforward to carry out the proofs of all the basic results in real and complex linear and polynomial algebra. For example, the fundamental theorem of algebra can be proved in ACA$_0$.

A *sequence of real numbers* is defined to be a double sequence of rational numbers $\langle q_{mn} : m, n \in \mathbb{N} \rangle$ such that for each m, $\langle q_{mn} : n \in \mathbb{N} \rangle$ is a real number. Such a sequence of real numbers is denoted $\langle x_m : m \in \mathbb{N} \rangle$, where $x_m = \langle q_{mn} : n \in \mathbb{N} \rangle$. Within ACA$_0$ we can prove that every bounded sequence of real numbers has a least upper bound. This is a very useful completeness property of the real number system. For instance, it implies that an infinite series of positive terms is convergent if and only if the finite partial sums are bounded. (Stronger completeness properties for the most part cannot be proved in ACA$_0$.)

We now turn to abstract algebra within ACA$_0$. Because of the restriction to the language of second order arithmetic, we cannot expect to obtain a good general theory of arbitrary (countable and uncountable) algebraic structures. However, we can develop *countable algebra*, i.e., the theory of countable algebraic structures, within ACA$_0$.

For instance, a *countable commutative ring* is defined within ACA$_0$ to be a structure $R, +_R, -_R, \cdot_R, 0_R, 1_R$, where $R \subseteq \mathbb{N}$, $+_R : R \times R \to R$, etc., and the usual commutative ring axioms are assumed. (We include $0 \neq 1$ among those axioms.) The subscript R is usually omitted. (An example is the ring of integers, $\mathbb{Z}, +_\mathbb{Z}, -_\mathbb{Z}, \cdot_\mathbb{Z}, 0_\mathbb{Z}, 1_\mathbb{Z}$, which was introduced above.) An *ideal* in R is a set $I \subseteq R$ such that $a \in I$ and $b \in I$ imply $a + b \in I$, $a \in I$ and $r \in R$ imply $a \cdot r \in I$, and $0 \in I$ and $1 \notin I$. We define an equivalence relation $=_I$ on R by $r =_I s$ if and only if $r - s \in I$. We let R/I be the set of $r \in R$ such that r is the $<_\mathbb{N}$-minimum element of its equivalence class under $=_I$. Thus R/I consists of one element of each $=_I$-equivalence class of elements of R. With the appropriate operations, R/I becomes a countable commutative ring, the quotient ring of R by I. The ideal I is said to be *prime* if R/I is an integral domain, and *maximal* if R/I is a field. With these definitions, the countable case of many basic results of commutative algebra can be proved in ACA$_0$. See §§III.5 and IV.6.

Other countable algebraic structures, e.g., countable groups, can be defined and discussed in a similar manner, within ACA$_0$. Countable fields are discussed in §§II.9, IV.4 and IV.5, and countable vector spaces are discussed in §III.4. It turns out that part of the theory of countable Abelian groups can be developed in ACA$_0$, but other parts of the theory require stronger systems. See §§III.6, V.7 and VI.4.

Next we indicate how some basic concepts and results of analysis and topology can be developed within ACA$_0$.

DEFINITION I.4.5 (complete separable metric spaces). Within ACA$_0$, a (code for a) *complete separable metric space* is a nonempty set $A \subseteq \mathbb{N}$ together with a function $d : A \times A \to \mathbb{R}$ satisfying $d(a, a) = 0$, $d(a, b) =$

$d(b, a) \geq 0$, and $d(a, c) \leq d(a, b) + d(b, c)$ for all $a, b, c \in A$. (Formally, d is a sequence of real numbers, indexed by $A \times A$.) We define a *point of the complete separable metric space* $\widehat{A}$ to be a sequence $x = \langle a_n : n \in \mathbb{N} \rangle$, $a_n \in A$, satisfying

$$\forall \epsilon \, (\epsilon > 0 \rightarrow \exists m \, \forall n \, (m < n \rightarrow d(a_m, a_n) < \epsilon)).$$

The pseudometric d is extended from A to $\widehat{A}$ by

$$d(x, y) = \lim_n d(a_n, b_n)$$

where $x = \langle a_n : n \in \mathbb{N} \rangle$ and $y = \langle b_n : n \in \mathbb{N} \rangle$. We write $x = y$ if and only if $d(x, y) = 0$.

For example, $\mathbb{R} = \widehat{\mathbb{Q}}$ under the metric $d(q, q') = |q - q'|$.

The idea of the above definition is that a complete separable metric space $\widehat{A}$ is presented by specifying a countable dense set A together with the restriction of the metric to A. Then $\widehat{A}$ is defined as the completion of A under the restricted metric. Just as in the case of the real number system, several difficulties arise from the circumstance that ACA_0 is formalized in the language of second order arithmetic. First, there is no variable or term that can denote the set of all points in $\widehat{A}$ (although we can use notations such as $x \in \widehat{A}$, meaning that x is a point of $\widehat{A}$). Second, equality for points of $\widehat{A}$ is an equivalence relation other than the identity relation. These difficulties are minor and do not seriously affect the content of the mathematical development concerning complete separable metric spaces within ACA_0. They only affect the outward form of that development. A more important limitation is that, in the language of second order arithmetic, we cannot speak at all about nonseparable metric spaces. This remark is related to our remarks in §I.1 about set-theoretic versus "ordinary" or non-set-theoretic mathematics.

DEFINITION I.4.6 (continuous functions). Within ACA_0, if $\widehat{A}$ and $\widehat{B}$ are complete separable metric spaces, a (code for a) *continuous function* $\phi : \widehat{A} \to \widehat{B}$ is a set $\Phi \subseteq A \times \mathbb{Q}^+ \times B \times \mathbb{Q}^+$ satisfying the following coherence conditions:

1. $(a, r, b, s) \in \Phi$ and $(a, r, b', s') \in \Phi$ imply $d(b, b') < s + s'$;
2. $(a, r, b, s) \in \Phi$ and $d(b, b') + s < s'$ imply $(a, r, b', s') \in \Phi$;
3. $(a, r, b, s) \in \Phi$ and $d(a, a') + r' < r$ imply $(a', r', b, s) \in \Phi$.

Here a' ranges over A, b' ranges over B, and r' and s' range over

$$\mathbb{Q}^+ = \{q \in \mathbb{Q} : q > 0\},$$

the positive rational numbers. In addition we require: for all $x \in \widehat{A}$ and $\epsilon > 0$ there exists $(a, r, b, s) \in \Phi$ such that $d(a, x) < r$ and $s < \epsilon$.

We can prove in ACA_0 that for all $x \in \widehat{A}$ there exists $y \in \widehat{B}$ such that $d(b, y) \leq s$ for all $(a, r, b, s) \in \Phi$ such that $d(a, x) < r$. This y is unique up to equality of points in $\widehat{B}$, and we define $\phi(x) = y$. It can be shown that $x = x'$ implies $\phi(x) = \phi(x')$.

The idea of the above definition is that $(a, r, b, s) \in \Phi$ is a *neighborhood condition* giving us a piece of information about the continuous function $\phi \colon \widehat{A} \to \widehat{B}$. Namely, $(a, r, b, s) \in \Phi$ tells us that for all $x \in \widehat{A}$, $d(x, a) < r$ implies $d(\phi(x), b) \leq s$. The code Φ consists of sufficiently many neighborhood conditions so as to determine $\phi(x) \in \widehat{B}$ for all $x \in \widehat{A}$.

Taking $\widehat{A} = \mathbb{R}^n$ and $\widehat{B} = \mathbb{R}$ in the above definition, we obtain a concept of continuous real-valued function of n real variables. Using this, the theory of differential and integral equations, calculus of variations, etc., can be developed as usual, within ACA$_0$. For instance, the Ascoli lemma can be proved in ACA$_0$ and then used to obtain the Peano existence theorem for solutions of ordinary differential equations (see §§III.2 and IV.8).

DEFINITION I.4.7 (open sets). Within ACA$_0$, let $\widehat{A}$ be a complete separable metric space. A (code for an) *open set* in $\widehat{A}$ is any set $U \subseteq A \times \mathbb{Q}^+$. For $x \in \widehat{A}$ we write $x \in U$ if and only if $d(x, a) < r$ for some $(a, r) \in U$.

The idea of definition I.4.7 is that $(a, r) \in A \times \mathbb{Q}^+$ is a code for a *neighborhood* or *basic open set* $B(a, r)$ in $\widehat{A}$. Here $x \in B(a, r)$ if and only if $d(a, x) < r$. An open set U is then defined as a union of basic open sets.

With definitions I.4.6 and I.4.7, the usual proofs of fundamental topological results can be carried out within ACA$_0$, for the case of complete separable metric spaces. For instance, the Baire category theorem and the Tietze extension theorem go through in this setting (see §§II.5, II.6, and II.7).

A *separable Banach space* is defined within ACA$_0$ to be a complete separable metric space $\widehat{A}$ arising from a countable pseudonormed vector space A over the rational field $\mathbb{Q}$. For example, let $A = \mathbb{Q}[x]$ be the ring of polynomials in one variable x over $\mathbb{Q}$. With the metric

$$d(f, g) = \left[\int_0^1 |f(x) - g(x)|^p \, dx \right]^{1/p},$$

$1 \leq p < \infty$, we have $\widehat{A} = \mathrm{L}_p[0, 1]$. Similarly, with the metric

$$d(f, g) = \sup_{0 \leq x \leq 1} |f(x) - g(x)|,$$

we have $\widehat{A} = \mathrm{C}[0, 1]$. As suggested by these examples, the basic theory of separable Banach and Frechet spaces can be developed formally within ACA$_0$. In particular, the Hahn/Banach theorem, the open mapping theorem, and the Banach/Steinhaus uniform boundedness principle can be proved in this setting (see §§II.10, IV.9, X.2).

Remark I.4.8. As in remark I.4.4, the above definitions of complete separable metric space, continuous function, open set, and separable Banach space are not the ones which we shall actually use in our detailed development in chapters II, III, and IV. However, the two sets of definitions are equivalent in ACA_0.

Notes for §I.4. The observation that a great deal of ordinary mathematics can be developed formally within a system something like ACA_0 goes back to Weyl [274]; see also definition X.3.2. See also Takeuti [260] and Zahn [281].

I.5. Π_1^1-CA_0 and Stronger Systems

In this section we introduce Π_1^1-CA_0 and some other subsystems of Z_2. These systems are much stronger than ACA_0.

Definition I.5.1 (Π_1^1 formulas). A formula φ is said to be Π_1^1 if it is of the form $\forall X\,\theta$, where X is a set variable and θ is an arithmetical formula. A formula φ is said to be Σ_1^1 if it is of the form $\exists X\,\theta$, where X is a set variable and θ is an arithmetical formula.

More generally, for $0 \le k \in \omega$, a formula φ is said to be Π_k^1 if it is of the form

$$\forall X_1\,\exists X_2\,\forall X_3 \cdots X_k\,\theta,$$

where $X_1,\dots,X_k$ are set variables and θ is an arithmetical formula. A formula φ is said to be Σ_k^1 if it is of the form

$$\exists X_1\,\forall X_2\,\exists X_3 \cdots X_k\,\theta,$$

where $X_1,\dots,X_k$ are set variables and θ is an arithmetical formula. In both cases, φ consists of k alternating set quantifiers followed by a formula with no set quantifiers. In the Π_k^1 case, the first set quantifier is universal, while in the Σ_k^1 case it is existential (assuming $k \ge 1$). Thus for instance a Π_2^1 formula is of the form $\forall X\,\exists Y\,\theta$, and a Σ_2^1 formula is of the form $\exists X\,\forall Y\,\theta$, where θ is arithmetical. A Π_0^1 or Σ_0^1 formula is the same thing as an arithmetical formula.

The equivalences $\neg\forall X\,\varphi \equiv \exists X\,\neg\varphi$, $\neg\exists X\,\varphi \equiv \forall X\,\neg\varphi$, and $\neg\neg\varphi \equiv \varphi$ imply that any Π_k^1 formula is logically equivalent to the negation of a Σ_k^1 formula, and vice versa. Moreover, using Π_k^1 (respectively Σ_k^1) to denote the class of formulas logically equivalent to a Π_k^1 formula (respectively a Σ_k^1 formula), we have

$$\Pi_k^1 \cup \Sigma_k^1 \subseteq \Pi_{k+1}^1 \cap \Sigma_{k+1}^1$$

for all $k \in \omega$. (This is proved by introducing dummy quantifiers.)

The hierarchy of L$_2$-formulas Π^1_k, $k \in \omega$, is closely related to the projective hierarchy in descriptive set theory.

DEFINITION I.5.2 (Π^1_1 and Π^1_k comprehension). Π^1_1-CA$_0$ is the subsystem of Z$_2$ whose axioms are the basic axioms I.2.4(i), the induction axiom I.2.4(ii), and the comprehension scheme I.2.4(iii) restricted to L$_2$-formulas $\varphi(n)$ which are Π^1_1. Thus we have the universal closure of

$$\exists X \, \forall n \, (n \in X \leftrightarrow \varphi(n))$$

for all Π^1_1 formulas $\varphi(n)$ in which X does not occur freely.

The systems Π^1_k-CA$_0$, $k \in \omega$, are defined similarly, with Π^1_k replacing Π^1_1. In particular Π^1_0-CA$_0$ is just ACA$_0$, and for all $k \in \omega$ we have

$$\Pi^1_k\text{-CA}_0 \subseteq \Pi^1_{k+1}\text{-CA}_0.$$

It is also clear that

$$Z_2 = \bigcup_{k \in \omega} \Pi^1_k\text{-CA}_0.$$

For this reason, Z$_2$ is sometimes denoted Π^1_∞-CA$_0$.

It would be possible to introduce systems Σ^1_k-CA$_0$, $k \in \omega$, but they would be superfluous, because a simple argument shows that Σ^1_k-CA$_0$ and Π^1_k-CA$_0$ are equivalent, i.e., they have the same theorems.

[Namely, given a Σ^1_k formula $\varphi(n)$, there is a logically equivalent formula $\neg\psi(n)$ where $\psi(n)$ is Π^1_k. Reasoning within Π^1_k-CA$_0$ and applying Π^1_k comprehension, we see that there exists a set Y such that

$$\forall n \, (n \in Y \leftrightarrow \psi(n)).$$

Applying arithmetical comprehension with Y as a parameter, there exists a set X such that

$$\forall n \, (n \in X \leftrightarrow n \notin Y).$$

Then clearly

$$\forall n \, (n \in X \leftrightarrow \varphi(n)).$$

This shows that all the axioms of Σ^1_k-CA$_0$ are theorems of Π^1_k-CA$_0$. The converse is proved similarly.]

We now discuss models of Π^1_k-CA$_0$, $1 \leq k \leq \infty$.

As explained in §I.3 above, ACA$_0$ has a minimum ω-model, and this model is very natural from both the recursion-theoretic and the model-theoretic points of view. It is therefore reasonable to ask about minimum ω-models of Π^1_k-CA$_0$. It turns out that, for $1 \leq k \leq \infty$, there is no minimum (or even minimal) ω-model of Π^1_k-CA$_0$. These negative results will be proved in §VIII.6. However, we can obtain a positive result by considering β-models instead of ω-models. The relevant definition is as follows.

DEFINITION I.5.3 (β-models). A β-*model* is an ω-model $\mathcal{S} \subseteq P(\omega)$ with the following property. If σ is any Π_1^1 or Σ_1^1 sentence with parameters from $\mathcal{S}$, then $(\omega, \mathcal{S}, +, \cdot, 0, 1, <)$ satisfies σ if and only if the intended model

$$(\omega, P(\omega), +, \cdot, 0, 1, <)$$

satisfies σ.

If T is any subsystem of Z_2, a β-*model of* T is any β-model satisfying the axioms of T. Chapter VII is a thorough study of β-models of subsystems of Z_2.

REMARK I.5.4 (β-models of Π_1^1-CA$_0$). For readers who are familiar with some basic concepts of hyperarithmetical theory, the β-models of Π_1^1-CA$_0$ can be characterized as follows. $\mathcal{S} \subseteq P(\omega)$ is a β-model of Π_1^1-CA$_0$ if and only if

  (i) $\mathcal{S} \neq \emptyset$;
  (ii) $A \in \mathcal{S}$ and $B \in \mathcal{S}$ imply $A \oplus B \in \mathcal{S}$;
  (iii) $A \in \mathcal{S}$ and $B \leq_H A$ imply $B \in \mathcal{S}$;
  (iv) $A \in \mathcal{S}$ implies HJ(A) $\in \mathcal{S}$.

Here $B \leq_H A$ means that B is hyperarithmetical in A, and HJ(A) denotes the hyperjump of A. In particular, there is a minimum (i.e., unique smallest) β-model of Π_1^1-CA$_0$, namely

$$\{A \in P(\omega) \colon \exists n \in \omega \; A \leq_H \text{HJ}(n, \emptyset)\}$$

where $\text{HJ}(0, X) = X$, $\text{HJ}(n + 1, X) = \text{HJ}(\text{HJ}(n, X))$. These results will be proved in §VII.1.

REMARK I.5.5 (minimum β-models of Π_k^1-CA$_0$). More generally, for each k in the range $1 \leq k \leq \infty$, it can be shown that there exists a minimum β-model of Π_k^1-CA$_0$. These models can be described in terms of Gödel's theory of constructible sets. For any ordinal number α, let L_α be the αth level of the constructible hierarchy. Then the minimum β-model of Π_k^1-CA$_0$ is of the form $L_\alpha \cap P(\omega)$, where $\alpha = \alpha_k$ is a countable ordinal number depending on k. Moreover, $\alpha_1 < \alpha_2 < \cdots < \alpha_\infty$, and the β-models $L_{\alpha_k} \cap P(\omega)$, $1 \leq k \leq \infty$, are all distinct. (These results are proved in §§VII.5 and VII.7.) It follows that, for each k, Π_{k+1}^1-CA$_0$ is properly stronger than Π_k^1-CA$_0$.

The development of ordinary mathematics within Π_1^1-CA$_0$ and stronger systems is discussed in §I.6 and in chapters V and VI. Models of Π_1^1-CA$_0$ and some stronger systems, including but not limited to Π_k^1-CA$_0$ for $k \geq 2$, are discussed in §§VII.1, VII.5, VII.6, VII.7, VIII.6, and IX.4. Our treatment of constructible sets is in §VII.4. Our treatment of hyperarithmetical theory is in §VIII.3.

Notes for §I.5. For an exposition of Gödel's theory of constructible sets, see any good textbook of axiomatic set theory, e.g., Jech [130].

I.6. Mathematics within Π_1^1-CA$_0$

The system Π_1^1-CA$_0$ was introduced in the previous section. We now discuss the development of ordinary mathematics within Π_1^1-CA$_0$. The material presented here will be restated and greatly refined and expanded in chapters V and VI.

We have seen in §I.4 that a large part of ordinary mathematics can already be developed in ACA$_0$, a subsystem of Z_2 which is much weaker than Π_1^1-CA$_0$. However, there are certain exceptional theorems of ordinary mathematics which can be proved in Π_1^1-CA$_0$ but cannot be proved in ACA$_0$. The exceptional theorems come from several branches of mathematics including countable algebra, the topology of the real line, countable combinatorics, and classical descriptive set theory.

What many of these exceptional theorems have in common is that they directly or indirectly involve countable ordinal numbers. The relevant definition is as follows.

DEFINITION I.6.1 (countable ordinal numbers). Within ACA$_0$ we define a *countable linear ordering* to be a structure $A, <_A$, where $A \subseteq \mathbb{N}$ and $<_A \subseteq A \times A$ is an irreflexive linear ordering of A, i.e., $<_A$ is transitive and, for all $a, b \in A$, exactly one of $a = b$ or $a <_A b$ or $b <_A a$ holds. The countable linear ordering $A, <_A$ is called a *countable well ordering* if there is no sequence $\langle a_n : n \in \mathbb{N} \rangle$ of elements of A such that $a_{n+1} <_A a_n$ for all $n \in \mathbb{N}$. We view a countable well ordering $A, <_A$ as a code for a countable ordinal number, α, which is intuitively just the order type of $A, <_A$. Two countable well orderings $A, <_A$ and $B, <_B$ are said to encode the same countable ordinal number if and only if they are isomorphic. Two countable well orderings $A, <_A$ and $B, <_B$ are said to be *comparable* if they are isomorphic or if one of them is isomorphic to a proper initial segment of the other. (Letting α and β be the corresponding countable ordinal numbers, this means that either $\alpha = \beta$ or $\alpha < \beta$ or $\beta < \alpha$.)

REMARK I.6.2. The fact that any two countable well orderings are comparable turns out to be provable in Π_1^1-CA$_0$ but not in ACA$_0$ (see theorem I.11.5.1 and §V.6). Thus Π_1^1-CA$_0$, but not ACA$_0$, is strong enough to develop a good theory of countable ordinal numbers. Because of this, Π_1^1-CA$_0$ is strong enough to prove several important theorems of ordinary mathematics which are not provable in ACA$_0$. We now present several examples of this phenomenon.

EXAMPLE I.6.3 (Ulm's theorem). Consider the well known structure theory for countable Abelian groups. Let $G, +_G, -_G, 0_G$ be a countable Abelian group. We say that G is *divisible* if for all $a \in G$ and $n > 0$ there exists $b \in G$ such that $nb = a$. We say that G is *reduced* if G has no nontrivial divisible subgroup. Within Π_1^1-CA$_0$, but not within ACA$_0$, one can prove that every countable Abelian group is the direct sum of a divisible group and a reduced group. Now assume that G is a countable Abelian p-group. (This means that for every $a \in G$ there exists $n \in \mathbb{N}$ such that $p^n a = 0$. Here p is a fixed prime number.) One defines a transfinite sequence of subgroups $G_0 = G$, $G_{\alpha+1} = pG_\alpha$, and for limit ordinals δ, $G_\delta = \bigcap_{\alpha < \delta} G_\alpha$. Thus G is reduced if and only if $G_\infty = 0$. The *Ulm invariants* of G are the numbers $\dim(P_\alpha / P_{\alpha+1})$, where $P_\alpha = \{a \in G_\alpha : pa = 0\}$ and the dimension is taken over the integers modulo p. Each Ulm invariant is either a natural number or ∞. *Ulm's theorem* states that two countable reduced Abelian p-groups are isomorphic if and only if their Ulm invariants are the same. Using the theory of countable ordinal numbers which is available in Π_1^1-CA_0, one can carry out the construction of the Ulm invariants and the usual proof of Ulm's theorem within Π_1^1-CA_0. Thus Ulm's theorem is a result of classical algebra which can be proved in Π_1^1-CA_0 but not in ACA_0. More on this topic is in §§V.7 and VI.4.

Example I.6.4 (the Cantor/Bendixson theorem). Next we consider a theorem concerning closed sets in n-dimensional Euclidean space. A *closed set* in $\mathbb{R}^n$ is defined to be the complement of an open set. (Open sets were discussed in definition I.4.7.)

If C is a closed set in $\mathbb{R}^n$, an *isolated point of* C is a point $x \in C$ such that $\{x\} = C \cap U$ for some open set U. Clearly C has at most countably many isolated points. We say that C is *perfect* if C has no isolated points. For any closed set C, the *derived set* of C is a closed set C' consisting of all points of C which are not isolated. Thus $C \setminus C'$ is countable, and $C' = C$ if and only if C is perfect. Given a closed set C, the derived sequence of C is a transfinite sequence of closed subsets of C, defined by $C_0 = C$, $C_{\alpha+1}$ = the derived set of C_α, and for limit ordinals δ, $C_\delta = \bigcap_{\alpha < \delta} C_\alpha$. Within Π_1^1-CA_0 we can prove that for all countable ordinal numbers α, the closed set C_α exists. Furthermore $C_{\beta+1} = C_\beta$ for some countable ordinal number β. In this case we clearly have $C_\beta = C_\alpha$ for all $\alpha > \beta$, so we write $C_\beta = C_\infty$. Clearly C_∞ is a perfect closed set. In fact, C_∞ can be characterized as the largest perfect closed subset of C, and C_∞ is therefore known as the *perfect kernel* of C.

In summary, for any closed set C we have $C = K \cup S$ where K is a perfect closed set (namely $K = C_\infty$) and S is a countable set (namely $S =$ the union of the sets $C_\alpha \setminus C_{\alpha+1}$ for all countable ordinal numbers α). If K happens to be the empty set, then C is itself countable.

The fact that every closed set in $\mathbb{R}^n$ is the union of a perfect closed set and a countable set is known as the *Cantor/Bendixson theorem*. It can be shown that the Cantor/Bendixson theorem is provable in Π_1^1-CA_0 but not in weaker systems such as ACA_0. This example is particularly striking because, although the proof of the Cantor/Bendixson theorem uses countable ordinal numbers, the statement of the theorem does not mention them. For details see §§VI.1 and V.4.

The Cantor/Bendixson theorem also applies more generally, to complete separable metric spaces other than $\mathbb{R}^n$. An important special case is the Baire space $\mathbb{N}^{\mathbb{N}}$. Note that points of $\mathbb{N}^{\mathbb{N}}$ may be identified with functions $f : \mathbb{N} \to \mathbb{N}$. The Cantor/Bendixson theorem for $\mathbb{N}^{\mathbb{N}}$ is closely related to the analysis of trees:

DEFINITION I.6.5 (trees). Within ACA$_0$ we let

$$\mathrm{Seq} = \mathbb{N}^{<\mathbb{N}} = \bigcup_{k \in \mathbb{N}} \mathbb{N}^k$$

denote the set of (codes for) finite sequences of natural numbers. For $\sigma, \tau \in \mathbb{N}^{<\mathbb{N}}$ there is $\sigma \,\widehat{}\, \tau \in \mathbb{N}^{<\mathbb{N}}$ which is the *concatenation*, σ followed by τ. A *tree* is a set $T \subseteq \mathbb{N}^{<\mathbb{N}}$ such that any initial segment of a sequence in T belongs to T. A *path* or *infinite path* through T is a function $f : \mathbb{N} \to \mathbb{N}$ such that for all $k \in \mathbb{N}$, the initial sequence

$$f[k] = \langle f(0), f(1), \ldots, f(k-1) \rangle$$

belongs to T. The set of paths through T is denoted $[T]$. Thus T may be viewed as a code for the closed set $[T] \subseteq \mathbb{N}^{\mathbb{N}}$. If T has no infinite path, we say that T is *well founded*. An *end node* of T is a sequence $\tau \in T$ which has no proper extension in T.

DEFINITION I.6.6 (perfect trees). Two sequences in $\mathbb{N}^{<\mathbb{N}}$ are said to be *compatible* if they are equal or one is an initial segment of the other. Given a tree $T \subseteq \mathbb{N}^{<\mathbb{N}}$ and a sequence $\sigma \in T$, we denote by T_σ the set of $\tau \in T$ such that σ is compatible with τ. Given a tree T, there is a *derived tree* $T' \subseteq T$ consisting of all $\sigma \in T$ such that T_σ contains a pair of incompatible sequences. We say that T is *perfect* if $T' = T$, i.e., every $\sigma \in T$ has a pair of incompatible extensions $\tau_1, \tau_2 \in T$.

Given a tree T, we may consider a transfinite sequence of trees defined by $T_0 = T$, $T_{\alpha+1} =$ the derived tree of T_α, and for limit ordinals δ, $T_\delta = \bigcap_{\alpha < \delta} T_\alpha$. We write $T_\infty = T_\beta$ where β is an ordinal such that $T_\beta = T_{\beta+1}$. Thus T_∞ is the largest perfect subtree of T. These notions concerning trees are analogous to example I.6.4 concerning closed sets. Indeed, the closed set $[T_\infty]$ is the perfect kernel of the closed set $[T]$ in the Baire space $\mathbb{N}^{\mathbb{N}}$. As in example I.6.4, it turns out that the existence of T_∞ is provable in Π_1^1-CA$_0$ but not in weaker systems such as ACA$_0$. This result will be proved in §VI.1.

Turning to another topic in mathematics, we point out that Π_1^1-CA$_0$ is strong enough to prove many of the basic results of classical descriptive set theory. By *classical descriptive set theory* we mean the study of Borel and analytic sets in complete separable metric spaces. The relevant definitions within ACA$_0$ are as follows.

DEFINITION I.6.7 (Borel sets). Let $\widehat{A}$ be a complete separable metric space. A (code for a) *Borel set B* in $\widehat{A}$ is defined to be a set $B \subseteq \mathbb{N}^{<\mathbb{N}}$ such that

(i) B is a well founded tree;
(ii) for any end node $\langle m_0, m_1, \ldots, m_k \rangle$ of B, we have $m_k = (a, r)$ for some $(a, r) \in A \times \mathbb{Q}^+$;
(iii) B contains exactly one sequence $\langle m_0 \rangle$ of length 1.

In particular, for each $a \in A$ and $r \in \mathbb{Q}^+$ there is a Borel code $\langle (a, r) \rangle$. We take $\langle (a, r) \rangle$ to be a code for the basic open neighborhood $B(a, r)$ as in definition I.4.7. Thus for all points $x \in \widehat{A}$ we have, by definition, $x \in B(a, r)$ if and only if $d(a, x) < r$. If B is a Borel code which is not of the form $\langle (a, r) \rangle$, then for each $\langle m_0, n \rangle \in B$ we have another Borel code

$$B_n = \{\langle \rangle\} \cup \{\langle n \rangle^\frown \tau : \langle m_0, n \rangle^\frown \tau \in B\}.$$

We use transfinite recursion to define the notion of a point $x \in \widehat{A}$ belonging to (the Borel set coded by) B, in such a way that $x \in B$ if and only if either m_0 is odd and $x \in B_n$ for some n, or m_0 is even and $x \notin B_n$ for some n. This recursion can be carried out in Π^1_1-CA$_0$; see §V.3.

Thus the Borel sets form a σ-algebra containing the basic open sets and closed under countable union, countable intersection, and complementation.

DEFINITION I.6.8 (analytic sets). Let $\widehat{A}$ be a complete separable metric space. A (code for an) *analytic set* $S \subseteq \widehat{A}$ is defined to be a (code for a) continuous function $\phi \colon \mathbb{N}^\mathbb{N} \to \widehat{A}$. We put $x \in S$ if and only if

$$\exists f (f \in \mathbb{N}^\mathbb{N} \wedge \phi(f) = x).$$

It can be proved in ACA$_0$ that a set is analytic if and only if it is defined by a Σ^1_1 formula with parameters.

EXAMPLE I.6.9 (classical descriptive set theory). Within Π^1_1-CA$_0$ we can emulate the standard proofs of some well known classical results on Borel and analytic sets. This is possible because Π^1_1-CA$_0$ includes a good theory of countable well orderings and countable well founded trees. In particular Souslin's theorem ("a set S is Borel if and only if S and its complement are analytic"), Lusin's theorem ("any two disjoint analytic sets can be separated by a Borel set"), and Kondo's theorem (coanalytic uniformization) are provable in Π^1_1-CA$_0$ but not in ACA$_0$. For details, see §§V.3 and VI.2.

With the above examples, Π^1_1-CA$_0$ emerges as being of considerable interest with respect to the development of ordinary mathematics. Other examples of ordinary mathematical theorems which are provable in Π^1_1-CA$_0$ are: determinacy of open sets in $\mathbb{N}^\mathbb{N}$ (see §V.8), and the Ramsey property for open sets in $[\mathbb{N}]^\mathbb{N}$ (see §V.9). These theorems, like Ulm's theorem and the Cantor/Bendixson theorem, are exceptional in that they are not provable in ACA$_0$.

REMARK I.6.10 (Friedman-style independence results).   There are a small number of even more exceptional theorems which, for instance, are provable in ZFC (i.e., Zermelo/Fraenkel set theory with the axiom of choice) but not in full Z$_2$. As an example, consider the following corollary, due to Friedman [71], of a theorem of Martin [177, 178]: Given a symmetric Borel set $B \subseteq I \times I$, $I = [0, 1]$, there exists a Borel function $\phi: I \rightarrow I$ such that the graph of ϕ is either included in or disjoint from B. Friedman [71] has shown that this result is not provable in Z$_2$ or even in simple type theory. This is related to Friedman's earlier result [66, 71] that Borel determinacy is not provable in simple type theory. More results of this kind are in [72] and in the Friedman volume [102].

Notes for §I.6. Chapters V and VI of this book deal with the development of mathematics in Π_1^1-CA$_0$. The crucial role of comparability of countable well orderings (remark I.6.2) was pointed out by Friedman [62, chapter II] and Steel [256, chapter I]; recent refinements are due to Friedman/Hirst [74] and Shore [223]. The impredicative nature of the Cantor/Bendixson theorem and Ulm's theorem was noted by Kreisel [149] and Feferman [58], respectively. An up-to-date textbook of classical descriptive set theory is Kechris [138].   Friedman has discovered a number of mathematically natural statements whose proofs require strong set existence axioms; see the Friedman volume [102] and recent papers such as [73].

I.7. The System RCA$_0$

In this section we introduce RCA$_0$, an important subsystem of Z$_2$ which is much weaker than ACA$_0$.

The acronym RCA stands for recursive comprehension axiom. This is because RCA$_0$ contains axioms asserting the existence of any set A which is recursive in given sets $B_1, \ldots, B_k$ (i.e., such that the characteristic function of A is computable assuming oracles for the characteristic functions of $B_1, \ldots, B_k$). As in ACA$_0$ and Π_1^1-CA$_0$, the subscript 0 in RCA$_0$ denotes restricted induction. The axioms of RCA$_0$ include Σ_1^0 induction, a form of induction which is weaker than arithmetical induction (as defined in §I.3) but stronger than the induction axiom I.2.4(ii).

We now proceed to the definition of RCA$_0$.

Let n be a number variable, let t be a numerical term not containing n, and let φ be a formula of L$_2$. We use the following abbreviations:

$$\forall n < t \, \varphi \equiv \forall n \, (n < t \rightarrow \varphi),$$
$$\exists n < t \, \varphi \equiv \exists n \, (n < t \wedge \varphi).$$

Thus $\forall n < t$ means "for all n less than t", and $\exists n < t$ means "there exists n less than t such that". We may also write $\forall n \le t$ instead of $\forall n < t + 1$, and $\exists n \le t$ instead of $\exists n < t + 1$.

The expressions $\forall n < t$, $\forall n \le t$, $\exists n < t$, $\exists n \le t$ are called *bounded number quantifiers*, or simply *bounded quantifiers*. A *bounded quantifier formula* is a formula φ such that all of the quantifiers occurring in φ are bounded number quantifiers. Thus the bounded quantifier formulas are a subclass of the arithmetical formulas. Examples of bounded quantifier formulas are

$$\exists m \le n \, (n = m + m),$$

asserting that n is even, and

$$\forall m < 2n \, (m \in X \leftrightarrow \exists k < m \, (m = 2k + 1)),$$

asserting that the first n elements of X are $1, 3, 5, \ldots, 2n - 1$.

DEFINITION I.7.1 (Σ_1^0 and Π_1^0 formulas). An L_2-formula φ is said to be Σ_1^0 if it is of the form $\exists m \, \theta$, where m is a number variable and θ is a bounded quantifier formula. An L_2-formula φ is said to be Π_1^0 if it is of the form $\forall m \, \theta$, where m is a number variable and θ is a bounded quantifier formula.

It can be shown that Σ_1^0 formulas are closely related to the notion of relative recursive enumerability in recursion theory. Namely, for $A, B \in P(\omega)$, A is recursively enumerable in B if and only if A is definable over some or any ω-model $(\omega, S, +, \cdot, 0, 1, <)$, $B \in S \subseteq P(\omega)$, by a Σ_1^0 formula with B as a parameter. (See also remarks I.3.4 and I.7.5.)

DEFINITION I.7.2 (Σ_1^0 induction). The Σ_1^0 *induction scheme*, Σ_1^0-IND, is the restriction of the second order induction scheme (as defined in §I.2) to L_2-formulas $\varphi(n)$ which are Σ_1^0. Thus we have the universal closure of

$$(\varphi(0) \wedge \forall n \, (\varphi(n) \rightarrow \varphi(n + 1))) \rightarrow \forall n \, \varphi(n)$$

where $\varphi(n)$ is any Σ_1^0 formula of L_2.

The Π_1^0 *induction scheme*, Π_1^0-IND, is defined similarly. It can be shown that Σ_1^0-IND and Π_1^0-IND are equivalent (in the presence of the basic axioms I.2.4(i)). This easy but useful result is proved in §II.3.

DEFINITION I.7.3 (Δ_1^0 comprehension). The Δ_1^0 *comprehension scheme* consists of (the universal closures of) all formulas of the form

$$\forall n \, (\varphi(n) \leftrightarrow \psi(n)) \rightarrow \exists X \, \forall n \, (n \in X \leftrightarrow \varphi(n)),$$

where $\varphi(n)$ is any Σ_1^0 formula, $\psi(n)$ is any Π_1^0 formula, n is any number variable, and X is a set variable which does not occur freely in $\varphi(n)$.

In the Δ_1^0 comprehension scheme, note that $\varphi(n)$ and $\psi(n)$ may contain parameters, i.e., free set variables and free number variables in addition to n. Thus an L$_2$-structure M satisfies Δ_1^0 comprehension if and only if $\mathcal{S}_M$ contains all subsets of $|M|$ which are both Σ_1^0 and Π_1^0 definable over M allowing parameters from $|M| \cup \mathcal{S}_M$.

DEFINITION I.7.4 (definition of RCA$_0$). RCA$_0$ is the subsystem of Z$_2$ consisting of the basic axioms I.2.4(i), the Σ_1^0 induction scheme I.7.2, and the Δ_1^0 comprehension scheme I.7.3.

REMARK I.7.5 (ω-models of RCA$_0$). In remark I.3.4, we characterized the ω-models of ACA$_0$ in terms of recursion theory. We can characterize the ω-models of RCA$_0$ in similar terms, as follows. $\mathcal{S} \subseteq P(\omega)$ is an ω-model of RCA$_0$ if and only if

(i) $\mathcal{S} \neq \emptyset$;
(ii) $A \in \mathcal{S}$ and $B \in \mathcal{S}$ imply $A \oplus B \in \mathcal{S}$;
(iii) $A \in \mathcal{S}$ and $B \leq_T A$ imply $B \in \mathcal{S}$.

(This result is proved in §VIII.1.) In particular, RCA$_0$ has a minimum (i.e., unique smallest) ω-model, namely

$$\mathrm{REC} = \{A \in P(\omega): A \text{ is recursive}\}.$$

More generally, given a set $B \in P(\omega)$, there is a unique smallest ω-model of RCA$_0$ containing B, consisting of all sets $A \in P(\omega)$ which are recursive in B.

The system RCA$_0$ plays two key roles in this book and in foundational studies generally. First, as we shall see in chapter II, the development of ordinary mathematics within RCA$_0$ corresponds roughly to the positive content of what is known as "computable mathematics" or "recursive analysis". Thus RCA$_0$ is a kind of formalized recursive mathematics. Second, RCA$_0$ frequently plays the role of a weak base theory in Reverse Mathematics. Most of the results of Reverse Mathematics in chapters III, IV, V, and VI will be stated formally as theorems of RCA$_0$.

REMARK I.7.6 (first order part of RCA$_0$). By remark I.3.3, the first order part of ACA$_0$ is first order arithmetic, PA. In a similar vein, we can characterize the first order part of RCA$_0$. Namely, let Σ_1^0-PA be PA with induction restricted to Σ_1^0 formulas. (Thus Σ_1^0-PA is a formal system whose language is L$_1$ and whose axioms are the basic axioms I.2.4(i) plus the universal closure of

$$(\varphi(0) \wedge \forall n \, (\varphi(n) \rightarrow \varphi(n+1))) \rightarrow \forall n \, \varphi(n)$$

for any formula $\varphi(n)$ of L$_1$ which is Σ_1^0.) Clearly the axioms of Σ_1^0-PA are included in those of RCA$_0$. Conversely, given any model

$$(|M|, +_M, \cdot_M, 0_M, 1_M, <_M) \tag{2}$$

of Σ_1^0-PA, it can be shown that there exists $\mathcal{S}_M \subseteq P(|M|)$ such that

$$(|M|, \mathcal{S}_M, +_M, \cdot_M, 0_M, 1_M, <_M)$$

is a model of RCA_0. (Namely, we can take $\mathcal{S}_M = \Delta_1^0\text{-Def}(M) =$ the set of all $A \subseteq |M|$ such that A is both Σ_1^0 and Π_1^0 definable over (2) allowing parameters from $|M|$.) It follows that, for any sentence σ in the language of first order arithmetic, σ is a theorem of RCA_0 if and only if σ is a theorem of Σ_1^0-PA. In other words, Σ_1^0-PA is the first order part of RCA_0. (These results are proved in §IX.1.)

Models of RCA_0 are discussed further in §§VIII.1, IX.1, IX.2, and IX.3. The development of ordinary mathematics within RCA_0 is outlined in §I.8 and is discussed thoroughly in chapter II.

REMARK I.7.7 (Σ_1^0 comprehension). It would be possible to define a system Σ_1^0-CA_0 consisting of the basic axioms I.2.4(i), the induction axiom I.2.4(ii), and the Σ_1^0 *comprehension scheme*, i.e., the universal closure of

$$\exists X \, \forall n \, (n \in X \leftrightarrow \varphi(n))$$

for all Σ_1^0 formulas $\varphi(n)$ of L_2 in which X does not occur freely. However, the introduction of Σ_1^0-CA_0 as a distinct subsystem of Z_2 is unnecessary, because it turns out that Σ_1^0-CA_0 is equivalent to ACA_0. This easy but important result will be proved in §III.1.

Generalizing the notion of Σ_1^0 and Π_1^0 formulas, we have:

DEFINITION I.7.8 (Σ_k^0 and Π_k^0 formulas). For $0 \leq k \in \omega$, an L_2-formula φ is said to be Σ_k^0 (respectively Π_k^0) if it is of the form

$$\exists n_1 \, \forall n_2 \, \exists n_3 \cdots n_k \, \theta$$

(respectively $\forall n_1 \exists n_2 \forall n_3 \cdots n_k \, \theta$), where $n_1, \ldots, n_k$ are number variables and θ is a bounded quantifier formula. In both cases, φ consists of k alternating unbounded number quantifiers followed by a formula containing only bounded number quantifiers. In the Σ_k^0 case, the first unbounded number quantifier is existential, while in the Π_k^0 case it is universal (assuming $k \geq 1$). Thus for instance a Π_2^0 formula is of the form $\forall m \, \exists n \, \theta$, where θ is a bounded quantifier formula. A Σ_0^0 or Π_0^0 formula is the same thing as a bounded quantifier formula.

Clearly any Σ_k^0 formula is logically equivalent to the negation of a Π_k^0 formula, and vice versa. Moreover, up to logical equivalence of formulas, we have $\Sigma_k^0 \cup \Pi_k^0 \subseteq \Sigma_{k+1}^0 \cap \Pi_{k+1}^0$, for all $k \in \omega$.

REMARK I.7.9 (induction and comprehension schemes). Generalizing definition I.7.2, we can introduce induction schemes Σ_k^i-IND and Π_k^i-IND, for all $k \in \omega$ and $i \in \{0, 1\}$. Clearly Σ_∞^0-IND $= \bigcup_{k \in \omega} \Sigma_k^0$-IND is equivalent to arithmetical induction, and Σ_∞^1-IND $= \bigcup_{k \in \omega} \Sigma_k^1$-IND is equivalent to the full second order induction scheme. It can be shown that, for all

$k \in \omega$ and $i \in \{0, 1\}$, Σ_k^i-IND is equivalent to Π_k^i-IND and is properly weaker than Σ_{k+1}^i-IND. As for comprehension schemes, it follows from remark I.7.7 that the systems Σ_k^0-CA$_0$ and Π_k^0-CA$_0$, $1 \leq k \in \omega$, are all equivalent to each other and to ACA$_0$, i.e., Π_0^1-CA$_0$. On the other hand, we have remarked in §I.5 that, for each $k \in \omega$, Π_k^1-CA$_0$ is equivalent to Σ_k^1-CA$_0$ and is properly weaker than Π_{k+1}^1-CA$_0$. In chapter VII we shall introduce the systems Δ_k^1-CA$_0$, $1 \leq k \in \omega$, and we shall show that Δ_k^1-CA$_0$ is properly stronger than Π_{k-1}^1-CA$_0$ and properly weaker than Π_k^1-CA$_0$.

Notes for §I.7. In connection with remark I.7.5, note that the literature of recursion theory sometimes uses the term *Turing ideals* referring to what we call ω-models of RCA$_0$. See for instance Lerman [161, page 29]. The system RCA$_0$ was first introduced by Friedman [69] (in an equivalent form, using a somewhat different language and axioms). The system Σ_1^0-PA was first studied by Parsons [201]. For a thorough discussion of Σ_1^0-PA and other subsystems of first order arithmetic, see Hájek/Pudlák [100] and Kaye [137].

I.8. Mathematics within RCA$_0$

In this section we sketch how some concepts and results of ordinary mathematics can be developed in RCA$_0$. This portion of ordinary mathematics is roughly parallel to the positive content of recursive analysis and recursive algebra. We shall also give some recursive counterexamples showing that certain other theorems of ordinary mathematics are recursively false and hence, although provable in ACA$_0$, cannot be proved in RCA$_0$.

As already remarked in I.4.4 and I.4.8, the strictures of RCA$_0$ require us to modify our definitions of "real number" and "point of a complete separable metric space". The needed modifications are as follows:

DEFINITION I.8.1 (partially replacing I.4.2). Within RCA$_0$, a (code for a) *real number* $x \in \mathbb{R}$ is defined to be a sequence of rational numbers $x = \langle q_n : n \in \mathbb{N} \rangle$, $q_n \in \mathbb{Q}$, such that

$$\forall m \, \forall n \, (m < n \to |q_m - q_n| < 1/2^m).$$

For real numbers x and y we have $x =_{\mathbb{R}} y$ if and only if

$$\forall m \, (|q_m - q_m'| \leq 1/2^{m-1}),$$

and $x <_{\mathbb{R}} y$ if and only if

$$\exists m \, (q_m + 1/2^m < q_m').$$

Note that with definition I.8.1 we now have that the predicate $x < y$ is Σ_1^0, and the predicates $x \leq y$ and $x = y$ are Π_1^0, for $x, y \in \mathbb{R}$. Thus real number comparisons have become easier, and therein lies the superiority of I.8.1 over I.4.2 within RCA_0.

DEFINITION I.8.2 (partially replacing I.4.5). Within RCA_0, a (code for a) complete separable metric space is defined as in I.4.5. However, a (code for a) *point of the complete separable metric space $\hat{A}$* is now defined in RCA_0 to be a sequence $x = \langle a_n : n \in \mathbb{N}\rangle$, $a_n \in A$, satisfying $\forall m \, \forall n$ $(m < n \to d(a_m, a_n) < 1/2^m)$. The extension of d to $\hat{A}$ is as in I.4.5.

Under definition I.8.2, the predicate $d(x, y) < r$ for $x, y \in \hat{A}$ and $r \in \mathbb{R}$ becomes Σ_1^0. This makes I.8.2 far more appropriate than I.4.5 for use in RCA_0. We shall also need to modify slightly our earlier definitions of "continuous function" in I.4.6 and "open set" in I.4.7; the modified definitions will be presented in II.6.1 and II.5.6.

With these new definitions, the development of mathematics within RCA_0 is broadly similar to the development within ACA_0 as already outlined in §I.4 above. For the most part, Δ_1^0 comprehension is an adequate substitute for arithmetical comprehension. Thus RCA_0 is strong enough to prove basic results of real and complex linear and polynomial algebra, up to and including the fundamental theorem of algebra, and basic properties of countable algebraic structures and of continuous functions on complete separable metric spaces. Also within RCA_0 we can introduce sequences of real numbers, sequences of continuous functions, and separable Banach spaces including examples such as $C[0, 1]$ and $L_p[0, 1]$, $1 \leq p < \infty$, just as in ACA_0 (§I.4). This detailed development within RCA_0 will be presented in chapter II.

In addition to basic results (e.g., the fact that the composition of two continuous functions is continuous), a number of nontrivial theorems are also provable in RCA_0. We have:

THEOREM I.8.3 (mathematics in RCA_0). *The following ordinary mathematical theorems are provable in* RCA_0:

1. *the Baire category theorem* (§§II.4, II.5);
2. *the intermediate value theorem* (§II.6);
3. *Urysohn's lemma and the Tietze extension theorem for complete separable metric spaces* (§II.7);
4. *the soundness theorem and a version of Gödel's completeness theorem in mathematical logic* (§II.8);
5. *existence of an algebraic closure of a countable field* (§II.9);
6. *existence of a unique real closure of a countable ordered field* (§II.9);
7. *the Banach/Steinhaus uniform boundedness principle* (§II.10).

On the other hand, a phenomenon of great interest for us is that many well known and important mathematical theorems which are routinely provable in ACA_0 turn out not to be provable at all in RCA_0. We now present an example of this phenomenon.

EXAMPLE I.8.4 (the Bolzano/Weierstraß theorem). Let us denote by BW the statement of the Bolzano/Weierstraß theorem: "Every bounded sequence of real numbers contains a convergent subsequence." It is straightforward to show that BW is provable in ACA$_0$.

We claim that BW is not provable in RCA$_0$.

To see this, consider the ω-model REC consisting of all recursive subsets of ω. We have seen in I.7.5 that REC is a model of RCA$_0$. We shall now show that BW is false in REC.

We use some basic results of recursive function theory. Let A be a recursively enumerable subset of ω which is not recursive. For instance, we may take $A = K = \{n : \{n\}(n) \text{ is defined}\}$. Let $f : \omega \to \omega$ be a one-to-one recursive function such that $A = $ the range of f. Define a bounded increasing sequence of rational numbers a_k, $k \in \omega$, by putting

$$a_k = \sum_{m=0}^{k} \frac{1}{2^{f(m)}}.$$

Clearly the sequence $\langle a_k \rangle_{k \in \omega}$, or more precisely its code, is recursive and hence is an element of REC. On the other hand, it can be shown that the real number

$$r = \sup_{k \in \omega} a_k = \sum_{m=0}^{\infty} \frac{1}{2^{f(m)}} = \sum_{n \in A} \frac{1}{2^n}$$

is not recursive, i.e. (any code of) r is not an element of REC. One way to see this would be to note that the characteristic function of the nonrecursive set A would be computable if we allowed (any code of) r as a Turing oracle.

Thus the ω-model REC satisfies "$\langle a_k \rangle_{k \in \mathbb{N}}$ is a bounded increasing sequence of rational numbers, and $\langle a_k \rangle_{k \in \mathbb{N}}$ has no least upper bound". In particular, REC satisfies "$\langle a_k \rangle_{k \in \mathbb{N}}$ is a bounded sequence of real numbers which has no convergent subsequence". Hence BW is false in the ω-model REC. Hence BW is not provable in RCA$_0$.

REMARK I.8.5 (recursive counterexamples). There is an extensive literature of what is known as "recursive analysis" or "computable mathematics", i.e., the systematic development of portions of ordinary mathematics within the particular ω-model REC. (See the notes at the end of this section.) This literature contains many so-called "recursive counterexamples", where methods of recursive function theory are used to show that particular mathematical theorems are false in REC. Such results are of great interest with respect to our Main Question, §I.1, because they imply that the set existence axioms of RCA$_0$ are not strong enough to prove the mathematical theorems under consideration. We have already presented one such recursive counterexample, showing that the Bolzano/Weierstraß theorem is false in REC, hence not provable in RCA_0. Other recursive counterexamples will be presented below.

EXAMPLE I.8.6 (the Heine/Borel covering lemma).   Let us denote by HB the statement of the Heine/Borel covering lemma: Every covering of the closed interval $[0, 1]$ by a sequence of open intervals has a finite subcovering. Again HB is provable in ACA_0. We shall exhibit a recursive counterexample showing that HB is false in REC, hence not provable in RCA_0.

Consider the well known Cantor middle third set $C \subseteq [0, 1]$ defined by

$$C = [0, 1] \setminus ((1/3, 2/3) \cup (1/9, 2/9) \cup (7/9, 8/9) \cup \ldots).$$

There is a well known obvious recursive homeomorphism $H : C \cong \{0, 1\}^\omega$, where $\{0, 1\}^\omega$ is the product of ω copies of the two-point discrete space $\{0, 1\}$. Points $h \in \{0, 1\}^\omega$ may be identified with functions $h : \omega \to \{0, 1\}$. For each $\varepsilon \in \{0, 1\}$ and $n \in \omega$, let U_n^ε be the union of 2^n effectively chosen rational open intervals such that

$$H(U_n^\varepsilon \cap C) = \{h \in \{0, 1\}^\omega : h(n) = \varepsilon\}.$$

For instance, corresponding to $\varepsilon = 0$ and $n = 2$ we could choose $U_2^0 = (-1, 1/18) \cup (1/6, 5/18) \cup (1/2, 13/18) \cup (5/6, 17/18)$.

Now let A, B be a disjoint pair of recursively inseparable, recursively enumerable subsets of ω. For instance, we could take $A = \{n : \{n\}(n) \simeq 0\}$ and $B = \{n : \{n\}(n) \simeq 1\}$. Since A and B are recursively inseparable, it follows that for any recursive point $h \in \{0, 1\}^\omega$ we have either $h(n) = 0$ for some $n \in A$, or $h(n) = 1$ for some $n \in B$. Let $f, g : \omega \to \omega$ be recursive functions such that $A = \text{rng}(f)$ and $B = \text{rng}(g)$. Then $U_{f(m)}^0$, $U_{g(m)}^1$, $m \in \omega$, give a recursive sequence of rational open intervals which cover the recursive reals in C but not all of C. Combining this with the middle third intervals $(1/3, 2/3)$, $(1/9, 2/9)$, $(7/9, 8/9)$, ..., we obtain a recursive sequence of rational open intervals which cover the recursive reals in $[0, 1]$ but not all of $[0, 1]$. Thus the ω-model REC satisfies "there exists a sequence of rational open intervals which is a covering of $[0, 1]$ but has no finite subcovering". Hence HB is false in REC. Hence HB is not provable in RCA_0.

EXAMPLE I.8.7 (the maximum principle).   Another ordinary mathematical theorem not provable in RCA_0 is the maximum principle: Every continuous real-valued function on $[0, 1]$ attains a supremum. To see this, let C, f, g, U_n^ε, $\varepsilon \in \{0, 1\}$, $n \in \omega$ be as in I.8.6, and let r, a_k, $k \in \omega$ be as in I.8.4. It is straightforward to construct a recursive code Φ for a function ϕ such that REC satisfies "$\phi : C \to \mathbb{R}$ is continuous and, for all $x \in C$, $\phi(x) = a_k$ where $k = $ the least m such that $x \in U_{f(m)}^0 \cup U_{g(m)}^1$". Thus $\sup\{\phi(x) : x \in C \cap \text{REC}\} = \sup_{k \in \omega} a_k = r$ is a nonrecursive real number, so REC satisfies "$\sup_{x \in C} \phi(x)$ does not exist". Since $0 < a_k < 2$

for all k, we actually have $\phi\colon C \rightarrow [0, 2]$ in REC. Also, we can extend ϕ uniquely to a continuous function $\psi\colon [0, 1] \rightarrow [0, 2]$ which is linear on intervals disjoint from C. Thus REC satisfies "$\psi\colon [0, 1] \rightarrow [0, 2]$ is continuous and $\sup_{x \in C} \psi(x)$ does not exist". Hence the maximum principle is false in REC and therefore not provable in RCA$_0$.

EXAMPLE I.8.8 (König's lemma). Recall our notion of tree as defined in I.6.5. A tree T is said to be *finitely branching* if for each $\sigma \in T$ there are only finitely many n such that $\sigma^\frown \langle n \rangle \in T$. *König's lemma* is the following statement: every infinite, finitely branching tree has an infinite path.

We claim that König's lemma is provable in ACA$_0$. An outline of the argument within ACA$_0$ is as follows. Let $T \subseteq \mathbb{N}^{<\mathbb{N}}$ be an infinite, finitely branching tree. By arithmetical comprehension, there is a subtree $T^* \subseteq T$ consisting of all $\sigma \in T$ such that T_σ (see definition I.6.6) is infinite. Since T is infinite, the empty sequence $\langle \rangle$ belongs to T^*. Moreover, by the pigeonhole principle, T^* has no end nodes. Define $f\colon \mathbb{N} \rightarrow \mathbb{N}$ by primitive recursion by putting $f(m) = $ the least n such that $f[m]^\frown \langle n \rangle \in T^*$, for all $m \in \mathbb{N}$. Then f is a path through T^*, hence through T, Q.E.D.

We claim that König's lemma is not provable in RCA$_0$. To see this, let A, B, f, g be as in I.8.6. Let $\{0, 1\}^{<\omega}$ be the full binary tree, i.e., the tree of finite sequences of 0's and 1's. Let T be the set of all $\tau \in \{0, 1\}^{<\omega}$ such that, if $k = $ the length of τ, then for all $m, n < k$, $f(m) = n$ implies $\tau(n) = 1$, and $g(m) = n$ implies $\tau(n) = 0$. Note that T is recursive. Moreover, $h \in \{0, 1\}^{\omega}$ is a path through T if and only if h separates A and B, i.e., $h(n) = 1$ for all $n \in A$ and $h(n) = 0$ for all $n \in B$. Thus T is an infinite, recursive, finitely branching tree with no recursive path. Hence we have a recursive counterexample to König's lemma, showing that König's lemma is false in REC, hence not provable in RCA$_0$.

The recursive counterexamples presented above show that, although RCA$_0$ is able to accommodate a large and significant portion of ordinary mathematical practice, it is also subject to some severe limitations. We shall eventually see that, in order to prove ordinary mathematical theorems such as the Bolzano/Weierstraß theorem, the Heine/Borel covering lemma, the maximum principle, and König's lemma, it is necessary to pass to subsystems of Z_2 that are considerably stronger than RCA$_0$. This investigation will lead us to another important theme: Reverse Mathematics (§§I.9, I.10, I.11, I.12).

REMARK I.8.9 (constructive mathematics). In some respects, our formal development of ordinary mathematics within RCA$_0$ resembles the practice of Bishop-style constructivism [20]. However, there are some substantial differences (see also the notes below):

1. The constructivists believe that mathematical objects are purely mental constructions, while we make no such assumption.

2. The meaning which the constructivists assign to the propositional connectives and quantifiers is incompatible with our classical interpretation.

3. The constructivists assume unrestricted induction on the natural numbers, while in RCA_0 we assume only Σ^0_1 induction.

4. We always assume the law of the excluded middle, while the constructivists deny it.

5. The typical constructivist response to a nonconstructive mathematical theorem is to modify the theorem by adding hypotheses or "extra data". In contrast, our approach in this book is to analyze the provability of mathematical theorems as they stand, passing to stronger subsystems of Z_2 if necessary. See also our discussion of Reverse Mathematics in §I.9.

Notes for §I.8. Some references on recursive and constructive mathematics are Aberth [2], Beeson [17], Bishop/Bridges [20], Demuth/Kučera [46], Mines/Richman/Ruitenburg [189], Pour-El/Richards [203], and Troelstra/van Dalen [268]. The relationship between Bishop-style constructivism and RCA_0 is discussed in [78, §0]. Chapter II of this book is devoted to the development of mathematics within RCA_0. Some earlier literature presenting some of this development in a less systematic manner is Simpson [236], Friedman/Simpson/Smith [78], Brown/Simpson [27].

I.9. Reverse Mathematics

We begin this section with a quote from Aristotle.

> Reciprocation of premises and conclusion is more frequent in mathematics, because mathematics takes definitions, but never an accident, for its premises—a second characteristic distinguishing mathematical reasoning from dialectical disputations. Aristotle, *Posterior Analytics* [184, 78a10].

The purpose of this section is to introduce one of the major themes of this book: Reverse Mathematics.

In order to motivate Reverse Mathematics from a foundational standpoint, consider the Main Question as defined in §I.1, concerning the role of set existence axioms. In §§I.4 and I.6, we have sketched an approximate answer to the Main Question. Namely, we have suggested that most theorems of ordinary mathematics can be proved in ACA_0, and that of the exceptions, most can be proved in $\Pi^1_1\text{-}CA_0$.

Consider now the following sharpened form of the Main Question: *Given a theorem τ of ordinary mathematics, what is the weakest natural subsystem $S(\tau)$ of Z_2 in which τ is provable?*

Surprisingly, it turns out that for many specific theorems τ this question has a precise and definitive answer. Furthermore, $S(\tau)$ often turns out to be one of five specific subsystems of Z_2. For convenience we shall now list these systems as S_1, S_2, S_3, S_4 and S_5 in order of increasing ability to accommodate ordinary mathematical practice. The odd numbered systems S_1, S_3 and S_5 have already been introduced as RCA_0, ACA_0 and $\Pi_1^1\text{-}\mathsf{CA}_0$ respectively. The even numbered systems S_2 and S_4 are intermediate systems which will be introduced in §§I.10 and I.11 below.

Our method for establishing results of the form $S(\tau) = S_j$, $2 \leq j \leq 5$ is based on the following empirical phenomenon: "When the theorem is proved from the right axioms, the axioms can be proved from the theorem." (Friedman [68].) Specifically, let τ be an ordinary mathematical theorem which is not provable in the weak base theory $S_1 = \mathsf{RCA}_0$. Then very often, τ turns out to be equivalent to S_j for some $j = 2, 3, 4$ or 5. The equivalence is provable in S_i for some $i < j$, usually $i = 1$.

For example, let $\tau = \mathsf{BW} = $ the Bolzano/Weierstraß theorem: every bounded sequence of real numbers has a convergent subsequence. We have seen in I.8.4 that BW is false in the ω-model REC. An adaptation of that argument gives the following result:

THEOREM I.9.1. BW *is equivalent to* ACA_0, *the equivalence being provable in* RCA_0.

PROOF. Note first that $\mathsf{ACA}_0 = \mathsf{RCA}_0$ plus arithmetical comprehension. Thus the forward direction of our theorem is obtained by observing that the usual proof of BW goes through in ACA_0, as already remarked in §I.4.

For the reverse direction (i.e., the converse), we reason within RCA_0 and assume BW. We are trying to prove arithmetical comprehension. Recall that, by relativization, arithmetical comprehension is equivalent to Σ_1^0 comprehension (see remark I.7.7). So let $\varphi(n)$ be a Σ_1^0 formula, say $\varphi(n) \equiv \exists m\, \theta(m, n)$ where θ is a bounded quantifier formula. For each $k \in \mathbb{N}$ define

$$c_k = \sum \{2^{-n} : n < k \wedge (\exists m < k)\, \theta(m, n)\}.$$

Then $\langle c_k : k \in \mathbb{N} \rangle$ is a bounded increasing sequence of rational numbers. This sequence exists by Δ_1^0 comprehension, which is available to us since we are working in RCA_0. Now by BW the limit $c = \lim_k c_k$ exists. Then we have

$$\forall n\, (\varphi(n) \leftrightarrow \forall k\, (|c - c_k| < 2^{-n} \rightarrow (\exists m < k)\, \theta(m, n))).$$

This gives the equivalence of a Σ_1^0 formula with a Π_1^0 formula. Hence by Δ_1^0 comprehension we conclude $\exists X\, \forall n\, (n \in X \leftrightarrow \varphi(n))$. This proves Σ_1^0 comprehension and hence arithmetical comprehension. $\square$

REMARK I.9.2 (on Reverse Mathematics). Theorem I.9.1 implies that $S_3 = \mathsf{ACA}_0$ is the weakest natural subsystem of Z_2 in which $\tau = \mathsf{BW}$ is provable. Thus, for this particular case involving the Bolzano/Weierstraß theorem, I.9.1 provides a definitive answer to our sharpened form of the Main Question.

Note that the proof of theorem I.9.1 involved the deduction of a set existence axiom (namely arithmetical comprehension) from an ordinary mathematical theorem (namely BW). This is the opposite of the usual pattern of ordinary mathematical practice, in which theorems are deduced from axioms. The deduction of axioms from theorems is known as *Reverse Mathematics*. Theorem I.9.1 illustrates how Reverse Mathematics is the key to obtaining precise answers for instances of the Main Question. This point will be discussed more fully in §I.12.

We shall now state a number of results, similar to I.9.1, showing that particular ordinary mathematical theorems are equivalent to the axioms needed to prove them. These Reverse Mathematics results with respect to ACA_0 and $\Pi_1^1\text{-}\mathsf{CA}_0$ will be summarized in theorems I.9.3 and I.9.4 and proved in chapters III and VI, respectively.

THEOREM I.9.3 (Reverse Mathematics for ACA_0). *Within* RCA_0 *one can prove that* ACA_0 *is equivalent to each of the following ordinary mathematical theorems*:

1. *Every bounded, or bounded increasing, sequence of real numbers has a least upper bound* (§III.2).
2. *The Bolzano/Weierstraß theorem: Every bounded sequence of real numbers, or of points in $\mathbb{R}^n$, has a convergent subsequence* (§III.2).
3. *Every sequence of points in a compact metric space has a convergent subsequence* (§III.2).
4. *The Ascoli lemma: Every bounded equicontinuous sequence of real-valued continuous functions on a bounded interval has a uniformly convergent subsequence* (§III.2).
5. *Every countable commutative ring has a maximal ideal* (§III.5).
6. *Every countable vector space over $\mathbb{Q}$, or over any countable field, has a basis* (§III.4).
7. *Every countable field (of characteristic 0) has a transcendence basis* (§III.4).
8. *Every countable Abelian group has a unique divisible closure* (§III.6).
9. *König's lemma: Every infinite, finitely branching tree has an infinite path* (§III.7).
10. *Ramsey's theorem for colorings of $[\mathbb{N}]^3$, or of $[\mathbb{N}]^4$, $[\mathbb{N}]^5$, ...* (§III.7).

THEOREM I.9.4 (Reverse Mathematics for $\Pi_1^1\text{-}\mathsf{CA}_0$). *Within* RCA_0 *one can prove that* $\Pi_1^1\text{-}\mathsf{CA}_0$ *is equivalent to each of the following ordinary mathematical statements*:

1. *Every tree has a largest perfect subtree* (§VI.1).
2. *The Cantor/Bendixson theorem*: *Every closed subset of* $\mathbb{R}$, *or of any complete separable metric space, is the union of a countable set and a perfect set* (§VI.1).
3. *Every countable Abelian group is the direct sum of a divisible group and a reduced group* (§VI.4).
4. *Every difference of two open sets in the Baire space* $\mathbb{N}^{\mathbb{N}}$ *is determined* (§VI.5).
5. *Every* G_δ *set in* $[\mathbb{N}]^{\mathbb{N}}$ *has the Ramsey property* (§VI.6).
6. *Silver's theorem*: *For every Borel (or coanalytic, or* F_σ*) equivalence relation with uncountably many equivalence classes, there exists a nonempty perfect set of inequivalent elements* (§VI.3).

More Reverse Mathematics results will be stated in §§I.10 and I.11 and proved in chapters IV and V, respectively. The significance of Reverse Mathematics for our Main Question will be discussed in §I.12.

Notes for §I.9. Historically, Reverse Mathematics may be viewed as a spin-off of Friedman's work [65, 66, 71, 72, 73] attempting to demonstrate the necessary use of higher set theory in mathematical practice. The theme of Reverse Mathematics in the context of subsystems of Z_2 first appeared in Steel's thesis [256, chapter I] (an outcome of Steel's reading of Friedman's thesis [62, chapter II] under Simpson's supervision [230]) and in Friedman [68, 69]; see also Simpson [238]. This theme was taken up by Simpson and his collaborators in numerous studies [236, 241, 76, 235, 234, 78, 79, 250, 243, 246, 245, 21, 27, 28, 280, 80, 113, 112, 247, 127, 128, 26, 93, 248] which established it as a subject. The slogan "Reverse Mathematics" was coined by Friedman during a special session of the American Mathematical Society organized by Simpson.

I.10. The System WKL₀

In this section we introduce WKL₀, a subsystem of Z_2 consisting of RCA₀ plus a set existence axiom known as *weak König's lemma*. We shall see that, in the notation of §I.9, WKL₀ = S_2 is intermediate between RCA₀ = S_1 and ACA₀ = S_3. We shall also state several results of Reverse Mathematics with respect to WKL₀ (theorem I.10.3 below).

In order to motivate WKL₀ in terms of foundations of mathematics, consider our Main Question (§I.1) as it applies to three specific theorems of ordinary mathematics: the Bolzano/Weierstraß theorem, the Heine/Borel covering lemma, the maximum principle. We have seen in I.8.4, I.8.6, I.8.7 that these three theorems are not provable in RCA₀. However, we have definitively answered the Main Question only for the

Bolzano/Weierstraß theorem, not for the other two. We have seen in I.9.1 that Bolzano/Weierstraß is equivalent to ACA_0 over RCA_0.

It will turn out (theorem I.10.3) that the Heine/Borel covering lemma, the maximum principle, and many other ordinary mathematical theorems are equivalent to each other and to weak König's lemma, over RCA_0. Thus WKL_0 is the weakest natural subsystem of Z_2 in which these ordinary mathematical theorems are provable. Thus WKL_0 provides the answer to these instances of the Main Question.

It will also turn out that WKL_0 is sufficiently strong to accommodate a large portion of mathematical practice, far beyond what is available in RCA_0, including many of the best-known non-constructive theorems. This will become clear in chapter IV.

We now present the definition of WKL_0.

DEFINITION I.10.1 (weak König's lemma). The following definitions are made within RCA_0. We use $\{0, 1\}^{<\mathbb{N}}$ or $2^{<\mathbb{N}}$ to denote the full binary tree, i.e., the set of (codes for) finite sequences of 0's and 1's. *Weak König's lemma* is the following statement: Every infinite subtree of $2^{<\mathbb{N}}$ has an infinite path. (Compare definition I.6.5 and example I.8.8.)

WKL_0 is defined to be the subsystem of Z_2 consisting of RCA_0 plus weak König's lemma.

REMARK I.10.2 (ω-models of WKL_0). By example I.8.8, the ω-model REC consisting of all recursive subsets of ω does not satisfy weak König's lemma. Hence REC is not a model of WKL_0. Since REC is the minimum ω-model of RCA_0 (remark I.7.5), it follows that RCA_0 is a proper subsystem of WKL_0. In addition, I.8.8 implies that WKL_0 is a subsystem of ACA_0. That it is a proper subsystem is not so obvious, but we shall see this in §VIII.2, where it is shown for instance that REC is the intersection of all ω-models of WKL_0. Thus we have

$$RCA_0 \subsetneq WKL_0 \subsetneq ACA_0$$

and there are ω-models for the independence.

We now list several results of Reverse Mathematics with respect to WKL_0. These results will be proved in chapter IV.

THEOREM I.10.3 (Reverse Mathematics for WKL_0). *Within RCA_0 one can prove that WKL_0 is equivalent to each of the following ordinary mathematical statements:*

1. *The Heine/Borel covering lemma: Every covering of the closed interval $[0, 1]$ by a sequence of open intervals has a finite subcovering* (§IV.1).
2. *Every covering of a compact metric space by a sequence of open sets has a finite subcovering* (§IV.1).
3. *Every continuous real-valued function on $[0, 1]$, or on any compact metric space, is bounded* (§IV.2).

4. *Every continuous real-valued function on* $[0, 1]$, *or on any compact metric space, is uniformly continuous* (§IV.2).
5. *Every continuous real-valued function on* $[0, 1]$ *is Riemann integrable* (§IV.2).
6. *The maximum principle: Every continuous real-valued function on* $[0, 1]$, *or on any compact metric space, has, or attains, a supremum* (§IV.2).
7. *The local existence theorem for solutions of (finite systems of) ordinary differential equations* (§IV.8).
8. *Gödel's completeness theorem: every finite, or countable, set of sentences in the predicate calculus has a countable model* (§IV.3).
9. *Every countable commutative ring has a prime ideal* (§IV.6).
10. *Every countable field (of characteristic* 0) *has a unique algebraic closure* (§IV.5).
11. *Every countable formally real field is orderable* (§IV.4).
12. *Every countable formally real field has a (unique) real closure* (§IV.4).
13. *Brouwer's fixed point theorem: Every uniformly continuous function* $\phi : [0, 1]^n \to [0, 1]^n$ *has a fixed point* (§IV.7).
14. *The separable Hahn/Banach theorem: If* f *is a bounded linear functional on a subspace of a separable Banach space, and if* $\|f\| \leq 1$, *then* f *has an extension* $\tilde{f}$ *to the whole space such that* $\|\tilde{f}\| \leq 1$ (§IV.9).

REMARK I.10.4 (mathematics within WKL$_0$). Theorem I.10.3 illustrates how WKL$_0$ is much stronger than RCA$_0$ from the viewpoint of mathematical practice. In fact, WKL$_0$ is strong enough to prove many well known nonconstructive theorems that are extremely important for mathematical practice but not true in the ω-model REC, hence not provable in RCA$_0$ (see §I.8).

REMARK I.10.5 (first order part of WKL$_0$). We have seen that WKL$_0$ is much stronger than RCA$_0$ with respect to both ω-models (remark I.10.2) and mathematical practice (theorem I.10.3, remark I.10.4). Nevertheless, it can be shown that WKL$_0$ is of the same strength as RCA$_0$ in a proof-theoretic sense. Namely, the first order part of WKL$_0$ is the same as that of RCA$_0$, *viz.* Σ_1^0-PA. (See also remark I.7.6.) In fact, given any model M of RCA$_0$, there exists a model $M' \supseteq M$ of WKL$_0$ having the same first order part as M. This model-theoretic conservation result will be proved in §IX.2.

Another key conservation result is that WKL$_0$ is conservative over the formal system known as PRA or *primitive recursive arithmetic*, with respect to Π_2^0 sentences. In particular, given a Σ_1^0 formula $\varphi(m, n)$ and a proof of $\forall m \, \exists n \, \varphi(m, n)$ in WKL$_0$, we can find a primitive recursive function $f : \omega \to \omega$ such that $\varphi(m, f(m))$ holds for all $m \in \omega$. This interesting and important result will be proved in §IX.3.

REMARK I.10.6 (Hilbert's program). The results of chapters IV and IX
are of great importance with respect to the foundations of mathematics,
specifically Hilbert's program. Hilbert's intention [114] was to justify all
of mathematics (including infinitistic, set-theoretic mathematics) by re-
ducing it to a restricted form of reasoning known as finitism. Gödel's
[94, 115, 55, 222] limitative results show that there is no hope of realizing
Hilbert's program completely. However, results along the lines of theorem
I.10.3 and remark I.10.5 show that a large portion of infinitistic mathematical practice is in fact finitistically reducible, because it can be carried
out in WKL_0. Thus we have a significant partial realization of Hilbert's
program of finitistic reductionism. See also remark IX.3.18.

Notes for §I.10. The formal system WKL_0 was first introduced by Friedman [69]. In the model-theoretic literature, ω-models of WKL_0 are sometimes known as *Scott systems*, referring to Scott [217]. Chapter IV of this
book is devoted to the development of mathematics within WKL_0 and Reverse Mathematics for WKL_0. Models of WKL_0 are discussed in §§VIII.2,
IX.2, and IX.3 of this book. The original paper on Hilbert's program
is Hilbert [114]. The significance of WKL_0 and Reverse Mathematics for
partial realizations of Hilbert's program is expounded in Simpson [246].

I.11. The System ATR_0

In this section we introduce and discuss ATR_0, a subsystem of Z_2 consisting of ACA_0 plus a set existence axiom known as *arithmetical transfinite
recursion*. Informally, arithmetical transfinite recursion can be described
as the assertion that the Turing jump operator can be iterated along any
countable well ordering starting at any set. The precise statement is given
in definition I.11.1 below.

From the standpoint of foundations of mathematics, the motivation for
ATR_0 is similar to the motivation for WKL_0, as explained in §I.10. (See
also the analogy in I.11.7 below.) Using the notation of §I.9, $\mathsf{ATR}_0 = S_4$
is intermediate between $\mathsf{ACA}_0 = S_3$ and $\Pi_1^1\text{-}\mathsf{CA}_0 = S_5$. It turns out that
ATR_0 is equivalent to several theorems of ordinary mathematics which are
provable in $\Pi_1^1\text{-}\mathsf{CA}_0$ but not in ACA_0.

As an example, consider the *perfect set theorem*: Every uncountable
closed set (or analytic set) has a perfect subset. We shall see that ATR_0 is
equivalent over RCA_0 to (either form of) the perfect set theorem. Thus
ATR_0 is the weakest natural subsystem of Z_2 in which the perfect set
theorem is provable. Actually, ATR_0 provides the answer not only to this
instance of the Main Question (§I.9) but also to many other instances
of it; see theorem I.11.5 below. Moreover, ATR_0 is sufficiently strong
to accommodate a large portion of mathematical practice beyond ACA_0,

including many basic theorems of infinitary combinatorics and classical descriptive set theory.

We now proceed to the definition of ATR$_0$.

DEFINITION I.11.1 (arithmetical transfinite recursion).   Consider an arithmetical formula $\theta(n, X)$ with a free number variable n and a free set variable X. Note that $\theta(n, X)$ may also contain parameters, i.e., additional free number and set variables. Fixing these parameters, we may view θ as an "arithmetical operator" $\Theta: P(\mathbb{N}) \to P(\mathbb{N})$, defined by

$$\Theta(X) = \{n \in \mathbb{N}: \theta(n, X)\}.$$

Now let $A, <_A$ be any countable well ordering (definition I.6.1), and consider the set $Y \subseteq \mathbb{N}$ obtained by transfinitely iterating the operator Θ along $A, <_A$. This set Y is defined by the following conditions: $Y \subseteq \mathbb{N} \times A$ and, for each $a \in A$, $Y_a = \Theta(Y^a)$, where $Y_a = \{m: (m, a) \in Y\}$ and $Y^a = \{(n, b): n \in Y_b \wedge b <_A a\}$. Thus, for each $a \in A$, Y^a is the result of iterating Θ along the initial segment of $A, <_A$ up to but not including a, and Y_a is the result of applying Θ one more time.

Finally, *arithmetical transfinite recursion* is the axiom scheme asserting that such a set Y exists, for every arithmetical operator Θ and every countable well ordering $A, <_A$. We define ATR$_0$ to consist of ACA$_0$ plus the scheme of arithmetical transfinite recursion. It is easy to see that ATR$_0$ is a subsystem of Π^1_1-CA$_0$, and we shall see below that it is a proper subsystem.

EXAMPLE I.11.2 (the ω-model ARITH).   Recall the ω-model

$$\text{ARITH} = \text{Def}((\omega, +, \cdot, 0, 1, <))$$
$$= \{X \subseteq \omega: \exists n \in \omega \; X \leq_T \text{TJ}(n, \emptyset)\}$$

consisting of all arithmetically definable subsets of ω (remarks I.3.3 and I.3.4). We have seen that ARITH is the minimum ω-model of ACA$_0$. Trivially for each $n \in \omega$ we have $\text{TJ}(n, \emptyset) \in \text{ARITH}$; here $\text{TJ}(n, \emptyset)$ is the result of iterating the Turing jump operator n times, i.e., along a finite well ordering of order type n. On the other hand, ARITH does not contain $\text{TJ}(\omega, \emptyset)$, the result of iterating the Turing jump operator ω times, i.e., along the well ordering $(\omega, <)$. Thus ARITH fails to satisfy this instance of arithmetical transfinite recursion. Hence ARITH is not an ω-model of ATR$_0$.

EXAMPLE I.11.3 (the ω-model HYP).   Another important ω-model is

$$\text{HYP} = \{X \subseteq \omega: X \leq_H \emptyset\}$$
$$= \{X \subseteq \omega: X \text{ is hyperarithmetical}\}$$
$$= \{X \subseteq \omega: \exists \alpha < \omega_1^{CK} \; X \leq_T \text{TJ}(\alpha, \emptyset)\}.$$

Here α ranges over the recursive ordinals, i.e., the countable ordinals which are order types of recursive well orderings of ω. We use ω_1^{CK} to denote Church/Kleene ω_1, i.e., the least nonrecursive ordinal. Clearly HYP is much larger than ARITH, and HYP contains many sets which are defined by arithmetical transfinite recursion. However, as we shall see in §VIII.3, HYP does not contain enough sets to be an ω-model of ATR$_0$.

REMARK I.11.4 (ω-models of ATR$_0$). In §§VII.2 and VIII.6 we shall prove two facts: (1) every β-model is an ω-model of ATR$_0$; (2) the intersection of all β-models is HYP, the ω-model consisting of the hyperarithmetical sets. From this it follows that HYP, although not itself an ω-model of ATR$_0$, is the intersection of all such ω-models. Hence ATR$_0$ does not have a minimum ω-model or a minimum β-model. Combining these observations with what we already know about ω-models of ACA$_0$ and Π_1^1-CA$_0$ (remarks I.3.4 and I.5.4), we see that

$$\mathsf{ACA}_0 \subsetneqq \mathsf{ATR}_0 \subsetneqq \Pi_1^1\text{-}\mathsf{CA}_0$$

and there are ω-models for the independence.

We now list several results of Reverse Mathematics with respect to ATR$_0$. These results will be proved in chapter V.

THEOREM I.11.5 (Reverse Mathematics for ATR$_0$). *Within* RCA$_0$ *one can prove that* ATR$_0$ *is equivalent to each of the following ordinary mathematical statements*:

1. *Any two countable well orderings are comparable* (§V.6).
2. *Ulm's theorem: Any two countable reduced Abelian p-groups which have the same Ulm invariants are isomorphic* (§V.7).
3. *The perfect set theorem: Every uncountable closed, or analytic, set has a perfect subset* (§V.4, V.5).
4. *Lusin's separation theorem: Any two disjoint analytic sets can be separated by a Borel set* (§§V.3, V.5).
5. *The domain of any single-valued Borel set in the plane is a Borel set* (§V.3, V.5).
6. *Every open, or clopen, subset of $\mathbb{N}^{\mathbb{N}}$ is determined* (§V.8).
7. *Every open, or clopen, subset of $[\mathbb{N}]^{\mathbb{N}}$ has the Ramsey property* (§V.9).

REMARK I.11.6 (mathematics within ATR$_0$). Theorem I.11.5 illustrates how ATR$_0$ is much stronger than ACA$_0$ from the viewpoint of mathematical practice. Namely, ATR$_0$ proves many well known ordinary mathematical theorems which fail in the ω-models ARITH and HYP and hence are not provable in ACA$_0$ (see §I.4) or even in somewhat stronger systems such as Σ_1^1-AC$_0$ (§VIII.4). A common feature of such theorems is that they require, implicitly or explicitly, a good theory of countable ordinal numbers.

Remark I.11.7 (Σ_1^0 and Σ_1^1 separation). From the viewpoint of mathematical practice, we have already noted an interesting analogy between WKL_0 and ATR_0, suggested by the following equation:

$$\frac{\text{WKL}_0}{\text{ACA}_0} \approx \frac{\text{ATR}_0}{\Pi_1^1\text{-CA}_0}.$$

We shall now extend this analogy by reformulating WKL_0 and ATR_0 in terms of separation principles.

Define Σ_1^0 *separation* to be the axiom scheme consisting of (the universal closures of) all formulas of the form

$$(\forall n \, \neg(\varphi_1(n) \wedge \varphi_2(n))) \to$$
$$\exists X \, (\forall n \, (\varphi_1(n) \to n \in X) \wedge \forall n \, (\varphi_2(n) \to n \notin X)),$$

where $\varphi_1(n)$ and $\varphi_2(n)$ are any Σ_1^0 formulas, n is any number variable, and X is a set variable which does not occur freely in $\varphi_1(n) \wedge \varphi_2(n)$. Define Σ_1^1 *separation* similarly, with Σ_1^1 formulas instead of Σ_1^0 formulas. It turns out that

$$\text{WKL}_0 \equiv \Sigma_1^0 \text{ separation},$$

and

$$\text{ATR}_0 \equiv \Sigma_1^1 \text{ separation},$$

over RCA_0. These equivalences, which will be proved in §§IV.4 and V.5 respectively, serve to strengthen the above-mentioned analogy between WKL_0 and ATR_0. They will also be used as technical tools for proving several of the reversals given by theorems I.10.3 and I.11.5.

Remark I.11.8. Another analogy in the same vein as that of I.11.7 is

$$\frac{\text{WKL}_0}{\text{RCA}_0} \approx \frac{\text{ATR}_0}{\Delta_1^1\text{-CA}_0}.$$

The system $\Delta_1^1\text{-CA}_0$ will be studied in §§VIII.3 and VIII.4, where we shall see that HYP is its minimum ω-model. Recall also (remark I.7.5) that REC is the minimum ω-model of

$$\text{RCA}_0 \equiv \Delta_1^0\text{-CA}_0.$$

Remark I.11.9 (first order part of ATR_0). It is known that the first order part of ATR_0 is the same as that of Feferman's system IR of predicative analysis; indeed, these two systems prove the same Π_1^1 sentences. Thus our development of mathematics within ATR_0 (theorem I.11.5, remark I.11.6, chapter V) may be viewed as contributions to a program of "predicative reductionism," analogous to Hilbert's program of finitistic reductionism (remark I.10.6, section IX.3). See also the proof of theorem IX.5.7 below.

Notes for §I.11. The formal system ATR_0 was first investigated by Friedman [68, 69] (see also Friedman [62, chapter II]) and Steel [256, chapter I]. A key reference for ATR_0 is Friedman/McAloon/Simpson [76]. Chapter V of this book is devoted to the development of mathematics within ATR_0 and Reverse Mathematics for ATR_0. Models of ATR_0 are discussed in §§VII.2, VII.3 and VIII.6. The basic reference for formal systems of predicative analysis is Feferman [56, 57]. The significance of ATR_0 for predicative reductionism has been discussed by Simpson [238, 246].

I.12. The Main Question, Revisited

The Main Question was introduced in §I.1. We now reexamine it in light of the results outlined in §§I.2 through I.11.

The Main Question asks which set existence axioms are needed to support ordinary mathematical reasoning. We take "needed" to mean that the set existence axioms are to be as weak as possible. When developing precise formal versions of the Main Question, it is natural also to consider formal languages which are as weak as possible. The language L_2 comes to mind because it is just adequate to define the majority of ordinary mathematical concepts and to express the bulk of ordinary mathematical reasoning. This leads in §I.2 to the consideration of subsystems of Z_2.

Two of the most obvious subsystems of Z_2 are ACA_0 and Π_1^1-CA_0, and in §§I.3–I.6 we outline the development of ordinary mathematics in these systems. The upshot of this is that a great many ordinary mathematical theorems are provable in ACA_0, and that of the exceptions, most are provable in Π_1^1-CA_0. The exceptions tend to involve countable ordinal numbers, either explicitly or implicitly. Another important subsystem of Z_2 is RCA_0, which is seen in §§I.7 and I.8 to embody a kind of formalized computable or constructive mathematics. Thus we have an approximate answer to the Main Question.

We then turn to a sharpened form of the Main Question, where we insist that the ordinary mathematical theorems should be logically equivalent to the set existence axioms needed to prove them. Surprisingly, this demand can be met in some cases; several ordinary mathematical theorems turn out to be equivalent over RCA_0 to either ACA_0 or Π_1^1-CA_0. This is our theme of Reverse Mathematics in §I.9. But the situation is not entirely satisfactory, because many ordinary mathematical theorems seem to fall into the gaps.

In order to improve the situation, we introduce two additional systems: WKL_0 lying strictly between RCA_0 and ACA_0, and analogously ATR_0 lying strictly between ACA_0 and Π_1^1-CA_0. These systems are introduced in §§I.10 and I.11 respectively. With this expanded complement of subsystems of Z_2, a certain stability is achieved; it now seems possible to "calibrate" a great many ordinary mathematical theorems, by showing that they are either provable in RCA_0 or equivalent over RCA_0 to WKL_0, ACA_0, ATR_0, or $\Pi_1^1\text{-}CA_0$.

Historically, the intermediate systems WKL_0 and ATR_0 were discovered in exactly in this way, as a response to the needs of Reverse Mathematics. See for example the discussion in Simpson [246, §§4,5].

From the above it is clear that the five basic systems RCA_0, WKL_0, ACA_0, ATR_0, $\Pi_1^1\text{-}CA_0$ arise naturally from investigations of the Main Question. The proof that these systems are mathematically natural is provided by Reverse Mathematics.

As a perhaps not unexpected byproduct, we note that these same five systems turn out to correspond to various well known, philosophically motivated programs in foundations of mathematics, as indicated in table 1. The foundational programs that we have in mind are: Bishop's program of constructivism [20] (see however remarks I.8.9 and IV.2.8); Hilbert's program of finitistic reductionism [114, 246] (see remarks I.10.6 and IX.3.18); Weyl's program of predicativity [274] as developed by Feferman [56, 57, 59]; predicative reductionism as developed by Friedman and Simpson [69, 76, 238, 247]; impredicativity as developed in Buchholz/Feferman/Pohlers/Sieg [29]. Thus, by studying the formalization of mathematics and Reverse Mathematics for the five basic systems, we can develop insight into the mathematical consequences of these philosophical proposals. Thus we can expect this book and other Reverse Mathematics studies to have a substantial impact on the philosophy of mathematics.

TABLE 1. Foundational programs and the five basic systems.

RCA_0	constructivism	Bishop
WKL_0	finitistic reductionism	Hilbert
ACA_0	predicativism	Weyl, Feferman
ATR_0	predicative reductionism	Friedman, Simpson
$\Pi_1^1\text{-}CA_0$	impredicativity	Feferman *et al.*

I.13. Outline of Chapters II through X

This section of our introductory chapter I consists of an outline of the remaining chapters.

The bulk of the material is organized in two parts. Part A consists of chapters II through VI and focuses on the development of mathematics within the five basic systems: RCA_0, WKL_0, ACA_0, ATR_0, $\Pi_1^1\text{-}CA_0$. A principal theme of Part A is Reverse Mathematics (see also §I.9). Part B, consisting of chapters VII through IX, is concerned with metamathematical properties of various subsystems of Z_2, including but not limited to the five basic systems. Chapters VII, VIII, and IX deal with β-models, ω-models, and non-ω-models, respectively. At the end of the book there is an appendix, chapter X, in which additional results are presented without proof but with references to the published literature. See also table 2.

TABLE 2. An overview of the entire book.

Introduction	Chapter I	introductory survey
	Chapter II	RCA_0
Part A	Chapter III	ACA_0
(mathematics within	Chapter IV	WKL_0
the 5 basic systems)	Chapter V	ATR_0
	Chapter VI	$\Pi_1^1\text{-}CA_0$
Part B	Chapter VII	β-models
(models of	Chapter VIII	ω-models
various systems)	Chapter IX	non-ω-models
Appendix	Chapter X	additional results

Part A: Mathematics Within Subsystems of Z_2. Part A consists of a key chapter II on the development of ordinary mathematics within RCA_0, followed by chapters III, IV, V, and VI on ordinary mathematics within the other four basic systems: ACA_0, WKL_0, ATR_0, and $\Pi_1^1\text{-}CA_0$, respectively. These chapters present many results of Reverse Mathematics showing that particular set existence axioms are necessary and sufficient to prove particular ordinary mathematical theorems. Table 3 indicates in more detail exactly where some of these results may be found. Table 3 may serve as a guide or road map concerning the role of set existence axioms in ordinary mathematical reasoning.

Chapter II: RCA_0. In §II.1 we define the formal system RCA_0 consisting of Δ_1^0 comprehension and Σ_1^0 induction. After that, the rest of chapter II is concerned with the development of ordinary mathematics within RCA_0. Although chapter II does not itself contain any Reverse Mathematics, it is necessarily a prerequisite for all of the Reverse Mathematics results to be presented in later chapters. This is because RCA_0 serves as our weak base theory (see §I.9 above).

TABLE 3. Ordinary mathematics within the five basic systems.

	RCA₀	WKL₀	ACA₀	ATR₀	Π¹₁-CA₀
analysis (separable):					
differential equations	IV.8	IV.8			
continuous functions	II.6, II.7	IV.2, IV.7	III.2		
completeness, etc.	II.4	IV.1	III.2		
Banach spaces	II.10	IV.9, X.2			X.2
open and closed sets	II.5	IV.1		V.4, V.5	VI.1
Borel and analytic sets	V.1			V.1, V.3	VI.2, VI.3
algebra (countable):					
countable fields	II.9	IV.4, IV.5	III.3		
commutative rings	III.5	IV.6	III.5		
vector spaces	III.4		III.4		
Abelian groups	III.6		III.6	V.7	VI.4
miscellaneous:					
mathematical logic	II.8	IV.3			
countable ordinals	V.1		V.6.10	V.1, V.6	
infinite matchings		X.3	X.3	X.3	
the Ramsey property			III.7	V.9	VI.6
infinite games			V.8	V.8	VI.5

In §II.2 we employ a device reminiscent of *Gödel's beta function* to prove within RCA₀ that finite sequences of natural numbers can be encoded as single numbers. This encoding is essential for §II.3, where we prove within RCA₀ that the class of functions from $f : \mathbb{N}^k \to \mathbb{N}$, $k \in \mathbb{N}$, is closed under *primitive recursion*. Another key technical result of §II.3 is that RCA₀ proves *bounded Σ_1^0 comprehension*, i.e., the existence of bounded subsets of $\mathbb{N}$ defined by Σ_1^0 formulas.

Armed with these preliminary results from §§II.2 and II.3, we begin the development of mathematics proper in §II.4 by discussing the *number systems* $\mathbb{N}$, $\mathbb{Z}$, $\mathbb{Q}$, and $\mathbb{R}$. Also in §II.4 we present an important completeness property of the real number system, known as *nested interval completeness*. An RCA₀ version of the *Baire category theorem* for k-dimensional Euclidean spaces $\mathbb{R}^k$, $k \in \mathbb{N}$, is stated; the proof is postponed to §II.5.

Sections II.5, II.6, and II.7 discuss *complete separable metric spaces* in RCA₀. Among the notions introduced (in a form appropriate for RCA₀) are *open sets*, *closed sets*, and *continuous functions*. We prove the following important technical result: An open set in a complete separable metric space $\widehat{A}$ is the same thing as a set in $\widehat{A}$ defined by a Σ_1^0 formula with an extensionality property (II.5.7). Nested interval completeness is used to prove the *intermediate value property* for continuous functions $\phi : \mathbb{R} \to \mathbb{R}$ in RCA$_0$ (II.6.6). A number of basic topological results for complete separable metric spaces are shown to be provable in RCA$_0$. Among these are *Urysohn's lemma* (II.7.3), the *Tietze extension theorem* (II.7.5), the *Baire category theorem* (II.5.8), and *paracompactness* (II.7.2).

Sections II.8 and II.9 deal with *mathematical logic* and *countable algebra*, respectively. We show in §II.8 that some surprisingly strong versions of basic results of mathematical logic can be proved in RCA$_0$. Among these are *Lindenbaum's lemma*, the *Gödel completeness theorem*, and the *strong soundness theorem*, via *cut elimination*. To illustrate the power of these results, we show that RCA$_0$ proves the consistency of *elementary function arithmetic*, EFA. In §II.9 we apply the results of §§II.3 and II.8 in a discussion of countable algebraically closed and real closed fields in RCA$_0$. We use *quantifier elimination* to prove within RCA$_0$ that every countable field has an *algebraic closure*, and that every countable ordered field has a *unique real closure*. (Uniqueness of algebraic closure is discussed later, in §IV.5.)

Section II.10 presents some basic concepts and results of the theory of *separable Banach spaces* and *bounded linear operators*, within RCA$_0$. It is shown that the standard proof of the *Banach/Steinhaus uniform boundedness principle*, via the Baire category theorem, goes through in this setting.

Chapter III: ACA$_0$. Chapter III is concerned with ACA$_0$, the formal system consisting of RCA$_0$ plus arithmetical comprehension. The focus of chapter III is Reverse Mathematics with respect to ACA$_0$. (See also §§I.4, I.3, and I.9.)

In §III.1 we define ACA$_0$ and show that it is equivalent over RCA$_0$ to Σ_1^0 comprehension and to the principle that for any function $f : \mathbb{N} \to \mathbb{N}$, the range of f exists. This equivalence is used to establish all of the Reverse Mathematics results which occupy the rest of the chapter. For example, it is shown in §III.2 that ACA$_0$ is equivalent to the *Bolzano/Weierstraß theorem*, i.e., sequential compactness of the closed unit interval. Also in §III.2 we introduce the notion of *compact metric space*, and we show that ACA$_0$ is equivalent to the principle that any sequence of points in a compact metric space has a convergent subsequence. We end §III.2 by showing that ACA$_0$ is equivalent to the *Ascoli lemma* concerning bounded equicontinuous families of continuous functions.

Sections III.3, III.4, III.5 and III.6 are concerned with countable algebra in ACA$_0$. It is perhaps interesting to note that chapter III has much more to say about algebra than about analysis.

We begin in §III.3 by reexamining the notion of an algebraic closure $h: K \to \widetilde{K}$ of a countable field K. We define a notion of *strong algebraic closure*, i.e., an algebraic closure with the additional property that the range of the embedding h exists as a set. Although the existence of algebraic closures is provable in RCA$_0$, we show in §III.3 that the existence of strong algebraic closures is equivalent to ACA$_0$. Similarly, although it is provable in RCA$_0$ that any countable ordered field has a real closure, we show in §III.3 that ACA$_0$ is required to prove the existence of a *strong real closure*.

In §III.4 we show that ACA$_0$ is equivalent to the theorem that every countable *vector space* over a countable field (or over the rational field $\mathbb{Q}$) has a basis. We then refine this result (following Metakides/Nerode [187]) by showing that ACA$_0$ is also equivalent to the assertion that every countable, infinite dimensional vector space over $\mathbb{Q}$ has an infinite linearly independent set. We also obtain similar results for *transcendence bases* of countable fields.

In §III.5 we turn to countable commutative rings. We use localization to show that ACA$_0$ is equivalent to the assertion that every countable commutative ring has a *maximal ideal*. In §III.6 we discuss *countable Abelian groups*. We show that ACA$_0$ is equivalent to the assertion that, for every countable Abelian group G, the *torsion subgroup* of G exists. We also show that, although the existence of *divisible closures* is provable in RCA$_0$, the uniqueness requires ACA$_0$

In §III.7 we consider *Ramsey's theorem*. We define RT(k) to be Ramsey's theorem for exponent k, i.e., the assertion that for every coloring of the k-element subsets of $\mathbb{N}$ with finitely many colors, there exists an infinite subset of $\mathbb{N}$ all of whose k-element subsets have the same color. We show that ACA$_0$ is equivalent to RT(k) for each "standard integer" $k \in \omega$, $k \geq 3$. From the viewpoint of Reverse Mathematics, the case $k = 2$ turns out to be anomalous: RT(2) is provable in ACA$_0$ but neither equivalent to ACA$_0$ nor provable in WKL$_0$. See also the notes at the end of §III.7. Another somewhat annoying anomaly is that the general assertion of Ramsey's theorem, $\forall k$ RT(k), is slightly stronger than ACA$_0$, due to the fact that ACA$_0$ lacks full induction.

An interesting technical result of §III.7 is that ACA$_0$ is equivalent to *König's lemma*: every infinite, finitely branching tree $T \subseteq \mathbb{N}^{<\mathbb{N}}$ has an infinite path. It turns out that ACA$_0$ is also equivalent to a much weaker sounding statement, namely König's lemma restricted to *binary trees*. (A tree $T \subseteq \mathbb{N}^{<\mathbb{N}}$ is defined to be binary if each node of T has at most two immediate successors.) The binary tree version of König's lemma is to be contrasted with its special case, *weak König's lemma*: every infinite tree $T \subseteq 2^{<\mathbb{N}}$ has an infinite path. It is important to understand that, in terms of set existence axioms and Reverse Mathematics, weak König's lemma is much weaker than König's lemma for binary trees. These observations provide a transition to the next chapter, which is concerned only with weak König's lemma and not at all with König's lemma for binary trees.

Chapter IV: WKL$_0$. Chapter IV focuses on Reverse Mathematics with respect to the formal system WKL$_0$ consisting of RCA$_0$ plus weak König's lemma. (See also the previous paragraph and §I.10.)

We begin in §IV.1 by showing that weak König's lemma is equivalent over RCA$_0$ to the *Heine/Borel covering lemma*: every covering of the closed unit interval [0, 1] by a sequence of open intervals has a finite subcovering. We then generalize this result by showing that WKL$_0$ proves a Heine/Borel covering property for arbitrary *compact metric spaces*. In order to obtain this generalization, we first prove a technical result: WKL$_0$ proves *bounded König's lemma*, i.e., König's lemma for subtrees of $\mathbb{N}^{<\mathbb{N}}$ which are bounded. (A tree $T \subseteq \mathbb{N}^{<\mathbb{N}}$ is said to be *bounded* if there exists a function $g : \mathbb{N} \to \mathbb{N}$ such that $\tau(m) < g(m)$ for all $\tau \in T$, $m < \mathrm{lh}(\tau)$.) We also develop some additional technical results which are needed in later sections.

Section IV.2 shows that various properties of continuous functions on compact metric spaces are provable in WKL$_0$ and in fact equivalent to weak König's lemma over RCA$_0$. Among the properties considered are *uniform continuity, Riemann integrability,* the *Weierstraß polynomial approximation theorem,* and the *maximum principle.* A key technical notion here is that of *modulus of uniform continuity* (definition IV.2.1).

In §IV.3 we return to mathematical logic. We show that several well known theorems of mathematical logic, such as the *completeness theorem* and the *compactness theorem* for both propositional logic and predicate calculus, are each equivalent to weak König's lemma over RCA$_0$. Our results here in §IV.3 are to be contrasted with those of §II.8.

Sections IV.4, IV.5 and IV.6 deal with countable algebra in WKL$_0$. We show in §IV.5 that weak König's lemma is equivalent to the assertion that every countable field has a *unique algebraic closure*. (We have already seen in §II.9 that the *existence* of algebraic closures is provable in RCA$_0$.) In §IV.4 we discuss *formally real fields,* i.e., fields in which -1 cannot be written as a sum of squares. We show that weak König's lemma is equivalent over RCA$_0$ to the assertion that every countable formally real field is *orderable,* and to the assertion that every countable formally real field has a *real closure.* In order to prove these results of Reverse Mathematics, we first prove a technical result characterizing WKL$_0$ in terms of Σ_1^0 *separation*; see also §I.11.

In §IV.6 we show that WKL$_0$ proves the existence of *prime ideals* in countable commutative rings. The argument for this result is somewhat interesting in that it involves not only two applications of weak König's lemma but also bounded Σ_1^0 comprehension. In addition, we obtain reversals showing that weak König's lemma is equivalent over RCA$_0$ to the existence of prime ideals, or even of radical ideals, in countable commutative rings. These results stand in contrast to §III.5, where we saw that ACA_0 is needed to prove the existence of *maximal ideals* in countable commutative rings. Thus it emerges that the usual textbook proof of the existence of prime ideals, via maximal ideals, is far from optimal with respect to its use of set existence axioms.

Sections IV.7, IV.8 and IV.9 are concerned with certain advanced topics in analysis. We begin in §IV.7 by showing that the well known *fixed point theorems* of Brouwer and Schauder are provable in WKL_0. In §IV.8 we use a fixed point technique to prove *Peano's existence theorem for solutions of ordinary differential equations*, in WKL_0. We also obtain reversals showing weak König's lemma is needed to prove the Brouwer and Schauder fixed point theorems and Peano's existence theorem. On the other hand, we note that the more familiar *Picard existence and uniqueness theorem*, assuming a Lipschitz condition, is already provable in RCA_0 alone.

Section IV.9 is concerned with Banach space theory in WKL_0. We build on the concepts and results of §§II.10 and IV.7. We begin by showing that yet another fixed point theorem, the *Markov/Kakutani theorem* for commutative families of affine maps, is provable in WKL_0. We then use this result to show that WKL_0 proves a version of the *Hahn/Banach extension theorem* for bounded linear functionals on separable Banach spaces. A reversal is also obtained.

Chapter V: ATR_0. Chapter V deals with mathematics in ATR_0, the formal system consisting of ACA_0 plus arithmetical transfinite recursion. (See also §I.11.) Many of the ordinary mathematical theorems considered in chapters V and VI are in the areas of countable combinatorics and classical descriptive set theory. The first few sections of chapter V focus on proving ordinary mathematical theorems in ATR_0. Reverse Mathematics with respect to ATR_0 is postponed to §V.5.

Chapter V begins with a preliminary §V.1 whose purpose is to elucidate the relationships among Σ_1^1 formulas, analytic sets, countable well orderings, and trees. An important tool is the *Kleene/Brouwer ordering* $KB(T)$ of an arbitrary tree $T \subseteq \mathbb{N}^{<\mathbb{N}}$. Key properties of the Kleene/Brouwer construction are: (1) $KB(T)$ is always a linear ordering; (2) $KB(T)$ is a well ordering if and only if T is well founded. The Kleene normal form theorem is proved in ACA_0 and is then used to show that any Π_1^1 assertion ψ can be expressed in ACA_0 by saying that an appropriately chosen tree T_ψ is well founded, or equivalently, $KB(T_\psi)$ is a well ordering.

In §V.2 we define the formal system ATR_0 and observe that it is strong enough to accommodate a good theory of *countable ordinal numbers*, encoded by countable well orderings. In §V.3 we show that ATR_0 is also strong enough to accommodate a good theory of *Borel and analytic sets* in the Cantor space $2^{\mathbb{N}}$. In this setting, the well known theorems of Souslin ("B is Borel if and only if B and its complement are analytic") and Lusin ("any two disjoint analytic sets can be separated by a Borel set") are proved, along with a lesser known closure property of Borel sets ("the domain of a single-valued Borel relation is Borel"). In §V.4 we advance our examination of classical descriptive set theory by showing that the *perfect set theorem* ("every uncountable analytic set has a nonempty perfect subset") is provable in ATR$_0$. This last result uses an interesting technique known as the *method of pseudohierarchies*, or "nonstandard H-sets", i.e., arithmetical transfinite recursion along countable linear orderings which are not well orderings.

In §V.5, most of the descriptive set-theoretic theorems mentioned in §§V.3 and V.4 are reversed, i.e., shown to be equivalent over RCA$_0$ to ATR$_0$. The reversals are based on our characterization of ATR$_0$ in terms of Σ^1_1 *separation*. See also §I.11. We also present the following alternative characterization: ATR$_0$ is equivalent to the assertion that, for any sequence of trees $\langle T_i : i \in \mathbb{N} \rangle$, if each T_i has at most one path, then the set $\{i : T_i$ has a path$\}$ exists. This equivalence is based on a sharpening of the Kleene normal form theorem.

We have already observed that the development of mathematics within ATR$_0$ seems to go hand in hand with a good theory of countable ordinal numbers. In §V.6 we sharpen this observation by showing that ATR$_0$ is actually equivalent over RCA$_0$ to a certain statement which is obviously indispensable for any such theory. The statement in question is, "any two countable well orderings are comparable", abbreviated CWO. The proof that CWO implies ATR$_0$ is rather technical and uses what are called *double descent trees*.

In §V.7 we return to the study of countable Abelian groups (see also §§III.6 and VI.4). We show that ATR$_0$ is needed to prove *Ulm's theorem* for reduced Abelian p-groups, as well as some consequences of Ulm's theorem. The reversals use the fact that ATR$_0$ is equivalent to CWO. Ulm's theorem is of interest with respect to our Main Question, because it seems to be one of the few places in analysis or algebra where transfinite recursion plays an apparently indispensable role.

In §§V.8 and V.9 we consider two other topics in ordinary mathematics where strong set existence axioms arise naturally. These are (1) infinite game theory, and (2) the Ramsey property.

The games considered in §V.8 are Gale/Stewart games, i.e., infinite games with perfect information. A payoff set $S \subseteq \mathbb{N}^{\mathbb{N}}$ is specified. Two players take turns choosing nonnegative integers $m_1, n_1, m_2, n_2, \ldots$, with full disclosure. The first player is declared the winner if the infinite sequence $\langle m_1, n_1, m_2, n_2, \ldots \rangle$ belongs to S. Otherwise the second player is declared the winner. Such a game is said to be *determined* if one player or the other has a winning strategy. Letting $\mathcal{S}$ be any class of payoff sets, $\mathcal{S}$-*determinacy* is the assertion that all games of this class are determined. It is well known that strong set existence axioms are correlated to determinacy for large classes of games. A striking result of this kind is due to Friedman

[66, 71], who showed that Borel determinacy requires $\aleph_1$ applications of the power set axiom.

We show in §V.8 that ATR_0 proves *open determinacy*, i.e., determinacy for all games in which the payoff set $S \subseteq \mathbb{N}^{\mathbb{N}}$ is open. This result uses pseudohierarchies, just as for the perfect set theorem. We also obtain a reversal, showing that open determinacy or even *clopen determinacy* is equivalent to ATR_0 over RCA_0. Our argument for the reversal proceeds via CWO. Along the way we obtain the following preliminary result: *determinacy for games of length* 3 is equivalent to ACA_0 over RCA_0.

As a consequence of open determinacy in ATR_0, we obtain the following interesting theorem: ATR_0 proves the Σ^1_1 *axiom of choice*. (More information on Σ^1_1 choice is in §VIII.4.)

In §V.9 we deal with a well known topological generalization of Ramsey's theorem. Let $[\mathbb{N}]^{\mathbb{N}}$ be the *Ramsey space*, i.e., the space of all infinite subsets of $\mathbb{N}$. Note that $[\mathbb{N}]^{\mathbb{N}}$ is canonically homeomorphic to the Baire space $\mathbb{N}^{\mathbb{N}}$ via $\Phi \colon [\mathbb{N}]^{\mathbb{N}} \cong \mathbb{N}^{\mathbb{N}}$ defined by

$$\Phi^{-1}(f) = \{f(0) + 1 + \cdots + 1 + f(n) \colon n \in \mathbb{N}\}.$$

A set $S \subseteq [\mathbb{N}]^{\mathbb{N}}$ is said to have the *Ramsey property* if there exists $X \in [\mathbb{N}]^{\mathbb{N}}$ such that either $[X]^{\mathbb{N}} \subseteq S$ or $[X]^{\mathbb{N}} \cap S = \emptyset$. (Here $[X]^{\mathbb{N}}$ denotes the set of infinite subsets of X.) The main result of §V.9 is that ATR_0 is equivalent over RCA_0 to the *open Ramsey theorem*, i.e., the assertion that every open subset of $[\mathbb{N}]^{\mathbb{N}}$ has the Ramsey property. The *clopen Ramsey theorem* is also seen to be equivalent over RCA_0 to ATR_0.

Chapter VI: $\Pi^1_1\text{-}\mathsf{CA}_0$. Chapter VI is concerned with mathematics and Reverse Mathematics with respect to the formal system $\Pi^1_1\text{-}\mathsf{CA}_0$, consisting of ACA_0 plus Π^1_1 comprehension. We show that $\Pi^1_1\text{-}\mathsf{CA}_0$ is just strong enough to prove several theorems of ordinary mathematics. It is interesting to note that several of these ordinary mathematical theorems, which are equivalent to Π^1_1 comprehension, have "ATR_0 counterparts" which are equivalent to arithmetical transfinite recursion. Thus chapter VI on $\Pi^1_1\text{-}\mathsf{CA}_0$ goes hand in hand with chapter V on ATR_0.

In §§VI.1 through VI.3 we consider several well known theorems of *classical descriptive set theory* in $\Pi^1_1\text{-}\mathsf{CA}_0$. We begin in §VI.1 by showing that the *Cantor/Bendixson theorem* ("every closed set consists of a perfect set plus a countable set") is equivalent to Π^1_1 comprehension. This result for the Baire space $\mathbb{N}^{\mathbb{N}}$ and the Cantor space $2^{\mathbb{N}}$ is closely related to an analysis of trees in $\mathbb{N}^{<\mathbb{N}}$ and $2^{<\mathbb{N}}$, respectively. The ATR_0 counterpart of the Cantor/Bendixson theorem is, of course, the perfect set theorem (§V.4).

In §VI.2 we show that *Kondo's theorem* (coanalytic uniformization) is provable in $\Pi^1_1\text{-}\mathsf{CA}_0$ and in fact equivalent to Π^1_1 comprehension over ATR_0. The reversal uses an ATR_0 formalization of *Suzuki's theorem* on Π^1_1 singletons.

In §VI.3 we consider *Silver's theorem*: For any coanalytic equivalence relation with uncountably many equivalence classes, there exists a nonempty perfect set of inequivalent elements. We show that a certain carefully stated reformulation of Silver's theorem is provable in ATR_0. (See lemma VI.3.1. The proof of this lemma is somewhat technical and uses *formalized hyperarithmetical theory* (§VIII.3) as well as *Gandy forcing* over countable coded ω-models.) We then use this ATR_0 result to show that Silver's theorem itself is provable in Π^1_1-CA_0. We also present a reversal showing that Silver's theorem specialized to Δ^0_2 equivalence relations is equivalent to Π^1_1 comprehension over RCA_0 (theorem VI.3.6).

In §VI.4 we resume our study of countable algebra. We show that Π^1_1 comprehension is equivalent over RCA_0 to the assertion that every countable Abelian group can be written as the direct sum of a divisible group and a reduced group. The ATR_0 counterpart of this assertion is Ulm's theorem (§V.7). Combining these results, we see that Π^1_1-CA_0 is just strong enough to develop the classical *structure theory of countable Abelian groups* as presented in, for instance, Kaplansky [136].

In §§VI.5 and VI.6 we resume our study of determinacy and the Ramsey property. We show that Π^1_1 comprehension is just strong enough to prove $\Sigma^0_1 \wedge \Pi^0_1$ *determinacy* and the Δ^0_2 *Ramsey theorem*. The ATR_0 counterparts of these results are, of course, Σ^0_1 determinacy (i.e., open determinacy) and the Σ^0_1 Ramsey theorem (i.e., the open Ramsey theorem). Our proof technique in §VI.6 uses countable coded β-models (§VII.2).

Section VI.7 serves as an appendix to §§VI.5 and VI.6. In it we remark that stronger forms of Ramsey's theorem and determinacy require *stronger set existence axioms*. For instance, the Δ^1_1 *Ramsey theorem* (i.e., the Galvin/Prikry theorem) and Δ^0_2 *determinacy* each require Π^1_1 *transfinite recursion* (theorem VI.7.3). Moreover, there are yet stronger forms of Ramsey's theorem and determinacy which go beyond Z_2 (remarks VI.7.6 and VI.7.7).

Note: The results in §VI.7 are stated without proof but with appropriate references to the published literature.

This completes our summary of Part A.

Part B: Models of Subsystems of Z_2. Part B is a fairly thorough study of metamathematical properties of subsystems of Z_2. We consider not only the five basic systems RCA_0, WKL_0, ACA_0, ATR_0, and Π^1_1-CA_0 but also many other systems, including Δ^1_k-CA_0 (Δ^1_k comprehension), Π^1_k-CA_0 (Π^1_k comprehension), Σ^1_k-AC_0 (Σ^1_k choice), Σ^1_k-DC_0 (Σ^1_k dependent choice), Π^1_k-TR_0 (Π^1_k transfinite recursion), and Π^1_k-TI_0 (Π^1_k transfinite induction), for arbitrary k in the range $1 \leq k \leq \infty$. Table 4 lists these systems in order of increasing *logical strength*, also known as *consistency strength*.

We have found it convenient to divide the metamathematical material of Part B into three chapters dealing with β-models, ω-models, and *non-ω-models* respectively. This threefold partition is perhaps somewhat misleading, and there are many cross-connections among the three chapters. This is mostly because the chapters which are ostensibly about β- and ω-models actually present their results in greater generality, so as to apply also to β- and ω-*submodels* of a given model, which need not itself be a β- or ω-model. Table 4 indicates where the main results concerning β-, ω- and non-ω-models of the various systems may be found.

Chapter VII: β-models. Recall from definition I.5.3 that a *β-model* is an ω-model M such that for any arithmetical formula $\theta(X)$ with parameters from M, if $\exists X \theta(X)$ then $(\exists X \in M)\theta(X)$. Such models are of importance because the concept of well ordering is absolute to them.

Throughout chapter VII, we find it convenient to consider a more general notion: M is a *β-submodel of* M' if M is a submodel of M' and, for all arithmetical formulas $\theta(X)$ with parameters from M, $M \models \exists X \theta(X)$ if and only if $M' \models \exists X \theta(X)$. Thus a β-model is the same thing as a β-submodel of the intended model $P(\omega)$.

Section VII.1 is introductory in nature. In it we characterize β-models of Π^1_1-CA$_0$ in terms of familiar recursion-theoretic notions. Namely, M is a β-model of Π^1_1-CA$_0$ if and only if M is closed under *relative recursiveness* and the *hyperjump*. We also obtain the obvious generalization to β-submodels. This is based on a formalized ACA$_0$ version of the *Kleene basis theorem*, according to which the sets recursive in HJ(X) form a basis for predicates which are arithmetical in X, provided HJ(X) exists.

In §VII.2 we consider *countable coded β-models*, i.e., β-models of the form $M = \{(W)_n : n \in \mathbb{N}\}$ where $W \subseteq \mathbb{N}$ and $(W)_n = \{m : (m, n) \in W\}$. Within ACA$_0$ we define the notion of *satisfaction* for such models, and we prove within ACA$_0$ that every such model satisfies ATR$_0$ and all instances of the *transfinite induction scheme*, Π^1_∞-TI$_0$, given by

$$\forall X (\mathrm{WO}(X) \to \mathrm{TI}(X, \varphi))$$

where φ is an arbitrary L$_2$-formula. Here WO(X) says that X is a countable well ordering, and TI(X, φ) expresses transfinite induction along X with respect to φ. We also prove within ACA$_0$ that if HJ(X) exists then there is a countable coded β-model $M \leq_T$ HJ(X) such that $X \in M$. These considerations have a number of interesting consequences: (1) Π^1_∞-TI$_0$ includes ATR$_0$; (2) Π^1_∞-TI$_0$ is not finitely axiomatizable; (3) there exists a β-model of Π^1_∞-TI$_0$ which is not a model of Π^1_1-CA$_0$; (4) Π^1_1-CA$_0$ proves the consistency of Π^1_∞-TI$_0$. We also obtain some technical results characterizing Π^1_2 sentences that are provable in Π^1_∞-TI$_0$ and in Π^1_2-TI$_0$.

In §VII.3 we introduce set-theoretic methods. We employ the language L$_{\mathrm{set}} = \{\in, =\}$ of Zermelo/Fraenkel set theory. Of key importance is an L$_{\mathrm{set}}$-theory ATR$_0^{\mathrm{set}}$, among whose axioms are the *Axiom of Countability*, asserting that all sets are hereditarily countable, and *Axiom Beta*, asserting that for any regular (i.e., well founded) binary relation r there exists a

TABLE 4. Models of subsystems of Z_2.

	β-models	ω-models	non-ω-models
RCA_0		VIII.1	IX.1
WKL_0		VIII.2; see note 1	IX.2–IX.3
Π^0_1-AC_0		,,	,,
Π^0_1-DC_0		,,	,,
strong Π^0_1-DC_0		,,	,,
ACA_0		VIII.1; see note 2	IX.1, IX.4.3–IX.4.6
Δ^1_1-CA_0		VIII.4; see note 2	IX.4.3–IX.4.6
Σ^1_1-AC_0		,,	,,
Σ^1_1-DC_0		VIII.4–VIII.5; notes 2, 3	
Π^1_1-TI_0		,,	
ATR_0	VII.2–VII.3, VIII.6	VIII.5–VIII.6; note 2	IX.4.7
Π^1_2-TI_0	VII.2.26–VII.2.32	see note 2	
Π^1_∞-TI_0	VII.2.14–VII.2.25	VIII.5.1–VIII.5.10; note 2	
strong Σ^1_1-DC_0	VII.6–VII.7	see notes 2 and 4	IX.4.8–IX.4.10
Π^1_1-CA_0	VII.1–VII.5, VII.7	,,	,,
Δ^1_2-CA_0	VII.5–VII.7	,,	,,
Σ^1_2-AC_0	VII.6	,,	,,
Σ^1_2-DC_0	,,	,,	
Π^1_1-TR_0	VII.1.18, VII.5.20, VII.7.12	VIII.4.24; see note 2	
strong Σ^1_2-DC_0	VII.6–VII.7	see notes 2 and 4	IX.4.8–IX.4.14
Π^1_{k+2}-CA_0	VII.5–VII.7	see note 2	,,
Δ^1_{k+3}-CA_0	,,	,,	,,
Σ^1_{k+3}-AC_0	VII.6	,,	,,
Σ^1_{k+3}-DC_0	,,	,,	
Π^1_{k+2}-TR_0	VII.5.20, VII.7.12	VIII.4.24; see note 2	
strong Σ^1_{k+3}-DC_0	VII.6–VII.7	see note 2	IX.4.8–IX.4.14
Π^1_∞-CA_0	VII.5–VII.7	,,	
Σ^1_∞-AC_0	VII.6–VII.7	,,	
Σ^1_∞-DC_0	,,	,,	

Notes:

1. Each of Π^0_1-AC_0 and Π^0_1-DC_0 and strong Π^0_1-DC_0 is equivalent to WKL_0. See lemma VIII.2.5.

Notes (cont.):

2. The ω-model incompleteness theorem VIII.5.6 applies to any system $S \supseteq \mathsf{ACA}_0$. The ω-model hard core theorem VIII.6.6 applies to any system $S \supseteq$ weak $\Sigma_1^1\text{-}\mathsf{AC}_0$. Quinsey's theorem VIII.6.12 applies to any system $S \supseteq \mathsf{ATR}_0$.

3. $\Pi_1^1\text{-}\mathsf{TI}_0$ is equivalent to $\Sigma_1^1\text{-}\mathsf{DC}_0$. See theorem VIII.5.12.

4. $\Sigma_2^1\text{-}\mathsf{AC}_0$ is equivalent to $\Delta_2^1\text{-}\mathsf{CA}_0$. $\Sigma_2^1\text{-}\mathsf{DC}_0$ is equivalent to $\Delta_2^1\text{-}\mathsf{CA}_0$ plus Σ_2^1 induction. Strong $\Sigma_1^1\text{-}\mathsf{DC}_0$ and strong $\Sigma_2^1\text{-}\mathsf{DC}_0$ are equivalent to $\Pi_1^1\text{-}\mathsf{CA}_0$ and $\Pi_2^1\text{-}\mathsf{CA}_0$, respectively. See remarks VII.6.3–VII.6.5 and theorem VII.6.9.

collapsing function, i.e., a function f such that $f(u) = \{f(v): \langle v, u \rangle \in r\}$ for all $u \in \mathrm{field}(r)$. By using well founded trees to encode hereditarily countable sets, we define a close relationship of mutual interpretability between ATR_0 and $\mathsf{ATR}_0^{\mathrm{set}}$. Under this interpretation, Σ_{k+1}^1 formulas of L_2 correspond to Σ_k^{set} formulas of L_{set} (theorem VII.3.24). Thus any formal system $T_0 \supseteq \mathsf{ATR}_0$ in L_2 is seen to have a *set-theoretic counterpart* T_0^{set} in L_{set} (definition VII.3.33). We point out that several familiar subsystems of Z_2 have elegant characterizations in terms of their set-theoretic counterparts. For instance, the principal axiom of $\Pi_\infty^0\text{-}\mathsf{TI}_0^{\mathrm{set}}$ is the $\in$-induction scheme, and the principal axiom of $\Sigma_2^1\text{-}\mathsf{AC}_0^{\mathrm{set}}$ is Σ_1^{set} collection.

In §VII.4 we explore Gödel's theory of *constructible sets* in a form appropriate for the study of subsystems of Z_2. We begin by defining within $\mathsf{ATR}_0^{\mathrm{set}}$ the inner model L^u of sets constructible from u, where u is any given nonempty transitive set. After that, we turn to *absoluteness results*. We prove within $\Pi_1^1\text{-}\mathsf{CA}_0^{\mathrm{set}}$ that the formula "r is a regular relation" is absolute to L^u. This fact is used to prove $\Pi_1^1\text{-}\mathsf{CA}_0^{\mathrm{set}}$ versions of the well known absoluteness theorems of Shoenfield and Lévy. We consider the inner models $L(X)$ and $\mathrm{HCL}(X)$ of sets that are constructible from X and *hereditarily constructibly countable* from X, respectively, where $X \subseteq \omega$. We prove within $\Pi_1^1\text{-}\mathsf{CA}_0^{\mathrm{set}}$ that $\mathrm{HCL}(X)$ satisfies $\Pi_1^1\text{-}\mathsf{CA}_0^{\mathrm{set}}$ plus $V = \mathrm{HCL}(X)$, and that Σ_2^1 and Σ_1^{set} formulas are absolute to $\mathrm{HCL}(X)$. We prove within $\mathsf{ATR}_0^{\mathrm{set}}$ that if $\mathrm{HCL}(X) \neq L(X)$ then $\mathrm{HCL}(X)$ satisfies $\Pi_\infty^1\text{-}\mathsf{CA}_0^{\mathrm{set}}$.

In §§VII.5, VII.6 and VII.7 we apply our results on constructible sets to the study of β-models of subsystems of second order arithmetic which are stronger than $\Pi_1^1\text{-}\mathsf{CA}_0$.

Section VII.5 is concerned with *strong comprehension schemes*. The main result is that if T_0 is any one of the systems $\Pi_1^1\text{-}\mathsf{CA}_0$, $\Delta_2^1\text{-}\mathsf{CA}_0$, $\Pi_2^1\text{-}\mathsf{CA}_0$, $\Delta_3^1\text{-}\mathsf{CA}_0$, ... , then T_0 implies its own relativization to the inner models $L(X) \cap P(\mathbb{N})$, $X \subseteq \mathbb{N}$. This has several interesting consequences: (1) $T_0 + \exists X \, \forall Y \, (Y \in L(X))$ is conservative over T_0 for Π_4^1 sentences; (2) T_0 has a *minimum β-model*, and this minimum β-model is of the form $L_\alpha \cap P(\omega)$ where α is an appropriately chosen countable ordinal. (These minimum β-models and their corresponding ordinals turn out to be distinct from one another; see §VII.7.) We also present generalizations involving minimum β-submodels of a given model.

Section VII.6 is concerned with several *strong choice schemes*, i.e., instances of the axiom of choice expressible in the language of second order arithmetic. Among the schemes considered are Σ_k^1 *choice*

$$\forall n \, \exists Y \eta(n, Y) \to \exists Z \, \forall n \, \eta(n, (Z)_n),$$

Σ_k^1 *dependent choice*

$$\forall n \, \forall X \, \exists Y \eta(n, Y) \to \exists Z \, \forall n \, \eta(n, (Z)^n, (Z)_n),$$

and *strong* Σ_k^1 *dependent choice*

$$\exists Z \, \forall n \, \forall Y (\eta(n, (Z)^n, Y) \to \eta(n, (Z)^n, (Z)_n)).$$

The corresponding formal systems are known as Σ_k^1-AC_0, Σ_k^1-DC_0, and strong Σ_k^1-DC_0, respectively. The case $k = 2$ is somewhat special. We show that Δ_2^1 comprehension implies Σ_2^1 choice, and even Σ_2^1 dependent choice provided Σ_2^1 induction is assumed. We also show that strong Σ_2^1 dependent choice is equivalent to Π_2^1 comprehension. These equivalences for $k = 2$ are based on the fact that Σ_2^1 *uniformization* is provable in Π_1^1-CA_0. Two proofs of this fact are given, one via Kondo's theorem and the other via Shoenfield absoluteness.

For $k \geq 3$ we obtain similar equivalences under the additional assumption $\exists X \, \forall Y \, (Y \in \mathrm{L}(X))$, via Σ_k^1 *uniformization*. We then apply our conservation theorems of the previous section to see that, for each $k \geq 3$, Σ_k^1 choice and strong Σ_k^1 dependent choice are conservative for Π_4^1 sentences over Δ_k^1 comprehension and Π_k^1 comprehension, respectively. Other results of a similar character are obtained. The case $k = 1$ is of a completely different character, and its treatment is postponed to §VIII.4.

Section VII.7 begins by generalizing the concept of β-model to β_k-*model*, i.e., an ω-model M such that all Σ_k^1 formulas with parameters from M are absolute to M. (Thus a β_1-model is the same thing as a β-model.) It is shown that, for each $k \geq 1$,

$$\forall X \, \exists M \, (X \in M \wedge M \text{ is a countable coded } \beta_k\text{-model})$$

is equivalent to strong Σ_k^1 dependent choice. This implies a kind of β_k-*model reflection principle* (theorem VII.7.6). Combining this with the results of §§VII.5 and VII.6, we obtain several noteworthy corollaries, e.g., the fact that Δ_{k+1}^1-CA_0 proves the existence of a countable coded β-model of Π_k^1-CA_0 which in turn proves the existence of a countable coded β-model of Δ_k^1-CA_0. From this it follows that the minimum β-models of Π_1^1-CA_0, Δ_2^1-CA_0, Π_2^1-CA_0, Δ_3^1-CA_0, ... are all distinct.

Chapter VIII: ω-models. The purpose of chapter VIII is to study ω-models of various subsystems of Z_2. We focus primarily on the five basic systems: RCA_0, WKL_0, ACA_0, ATR_0, Π_1^1-CA_0. We note that each of these systems is finitely axiomatizable. We also obtain some general results about fairly arbitrarily L_2-theories, which may be stronger than Π_1^1-CA_0

and need not be finitely axiomatizable. Many of our results on ω-models are formulated more generally, so as to apply also to ω-submodels of a given non-ω-model.

Section VIII.1 is introductory in nature. We characterize models of RCA_0 and ACA_0 in terms of Turing reducibility and the Turing jump operator. We show that the *minimum ω-models of* RCA_0 *and* ACA_0 are $\mathsf{REC} = \{X \colon X \text{ is recursive}\}$ and $\mathsf{ARITH} = \{X \colon X \text{ is arithmetical}\}$ respectively. We apply the strong soundness theorem and countable coded ω-models to show that ATR_0 proves the *consistency of* ACA_0, which in turn proves the *consistency of* RCA_0.

In §VIII.2 we consider models of WKL_0. We begin by showing that WKL_0 proves *strong Π_1^0 dependent choice*, which in turn implies the existence of a countable coded *strict β-model*. Such a model necessarily satisfies WKL_0, so we are surprisingly close to asserting that WKL_0 proves its own consistency (see however remark VIII.2.14). In particular, ACA_0 actually does prove the *consistency of* WKL_0, via countable coded ω-models (corollary VIII.2.12). Moreover, WKL_0 *has no minimal ω-model* (corollary VIII.2.8).

The rest of §VIII.2 is concerned with the *basis problem*: Given an infinite recursive tree $T \subseteq 2^{<\omega}$, to find a path through T which is in some sense "close to being recursive." We obtain three results, the *low basis theorem*, the *almost recursive basis theorem*, and the *GKT basis theorem*, which provide various solutions of the basis problem. They also imply the existence of *countable ω-models of* WKL_0 with various properties (theorems VIII.2.17, VIII.2.21, VIII.2.24). In particular, REC *is the intersection of all ω-models of* WKL_0 (corollary VIII.2.27).

In §VIII.3 we develop the technical machinery of *formalized hyperarithmetical theory*. We define the H-*sets* H_a^X for $X \subseteq \mathbb{N}$ and $a \in \mathcal{O}^X$. We note that ATR_0 is equivalent to $\forall X \,\forall a \,(\mathcal{O}(a, X) \to \mathsf{H}_a^X \text{ exists})$. We prove ATR_0 versions of the major classical results: *invariance of Turing degree* (VIII.3.13); $\Delta_1^1 = \mathrm{HYP}$ (VIII.3.19); the theorem on *hyperarithmetical quantifiers* (VIII.3.20, VIII.3.27). The latter result involves *pseudohierarchies*. An unorthodox feature of our exposition is that we do not use the recursion theorem.

In §VIII.4 we use the machinery of §VIII.3 to study ω-models of the systems $\Delta_1^1\text{-}\mathsf{CA}_0$, $\Sigma_1^1\text{-}\mathsf{AC}_0$, and $\Sigma_1^1\text{-}\mathsf{DC}_0$. We also consider a closely related system known as *weak $\Sigma_1^1\text{-}\mathsf{AC}_0$*. We show that $\mathrm{HYP} = \{X \colon X \text{ is hyperarithmetical}\}$ is the *minimum ω-model* of each of these four systems. The proof of this result uses Π_1^1 *uniformization*. Although the main results of classical hyperarithmetical theory are provable in ATR_0 (§VIII.3), the existence of the ω-model HYP is not (remark VIII.4.4). Nevertheless, we show that ATR_0 proves the existence of countable coded ω-models of $\Sigma_1^1\text{-}\mathsf{AC}_0$ etc. (theorem VIII.4.20). Indeed, ATR_0 proves that HYP is the intersection of all such ω-models (theorem VIII.4.23). In particular, ATR_0 proves the *consistency of* Σ_1^1-AC_0 etc.

In §VIII.5 we present two surprising theorems of Friedman which apply to fairly arbitrary L_2-theories $S \supseteq ACA_0$. They are: (1) If S is recursively axiomatizable and has an ω-model, then so does $S \wedge \neg \exists$ countable coded ω-model of S. (2) If S is finitely axiomatizable, then Π_∞^1-TI_0 proves $S \to \exists$ a countable coded ω-model of S. Note that (1) is an *ω-model incompleteness theorem*, while (2) is an *ω-model reflection principle*. Combining (1) and (2), we see that if S is finitely axiomatizable and has an ω-model, then there exists an ω-model of S which is does not satisfy Π_∞^1-TI_0 (corollary VIII.5.8).

At the end of §VIII.5 we prove that Π_1^1 *transfinite induction* is equivalent to ω-*model reflection for* Σ_3^1 *formulas*, which is equivalent to Σ_1^1 *dependent choice* (theorem VIII.5.12). From this it follows that there exists an ω-model of ATR_0 in which Σ_1^1-DC_0 fails (theorem VIII.5.13). This is in contrast to the fact that ATR_0 implies Σ_1^1-AC_0 (theorem V.8.3).

Section VIII.6 presents several *hard core theorems*. We show that any model M of ATR_0 has a proper β-submodel; indeed, by corollary VIII.6.10, HYP^M is the intersection of all such submodels. We also prove the following theorem of Quinsey: if M is any ω-model of a recursively axiomatizable L_2-theory $S \supseteq ATR_0$, then M has a proper submodel which is again a model of S (theorem VIII.6.12). Indeed, HYP^M is the intersection of all such submodels (exercise VIII.6.23). In particular, no such S has a minimal ω-model.

Chapter IX: non-ω-models. In chapter IX we study non-ω-models of various subsystems of Z_2. Section IX.1 deals with RCA_0 and ACA_0. Sections IX.2 and IX.3 are concerned with WKL_0. Section IX.4 is concerned with various systems including Π_k^1-CA_0 and Σ_k^1-AC_0, $k \geq 0$. For most of the results of chapter IX, it is essential that our systems contain only restricted induction and not full induction. Many of the results can be phrased as conservation theorems. The methods of §§IX.3 and IX.4 depend crucially on the existence of nonstandard integers.

We begin in §IX.1 by showing that every model M of PA can be expanded to a model of ACA_0. The expansion is accomplished by letting $S_M = \mathrm{Def}(M) = \{X \subseteq |M| : X$ is first order definable over M allowing parameters from $M\}$. From this it follows that PA is the first order part of ACA_0, and that ACA_0 has the same consistency strength as PA. We then prove analogous results for RCA_0. Namely, every model M of Σ_1^0-PA can be expanded to a model of RCA_0; the expansion is accomplished by letting $S_M = \Delta_1^0$-$\mathrm{Def}(M) = \{X \subseteq |M| : X$ is Δ_1^0 definable over M allowing parameters from $M\}$. The delicate point of this argument is to show that the expansion preserves Σ_1^0 induction. It follows that Σ_1^0-PA is the first order part of RCA_0, and that RCA_0 has the same consistency strength as Σ_1^0-PA.

In §IX.2 we show that WKL_0 has the same first order part and consistency strength as RCA_0. This is based on the following model-theoretic result due to Harrington: Given a countable model M of RCA_0, we can construct a countable model M' of WKL_0 such that M is an ω-submodel of M'. The model M' is obtained from M by iterated forcing, where at each stage we force with trees to add a generic path through a tree. Again, the delicate point is to verify that Σ_1^0 induction is preserved. This model-theoretic result implies that WKL_0 is conservative over RCA_0 for Π_1^1 sentences.

In §IX.3 we introduce the well known formal system PRA of *primitive recursive arithmetic*. This theory of primitive recursive functions contains a function symbol and defining axioms for each such function. We prove the following result of Friedman: WKL_0 has the same consistency strength as PRA and is conservative over PRA for Π_2^0 sentences. Our proof uses a model-theoretic method due to Kirby and Paris, involving *semiregular cuts*. The foundational significance of PRA is that it embodies *Hilbert's concept of finitism*. Therefore, Friedman's theorem combined with the mathematical work of chapters II and IV shows that a significant portion of mathematical practice is finitistically reducible. Thus we have a *partial realization of Hilbert's program*; see also remark IX.3.18.

In §IX.4 we use *recursively saturated models* to prove some surprising conservation theorems for various subsystems of Z_2. The main results may be summarized as follows: For each $k \geq 0$, Σ_{k+1}^1-AC_0 has the same consistency strength as Π_k^1-CA_0 and is conservative over Π_k^1-CA_0 for Π_l^1 sentences, $l = \min(k + 2, 4)$. These results are due to Barwise/Schlipf, Feferman, Friedman, and Sieg. We also obtain a number of related results.

Section IX.5 is a very brief discussion of Gentzen-style proof theory, with emphasis on provable ordinals of subsystems of Z_2.

This completes our summary of Part B.

Appendix: Chapter X: Additional Results. Chapter X is an appendix in which some additional Reverse Mathematics results and problems are presented without proof but with references to the published literature.

In §X.1 we consider *measure theory* in subsystems of Z_2. We introduce the formal system WWKL_0 consisting of RCA_0 plus *weak weak König's lemma* and show that it is just strong enough to prove several measure theoretic results, e.g., the *Vitali covering theorem*. We also consider measure theory in stronger systems such as ACA_0.

In §X.2 we mention some additional results on *separable Banach spaces* in subsystems of Z_2. We note that WKL_0 is just strong enough to prove *Banach separation*. We develop various notions related to the *weak-* topology* on X^*, the dual of a separable Banach space. We show that Π_1^1-CA_0 is just strong enough to prove the existence of the *weak-*-closed linear span* of a countable set Y in X^*.

In §X.3 we consider *countable combinatorics* in subsystems of Z_2. We note that *Hindman's theorem* lies between ACA_0 and a slightly stronger system, ACA_0^+. We mention a similar result for the closely related *Auslander/Ellis theorem* of topological dynamics. In the area of *matching theory*, we show that the *Podewski/Steffens theorem* ("every countable bipartite graph has a König covering") is equivalent to ATR_0. At the end of the section we consider *well quasiordering theory*, noting for instance that the *Nash-Williams transfinite sequence theorem* lies between ATR_0 and Π_1^1-CA_0.

In §X.4 we initiate a project of weakening the base theory for Reverse Mathematics. We introduce a system RCA_0^* which is essentially RCA_0 with Σ_1^0 induction weakened to Σ_0^0 induction. We also introduce a system WKL_0^* consisting of RCA_0^* plus weak König's lemma. We present some conservation results showing in particular that RCA_0^* and WKL_0^* have the same consistency strength as EFA, elementary function arithmetic. We note that several theorems of countable algebra are equivalent over RCA_0^* to Σ_1^0 induction. Among these are: (1) every polynomial over a countable field has an irreducible factor; (2) every finitely generated vector space over $\mathbb{Q}$ has a basis.

I.14. Conclusions

In this chapter we have presented and motivated the main themes of the book, including the Main Question (§§I.1, I.12) and Reverse Mathematics (§I.9). A detailed outline of the book is in section I.13. The five most important subsystems of second order arithmetic are RCA_0, WKL_0, ACA_0, ATR_0, Π_1^1-CA_0. Part A of the book consists of chapters II through VI and focuses on the development of mathematics in these five systems. Part B consists of chapters VII through IX and focuses on models of these and other subsystems of Z_2. Additional results are presented in an appendix, chapter X.

Part A

DEVELOPMENT OF MATHEMATICS WITHIN SUBSYSTEMS OF Z_2

Chapter II

RECURSIVE COMPREHENSION

II.1. The Formal System RCA_0

The purpose of this chapter is to study a certain subsystem of second order arithmetic known as RCA_0. RCA_0 is the weakest subsystem of Z_2 to be studied extensively in this book. It will play a key role in chapters III through VI as the "weak base theory" for Reverse Mathematics.

The acronym RCA stands for "recursive comprehension axiom." Roughly speaking, the set existence axioms of RCA_0 are only strong enough to prove the existence of recursive sets of natural numbers. However, these axioms do not rule out the existence of nonrecursive sets of natural numbers.

The purpose of this section is to present the axioms of RCA_0 and characterize the ω-models of RCA_0. In the rest of the chapter, we shall show that certain portions of ordinary mathematics can be developed within RCA_0. Some further results on models of RCA_0 will be presented in chapters VIII and IX.

In order to state the axioms of RCA_0 we shall need some definitions.

DEFINITION II.1.1. Let φ be a formula of L_2, let n be a number variable, and let t be a numerical term which does not contain n. We abbreviate $\forall n \, (n < t \to \varphi)$ as $(\forall n < t) \, \varphi$. We abbreviate $\exists n \, (n < t \land \varphi)$ as $(\exists n < t) \, \varphi$. The quantifiers $\forall n < t$ and $\exists n < t$ are known as *bounded quantifiers*.

DEFINITION II.1.2. An L_2-formula is said to be Σ_0^0 if it is built up from atomic formulas by means of propositional connectives and bounded number quantifiers. For $k \in \omega$, an L_2-formula is said to be Σ_k^0 (respectively Π_k^0) if it is of the form $\exists n_1 \, \forall n_2 \cdots n_k \, \theta$ (respectively $\forall n_1 \, \exists n_2 \cdots n_k \, \theta$) where θ is Σ_0^0.

In particular, a Σ_1^0 (respectively Π_1^0) formula is one of the form $\exists n \, \theta$ (respectively $\forall n \, \theta$) where θ is Σ_0^0. Note that although Σ_k^0 and Π_k^0 formulas contain no set quantifiers, they may contain free set variables.

DEFINITION II.1.3. For each $k \in \omega$, the scheme of Σ_k^0 *induction* consists of all axioms of the form

$$(\varphi(0) \wedge \forall n\, (\varphi(n) \to \varphi(n+1))) \to \forall n\, \varphi(n)$$

where $\varphi(n)$ is any Σ_k^0 formula of the language of second order arithmetic. The scheme of Π_k^0 *induction* is defined similarly.

DEFINITION II.1.4. The scheme of Δ_1^0 *comprehension* consists of all axioms of the form

$$\forall n\, (\varphi(n) \leftrightarrow \psi(n)) \to \exists X\, \forall n\, (n \in X \leftrightarrow \varphi(n))$$

where $\varphi(n)$ is Σ_1^0, $\psi(n)$ is Π_1^0, and X is not free in $\varphi(n)$.

We are now ready to define RCA$_0$.

DEFINITION II.1.5. RCA$_0$ is the formal system in the language L$_2$ whose axioms consist of the basic axioms (see definition I.2.4(i)) plus the schemes of Σ_1^0 induction and Δ_1^0 comprehension.

We now characterize the ω-models of RCA$_0$. We assume familiarity with the elements of recursive function theory (see e.g., Davis [44] or Rogers [208] or Cutland [43]).

LEMMA II.1.6. *Let X and Y be subsets of ω. The following are equivalent.*

(i) X *is recursively enumerable in Y;*
(ii) X *is definable (in the intended model of second order arithmetic) by a Σ_1^0 formula with parameter Y.*

PROOF. This is immediate from any one of several familiar characterizations of "recursively enumerable in".   □

THEOREM II.1.7. *Let S be a collection of subsets of ω. S is an ω-model of RCA$_0$ if and only if S enjoys the following closure properties:*

(i) S *is nonempty;*
(ii) *if $X, Y \in S$, then $X \oplus Y \in S$ where*

$$X \oplus Y = \{2n : n \in X\} \cup \{2n+1 : n \in Y\};$$

(iii) *if $X \leq_T Y$ and $Y \in S$, then $X \in S$. Here $\leq_T$ denotes Turing reducibility, i.e., $X \leq_T Y$ if and only if X is recursive in Y.*

(See also §§VII.1 and VIII.1.)

PROOF. Lemma II.1.6 implies that $X \leq_T (Y_1 \oplus \cdots \oplus Y_n)$ if and only if both X and $\omega \setminus X$ are definable by Σ_1^0 formulas with parameters $Y_1, \ldots, Y_n$. From this the theorem follows easily.   □

REMARK. Collections $S \subseteq P(\omega)$ satisfying (i), (ii) and (iii) are known as *Turing ideals*. Such collections have been studied extensively in the literature on degrees of unsolvability.

Corollary II.1.8. *The minimum ω-model of* RCA$_0$ *is the collection*

$$\text{REC} = \{X \subseteq \omega : X \text{ is recursive}\}.$$

For more information on models of RCA$_0$, see chapters VIII and IX.

Notes for §II.1. The system RCA$_0$ is due to Friedman [69]. Actually Friedman's axiomatization of RCA$_0$ is somewhat different from the one used here. For a thorough introduction to recursive function theory, see Davis [44] and Rogers [208]. For a survey of results on degrees of unsolvability, see Simpson [231].

II.2. Finite Sequences

It is well known that finite sequences of natural numbers can be encoded as single natural numbers. The purpose of this section is to show that one such coding method can be developed formally within RCA$_0$.

We begin with some elementary properties of the natural numbers. Within RCA$_0$ we define $\mathbb{N}$ to be the set of all natural numbers, i.e., the unique set X such that $\forall n \, (n \in X)$. The following lemma can be summarized by saying that the natural number system $\mathbb{N}, +, \cdot, 0, 1, <$ is a commutative ordered semiring with cancellation.

Lemma II.2.1. *The following are provable in* RCA$_0$.

(i) $(m + n) + p = m + (n + p)$
(ii) $0 + m = m$
(iii) $1 + m = m + 1$
(iv) $m + n = n + m$
(v) $m \cdot (n + p) = m \cdot n + m \cdot p$
(vi) $(m \cdot n) \cdot p = m \cdot (n \cdot p)$
(vii) $(m + n) \cdot p = m \cdot p + n \cdot p$
(viii) $0 \cdot m = 0$
(ix) $1 \cdot m = m$
(x) $m \cdot n = n \cdot m$
(xi) $(m < n \wedge n < p) \rightarrow m < p$
(xii) $m < n \rightarrow m + 1 < n + 1$
(xiii) $m + 1 < n + 1 \rightarrow m < n$
(xiv) $n \neq 0 \rightarrow 0 < n$
(xv) $m < n \vee m = n \vee n < m$
(xvi) $\neg n < n$
(xvii) $m < n \rightarrow m + p < n + p$
(xviii) $m + p < n + p \rightarrow m < n$
(xix) $m < m + n + 1$
(xx) $m + p = n + p \rightarrow m = n$
(xxi) $(p \neq 0 \wedge m < n) \rightarrow m \cdot p < n \cdot p$
(xxii) $(p \neq 0 \wedge m \cdot p < n \cdot p) \rightarrow m < n$

(xxiii) $(p \neq 0 \land m \cdot p = n \cdot p) \to m = n$
(xxiv) $m < n \to (\exists k < n) m + k + 1 = n$
(xxv) $n \neq 0 \to (\exists m < n) m + 1 = n$.

Proof. Each of statements (i)–(xxiii) is proved by a straightforward induction on the alphabetically last variable occurring in the statement. Previous statements may be used in the base step or the successor step. For example, here is the proof of (x) $m \cdot n = n \cdot m$. We proceed by induction on n. For $n = 0$ we have, using (viii) and one of the basic axioms, $m \cdot 0 = 0 = 0 \cdot m$. For $n + 1$ we have, using the induction hypothesis $m \cdot n = n \cdot m$ as well as (ix) and (vii) and one of the basic axioms, $m \cdot (n+1) = m \cdot n + m = n \cdot m + m = n \cdot m + 1 \cdot m = (n+1) \cdot m$. By induction on n it follows that $m \cdot n = n \cdot m$ for all n. (It is interesting to note that only quantifier-free induction is used in the proofs of (i)–(xxiii).)

We now prove (xxiv) by induction on n. For $n = 0$, we have by one of the basic axioms $\neg m < 0$ so there is nothing to prove. For $n + 1$, if $m < n + 1$ then by one of the basic axioms, either $m < n$ or $m = n$. If $m < n$ it follows by induction that $m + k + 1 = n$ for some $k < n$. Hence $m + (k + 1) + 1 = n + 1$ and by (xii) we have $k + 1 < n + 1$. If $m = n$ then $m + 0 + 1 = n + 1$ and by (xiv) and one of the basic axioms we have $0 < n + 1$. Statement (xxiv) follows by Σ_0^0 induction. Statement (xxv) is a special case of (xxiv) in view of (xiv). This completes the proof of lemma II.2.1. □

Within RCA_0 we define a *pairing map*

$$(i, j) = (i + j)^2 + i$$

where of course $k^2 = k \cdot k$. Part 2 of the following theorem says that the pairing map is a one-to-one mapping of $\mathbb{N} \times \mathbb{N}$ into $\mathbb{N}$.

Theorem II.2.2. *The following are provable in* RCA_0.
1. $i \leq (i, j)$ *and* $j \leq (i, j)$.
2. $(i, j) = (i', j') \to (i = i' \land j = j')$.

Proof. Part 1 is obvious from II.2.1(xix). For part 2, given $k = (i, j) = (i + j)^2 + i$, we claim that there exists a unique m such that $m^2 \leq k < (m + 1)^2$. Existence of m is obvious by taking $m = i + j$, and uniqueness follows from the fact that $m < n \to m^2 < n^2$. It now follows that $i = k - m^2$ and $j = m - i$. This proves part 2. □

Our next goal is to show that finite sets of natural numbers can be encoded as single natural numbers. This requires us to develop a little bit of elementary number theory within RCA_0.

From now on we write $mn = m \cdot n$. We say that m *divides* n (written $m \mid n$) if $\exists q\, (mq = n)$. For m_1 and m_2 both nonzero, we say that m_1 is *prime relative to* m_2 if $\forall n\, (m_2 \mid m_1 n \to m_2 \mid n)$.

Lemma II.2.3. *The following is provable in* RCA_0. *If* m_1 *is prime relative to* m_2 *then* m_2 *is prime relative to* m_1.

Proof. Assume that m_1 is prime relative to m_2. Let n be given such that $m_1 \mid m_2 n$. Let q be such that $m_1 q = m_2 n$. Then $m_2 \mid m_1 q$. Since m_1 is prime relative to m_2 it follows that $m_2 \mid q$. Let r be such that $m_2 r = q$. Then $m_1 m_2 r = m_1 q = m_2 n$. Hence $m_1 r = n$. So $m_1 \mid n$. This completes the proof. □

Lemma II.2.4. *The following is provable in* RCA$_0$. (i) *Given k, there exists $m > 0$ such that $\forall i < k \ (i + 1 \ divides \ m)$.* (ii) *Let k and m be as in part* (i). *Then $m(i + 1) + 1$ and $m(j + 1) + 1$ are relatively prime to each other for all $i < j < k$.*

Proof. Part (i) is easily proved by Σ_1^0 induction on k. For part (ii) let $i < j < k$ be given. We shall show that $m(j + 1) + 1$ is prime relative to $m(i + 1) + 1$. First note that for all n, if m divides $m(i + 1)n + n$ then m divides n. Thus $m(i + 1) + 1$ is prime relative to m. Hence by lemma II.2.3, m is prime relative to $m(i + 1) + 1$. Now let n be given such that $m(i+1)+1$ divides $(m(j+1)+1)n$. Let l be such that $i+l+1 = j$. Then $(m(j + 1) + 1)n = (m(i + l + 1 + 1) + 1)n = (m(i + 1) + 1)n + m(l + 1)n$. Therefore $m(i + 1) + 1$ divides $m(l + 1)n$. Since m is prime relative to $m(i+1)+1$ it follows that $m(i+1)+1$ divides $(l+1)n$. Since $l+1$ divides m it follows that $m(i + 1) + 1$ divides mn. Using again the fact that m is prime relative to $m(i + 1) + 1$, we see that $m(i + 1) + 1$ divides n. This completes the proof that $m(j + 1) + 1$ is prime relative to $m(i + 1) + 1$. It follows by lemma II.2.3 that $m(i+1)+1$ is prime relative to $m(j+1)+1$. This completes the proof. □

Within RCA$_0$ we define a *finite set* to be a set X such that $\exists k \ \forall i \ (i \in X \to i < k)$. We now show that finite sets can be encoded as natural numbers.

Theorem II.2.5. *The following is provable in* RCA$_0$. *For any finite set $X \subseteq \mathbb{N}$ there exist k, m and $n \in \mathbb{N}$ such that*

$$\forall i \ (i \in X \leftrightarrow (i < k \land m(i + 1) + 1 \ divides \ n)). \tag{3}$$

The least number of the form $(k, (m, n))$ such that (3) holds is called the *code* of the finite set X. Thus each finite set of natural numbers has a unique code. This fact is extremely important.

Proof. Let k be such that $\forall i \ (i \in X \to i < k)$. By II.2.4 let m be such that the numbers $m(i + 1) + 1$ for $i < k$ are pairwise relatively prime. Let $\varphi(j)$ be a Σ_1^0 formula asserting that either $j > k$ or $\exists n \ \forall i < k \ (m(i+1)+1$ divides $n \leftrightarrow (i \in X \land i < j))$. We prove $\forall j \ \varphi(j)$ by induction on j. For $j = 0$ or $j > k$ there is nothing to prove. For $j' = j + 1 \leq k$ put $n' = n(m(j + 1) + 1)$ if $j \in X$, $n' = n$ if $j \notin X$. Then for each $i < k$ we see that $m(i + 1) + 1$ divides n' if and only if either $i = j \in X$ or $m(i + 1) + 1$ divides n. Hence by the induction hypothesis it follows that $\forall i < k \ (m(i + 1) + 1$ divides $n' \leftrightarrow (i \in X \land i < j + 1))$. This proves $\forall j \ \varphi(j)$. From $\varphi(k)$ we have the conclusion of the theorem. □

We can now present our coding method for finite sequences of natural numbers.

DEFINITION II.2.6. The following definitions are made in RCA$_0$. A *finite sequence of natural numbers* is a finite set X such that $\forall n\, (n \in X \rightarrow \exists i\, \exists j\, (n = (i,j)))$ and $\forall i\, \forall j\, \forall k\, (((i,j) \in X \wedge (i,k) \in X) \rightarrow j = k)$ and $\exists l\, \forall i\, (i < l \leftrightarrow \exists j\, ((i,j) \in X))$. Here (i,j) denotes the pairing map of theorem II.2.2. The number l is uniquely determined and is called the *length* of X. The *code* of the finite sequence X is just the code of X as a finite set (theorem II.2.5).

The set of all codes of finite sequences is denoted Seq or $\mathbb{N}^{<\mathbb{N}}$. This set exists by Σ^0_0 comprehension. If $s \in$ Seq is the code of the finite sequence X, we write lh(s) for the length of X, and if $i < $ lh(s) we write $s(i)$ for the unique j such that $(i,j) \in X$. We shall sometimes use notations such as

$$s = \langle s(0), s(1), \ldots, s(\mathrm{lh}(s) - 1)\rangle$$

or

$$s = \langle s(i) : i < \mathrm{lh}(s)\rangle.$$

Whenever convenient we shall identify a finite sequence of natural numbers with its code.

If $s, t \in$ Seq we denote *concatenation* by $\,\widehat{}\,$, i.e.,

$$s\widehat{}t = \langle s(0), \ldots, s(\mathrm{lh}(s) - 1), t(0), \ldots, t(\mathrm{lh}(t) - 1)\rangle$$

so that lh$(s\widehat{}t) = $ lh$(s) + $ lh(t). In particular

$$s\widehat{}\langle n\rangle = \langle s(0), \ldots, s(\mathrm{lh}(s) - 1), n\rangle$$

and lh$(s\widehat{}\langle n\rangle) = $ lh$(s) + 1$. We write $s \subseteq t$ to mean that s is an *initial segment* of t, i.e., lh$(s) \le$ lh$(t) \wedge (\forall i <$ lh$(s))\, s(i) = t(i)$. Note that the predicates lh$(s) = m$, $s(i) = n$, $s \subseteq t$, $s\widehat{}\langle n\rangle = t$, etc., are Σ^0_0.

We state here the following formal version of the well known *Kleene normal form theorem* for Σ^0_1 relations. This result will be used several times.

THEOREM II.2.7 (normal form theorem). *Let* $\varphi(X)$ *be a* Σ^0_1 *formula. Then we can find a* Σ^0_0 *formula* $\theta(s)$ *such that* RCA$_0$ *proves*

$$\forall X\, (\varphi(X) \leftrightarrow \exists m\, \theta(X[m])).$$

Here we write $X[m] = \langle \xi_0, \xi_1, \ldots, \xi_{m-1}\rangle$ where $\xi_i = 1$ if $i \in X$, 0 if $i \notin X$. Thus $X[m]$ is the finite initial sequence of length m of the characteristic function of X. Note that $\varphi(X)$ may contain free variables other than X. If this is the case, then $\theta(s)$ will also contain those free variables.

PROOF. The proof is obtained by straightforwardly formalizing the Kleene normal form theorem in RCA$_0$, using the methods of §II.3. See also Kleene [142] or Rogers [208]. See also the last part of the proof of lemma IX.2.4 below.   □

Notes for §II.2. Our method of encoding finite sequences (II.2.5, II.2.6) is adapted from Shoenfield [222, page 115].

II.3. Primitive Recursion

In this section we prove within RCA$_0$ that the universe of total number-theoretic functions is closed under composition, primitive recursion, and the least number operator. As an application of these results we show that the Σ_1^0 induction scheme of RCA$_0$ is equivalent to a certain set existence principle known as bounded Σ_1^0 comprehension.

DEFINITION II.3.1 (functions). The following definitions are made in RCA$_0$. Let X and Y be sets of natural numbers. We write $X \subseteq Y$ to mean $\forall n \, (n \in X \to n \in Y)$. We define $X \times Y$ to be the set of all k such that $\exists i \le k \, \exists j \le k \, (i \in X \land j \in Y \land (i,j) = k)$. This set exists by Σ_0^0 comprehension; (i,j) denotes the pairing map of theorem II.2.2. We define a function $f : X \to Y$ to be a set $f \subseteq X \times Y$ such that $\forall i \, \forall j \, \forall k \, (((i,j) \in f \land (i,k) \in f) \to j = k)$ and $\forall i \, \exists j \, (i \in X \to (i,j) \in f)$. If $f : X \to Y$ and $i \in X$ we denote by $f(i)$ the unique j such that $(i,j) \in f$.

THEOREM II.3.2 (composition). *The following is provable in RCA$_0$. If $f : X \to Y$ and $g : Y \to Z$ then there exists $h = gf : X \to Z$ defined by $h(i) = g(f(i))$.*

PROOF. We have $\exists j \, ((i,j) \in f \land (j,k) \in g) \leftrightarrow (i \in X \land \forall j \, ((i,j) \in f \to (j,k) \in g))$. Hence by Δ_1^0 comprehension there exists h such that $(i,k) \in h \leftrightarrow \exists j \, ((i,j) \in f \land (j,k) \in g)$. Clearly $h = gf$. □

DEFINITION II.3.3. The following definitions are made in RCA$_0$. The set of all $s \in \text{Seq}$ such that $\text{lh}(s) = k$ is denoted $\mathbb{N}^k$. This set exists by Σ_0^0 comprehension. If $f : \mathbb{N}^k \to \mathbb{N}$ and $s = \langle n_1, \dots, n_k \rangle \in \mathbb{N}^k$, we sometimes write $f(n_1, \dots, n_k)$ instead of $f(s)$.

The above definition permits us to discuss k-ary functions $f : \mathbb{N}^k \to \mathbb{N}$ for variable $k \in \mathbb{N}$, within RCA$_0$. We can also discuss finite sequences $\langle f_1, \dots, f_m \rangle$ of k-ary functions $f_i : \mathbb{N}^k \to \mathbb{N}$, $1 \le i \le m$. Such a sequence is identified in the obvious way with a single function $f : \mathbb{N}^k \to \mathbb{N}^m$. Thus theorem II.3.2 implies that the universe of functions is closed under generalized composition, i.e., given $f_i : \mathbb{N}^k \to \mathbb{N}$, $1 \le i \le m$, and $g : \mathbb{N}^m \to \mathbb{N}$, there exists $h : \mathbb{N}^k \to \mathbb{N}$ defined by $h(n_1, \dots, n_k) = g(f_1(n_1, \dots, n_k), \dots, f_m(n_1, \dots, n_k))$.

The next theorem says that the universe of k-ary functions, $k \in \mathbb{N}$, is closed under *primitive recursion*.

THEOREM II.3.4 (primitive recursion). *The following is provable in RCA$_0$. Given $f : \mathbb{N}^k \to \mathbb{N}$ and $g : \mathbb{N}^{k+2} \to \mathbb{N}$, there exists a unique $h : \mathbb{N}^{k+1} \to \mathbb{N}$*

defined by

$$h(0, n_1, \ldots, n_k) = f(n_1, \ldots, n_k),$$
$$h(m + 1, n_1, \ldots, n_k) = g(h(m, n_1, \ldots, n_k), m, n_1, \ldots, n_k).$$

PROOF. Let $\theta(s, m, \langle n_1, \ldots, n_k \rangle)$ say that $s \in$ Seq and $\mathrm{lh}(s) = m + 1$ and $s(0) = f(n_1, \ldots, n_k)$ and, for all $i < m$, $s(i + 1) = g(s(i), i, n_1, \ldots, n_k)$. The formula $\exists s\, \theta(s, m, \langle n_1, \ldots, n_k \rangle)$ is Σ_1^0 so for each fixed $\langle n_1, \ldots, n_k \rangle \in \mathbb{N}^k$ we can prove this formula by the obvious Σ_1^0 induction on m. Also, if $\theta(s, m, \langle n_1, \ldots, n_k \rangle)$ and $\theta(s', m, \langle n_1, \ldots, n_k \rangle)$ then we can prove $s(i) = s'(i)$ by induction on $i < m + 1$. It follows that for all $\langle n_1, \ldots, n_k \rangle \in \mathbb{N}^k$ and all m and j,

$$\exists s\, (\theta(s, m, \langle n_1, \ldots, n_k \rangle)) \wedge s(m) = j) \leftrightarrow$$
$$\forall s\, (\theta(s, m, \langle n_1, \ldots, n_k \rangle)) \to s(m) = j).$$

Hence by Δ_1^0 comprehension there exists $h \colon \mathbb{N}^{k+1} \to \mathbb{N}$ such that

$$h(m, n_1, \ldots, n_k) = j$$

if and only if $\exists s\, (\theta(s, m, \langle n_1, \ldots, n_k \rangle)) \wedge s(m) = j)$. Clearly h has the desired properties. $\qquad\square$

Next we show that the universe of functions is closed under the *least number operator*, i.e., *minimization*.

THEOREM II.3.5 (minimization). *The following is provable in* RCA$_0$. *Let* $f \colon \mathbb{N}^{k+1} \to \mathbb{N}$ *be such that for all* $\langle n_1, \ldots, n_k \rangle \in \mathbb{N}^k$ *there exists* $m \in \mathbb{N}$ *such that* $f(m, n_1, \ldots, n_k) = 1$. *Then there exists* $g \colon \mathbb{N}^k \to \mathbb{N}$ *defined by* $g(n_1, \ldots, n_k) = $ *least* m *such that* $f(m, n_1, \ldots, n_k) = 1$.

PROOF. By Σ_0^0 comprehension there exists $g \subseteq \mathbb{N}^k \times \mathbb{N}$ such that

$$(\langle n_1, \ldots, n_k \rangle, m) \in g$$

if and only if $(\langle m, n_1, \ldots, n_k \rangle, 1) \in f \wedge \neg (\exists j < m)\, (\langle j, n_1, \ldots, n_k \rangle, 1) \in f$. The hypothesis of the theorem implies that $g \colon \mathbb{N}^k \to \mathbb{N}$ and clearly g has the desired property. $\qquad\square$

We now present some important consequences of the above results.

LEMMA II.3.6. *The following is provable in* RCA$_0$. *For any infinite set* $X \subseteq \mathbb{N}$, *there exists a function* $\pi_X \colon \mathbb{N} \to \mathbb{N}$ *such that* $\forall k\, \forall m\, (k < m \to \pi_X(k) < \pi_X(m))$ *and* $\forall n\, (n \in X \leftrightarrow \exists m\, (\pi_X(m) = n))$.

PROOF. First define $\nu_X \colon \mathbb{N} \to \mathbb{N}$ by $\nu_X(m) = $ least n such that $n \in X$ and $n \geq m$. Then use primitive recursion (theorem II.3.4) to define $\pi_X \colon \mathbb{N} \to \mathbb{N}$ by $\pi_X(0) = \nu_X(0)$, $\pi_X(m + 1) = \nu_X(\pi_X(m) + 1)$. Using Σ_0^0 induction it follows easily that $k < m \to \pi_X(k) < \pi_X(m)$ and $n \in X \to (\exists m \leq n)\, \pi_X(m) = n$. $\qquad\square$

The next lemma is analogous to the well known fact that an infinite recursively enumerable set is the range of a one-to-one recursive function.

LEMMA II.3.7. *Let $\varphi(n)$ be a Σ_1^0 formula in which X and f do not occur freely. The following is provable in* RCA$_0$. *Either there exists a finite set X such that $\forall n \, (n \in X \leftrightarrow \varphi(n))$, or there exists a one-to-one function $f : \mathbb{N} \to \mathbb{N}$ such that $\forall n \, (\varphi(n) \leftrightarrow \exists m \, (f(m) = n))$.*

PROOF. Suppose that the first alternative fails. Write $\varphi(n)$ as $\exists j \, \theta(j, n)$ where $\theta(j, n)$ is Σ_0^0. By Σ_0^0 comprehension let Y be the set of all (j, n) such that $\theta(j, n) \wedge \neg(\exists i < j) \, \theta(i, n)$. Since the first alternative fails, Y is infinite. Hence by lemma II.3.6 let $\pi_Y : \mathbb{N} \to \mathbb{N}$ be the function which enumerates the elements of Y in strictly increasing order. By Σ_0^0 comprehension let $p_2 : \mathbb{N} \to \mathbb{N}$ be the *second projection function*, i.e., $p_2((j, n)) = n$ for all $j, n \in \mathbb{N}$. Let $f : \mathbb{N} \to \mathbb{N}$ be the composition defined by $f(m) = p_2(\pi_Y(m))$. The definition of Y implies that f is one-to-one, and clearly f enumerates exactly those $n \in \mathbb{N}$ such that $\varphi(n)$. This completes the proof. $\square$

DEFINITION II.3.8 (bounded Σ_k^0 comprehension). For each $k \in \omega$ the scheme of *bounded Σ_k^0 comprehension* consists of all axioms of the form

$$\forall n \, \exists X \, \forall i \, (i \in X \leftrightarrow (i < n \wedge \varphi(i)))$$

where $\varphi(i)$ is any Σ_k^0 formula in which X does not occur freely.

THEOREM II.3.9. RCA$_0$ *proves bounded Σ_1^0 comprehension.*

PROOF. We reason in RCA$_0$. Let $\varphi(i)$ be a Σ_1^0 formula in which X does not occur freely. Given n, suppose there is no finite set X such that $\forall i \, (i \in X \leftrightarrow (i < n \wedge \varphi(i)))$. Then by lemma II.3.7 there exists a one-to-one function $f : \mathbb{N} \to \mathbb{N}$ such that $\forall m \, (f(m) < n \wedge \varphi(f(m)))$. In particular the restriction of f to $\{0, \ldots, n-1, n\}$ is a finite one-to-one function from $\{0, \ldots, n-1, n\}$ into $\{0, \ldots, n-1\}$. But it is easy to prove (by Σ_0^0 induction on the codes of finite functions) that no finite function can have the mentioned properties. This contradiction completes the proof. $\square$

COROLLARY II.3.10. RCA$_0$ *proves the Π_1^0 induction scheme*

$$(\psi(0) \wedge \forall n \, (\psi(n) \to \psi(n+1))) \to \forall n \, \psi(n)$$

where $\psi(n)$ is any Π_1^0 formula.

PROOF. Reasoning within RCA$_0$, assume the hypothesis. Fix n. We must show that $\psi(n)$ holds. By bounded Σ_1^0 comprehension (theorem II.3.9) using n as a parameter, let X be such that $\forall m \, (m \in X \leftrightarrow (m \leq n \wedge \neg\psi(m)))$. By Δ_1^0 comprehension let Y be such that $\forall m \, (m \in Y \leftrightarrow m \notin X)$. By assumption we have $0 \in Y$ and $\forall m \, (m \in Y \to m+1 \in Y)$. Hence by the induction axiom I.2.4(ii) we have $\forall m \, (m \in Y)$, in particular $n \in Y$. Hence $\psi(n)$ holds. This completes the proof. $\square$

REMARK II.3.11. In chapter I we emphasized the role of set existence axioms. It is therefore interesting to note that, despite appearances, the Σ_1^0

induction axiom of RCA_0 can be considered to be a set existence axiom. Namely, Σ_1^0 induction is provably equivalent to bounded Σ_1^0 comprehension (in the presence of the basic axioms, the induction axiom, and Δ_1^0 comprehension).

[One half of this equivalence has already been established as theorem II.3.9. The other half is easily proved, as follows. Given $\varphi(0) \wedge \forall n\, (\varphi(n) \to \varphi(n+1))$ where $\varphi(n)$ is Σ_1^0, fix n and apply bounded Σ_1^0 comprehension to get a set X such that $\forall m \leq n\, (m \in X \leftrightarrow \varphi(m))$. Then apply Σ_0^0 comprehension to get a set Y such that $\forall m\, (m \in Y \leftrightarrow (m \in X \vee m > n))$. Clearly $0 \in Y \wedge \forall m\, (m \in Y \to m+1 \in Y)$. Hence by the induction axiom $\forall m\, (m \in Y)$. In particular $n \in Y$ so $\varphi(n)$ holds. Since n is arbitrary we have we have $\forall n\, \varphi(n)$. See also Simpson/Smith [250, lemma 2.5] and remark X.4.3.]

EXERCISE II.3.12. Show that, for each $k \in \omega$, RCA_0 proves

$$\Sigma_k^0 \text{ induction} \leftrightarrow \Pi_k^0 \text{ induction.}$$

EXERCISE II.3.13. Show that, for each $k \in \omega$, RCA_0 proves

$$\Sigma_k^0 \text{ induction} \leftrightarrow \text{bounded } \Sigma_k^0 \text{ comprehension.}$$

EXERCISE II.3.14. Show that RCA_0 proves the *strong Σ_1^0 bounding scheme*:

$$\forall m\, \exists n\, \forall i < m\, ((\exists j\, \varphi(i,j)) \to (\exists j < n)\, \varphi(i,j))$$

where $\varphi(i,j)$ is any Σ_1^0 formula in which n does not occur freely.

EXERCISE II.3.15. For each $k \in \omega$, define the *strong Σ_k^0 bounding principle* (in analogy with the previous exercise) and show that RCA_0 proves

$$\Sigma_k^0 \text{ induction} \leftrightarrow \text{strong } \Sigma_k^0 \text{ bounding.}$$

REMARK II.3.16. The main focus of this section has been our basic result on primitive recursion, theorem II.3.4. From the viewpoint of ordinary mathematics, the most important consequence of theorem II.3.4 is that *elementary number theory can be developed straightforwardly within* RCA_0. For instance, we can use theorem II.3.4 to prove the existence of the exponential function $f(m,n) = m^n$ defined by $f(m,0) = 1$, $f(m,n+1) = f(m,n) \cdot m$. We can then show that RCA_0 proves basic properties such as $(m_1 m_2)^n = m_1^n m_2^n$, $m^{n_1 + n_2} = m^{n_1} m^{n_2}$, $m^{n_1 n_2} = (m^{n_1})^{n_2}$. Also within RCA_0 we can straightforwardly state and prove fundamental results such as unique prime power factorization.

It appears that even the most intricate arguments of elementary number theory, finite combinatorics, and finite group theory can be transcribed into RCA_0. This holds so long as the arguments in question make no essential use of infinite sets. Indeed, such arguments can usually be developed within the much weaker theory EFA consisting of Σ_0^0 comprehension, the induction axiom, and the basic axioms augmented by $m^0 = 1$,

$m^{n+1} = m^n \cdot m$ where now exponentiation is treated as a primitive binary operation symbol.

In the rest of this chapter we shall turn to infinitary mathematics. We shall show that certain elementary portions of the theory of continuous functions, countable algebra, and mathematical logic can be developed within RCA_0.

Notes for §II.3. Friedman's original axiomatization of RCA_0 [69] was based on primitive recursion rather than Σ_1^0 induction. The results of this section are essentially due to Friedman (unpublished). For more information on the Σ_k^0 bounding principle, etc., see Kirby/Paris [141] and Hájek/Pudlák [100].

II.4. The Number Systems

In this section we begin the development of ordinary mathematics within RCA_0. We present what amounts to the usual Dedekind/Cauchy construction of the number systems.

We begin with the ring of integers $\mathbb{Z}$. The most basic properties of the natural number system $\mathbb{N}$ have already been developed (lemma II.2.1). We shall now define integers to be certain ordered pairs $(m, n) \in \mathbb{N} \times \mathbb{N}$. In order to do so, we first define the following operations and relations on $\mathbb{N} \times \mathbb{N}$:

$$(m, n) +_{\mathbb{Z}} (p, q) = (m + p, n + q),$$
$$(m, n) -_{\mathbb{Z}} (p, q) = (m + q, n + p),$$
$$(m, n) \cdot_{\mathbb{Z}} (p, q) = (m \cdot p + n \cdot q, m \cdot q + n \cdot p),$$
$$(m, n) <_{\mathbb{Z}} (p, q) \leftrightarrow m + q < n + p,$$
$$(m, n) =_{\mathbb{Z}} (p, q) \leftrightarrow m + q = n + p.$$

Clearly $=_{\mathbb{Z}}$ is an equivalence relation on $\mathbb{N} \times \mathbb{N}$. We define an *integer* to be any element of $\mathbb{N} \times \mathbb{N} \subseteq \mathbb{N}$ which is the least element of its equivalence class. (Here "least" refers to the ordering of $\mathbb{N}$.) We can prove in RCA_0 that the set $\mathbb{Z}$ of all integers exists. We can then define $+, -, \cdot, 0, 1, <$ on $\mathbb{Z}$ accordingly. (For instance, for all $a, b \in \mathbb{Z}$, we define $a + b =$ the unique $c \in \mathbb{Z}$ such that $c =_{\mathbb{Z}} a +_{\mathbb{Z}} b$.) We can then prove:

THEOREM II.4.1. *The following is provable in* RCA_0. *The system*

$$\mathbb{Z}, +, -, \cdot, 0, 1, <$$

is an ordered integral domain, is Euclidean, etc.

PROOF. We identify $m \in \mathbb{N}$ with $(m, 0) \in \mathbb{Z}$. Note that, under this identification, $(m, n) =_{\mathbb{Z}} m - n$. The proof of the basic properties of the ring of integers $\mathbb{Z}$ is straightforward using lemma II.2.1.   □

Next we introduce the field $\mathbb{Q}$ of rational numbers. Let $\mathbb{Z}^+$ be the set of positive integers. Rational numbers will be defined to be certain ordered pairs $(a, b) \in \mathbb{Z} \times \mathbb{Z}^+$. We define the following operations and relations on $\mathbb{Z} \times \mathbb{Z}^+$:

$$(a, b) +_{\mathbb{Q}} (c, d) = (a \cdot d + b \cdot c, b \cdot d),$$
$$(a, b) -_{\mathbb{Q}} (c, d) = (a \cdot d - b \cdot c, b \cdot d),$$
$$(a, b) \cdot_{\mathbb{Q}} (c, d) = (a \cdot c, b \cdot d),$$
$$(a, b) <_{\mathbb{Q}} (c, d) \leftrightarrow a \cdot d < b \cdot c,$$
$$(a, b) =_{\mathbb{Q}} (c, d) \leftrightarrow a \cdot d = b \cdot c.$$

Again $=_{\mathbb{Q}}$ is an equivalence relation on $\mathbb{Z} \times \mathbb{Z}^+$, and we define a *rational number* to be any element of $\mathbb{Z} \times \mathbb{Z}^+ \subseteq \mathbb{N}$ which is the least element of its equivalence class. (Here "least" refers to the ordering of $\mathbb{N}$.) The set of all rational numbers is denoted $\mathbb{Q}$ and we define $+, -, \cdot, 0, 1, <$ on $\mathbb{Q}$ accordingly. We then prove:

THEOREM II.4.2. *The following is provable in* RCA$_0$. *The system*

$$\mathbb{Q}, +, -, \cdot, 0, 1, <$$

is an ordered field.

PROOF. For all $r, s \in \mathbb{Q}$ with $s \neq 0$, we define $r/s =$ the unique $q \in \mathbb{Q}$ such that $q \cdot s = r$. We identify $a \in \mathbb{Z}$ with the unique $r \in \mathbb{Q}$ such that $r =_{\mathbb{Q}} (a, 1)$. Under this identification, for all $(a, b) \in \mathbb{Z} \times \mathbb{Z}^+$, $(a, b) = a/b$. The proof of theorem II.4.2 is straightforward using theorem II.4.1.   □

We now introduce the real number system. We use a modification of the usual definition via Cauchy sequences of rational numbers.

DEFINITION II.4.3. A *sequence of rational numbers* is defined in RCA$_0$ to be a function $f : \mathbb{N} \to \mathbb{Q}$. We usually denote such a sequence as $\langle q_k : k \in \mathbb{N} \rangle$ where $q_k = f(k)$.

DEFINITION II.4.4 (the real number system). A *real number* is defined in RCA$_0$ to be a sequence of rational numbers $\langle q_k : k \in \mathbb{N} \rangle$ such that $\forall k \, \forall i \, (|q_k - q_{k+i}| \leq 2^{-k})$. Here $|q|$ denotes the *absolute value* of a rational number $q \in \mathbb{Q}$, i.e., $|q| = q$ if $q \geq 0$, $-q$ otherwise. Two real numbers $\langle q_k : k \in \mathbb{N} \rangle$ and $\langle q_k' : k \in \mathbb{N} \rangle$ are said to be *equal* if $\forall k \, (|q_k - q_k'| \leq 2^{-k+1})$.

We shall often use special variables such as x, y, ... to range over real numbers. We then write $x = y$ to mean that the real numbers x and y are equal in the sense of definition II.4.4. When describing definitions or proofs within RCA$_0$, we shall sometimes use the symbol $\mathbb{R}$ informally to denote the set of all real numbers. Thus for instance $\forall x \in \mathbb{R} \ldots$ means $\forall x$ (if x is a real number then ...). Of course the set $\mathbb{R}$ does not formally exist within RCA$_0$, since RCA$_0$ is limited to the language L_2 of second order arithmetic.

REMARK. Note that we are taking equality between real numbers (the $=$ of definition II.4.4) to be an equivalence relation rather than true identity. This choice is dictated by our goal of developing mathematics within subsystems of second order arithmetic such as RCA_0 and ACA_0. One might consider alternative definitions under which a real number would be an equivalence class or a representative of an equivalence class. Both alternatives turn out to be inappropriate. Equivalence classes would require the language of third order arithmetic, and the use of representatives would demand a strong form of the axiom of choice which is not available even in full second order arithmetic, Z_2.

Working within RCA_0 we embed $\mathbb{Q}$ into $\mathbb{R}$ by identifying $q \in \mathbb{Q}$ with the real number $x_q = \langle q \rangle = \langle q_k : k \in \mathbb{N} \rangle$ where $q_k = q$ for all $k \in \mathbb{N}$. A real number x is said to be *rational* if $x = x_q$ for some $q \in \mathbb{Q}$. (Here $=$ is as in definition II.4.4.)

The sum of two real numbers $x = \langle q_k : k \in \mathbb{N} \rangle$ and $y = \langle q'_k : k \in \mathbb{N} \rangle$ is defined by

$$x + y = \langle q_{k+1} + q'_{k+1} : k \in \mathbb{N} \rangle.$$

We note that $|(q_{k+1} + q'_{k+1}) - (q_{k+i+1} + q'_{k+i+1})| \le |q_{k+1} - q_{k+i+1}| + |q'_{k+1} - q'_{k+i+1}| \le 2^{-k-1} + 2^{-k-1} = 2^{-k}$ so $x + y$ is a real number. Trivially $-x = \langle -q_k : k \in \mathbb{N} \rangle$ is also a real number. We define $x \le y$ if and only if $\forall k \, (q_k \le q'_k + 2^{-k+1})$. Clearly $x = y$ if and only if $x \le y$ and $y \le x$. We define $x < y$ if and only if $y \not\le x$. It is straightforward to verify in RCA_0 that the system $\mathbb{R}, +, -, 0, 1, <$ obeys all the axioms for an ordered Abelian group, for example

$$x < y \lor x = y \lor x > y,$$
$$x < y \leftrightarrow x + z < y + z,$$

etc.

Note that formulas such as $x \le y$, $x = y$, $x + y = z$, ... are Π^0_1 while $x < y$, $x \ne 0$, ... are Σ^0_1.

Multiplication of real numbers $x = \langle q_k : k \in \mathbb{N} \rangle$ and $y = \langle q'_k : k \in \mathbb{N} \rangle$ is defined by

$$x \cdot y = \langle q_{n+k} \cdot q'_{n+k} : k \in \mathbb{N} \rangle$$

where n is as small as possible such that $2^n \ge |q_0| + |q'_0| + 2$. We note that $x \cdot y$ is a real number since

$$|q_{n+k} \cdot q'_{n+k} - q_{n+k+i} \cdot q'_{n+k+i}|$$
$$\le |q_{n+k}| \cdot |q'_{n+k} - q'_{n+k+i}| + |q_{n+k} - q_{n+k+i}| \cdot |q'_{n+k+i}|$$
$$\le 2^{-n-k} (|q_0| + |q'_0| + 2)$$
$$\le 2^{-k}.$$

We can then prove straightforwardly:

Theorem II.4.5. *It is provable in* RCA_0 *that the real number system*

$$\mathbb{R}, +, -, \cdot, 0, 1, <, =$$

obeys all the axioms of an Archimedean ordered field.

It is natural now to ask whether the real number system is complete. In RCA_0 we cannot discuss arbitrary bounded subsets of $\mathbb{R}$. Thus we cannot even formulate the least upper bound principle in full generality. However, we can discuss sequences of elements of $\mathbb{R}$.

Definition II.4.6 (sequences of real numbers). Within RCA_0, a *sequence of real numbers* is a function $f: \mathbb{N} \times \mathbb{N} \to \mathbb{Q}$ such that for each $n \in \mathbb{N}$ the function $(f)_n: \mathbb{N} \to \mathbb{Q}$ defined by $(f)_n(k) = f((k, n))$ is a real number (in the sense of definition II.4.4). We shall employ notations such as $\langle x_n : n \in \mathbb{N} \rangle$ for the sequence f with $(f)_n = x_n$.

Using the previous definition we can discuss sequential convergence within RCA_0. We say that the sequence $\langle x_n : n \in \mathbb{N} \rangle$ *converges* to x, written $x = \lim_n x_n$, if $\forall \epsilon > 0 \, \exists n \, \forall i \, (|x - x_{n+i}| < \epsilon)$. The sequence $\langle x_n : n \in \mathbb{N} \rangle$ is said to be *convergent* if $\lim_n x_n$ exists.

Unfortunately, the axioms of RCA_0 are not even strong enough to prove that $\mathbb{R}$ is sequentially complete. This is shown by the following counterexample.

Example II.4.7. Let $f: \omega \to \omega$ be a one-to-one recursive function whose range is not recursive. For each $n \in \omega$ put

$$c_n = \sum_{i=0}^{n} 2^{-f(i)}.$$

Clearly $c_0 < c_1 < \cdots < c_n < \cdots < 2$ so $\langle c_n : n \in \omega \rangle$ is a recursive, bounded, increasing sequence of rational numbers. However, the real number $c = \lim_n c_n$ is clearly not recursive.

From the above counterexample, it follows that the least upper bound principle for sequences of real numbers is false in the ω-model REC $= \{X \subseteq \omega : X \text{ is recursive}\}$. Since REC $\models RCA_0$, it follows that the least upper bound principle for sequences of real numbers is not provable in RCA_0.

However, not all is lost. The following *nested interval completeness* property of $\mathbb{R}$ is provable in RCA_0 and suffices for many purposes.

Theorem II.4.8 (nested interval completeness). *The following is provable in* RCA_0. *Let* $\langle a_n : n \in \mathbb{N} \rangle$ *and* $\langle b_n : n \in \mathbb{N} \rangle$ *be sequences of real numbers such that for all* n, $a_n \leq a_{n+1} \leq b_{n+1} \leq b_n$, *and* $\lim_n |a_n - b_n| = 0$. *Then there exists a real number* x *such that* $x = \lim_n a_n = \lim_n b_n$.

Proof. Let $\langle q_{nk} : n, k \in \mathbb{N} \rangle$ and $\langle q'_{nk} : n, k \in \mathbb{N} \rangle$ be double sequences of rationals such that for all n, $a_n = \langle q_{nk} : k \in \mathbb{N} \rangle$ and $b_n = \langle q'_{nk} : k \in \mathbb{N} \rangle$. (Compare definitions II.4.3, II.4.4, and II.4.6.) Clearly for each k there exists n such that $n \geq k+2$ and $|q_{nn} - q'_{nn}| \leq 2^{-k-2}$. Let $f(k)$ be the least such n (theorem II.3.5) and put $x = \langle q''_k : k \in \mathbb{N} \rangle$ where $q''_k = q_{f(k),f(k)}$. It is straightforward to verify that x is a real number and that $a_n \leq x \leq b_n$ for all n. Thus $x = \lim_n a_n = \lim_n b_n$. This completes the proof. $\square$

Using nested interval completeness, we can prove that $\mathbb{R}$ is uncountable:

THEOREM II.4.9 (uncountability of $\mathbb{R}$). *The following is provable in* RCA$_0$. *For any sequence of real numbers* $\langle x_n : n \in \mathbb{N} \rangle$ *there exists a real number* y *such that* $\forall n \, (x_n \neq y)$.

PROOF. Let $\langle q_{nk} : n \in \mathbb{N}, k \in \mathbb{N} \rangle$ be a double sequence of rational numbers such that $x_n = \langle q_{nk} : k \in \mathbb{N} \rangle$ for each n. By primitive recursion (theorem II.3.4) define a sequence of rational intervals $\langle (a_n, b_n) : n \in \mathbb{N} \rangle$ as follows: $(a_0, b_0) = (0, 1)$;

$$(a_{n+1}, b_{n+1}) = \begin{cases} ((a_n + 3b_n)/4, b_n) & \text{if } q_{n.2n+3} \leq (a_n + b_n)/2, \\ (a_n, (3a_n + b_n)/4) & \text{otherwise.} \end{cases}$$

For each n we have $|a_n - b_n| = 2^{-2n}$ so $\lim_n |a_n - b_n| = 0$. By theorem II.4.8 let $y = \lim_n a_n = \lim_n b_n$. (Alternatively we could just define y to be the rational sequence $\langle a_n : n \in \mathbb{N} \rangle$ and note directly that y is a real number.) If $q_{n.2n+3} \leq \frac{1}{2}(a_n + b_n)$, we have $x_n \leq \frac{1}{2}(a_n + b_n) + 2^{-2n-3} < a_{n+1} \leq y$. In the other case we have $x_n \geq \frac{1}{2}(a_n + b_n) - 2^{-2n-3} > b_{n+1} \geq y$. Thus $\forall n \, (x_n \neq y)$. $\square$

In a similar vein we can prove the *Baire category theorem* for $\mathbb{R}^k$, $k \in \mathbb{N}$, within RCA$_0$. First we present the relevant definitions. Within RCA$_0$ we define a *point of* $\mathbb{R}^k$ to be a finite sequence of real numbers of length k. We use notations such as $\langle x_1, \ldots, x_k \rangle$ for points of $\mathbb{R}^k$. Within RCA$_0$ we define a (code for a) *basic open set* in $\mathbb{R}^k$ to be an ordered $2k$-tuple of rational numbers $\langle a_1, b_1, \ldots, a_k, b_k \rangle \in \mathbb{Q}^{2k}$ such that $a_i < b_i$ for all i, $1 \leq i \leq k$. A (code for an) *open set* in $\mathbb{R}^k$ is any set U of (codes for) basic open sets in $\mathbb{R}^k$. We then define $\langle x_1, \ldots, x_k \rangle \in U$ to mean that there exists $\langle a_1, b_1, \ldots, a_k, b_k \rangle \in U$ such that $a_i < x_i < b_i$ for all i, $1 \leq i \leq k$. An open set U in $\mathbb{R}^k$ is said to be *dense* if it contains points from each basic open set in $\mathbb{R}^k$. Using these definitions we have the Baire category theorem for $\mathbb{R}^k$:

THEOREM II.4.10 (Baire category theorem for $\mathbb{R}^k$). *The following is provable in* RCA$_0$. *Let* $\langle U_n : n \in \mathbb{N} \rangle$ *be a sequence of dense open sets in* $\mathbb{R}^k$. *Then there exists* $x \in \mathbb{R}^k$ *such that* $x \in U_n$ *for all* $n \in \mathbb{N}$.

PROOF. Similar to the proof of the previous theorem. $\square$

In the next section, theorems II.4.8, II.4.9 and II.4.10 as well as the definitions preceding theorem II.4.10 will be generalized to the context of complete separable metric spaces.

EXERCISE II.4.11 (real linear algebra). Show that RCA_0 is strong enough to develop the basics of real linear algebra, including Gaussian elimination, etc.

Hint: Given a generic system of linear equations

$$a_{11}x_1 + \cdots + a_{1n}x_n = b_1$$

$$\vdots$$

$$a_{m1}x_1 + \cdots + a_{mn}x_n = b_m$$

we can form a finite decision tree T_{mn} representing all of the possibilities for Gaussian elimination. Each node $v \in T_{mn}$ has at most two immediate successors which are distinguished by whether a certain integer polynomial p_v in the coefficients

$$a_{11}, \ldots, a_{1n}, \ldots, a_{m1}, \ldots, a_{mn} \tag{4}$$

is equal or unequal to 0. By bounded Σ_1^0 comprehension in RCA_0, any particular set of real coefficients (4) gives rise to a path through T_{mn} describing this instance of Gaussian elimination.

Notes for §II.4. Our treatment of the real number system in the context of RCA_0 is analogous to that of Aberth [2] in the somewhat different context of recursive analysis. For a constructivist treatment, see Bishop/Bridges [20]. In an early paper Simpson [236] developing mathematics within RCA_0, we defined a real number to be the set of smaller rational numbers. This alternative definition, although in a sense equivalent to definition II.4.4 above, turns out to be inappropriate for other reasons, as explained in Brown/Simpson [27, §3].

II.5. Complete Separable Metric Spaces

In the previous section we defined the real numbers, within RCA_0, to be the "completion" of the rational numbers. We shall now use the same idea to define a complete separable metric space $\widehat{A}$, within RCA_0, to be the "completion" of its countable dense subset A.

DEFINITION II.5.1 (complete separable metric spaces). A (code for a) *complete separable metric space* $\widehat{A}$ is defined in RCA_0 to be a nonempty set $A \subseteq \mathbb{N}$ together with a sequence of real numbers $d : A \times A \to \mathbb{R}$ such that $d(a, a) = 0$, $d(a, b) = d(b, a) \geq 0$, and $d(a, b) + d(b, c) \geq d(a, c)$ for all $a, b, c \in A$. A *point* of $\widehat{A}$ is a sequence $x = \langle a_k : k \in \mathbb{N} \rangle$ of elements of A, such that $\forall i \, \forall j \, (i < j \to d(a_i, a_j) \leq 2^{-i})$. We write $x \in \widehat{A}$ to mean that x is a point of $\widehat{A}$.

If $x = \langle a_k : k \in \mathbb{N} \rangle$ and $y = \langle b_k : k \in \mathbb{N} \rangle$ are points of $\widehat{A}$, we define $d(x, y) = \lim_k d(a_k, b_k)$. We define $x = y$ to mean that $d(x, y) = 0$. Note that the condition $x = y$ is Π_1^0 since it is equivalent to $\forall k \, (d(a_k, b_k) \leq 2^{-k+1})$.

Each $a \in A$ is identified with the point $x_a = \langle a : k \in \mathbb{N} \rangle \in \widehat{A}$. Thus by definition the countable set A is dense in $\widehat{A}$; indeed, for all $x \in \widehat{A}$ we have $d(x, a_k) \leq 2^{-k}$, where $x = \langle a_k : k \in \mathbb{N} \rangle$. This justifies our designation of $\widehat{A}$ as "separable." In order to justify our designation of $\widehat{A}$ as "complete," we present the following exercise generalizing our earlier discussion of nested interval completeness.

EXERCISE II.5.2. Within RCA$_0$, show that $\widehat{A}$ is complete in the following sense. Let $\langle x_n : n \in \mathbb{N} \rangle$ be a sequence of points of $\widehat{A}$. Assume that there exists a sequence of real numbers $\langle r_n : n \in \mathbb{N} \rangle$ such that $\forall m \, \forall n \, (m < n \rightarrow d(x_n, x_m) \leq r_m)$ and $\lim_n r_n = 0$. Then $\langle x_n : n \in \mathbb{N} \rangle$ is *convergent*, i.e., there exists a point $x \in \widehat{A}$ (unique up to $=$ as defined in II.5.1) such that $x = \lim_n x_n$.

We now give some examples and constructions of complete separable metric spaces, within RCA$_0$.

EXAMPLE II.5.3 (the real numbers). Within RCA$_0$, for $q, q' \in \mathbb{Q}$ define $d(q, q') = |q - q'|$. Then $\widehat{\mathbb{Q}} = \mathbb{R}$, i.e., the reals are the completion of the rationals. More generally, any closed (bounded or unbounded) interval of $\mathbb{R}$ is a complete separable metric space with the same metric. For example we have the *closed unit interval*

$$[0, 1] = \{x : 0 \leq x \leq 1\}.$$

EXAMPLE II.5.4 (finite product spaces). Within RCA$_0$ we can define the notion of a sequence of codes for complete separable metric spaces. Given a finite sequence of such codes A_i, $1 \leq i \leq m$, we can form the m-fold Cartesian product

$$A = A_1 \times \cdots \times A_m = \{\langle a_1, \ldots, a_m \rangle : a_i \in A_i\}$$

and define $d : A \times A \to \mathbb{R}$ by

$$d(\langle a_1, \ldots, a_m \rangle, \langle b_1, \ldots, b_m \rangle) = \sqrt{d_1(a_1, b_1)^2 + \cdots + d_m(a_m, b_m)^2}.$$

We can then prove within RCA$_0$ the following facts: (i) $\widehat{A}$ is a complete separable metric space; (ii) the points of $\widehat{A}$ can be identified with finite sequences $\langle x_1, \ldots, x_m \rangle$ with $x_i \in \widehat{A_i}$ for $1 \leq i \leq m$; and (iii) under this identification, the metric on $\widehat{A}$ is given by

$$d(\langle x_1, \ldots, x_m \rangle, \langle y_1, \ldots, y_m \rangle) = \sqrt{d_1(x_1, y_1)^2 + \cdots + d_m(x_m, y_m)^2}.$$

Thus we are justified in writing

$$\widehat{A} = \widehat{A}_1 \times \cdots \times \widehat{A}_m = \prod_{i=1}^{m} \widehat{A}_i.$$

In particular we have within RCA_0 the m-dimensional Euclidean spaces $\mathbb{R}^m$ for all $m \in \mathbb{N}$. The points of $\mathbb{R}^m$ can be identified with m-tuples $\langle x_1, \ldots, x_m \rangle$, $x_i \in \mathbb{R}$.

EXAMPLE II.5.5 (infinite product spaces). Given an infinite sequence of (codes for) complete separable metric spaces $\widehat{A}_i, i \in \mathbb{N}$, we can form the infinite product space $\widehat{A} = \prod_{i=0}^{\infty} \widehat{A}_i$ as follows. For each $i \in \mathbb{N}$, we let c_i be the smallest element of $A_i \subseteq \mathbb{N}$ (in the usual ordering of $\mathbb{N}$). We define

$$A = \bigcup_{m=0}^{\infty} (A_0 \times \cdots \times A_m) = \{\langle a_i : i \leq m \rangle : m \in \mathbb{N}, a_i \in A_i\}$$

and $d : A \times A \to \mathbb{R}$ by

$$d(\langle a_i : i \leq m \rangle, \langle b_i : i \leq n \rangle) = \sum_{i=0}^{\infty} \frac{d_i(a'_i, b'_i)}{1 + d_i(a'_i, b'_i)} \cdot \frac{1}{2^i}$$

where

$$a'_i = \begin{cases} a_i & \text{if } i \leq m, \\ c_i & \text{otherwise} \end{cases}$$

and

$$b'_i = \begin{cases} b_i & \text{if } i \leq m, \\ c_i & \text{otherwise.} \end{cases}$$

We can then prove within RCA_0 the following facts: (i) $\widehat{A}$ is a complete separable metric space; (ii) the points of $\widehat{A}$ can be identified with the sequences $\langle x_i : i \in \mathbb{N} \rangle$ where $x_i \in \widehat{A}_i$ for all $i \in \mathbb{N}$; and (iii) under this identification, the metric on $\widehat{A}$ is given by

$$d(\langle x_i : i \in \mathbb{N} \rangle, \langle y_i : i \in \mathbb{N} \rangle) = \sum_{i=0}^{\infty} \frac{d_i(x_i, y_i)}{1 + d_i(x_i, y_i)} \cdot \frac{1}{2^i}.$$

These three conditions define the usual textbook construction of the product of a sequence of complete separable metric spaces. Thus we are justified in writing

$$\widehat{A} = \prod_{i=0}^{\infty} \widehat{A}_i.$$

In particular, we have within RCA$_0$ the *Cantor space*

$$2^{\mathbb{N}} = \{0, 1\}^{\mathbb{N}} = \prod_{i=0}^{\infty}\{0, 1\},$$

the *Baire space*

$$\mathbb{N}^{\mathbb{N}} = \prod_{i=0}^{\infty}\mathbb{N},$$

and the *Hilbert cube*

$$[0, 1]^{\mathbb{N}} = \prod_{i=0}^{\infty}[0, 1].$$

The points of the Cantor space and the Baire space can be identified with functions $f : \mathbb{N} \to \{0, 1\}$ and $f : \mathbb{N} \to \mathbb{N}$ respectively. The points of the Hilbert cube can be identified with sequences $\langle x_i : i \in \mathbb{N}\rangle$, $0 \le x_i \le 1$.

We now begin our discussion of the topology of complete separable metric spaces. This discussion will be continued in §§II.6, II.7 and II.10 and in chapters II–VI.

DEFINITION II.5.6 (open sets). Within RCA$_0$, let $\widehat{A}$ be a complete separable metric space. A (code for an) *open set* U in $\widehat{A}$ is a set $U \subseteq \mathbb{N} \times A \times \mathbb{Q}^+$, where

$$\mathbb{Q}^+ = \{q \in \mathbb{Q} : q > 0\}.$$

A point $x \in \widehat{A}$ is said to *belong to* U (abbreviated $x \in U$) if

$$\exists n \, \exists a \, \exists r \, ((d(x, a) < r \wedge (n, a, r) \in U).$$

Note that the formula $x \in U$ is Σ_1^0.

We regard $(a, r) \in A \times \mathbb{Q}^+$ as a code for the basic open ball $\mathrm{B}(a, r)$ consisting of all points $x \in \widehat{A}$ such that $d(x, a) < r$. The idea of the preceding definition is that U encodes the open set which is the union of the balls $\mathrm{B}(a, r)$ such that $\exists n \, ((n, a, r) \in U)$. We shall sometimes use notations such as $(a, r) < U$ meaning that

$$\exists n \, \exists b \, \exists s \, (d(a, b) + r < s \wedge (n, b, s) \in U).$$

Note that this condition is Σ_1^0 and implies that the closure of $\mathrm{B}(a, r)$ is included in U. We write $(a, r) < (b, s)$ to mean $d(a, b) + r < s$.

The following lemma will provide many examples of open sets within RCA$_0$.

LEMMA II.5.7. *For any Σ_1^0 formula $\varphi(x)$, the following is provable in RCA$_0$. Let $\widehat{A}$ be a complete separable metric space. Assume that for all x and $y \in \widehat{A}$, $x = y$ and $\varphi(x)$ imply $\varphi(y)$. Then there exists an open set $U \subseteq \widehat{A}$ such that for all $x \in \widehat{A}$, $x \in U$ if and only if $\varphi(x)$.*

PROOF. By the normal form theorem for Σ_1^0 formulas (theorem II.2.7), there exists a Σ_0^0 formula $\theta(n)$ such that RCA$_0$ proves: for all $x = \langle a_k : k \in \mathbb{N} \rangle \in \widehat{A}$, $\varphi(x) \leftrightarrow \exists m\, \theta(\langle a_k : k \leq m \rangle)$. We reason within RCA$_0$. By Σ_0^0 comprehension, let U be the set of all $(n, a, r) \in \mathbb{N} \times A \times \mathbb{Q}^+$ such that, for some $m \in \mathbb{N}$, $n = \langle a_k : k \leq m \rangle \in A^{m+1}$ and $\theta(\langle a_k : k \leq m \rangle)$ holds and $a = a_m$ and $r = 2^{-m-1}$ and

$$(\forall i \leq m)\,(\forall j \leq m)\,(i < j \rightarrow d(a_i, a_j)_{m+1} \leq 2^{-i-1}).$$

(Here $d(a, b)_k$ denotes the kth rational approximation to the real number $d(a, b)$, i.e., $d(a, b)_k = q_k \in \mathbb{Q}$ where $d(a, b) = \langle q_k : k \in \mathbb{N} \rangle \in \mathbb{R}$.)

Thus U is (a code for) an open set in $\widehat{A}$. It remains to prove that, for all $y \in \widehat{A}$, $y \in U$ if and only if $\varphi(y)$.

Assume first that $y \in U$. Then $d(a, y) < r$ for some $(n, a, r) \in U$. Let $n = \langle a_k : k \leq m \rangle$ as above. Thus $a = a_m$, $r = 2^{-m-1}$, and $d(a_m, y) < 2^{-m-1}$. Write $y = \langle b_k : k \in \mathbb{N} \rangle$ and let $m' > m$ be so large that $d(a_m, b_j) \leq 2^{-m-1}$ for all $j \geq m'$. Then for all $i < m$ and $j \geq m'$ we have

$$\begin{aligned} d(a_i, b_j) &\leq d(a_i, a_m) + d(a_m, b_j) \\ &\leq d(a_i, a_m)_{m+1} + 2^{-m-1} + d(a_m, b_j) \\ &\leq 2^{-i-1} + 2^{-m-1} + 2^{-m-1} \\ &\leq 2^{-i}. \end{aligned}$$

Put $z = \langle a_0, a_1, \ldots, a_m \rangle ^\frown \langle b_{m'}, b_{m'+1}, \ldots \rangle$. Then by the previous inequality we have $z \in \widehat{A}$. Moreover $z = y$, and $\varphi(z)$ holds. Hence $\varphi(y)$ holds.

Conversely, assume that $\varphi(y)$ holds. Write $y = \langle b_k : k \in \mathbb{N} \rangle$. Put $x = \langle a_k : k \in \mathbb{N} \rangle$ where $a_k = b_{k+2}$ for all $k \in \mathbb{N}$. Then $x \in \widehat{A}$, and $x = y$, so $\varphi(x)$ holds. Moreover, for all $i < j \in \mathbb{N}$ we have $d(a_i, a_j) = d(b_{i+2}, b_{j+2}) \leq 2^{-i-2}$, hence $d(a_i, a_j)_k \leq 2^{-i-2} + 2^{-k} \leq 2^{-i-1}$ for all $k \geq i+2$. Let m be such that $\theta(\langle a_k : k \leq m \rangle)$ holds. Put $n = \langle a_k : k \leq m \rangle$, $a = a_m$, and $r = 2^{-m-1}$. Then $(n, a, r) \in U$. Also $d(a, y) = d(a_m, x) = \lim_j d(a_m, a_j) \leq 2^{-m-2} < 2^{-m-1} = r$, which implies that $y \in U$. This completes the proof. □

We shall now prove the following RCA$_0$ version of the *Baire category theorem*.

THEOREM II.5.8 (Baire category theorem). *The following is provable in* RCA$_0$. *Let* $\langle U_k : k \in \mathbb{N} \rangle$ *be a sequence of dense open sets in* $\widehat{A}$. *Then* $\bigcap_{k \in \mathbb{N}} U_k$ *is dense in* $\widehat{A}$.

PROOF. We reason within RCA$_0$. We wish to show that $\bigcap_{k \in \mathbb{N}} U_k$ is dense in $\widehat{A}$. Given $y \in \widehat{A}$ and $\epsilon > 0$, we must find $x \in \widehat{A}$ such that $d(x, y) < \epsilon$ and $x \in U_k$ for all $k \in \mathbb{N}$. We shall define the point $x = \langle a_k : k \in \mathbb{N} \rangle$

by recursion on k. Since U_0 is dense, we can find $(a_0, r_0) \in A \times \mathbb{Q}^+$ such that $(a_0, r_0) < (y, \epsilon)$, $(a_0, r_0) < U_0$, and $r_0 \leq 1/2$. Let $\varphi(k, a, r, b, s)$ be a Σ_1^0 formula which expresses the following: $(a, r) \in A \times \mathbb{Q}^+$, $(b, s) \in A \times \mathbb{Q}^+$, $(b, s) < (a, r)$, $(b, s) < U_k$, and $s \leq 2^{-k-1}$. From the density of U_k, it follows that for each $(k, a, r) \in \mathbb{N} \times A \times \mathbb{Q}^+$ there exists (b, s) such that $\varphi(k, a, r, b, s)$. Write

$$\varphi(k, a, r, b, s) \equiv \exists n \, \theta(k, a, r, b, s, n)$$

where θ is Σ_0^0. By minimization (theorem II.3.5), there exists a function

$$f : \mathbb{N} \times A \times \mathbb{Q}^+ \to \mathbb{N} \times A \times \mathbb{Q}^+$$

such that $f(k, a, r)$ is the least (n, b, s) such that $\theta(k, a, r, b, s, n)$ holds. By primitive recursion (theorem II.3.4), there exists a function $g : \mathbb{N} \to A \times \mathbb{Q}^+$ such that $g(0) = (a_0, r_0)$ and, for all $k \in \mathbb{N}$, $g(k+1) = (a_{k+1}, r_{k+1})$ where $f(k, a_k, r_k) = (n_k, a_{k+1}, r_{k+1})$. Hence $\varphi(k, a_k, r_k, a_{k+1}, r_{k+1})$ holds for all k. It is not hard to check that $x = \langle a_k : k \in \mathbb{N} \rangle$ is a point of $\widehat{A}$ and that $x \in U_k$ for all k, and $d(x, y) < \epsilon$. This completes the proof. □

COROLLARY II.5.9. *The following is provable in* RCA$_0$. *Let $\widehat{A}$ be a complete separable metric space with no isolated points. Then $\widehat{A}$ is uncountable, i.e., for all sequences of points $\langle x_k : k \in \mathbb{N} \rangle$, $x_k \in \widehat{A}$, there exists a point $y \in \widehat{A}$ such that $\forall k \, (x_k \neq y)$.*

PROOF. We reason within RCA$_0$. Let $\varphi(k, y)$ be a Σ_1^0 formula (with parameter $\langle x_k : k \in \mathbb{N} \rangle$) which says that $y \neq x_k$. By lemma II.5.7, RCA$_0$ proves the existence of a sequence of open sets $\langle U_k : k \in \mathbb{N} \rangle$ such that, for all $y \in \widehat{A}$ and $k \in \mathbb{N}$, $y \in U_k$ if and only if $\varphi(k, y)$. For each $k \in \mathbb{N}$, since x_k is not an isolated point, U_k is dense. The desired conclusion follows from the Baire category theorem II.5.8. □

EXERCISE II.5.10. In RCA$_0$ show that, given a sequence $\langle U_n : n \in \mathbb{N} \rangle$ of (codes for) open sets in $\widehat{A}$, we can effectively find (a code for) an open set U in $\widehat{A}$ such that for all points $x \in \widehat{A}$, $x \in U$ if and only if $\exists n \, (x \in U_n)$. Thus we are justified in writing $U = \bigcup_{n \in \mathbb{N}} U_n$ and in saying that the union of countably many open sets is open.

EXERCISE II.5.11. In RCA$_0$ show that, given a finite sequence $\langle U_k : k < n \rangle$ of (codes for) open sets in $\widehat{A}$, we can effectively find (a code for) an open set U in $\widehat{A}$ such that, for all points $x \in \widehat{A}$, $x \in U$ if and only if $x \in U_k$ for all $k < n$. Thus we are justified in writing $U = \bigcap_{k=0}^{n-1} U_k$ and in saying that the intersection of finitely many open sets is open.

DEFINITION II.5.12 (closed sets). Let $\widehat{A}$ be a complete separable metric space. A *closed set* in $\widehat{A}$ is defined in RCA$_0$ to be the complement of an open set in $\widehat{A}$. In other words, we define a code for a closed set C to be the same thing as a code for an open set U, and we define $x \in C$ if and only if $x \notin U$. Note that the formula $x \in C$ is Π_1^0.

EXERCISE II.5.13. Within RCA_0 show that, in any complete separable metric space $\widehat{A}$, the countable intersection and finite union of closed sets is closed.

EXERCISE II.5.14. Within RCA_0 show that the open unit interval

$$(0, 1) = \{x : 0 < x < 1\}$$

is a complete separable metric space under

$$d(x, y) = |x - y| + \left| \frac{1}{h(x)} - \frac{1}{h(y)} \right|$$

where

$$h(x) = \frac{1}{2} - \left| \frac{1}{2} - x \right|.$$

EXERCISE II.5.15. Show that the following is provable in RCA_0. If $\widehat{A}$ is a complete separable metric space and if U is a nonempty open set in $\widehat{A}$, there exists a complete separable metric space $\widehat{B}$ which is *homeomorphic to* U, i.e., there exist continuous functions $f : \widehat{B} \to U$, $g : U \to \widehat{B}$ such that $f(g(x)) = x$ for all $x \in U$. (Continuous functions will be defined in the next section.)

REMARK II.5.16. The previous exercise does not go through in RCA_0 if we replace the nonempty open set U by a nonempty closed set C. See exercise IV.2.11.

For more on complete separable metric spaces in RCA_0, see §§II.6, II.7 and II.10. See also chapters III and IV.

Notes for §II.5. Our definition of complete separable metric space within RCA_0 (II.5.1) comes from Brown/Simpson [27]. Lemma II.5.7 is due to Simpson, unpublished. Our RCA_0 version of the Baire category theorem (II.5.8) is due to Simpson, unpublished. A stronger version of the Baire category theorem is discussed in Brown/Simpson [28]; see also Mytilinaios/Slaman [194]. Alternative notions of closed set are considered in Brown [25] and Giusto/Simpson [93].

II.6. Continuous Functions

In this section we continue the work of the previous section. We show that certain portions of the theory of continuous functions on complete separable metric spaces can be developed within RCA_0. For more information on complete separable metric spaces and continuous functions, see the next section.

DEFINITION II.6.1 (continuous functions). Within RCA$_0$, let $\widehat{A}$ and $\widehat{B}$ be complete separable metric spaces. A (code for a) *continuous partial function* ϕ from $\widehat{A}$ to $\widehat{B}$ is a set of quintuples $\Phi \subseteq \mathbb{N} \times A \times \mathbb{Q}^+ \times B \times \mathbb{Q}^+$ which is required to have certain properties. We write $(a, r)\Phi(b, s)$ as an abbreviation for $\exists n\, ((n, a, r, b, s) \in \Phi)$. The properties which we require are:

1. if $(a, r)\Phi(b, s)$ and $(a, r)\Phi(b', s')$, then $d(b, b') \le s + s'$;
2. if $(a, r)\Phi(b, s)$ and $(a', r') < (a, r)$, then $(a', r')\Phi(b, s)$;
3. if $(a, r)\Phi(b, s)$ and $(b, s) < (b', s')$, then $(a, r)\Phi(b', s')$;

where the notation $(a', r') < (a, r)$ means that $d(a, a') + r' < r$.

The idea of the definition is that Φ encodes a partially defined, continuous function ϕ from $\widehat{A}$ to $\widehat{B}$. Recall from the previous section that B(a, r) denotes the basic open ball centered at a with radius r. Intuitively, $(a, r)\Phi(b, s)$ is a piece of information to the effect that $\phi(x) \in$ the closure of B(b, s) whenever $x \in$ B(a, r), provided $\phi(x)$ is defined. This is made precise in the following two paragraphs.

A point $x \in \widehat{A}$ is said to *belong to the domain of* ϕ, abbreviated $x \in$ dom(ϕ), provided the code Φ of ϕ contains sufficient information to evaluate ϕ at x. This means that for all $\epsilon > 0$ there exists $(a, r)\Phi(b, s)$ such that $d(x, a) < r$ and $s < \epsilon$. If $x \in$ dom(ϕ), we define the value $\phi(x)$ to be the unique point $y \in \widehat{B}$ such that $d(y, b) \le s$ for all $(a, r)\Phi(b, s)$ with $d(x, a) < r$. If $x \in$ dom(ϕ), we can use the code Φ and minimization (theorem II.3.5) to prove within RCA$_0$ that $\phi(x)$ exists. Then, using condition II.6.1.1, it is easy to prove within RCA$_0$ that $\phi(x)$ is unique (up to equality of points in $\widehat{B}$, as defined in II.5.1).

We write $\phi(x)$ *is defined* to mean that $x \in$ dom(ϕ). We say that ϕ is *totally defined on* $\widehat{A}$ if $\phi(x)$ is defined for all $x \in \widehat{A}$. We write $\phi : \widehat{A} \to \widehat{B}$ to mean that ϕ is a continuous, totally defined function from $\widehat{A}$ to $\widehat{B}$.

We now present some examples of (codes for) continuous functions within RCA$_0$.

LEMMA II.6.2. *Within* RCA$_0$, *let* $\widehat{A}$ *and* $\widehat{B}$ *be complete separable metric spaces.*

1. *The identity function* $\phi : \widehat{A} \to \widehat{A}$ *given by* $\phi(x) = x$ *is continuous.*
2. *For any* $y \in \widehat{B}$, *the constant function* $\phi : \widehat{A} \to \widehat{B}$, *given by* $\phi(x) = y$ *for all* $x \in \widehat{A}$, *is continuous.*
3. *The metric* $d : \widehat{A} \times \widehat{A} \to \mathbb{R}$ *is continuous.*

PROOF. For part 1, let $\varphi(a, r, b, s)$ be a Σ^0_1 formula which says that $(a, r), (b, s) \in A \times \mathbb{Q}^+$ and $(a, r) < (b, s)$. (Recall that $(a, r) < (b, s)$ is an abbreviation for $d(a, b) + r < s$.) Write $\varphi(a, r, b, s) \equiv \exists n\, \theta(n, a, r, b, s)$ where θ is Σ^0_0. By Σ^0_0 comprehension, let Φ be the set of (n, a, r, b, s) such that $\theta(n, a, r, b, s)$ holds. It is straightforward to check that Φ is a code for a continuous function $\phi : \widehat{A} \to \widehat{A}$, and that $\phi(x) = x$ for all $x \in \widehat{A}$. (The proof of part 1 should be studied carefully. The same idea will be used in all later constructions of continuous functions within RCA$_0$.)

For part 2, let Φ be such that $(a, r)\Phi(b, s)$ if and only if $(a, r) \in A \times \mathbb{Q}^+$, $(b, s) \in B \times \mathbb{Q}^+$, and $d(b, y) < s$. It is straightforward to check that Φ is a code for a continuous function $\phi \colon \widehat{A} \to \widehat{B}$, and that $\phi(x) = y$ for all $x \in \widehat{A}$.

For part 3, let Φ be such that $(a, r)\Phi(b, s)$ if and only if $a = (a_1, a_2) \in A \times A$, $r \in \mathbb{Q}^+$, $b \in \mathbb{Q}$, $s \in \mathbb{Q}^+$, and $|d(a_1, a_2) - b| + 2r < s$. It is not difficult to check that Φ is a code for a continuous function $\phi \colon \widehat{A} \times \widehat{A} \to \mathbb{R}$, and that $\phi(x_1, x_2) = d(x_1, x_2)$ for all $x_1, x_2 \in \widehat{A}$.

This completes the proof of lemma II.6.2. $\square$

LEMMA II.6.3. *The following is provable in* RCA$_0$. *Addition, subtraction, multiplication and division are continuous functions from* $\mathbb{R}$ *into* $\mathbb{R}$. *For any* $m \in \mathbb{N}$, *the functions* $\sum_{i=1}^{m} x_i$, $\prod_{i=1}^{m} x_i$ *and* $\max(x_1, \ldots, x_m)$ *are continuous functions from* $\mathbb{R}^m$ *into* $\mathbb{R}$.

PROOF. For example, let Φ be such that $(a, r)\Phi(b, s)$ if and only if $(a, r) \in \mathbb{Q} \times \mathbb{Q}^+$ and $(b, s) \in \mathbb{Q} \times \mathbb{Q}^+$ and $b - s < (a+r)^{-1} < (a-r)^{-1} < b + s$ and either $0 < a - r$ or $a + r < 0$. It is straightforward to check that Φ is a code for a continuous function ϕ from $\mathbb{R}$ into $\mathbb{R}$, that $\phi(x) = x^{-1}$ for all nonzero $x \in \mathbb{R}$, and that $\phi(0)$ is undefined. $\square$

LEMMA II.6.4. *The following is provable in* RCA$_0$. *If* $f \colon \widehat{A} \to \widehat{B}$ *and* $g \colon \widehat{B} \to \widehat{C}$ *are continuous, then so is the composition* $h = gf \colon \widehat{A} \to \widehat{C}$ *given by* $h(x) = g(f(x))$.

PROOF. Let F and G be the codes of $f \colon \widehat{A} \to \widehat{B}$ and $g \colon \widehat{B} \to \widehat{C}$ respectively. Let H be such that $(a, r)H(c, t)$ if and only if there exists (b, s) and $s' > s$ such that $(a, r)F(b, s)$ and $(b, s')G(c, t)$. It is straightforward to check that H is a code for a continuous function $h \colon \widehat{A} \to \widehat{C}$ and that $h(x) = g(f(x))$ for all $x \in \widehat{A}$. $\square$

From lemmas II.6.3 and II.6.4 we see that, in RCA$_0$, any polynomial $f(x_1, \ldots, x_m)$ in m indeterminates with coefficients from $\mathbb{R}$ gives rise to a continuous function $f \colon \mathbb{R}^m \to \mathbb{R}$. The following lemma can be used to show that functions defined by power series, such as e^x and $\sin x$, are also continuous.

LEMMA II.6.5. *The following is provable in* RCA$_0$. *Let* $\sum_{k=0}^{\infty} \alpha_k$ *be a convergent series of nonnegative real numbers* $\alpha_k \geq 0$. *Let* $\langle \phi_k \colon k \in \mathbb{N} \rangle$ *be a sequence of continuous functions* $\phi_k \colon \widehat{A} \to \mathbb{R}$ *such that* $|\phi_k(x)| \leq \alpha_k$ *for all* $k \in \mathbb{N}$ *and* $x \in \widehat{A}$. *Then* $\phi = \sum_{k=0}^{\infty} \phi_k \colon \widehat{A} \to \mathbb{R}$ *is continuous, and* $|\phi(x)| \leq \sum_{k=0}^{\infty} \alpha_k$ *for all* $x \in \widehat{A}$.

PROOF. We reason within RCA$_0$. Let Φ be such that $(a, r)\Phi(b, s)$ if and only if, for some $m \in \mathbb{N}$, there exist $(a, r)\Phi_k(b_k, s_k)$, $k < m$, such that

$b = \sum_{k<m} b_k$ and

$$\sum_{k=0}^{\infty} s_k + \sum_{k=m}^{\infty} \alpha_k < s.$$

It is straightforward to verify that Φ is a code for a continuous function $\phi \colon \widehat{A} \to \mathbb{R}$ as required.   □

We now specialize to the study of continuous functions on $\mathbb{R}$. We show that the *intermediate value theorem* can be proved within RCA$_0$. (See also exercise IV.2.12.)

THEOREM II.6.6 (intermediate value theorem). *The following is provable in RCA$_0$. If $\phi(x)$ is continuous on the unit interval $0 \le x \le 1$, and if $\phi(0) < 0 < \phi(1)$, then there exists x such that $0 < x < 1$ and $\phi(x) = 0$.*

PROOF. We may assume that $\phi(q) \ne 0$ for all rational numbers q with $0 < q < 1$. Then by Δ_1^0 comprehension there exists a set X consisting of all $q \in \mathbb{Q}$ such that $0 < q < 1$ and $\phi(q) < 0$. By primitive recursion using X as a parameter, define a nested sequence of rational intervals

$$(a_0, b_0) = (0, 1),$$

$$(a_{n+1}, b_{n+1}) = \begin{cases} ((a_n + b_n)/2, b_n) & \text{if } \phi((a_n + b_n)/2) < 0, \\ (a_n, (a_n + b_n)/2) & \text{if } \phi((a_n + b_n)/2) > 0. \end{cases}$$

By Σ_0^0 induction we see that $\phi(a_n) < 0 < \phi(b_n)$ for all $n \in \mathbb{N}$. Also $|a_n - b_n| = 2^{-n}$. Thus $x = \langle a_n : n \in \mathbb{N} \rangle = \langle b_n : n \in \mathbb{N} \rangle$ is a real number. We claim that $\phi(x) = 0$. Suppose not, say $\phi(x) < 0$. Let Φ be the code of ϕ. Let $(u, r)\Phi(v, s)$ be such that $|x - u| < r$ and $s < |\phi(x)|/2$. Since $|\phi(x) - v| \le s$, we have $v + s < 0$. Let n be so large that $|b_n - u| < r$. Then $|\phi(b_n) - v| \le s$, hence $\phi(b_n) \le v + s < 0$, a contradiction. This completes the proof.   □

COROLLARY II.6.7. *It is provable in RCA$_0$ that the ordered field of real numbers $\mathbb{R}, +, -, \cdot, 0, 1, <$ is real closed, i.e., has the intermediate value property for all polynomials.*

PROOF. This is immediate from theorem II.6.6 plus the fact, noted above, that polynomials give rise to continuous functions.   □

REMARK II.6.8. Given a continuous real-valued function $\phi(x)$ defined for $0 \le x \le 1$, it is natural to ask whether RCA$_0$ proves the *maximum principle*. We shall see later that RCA$_0$ is not even strong enough to prove that the values $\phi(x)$, $0 \le x \le 1$ are bounded above. Even if they are, one cannot prove in RCA$_0$ that $\sup \phi(x)$ exists. And even if $c = \sup \phi(x)$ exists, one cannot prove that this maximum value is attained, i.e., RCA$_0$ does not prove the existence of an x such that $\phi(x) = c$. See especially §IV.2.

Exercise II.6.9. Within RCA_0, let $\phi \colon \widehat{A} \to \widehat{B}$ be continuous. Show that, given (a code for) an open set $V \subseteq \widehat{B}$, we can effectively find (a code for) an open set $U \subseteq \widehat{A}$ such that for all points $x \in \widehat{A}$, $x \in U$ if and only if $\phi(x) \in V$. Thus we are justified in writing $U = \phi^{-1}(V)$ and in saying that the inverse image of an open set under a continuous function is open.

Exercise II.6.10. Within RCA_0, let $\phi \colon \mathbb{R} \to \mathbb{R}$ be continuous. Assume that the derivative

$$\phi'(x) = \lim_{\Delta x \to 0} \frac{\phi(x + \Delta x) - \phi(x)}{\Delta x}$$

exists and is $\leq M$ for all $x \in \mathbb{R}$. Show that

$$\frac{\phi(b) - \phi(a)}{b - a} \leq M$$

for all $x \in \mathbb{R}$.

Notes for §II.6. Our concept of continuous function within RCA_0 is the same as that of [236] and Brown/Simpson [27]. Another approach has been taken by Aberth [2] and Bishop, who define continuous functions on the real line to be uniformly continuous on bounded intervals. See for instance Bishop/Bridges [20, page 38]. Thus their approach relies on the fact that the real line is locally compact. Our approach works for complete separable metric spaces which are not required to be locally compact, e.g., the Baire space.

There are considerable differences between Bishop's constructive mathematics and our development of mathematics within the formal system RCA_0. One difference is that Bishop eschews the use of formal systems altogether. A major difference is that Bishop rejects the law of the excluded middle. As a consequence, the intermediate value theorem is not constructively valid in Bishop's sense, even though by II.6.6 it is provable in RCA_0.

Exercise II.6.10 is related to a result of Aberth [2] in recursive analysis. The other results of this section are due to Simpson, unpublished.

II.7. More on Complete Separable Metric Spaces

In this section we shall prove some additional theorems concerning the topology of complete separable metric spaces, within RCA_0. Throughout this section, we assume that $\widehat{A}$ is a complete separable metric space. For notational convenience, we write $X = \widehat{A}$.

Lemma II.7.1. *The following is provable in RCA_0.*

1. *Given (a code for) an open set $U \subseteq X$, we can effectively find a (code for a) continuous function $h_U \colon X \to [0, 1]$ such that for all $x \in X$, $x \in U$ if and only if $h_U(x) > 0$.*

2. *Conversely, given a (code for a) continuous function $f : X \to \mathbb{R}$, we can effectively find (a code for) an open set V such that for all $x \in X$, $x \in V$ if and only if $f(x) > 0$.*

PROOF. For part 1, we put $h_U = \sum_{k \in U} h_k$ where

$$h_k(x) = \frac{\max(0, r - d(a, x))}{r \cdot 2^{k+1}}$$

for all $k = (n, a, r) \in U$ and $x \in X$. The continuity of h_U follows from lemmas II.6.2–II.6.5 since $|h_k| \leq 2^{-k-1}$. It is obvious that $0 \leq h_U \leq 1$, and that $h_U(x) > 0$ if and only if $x \in U$.

For the converse, let F be the code of f. Let $\varphi(x)$ be a Σ_1^0 formula which says that $f(x) > 0$, i.e., there exists $(a, r)F(b, s)$ such that $d(a, x) < r$ and $b - s > 0$. By lemma II.5.7 we get an open set $U \subseteq X$ such that, for all $x \in X$, $x \in U$ if and only if $\varphi(x)$. This completes the proof. □

The following theorem expresses the well known fact that complete separable metric spaces are *paracompact* (see also the notes at end of this section). An *open covering* of X is defined to be a sequence of open sets $\langle U_n : n \in \mathbb{N} \rangle$ in X such that for all $x \in X$ there exists $n \in \mathbb{N}$ such that $x \in U_n$.

THEOREM II.7.2 (paracompactness). *The following is provable in RCA$_0$. Given an open covering $\langle U_n : n \in \mathbb{N} \rangle$, we can effectively find an open covering $\langle V_n : n \in \mathbb{N} \rangle$ such that $V_n \subseteq U_n$ for all n, and $\langle V_n : n \in \mathbb{N} \rangle$ is* locally finite, *i.e., for all $x \in X$ there exists an open set W such that $x \in W$ and $W \cap V_n = \emptyset$ for all but finitely many n.*

PROOF. We reason within RCA$_0$. Let $\langle U_n : n \in \mathbb{N} \rangle$ be an open covering of X. By lemma II.7.1.1, we can find a sequence of continuous functions $h_n : X \to [0, 1]$, $n \in \mathbb{N}$, such that for all $x \in X$ and $n \in \mathbb{N}$, $x \in U_n$ if and only if $h_n(x) > 0$. Put

$$g_n = \frac{h_n \cdot 2^{-n}}{\sum_{m \in \mathbb{N}} h_m \cdot 2^{-m}}.$$

Thus $0 \leq g_n \leq 1$, $\sum_{n \in \mathbb{N}} g_n = 1$, and $g_n(x) > 0$ if and only if $x \in U_n$. Thus $\langle g_n : n \in \mathbb{N} \rangle$ is a partition of unity. Put

$$f_n = \min \left(\frac{1}{2}, \sum_{m \leq n} g_m \right) - \min \left(\frac{1}{2}, \sum_{m < n} g_m \right).$$

Thus $0 \leq f_n \leq g_n$ and $\sum_{n \in \mathbb{N}} f_n = 1/2$. Furthermore, for any $x \in X$, $f_n(x) = 0$ for all n such that $\sum_{m < n} g_m(x) > 1/2$. Now to finish the proof, apply lemma II.7.1.2 to get a sequence of open sets $\langle V_n : n \in \mathbb{N} \rangle$ such that for all $x \in X$ and $n \in \mathbb{N}$, $x \in V_n$ if and only if $f_n(x) > 0$. Clearly $\langle V_n : n \in \mathbb{N} \rangle$ has the desired properties. □

The rest of this section is devoted to a proof of the Tietze extension theorem for complete separable metric spaces, within RCA_0. We start with the following version of Urysohn's Lemma.

LEMMA II.7.3 (Urysohn's lemma). *The following is provable in RCA_0. Given (codes for) disjoint closed sets C_0 and C_1 in X, we can effectively find a (code for a) continuous function $g: X \to [0,1]$ such that, for all $x \in X$ and $i \in \{0,1\}$, $x \in C_i$ if and only if $g(x) = i$.*

PROOF. Let C_0 and C_1 be given. By lemma II.7.1.1 we can effectively find continuous functions $h_i: X \to [0,1]$ such that for all $x \in X$ and $i \in \{0,1\}$, $x \in C_i$ if and only if $h_i(x) = 0$. Define a continuous function g on X by $g = h_0/(h_0 + h_1)$. It is easy to verify that g has the desired properties. $\qquad\qquad\square$

We shall need the following variant of the previous lemma.

LEMMA II.7.4. *The following is provable in RCA_0. Given a closed set $C \subseteq X$ and a continuous function $f: C \to [-1,1]$, we can effectively find a continuous function $g: X \to [-1/3, 1/3]$ such that $|f(x) - g(x)| \leq 2/3$ for all $x \in C$.*

PROOF. By lemma II.7.1.2, let U be an open set such that for all $x \in X$, $x \in U$ if and only if either $x \notin C$, or $x \in C$ and $f(x) > -1/3$. Similarly let V be an open set such that for all $x \in X$, $x \in V$ if and only if either $x \notin C$, or $x \in C$ and $f(x) < 1/3$. Letting $h_U, h_V: X \to [0,1]$ be as in lemma II.7.1.1, define g: $X \to [-1/3, 1/3]$ by

$$g = \frac{1}{3} \cdot \frac{h_U - h_V}{h_U + h_V}.$$

The denominator $h_U + h_V$ is everywhere nonzero since $X = U \cup V$. If $x \in C$ and $f(x) \leq -1/3$, then $x \notin U$ so $h_U(x) = 0$, hence $g(x) = -1/3$. Similarly if $x \in C$ and $f(x) \geq 1/3$, then $g(x) = 1/3$. This proves the lemma. $\qquad\qquad\square$

The following is our version of the Tietze extension theorem.

THEOREM II.7.5 (Tietze extension theorem). *The following is provable in RCA_0. Given a (code for a) closed set $C \subseteq X$ and a (code for a) continuous function $f: C \to [-1,1]$, we can effectively find a (code for a) continuous function $g: X \to [-1,1]$ such that $g(x) = f(x)$ for all $x \in C$.*

PROOF. We shall first give the construction of g, then indicate how to prove within RCA_0 that the construction works.

We begin with $f = f_0: C \to [-1,1]$. Apply lemma II.7.4 to get $g_0: X \to [-1/3, 1/3]$ such that $|f_0 - g_0| \leq 2/3$ on C. Set $f_1 = f_0 - g_0 = f - g_0: C \to [-2/3, 2/3]$. Apply lemma II.7.4 again to get $g_1: X \to [-2/9, 2/9]$ such that $|f_1 - g_1| \leq 4/9$ on C. Set $f_2 = f_1 - g_1 = f - (g_0 + g_1): C \to [-4/9, 4/9]. \ldots.$ In general, we have

$$f_n = f - (g_0 + g_1 + \cdots + g_{n-1}): C \to [-(2/3)^n, (2/3)^n],$$

and the inductive step consists of applying lemma II.7.4 to get $g_n: X \to [-2^n/3^{n+1}, 2^n/3^{n+1}]$ such that $|f_n - g_n| \leq (2/3)^{n+1}$ on C, then setting $f_{n+1} = f_n - g_n = f - (g_0 + g_1 + \cdots + g_n): C \to [-(2/3)^{n+1}, (2/3)^{n+1}]$. Finally we put $g = \sum_{n=0}^{\infty} g_n: X \to [-1, 1]$ which is continuous by lemma II.6.5. It is then clear that $f = g$ on C. This completes the construction of g.

Within RCA$_0$, the above construction is to be interpreted as a simultaneous enumeration of the codes of f_n and g_n, for all $n \in \mathbb{N}$. The key to showing that the construction works will be to prove the following claim: for all $x \in X$ and all n, $g_n(x)$ is defined. Our basic strategy is to try to prove this claim by induction on n. Unfortunately the claim is not obviously Σ_1^0 or Π_1^0 and so its proof does not obviously go through in RCA$_0$.

We resolve this difficulty as follows. Tracing back through the construction, we see that g_n is defined from $f_n = f - (g_0 + g_1 + \cdots + g_{n-1})$ by

$$U_n = (X \setminus C) \cup \{x \in C: f_n(x) > -(1/3)^{n+1}\},$$

$$V_n = (X \setminus C) \cup \{x \in C: f_n(x) < (1/3)^{n+1}\},$$

$$g_n = (1/3)^{n+1} \cdot \frac{h_{U_n} - h_{V_n}}{h_{U_n} + h_{V_n}}.$$

Thus $g_n(x)$ is defined provided the denominator $h_{U_n}(x) + h_{V_n}(x)$ is nonzero, i.e., provided x belongs to the open set $U_n \cup V_n$. And this holds provided either $x \notin C$ or $f_n(x) = f(x) - (g_0(x) + \cdots + g_{n-1}(x))$ is defined. The fact that U_n and V_n are (codes for) open sets is obvious at the outset and does not require a proof by induction on n.

So, to prove that $g_n(x)$ is defined for all $x \in X$ and $n \in \mathbb{N}$, we proceed as follows. Fix $x \in X$. Prove by induction on n that $x \in U_n \cup V_n$. This assertion is Σ_1^0 so we may carry out the inductive argument within RCA$_0$. The inductive step is as follows. Assume that $x \in U_k \cup V_k$ for all $k < n$. Then, as in the previous paragraph, $g_k(x)$ is defined for all $k < n$. Hence either $x \notin C$ or $f_n(x) = f(x) - (g_0(n) + \cdots + g_{n-1}(x))$ is defined. Hence $x \in U_n \cup V_n$.

This completes the proof.   □

Notes for §II.7. For general information on metric spaces (paracompactness, Tietze extension theorem, etc.), see e.g., Engelking [52]. The material in this section is due to Simpson, unpublished. For a somewhat more detailed treatment, see Brown [24]. Some alternative versions of the Tietze extension theorem are analyzed in Giusto/Simpson [93].

II.8. Mathematical Logic

The purpose of this section is to point out that weak versions of some basic results of mathematical logic can be formulated and proved in RCA_0.

A *language* is a set of relation, operation, and constant symbols. We work within RCA_0 and assume a fixed countable language L. Terms and formulas of first order logic (i.e., predicate calculus) are defined as usual. We identify terms and formulas with their Gödel numbers under a fixed Gödel numbering. Such a Gödel numbering can be constructed by primitive recursion (theorem II.3.4) using L as a parameter. We can also prove in RCA_0 that there exist sets Trm, Fml, Snt, and Axm consisting of all Gödel numbers of terms, formulas, sentences and logical axioms respectively. We assume that the logical axioms and rules have been set up so that the only logical rule is modus ponens. (See the notes at the end of the section.)

DEFINITION II.8.1 (provability predicate). The following definitions are made in RCA_0. For any set of formulas $X \subseteq$ Fml, let $Prf(X, p)$ be the Σ_0^0 formula which says that p is a *proof from* X, i.e., $p \in$ Seq $\wedge \forall k\, (k < lh(p) \rightarrow p(k) \in$ Fml$) \wedge \forall k\, (k < lh(p) \rightarrow (p(k) \in X \vee p(k) \in$ Axm $\vee (\exists i < k)\,(\exists j < k)\,(p(i) = (p(j) \rightarrow p(k)))))$. We say that φ is *provable from* X (written $Pbl(X, \varphi)$) if $\exists p\, (Prf(X, p) \wedge (\exists i < lh(p))\,(p(i) = \varphi))$. Note that the formula $Pbl(X, \varphi)$ is Σ_1^0.

DEFINITION II.8.2 (consistency, etc.). The following definitions are made in RCA_0. A set $X \subseteq$ Snt is *consistent* if $\neg \exists \varphi\, (Pbl(X, \varphi) \wedge Pbl(X, \neg\varphi))$. X is *closed under logical consequence* if $\forall \sigma\, ((\sigma \in$ Snt $\wedge Pbl(X, \sigma)) \rightarrow \sigma \in X)$. X is *complete* if $\forall \sigma\, (\sigma \in$ Snt $\rightarrow (Pbl(X, \sigma) \vee Pbl(X, \neg\sigma)))$.

DEFINITION II.8.3 (models). The following definition is made in RCA_0. A *countable model* is a function $M: T_M \cup S_M \rightarrow |M| \cup \{0, 1\}$. Here $|M| \subseteq \mathbb{N}$ is a set called the *universe* of M, and T_M and S_M are respectively the sets of closed terms and sentences of the expanded language $L_M = L \cup \{\underline{m}: m \in |M|\}$ with new constant symbols $\underline{a}$ for each element a of $|M|$. The function M is required to obey the familiar clauses of Tarski's truth definition:

1. $t \in T_M$ implies $M(t) \in |M|$;
2. $\sigma \in S_M$ implies $M(\sigma) \in \{0, 1\}$;
3. for any $t_1, t_1', \ldots, t_n, t_n' \in T_M$, if $M(t_i) = M(t_i')$, $1 \leq i \leq n$, then $M(\underline{R}(t_1, \ldots, t_n)) = M(\underline{R}(t_1', \ldots, t_n'))$ and $M(\underline{o}(t_1, \ldots, t_n)) = M(\underline{o}(t_1', \ldots, t_n'))$; here $\underline{R}$ is a relation symbol and $\underline{o}$ is a function symbol;
4. $M(\neg\sigma) = 1 - M(\sigma)$;
5. $M(\sigma_1 \wedge \sigma_2) = M(\sigma_1) \cdot M(\sigma_2)$;
6. $M(\forall v\, \varphi(v)) = \prod_{a \in |M|} M(\varphi(\underline{a}))$;

etc.

The following is a weak version of Gödel's completeness theorem.

THEOREM II.8.4 (weak completeness theorem). *The following is provable in* RCA$_0$. *Let* $X \subseteq$ Snt *be consistent and closed under logical consequence. Then there exists a countable model M such that $M(\sigma) = 1$ for all $\sigma \in X$.*

PROOF. We first prove a weak version of Lindenbaum's lemma.

LEMMA II.8.5 (weak Lindenbaum lemma). *The following is provable in* RCA$_0$. *Suppose $X \subseteq$ Snt is consistent and closed under logical consequence. Then there exists $X^* \subseteq$ Snt such that $X \subseteq X^*$ and X^* is consistent, complete, and closed under logical consequence.*

PROOF. Let $\langle \sigma_n : n \in \mathbb{N} \rangle$ be a one-to-one enumeration of Snt. Define a sequence of sentences $\langle \sigma_n^* : n \in \mathbb{N} \rangle$ by primitive recursion as follows: $\sigma_n^* = \sigma_n$ if $((\sigma_0^* \wedge \cdots \wedge \sigma_{n-1}^*) \rightarrow \sigma_n) \in X$; $\sigma_n^* = \neg\sigma_n$ otherwise. Let X^* be the set of all σ_n^*, $n \in \mathbb{N}$. Clearly X^* has the desired properties. The lemma is proved. □

Now to prove theorem II.8.4, let C be an infinite set of new constant symbols. Let $\langle c_n : n \in \mathbb{N} \rangle$ be a one-to-one enumeration of C and let $\langle \varphi_n(x) : n \in \mathbb{N} \rangle$ be an enumeration of all formulas with one free variable in the expanded language $L_1 = L \cup C$. We may safely assume that c_n does not occur in $\varphi_i(x)$, $i < n$. If τ is any L_1-sentence, we write

$$\tau^- \equiv \forall z_0 \cdots \forall z_n \, \tau(z_0/c_0, \ldots, z_n/c_n)$$

where

$$n = n_\tau = \sup\{n : c_n \text{ occurs in } \tau\}$$

and $z_0, \ldots, z_n$ are new variables. Thus τ^- is an L-sentence. Form Henkin axioms

$$\eta_n \equiv (\exists x \, \varphi_n(x)) \rightarrow \varphi_n(c_n)$$

and let X_1 be the set of all sentences of L_1 which are provable from X plus the Henkin axioms. X_1 exists by Δ_1^0 comprehension since

$$\sigma \in X_1 \leftrightarrow ((\eta_0 \wedge \cdots \wedge \eta_n) \rightarrow \sigma)^- \in X$$

where $n = n_\sigma$. Clearly X_1 is consistent and closed under logical consequence, so by lemma II.8.5 let X_1^* be a completion of X_1. A countable model M can be read off from X_1^* in the usual way. Namely, let $|M|$ be the set of all $c_n \in C$ such that $\neg\exists m \, (m < n \wedge (c_m = c_n) \in X_1^*)$. For all $\sigma \in S_M$ put $M(\sigma) = 1$ if and only if $\sigma \in X_1^*$. This completes the proof of theorem II.8.4. □

COROLLARY II.8.6. *The following is provable in* RCA$_0$. *If $X \subseteq$ Snt is consistent and complete, then there exists a countable model M such that $M(\sigma) = 1$ for all $\sigma \in X$.*

Proof. The hypotheses on X imply that for all $\sigma \in$ Snt,

$$\text{Pbl}(X, \sigma) \leftrightarrow \neg\text{Pbl}(X, \neg\sigma).$$

Hence by Δ_1^0 comprehension there exists a set Pbl_X consisting of all sentences which are provable from X. The corollary is proved by applying theorem II.8.5 to Pbl_X.   □

Remark II.8.7. In connection with the above theorem and corollary, note that it is not provable in RCA$_0$ that every consistent set of sentences can be extended to a consistent set of sentences which is closed under logical consequence. For example, let Q be the set of axioms of Robinson's system. (See the notes at end of section.) Then Q is finite but, as is well known, there is no recursive consistent set of sentences which contains Q and is closed under logical consequence. Thus the ω-model REC satisfies "Q is consistent but has no countable model." In chapter IV we shall see that WKL$_0$ is strong enough to prove the full Gödel completeness theorem: Every consistent set of sentences in a countable language has a countable model.

We now consider converses of the Gödel completeness theorem. The following version of the soundness theorem is easy to prove.

Theorem II.8.8 (soundness theorem). *The following is provable in* RCA$_0$. *If $X \subseteq$ Snt and there exists a countable model M such that $M(\sigma) = 1$ for all $\sigma \in X$, then X is consistent.*

Proof. For any formula φ let $\overline{\varphi}$ be the *universal closure* of φ, i.e., the sentence obtained by prefixing φ with universal quantifiers. Given p such that $\text{Prf}(X, p)$, it is straightforward to prove by induction on $k < \text{lh}(p)$ that $M\left(\overline{p(k)}\right) = 1$. This implies the theorem.   □

For later use we prove the following stronger version of the soundness theorem.

Definition II.8.9 (weak models). Within RCA$_0$, let $X \subseteq$ Snt be a set of sentences. A *weak countable model of X* is a function $M : \text{T}_M \cup \text{S}_M^X \to |M| \cup \{0, 1\}$. Here $|M|$ and T_M are as in definition II.8.3, and S_M^X is the set of all $\sigma \in \text{S}_M$ such that σ is a propositional combination of substitution instances of subformulas of elements of X. We require M to obey the clauses of definition II.8.3 except that the clause involving $\forall v\, \varphi(v)$ applies only when $\forall v\, \varphi(v) \in \text{S}_M^X$. We also require that $M(\sigma) = 1$ whenever $\sigma \in X$.

Theorem II.8.10 (strong soundness theorem). *The following is provable in* RCA$_0$. *If there exists a weak countable model of $X \subseteq$ Snt, then X is consistent.*

Proof. Consider a cut-free system of axioms and rules for logic (see notes at end of section). In RCA$_0$ we can carry out the usual syntactical proof that if φ is provable from the empty set (in the sense of definition II.8.1), then there exists a cut-free proof of φ. This cut-free proof has the property that each formula occurring in it is a substitution instance of a subformula of φ.

Assume now that M is a weak countable model of $X \subseteq$ Snt, but X is not consistent. Then there exists $\sigma_1, \ldots, \sigma_n \in X$ such that $\neg(\sigma_1 \wedge \cdots \wedge \sigma_n)$ is provable from the empty set. Let p be a cut-free proof of $\neg(\sigma_1 \wedge \cdots \wedge \sigma_n)$. Then S_M^X contains all $\sigma \in S_M$ which are substitution instances of formulas in p. By Π_1^0 induction on the length of p we can prove that $M(\sigma) = 1$ for all such σ. In particular $M(\neg(\sigma_1 \wedge \cdots \wedge \sigma_n)) = 1$, but this is impossible since $M(\sigma_1) = \cdots = M(\sigma_n) = 1$. The proof is complete. $\square$

In order to illustrate the significance of the above result, we present the following application. Let $L = L_1(\exp)$ be the language of first order arithmetic $+, \cdot, 0, 1, <, =$ augmented by a binary operation symbol $\exp(m, n) = m^n$ intended to denote exponentiation. Let EFA (elementary function arithmetic) consist of the basic axioms (definition I.2.4) augmented by

$$m^0 = 1, \quad m^{n+1} = m^n \cdot m,$$

plus Σ_0^0 induction.

THEOREM II.8.11 (consistency of EFA). RCA$_0$ *proves the consistency of* EFA.

PROOF. We reason within RCA$_0$. Let EFA$'$ be the same as EFA with the Σ_0^0 induction scheme

$$(\theta(0) \wedge \forall n\, (\theta(n) \rightarrow \theta(n+1))) \rightarrow \forall n\, \theta(n)$$

replaced by the equivalent scheme

$$\forall n\, ((\theta(0) \wedge \forall k < n\, (\theta(k) \rightarrow \theta(k+1))) \rightarrow \theta(n)).$$

Here $\theta(n)$ denotes an arbitrary Σ_0^0 formula in the language of EFA. Let X be the set of all universal closures of axioms of EFA$'$. In order to show that EFA is consistent, it suffices to prove the consistency of EFA$'$, i.e., of X. And for this it suffices by theorem II.8.10 to construct a weak countable model M of X.

We begin by letting $|M| = \mathbb{N}$. Note that X consists of Π_1^0 sentences; this is why we switched from EFA to EFA$'$. Let T_M and S_M^X be as in definition II.8.9. Let S_M^- be the set of all Σ_0^0 sentences in the language of EFA with parameters from $|M|$. Note that $S_M^- \subseteq S_M^X$. Using primitive recursion (theorem II.3.4), it is straightforward to prove the existence of a function

$$M^-: T_M \cup S_M^- \rightarrow |M| \cup \{0, 1\}$$

obeying the Tarski clauses. Since X consists of Π_1^0 sentences, it is trivial to extend M^- to a function

$$M: T_M \cup S_M^X \rightarrow |M| \cup \{0, 1\}$$

which also obeys the Tarski clauses. It is then easy to check that $M(\sigma) = 1$ for each $\sigma \in X$. This completes the proof.   □

Notes for §II.8. In definition II.8.1 we assumed that the logical axioms and rules had been set up so that the only rule is modus ponens. For one way to do this, see Enderton [51, §2.4]. Theorem II.8.4 applied to the ω-model REC implies that every recursively decidable theory has a recursive model with a recursive satisfaction predicate. This result is originally due to Morley and is the beginning of a subject known as *recursive model theory*. For a recent survey of recursive model theory, see [53]. The original source of Robinson's system Q is Tarski/Mostowski/Robinson [266]. The proof of theorem II.8.10 used a cut-free system of logical axioms and rules, for which see e.g., Kleene [142].

The material in this section is due to Simpson, unpublished.

II.9. Countable Fields

In this section we show that some of the usual constructions of countable algebraic structures can be carried out in RCA_0.

DEFINITION II.9.1 (fields). The following definitions are made in RCA_0. A *countable field* K consists of a set $|K| \subseteq \mathbb{N}$ together with binary operations $+_K$, $\cdot_K$ and a unary operation $-_K$ and distinguished elements 0_K, 1_K such that the system $|K|, +_K, -_K, \cdot_K, 0_K, 1_K$ obeys the usual field axioms, e.g., $\forall x \forall y \, (x \cdot y = y \cdot x)$ and $\forall x \, (x \neq 0 \rightarrow \exists y \, (x \cdot y = 1))$. The *polynomial ring* $K[x]$ consists of all finite sequences $\langle a_0, \ldots, a_n \rangle$ of elements of $|K|$ such that $n = 0$ or $a_n \neq 0$. We denote $\langle a_0, \ldots, a_n \rangle$ by $\sum_{i=0}^{n} a_i x^i$.

The theory of finite extensions of a countable field can be developed as usual within RCA_0. As usual, a countable field K is said to be *algebraically closed* if for all nonconstant polynomials $f(x) \in K[x]$ there exists $a \in K$ such that $f(a) = 0$.

DEFINITION II.9.2 (algebraic closure). The following definition is made in RCA_0. Let K be a countable field. An *algebraic closure* of K consists of an algebraically closed countable field $\widetilde{K}$ together with a monomorphism $h \colon K \to \widetilde{K}$ such that for all $b \in \widetilde{K}$ there exists a nonzero polynomial $f(x) \in K[x]$ such that $h(f)(b) = 0$. Here for $f(x) = \sum_{i=0}^{n} a_i x^i \in K[x]$ we write $h(f)(x) = \sum_{i=0}^{n} h(a_i) x^i \in \widetilde{K}[x]$. (Caution: We cannot prove in RCA_0 that there exists a set which is the image of the monomorphism $h \colon K \to \widetilde{K}$. See §III.3.)

In order to prove that every countable field has an algebraic closure, we shall invoke the model-theoretic results which were presented in the previous section. Let L be the language of fields with symbols $+, -, \cdot, 0, 1$. Let AF be the usual set of field axioms, e.g., $\forall x\, (x \neq 0 \rightarrow \exists y\, (x \cdot y = 1))$. Let ACF be the usual set of axioms for an algebraically closed field: ACF consists of AF plus the infinite set of axioms

$$\forall x_0 \cdots \forall x_{n-1} \exists y\, (y^n + x_{n-1} y^{n-1} + \cdots + x_1 y + x_0 = 0)$$

for all $n \in \mathbb{N}$, $n \geq 1$.

LEMMA II.9.3. *The following facts are provable in* RCA$_0$. (i) ACF *admits elimination of quantifiers, i.e., for any formula φ there exists a quantifier-free formula φ^* such that* ACF *proves* $\varphi \leftrightarrow \varphi^*$. (ii) *For any quantifier-free formula φ, if* ACF *proves φ then* AF *proves φ.*

PROOF. These well known results have purely syntactical proofs which can be transcribed into RCA$_0$, using the availability of various primitive recursive functions and predicates. (See the notes at end of this section.) □

THEOREM II.9.4 (existence of algebraic closure). *It is provable in* RCA$_0$ *that every countable field K has an algebraic closure.*

PROOF. Let Δ_K be the *quantifier-free diagram* of K, i.e., the set of all quantifier-free sentences of L_K which are true in K. Clearly K can be expanded to a weak countable model of $\Delta_K \cup$ AF. Hence by theorem II.8.10, $\Delta_K \cup$ AF is consistent. It follows by lemma II.9.3(ii) that $\Delta_K \cup$ ACF is consistent. Also $\Delta_K \cup$ ACF is complete by lemma II.9.3(i). Hence by corollary II.8.6 there exists a countable model M of $\Delta_K \cup$ ACF. Clearly M may be viewed as a countable algebraically closed field and there is a canonical embedding $k\colon K \rightarrow M$. Let $\varphi(b)$ be a Σ^0_1 formula saying that $b \in |M|$ and there exists a nonconstant $f(x) \in K[x]$ such that $k(f)(b) = 0$. By lemma II.3.7 there exists a one-to-one function $g\colon \mathbb{N} \rightarrow |M|$ such that for all $b \in |M|$, $\varphi(b)$ if and only if $\exists j\, (g(j) = b)$. Put $|\widetilde{K}| = \mathbb{N}$ and define the field operations of $\widetilde{K}$ by pulling back via g, e.g., $i +_{\widetilde{K}} j = g^{-1}(g(i) +_M g(j))$. Clearly $\widetilde{K}$ is an algebraic closure of K with the monomorphism $h\colon K \rightarrow \widetilde{K}$ given by $h(a) = g^{-1}(k(a))$. This completes the proof. □

DEFINITION II.9.5 (real closure). The following definitions are made in RCA$_0$. A *countable ordered field* consists of a countable field K together with a binary relation $<_K \subseteq |K|^2$ such that $K, <_K$ obeys the usual ordered field axioms, e.g., $\forall x \forall y\, (x < y \lor x = y \lor y < x)$ and $(x < y \leftrightarrow x + z < y + z)$. A countable ordered field is said to be *real closed* if it has the intermediate value property for polynomials, i.e., for all $g(x) \in K[x]$ and $a, b \in K$, if $g(a) < 0 < g(b)$ then there exists $c \in K$ between a and b such that $g(c) = 0$. A *real closure* of a countable ordered field K consists of a countable real closed ordered field $\overline{K}$ together with a monomorphism $h \colon K \to \overline{K}$ such that for each $b \in \overline{K}$ there exists a nonconstant $f(x) \in K[x]$ such that $h(f)(b) = 0$.

Our proof that every countable ordered field has a real closure will be similar to the above proof of the corresponding result for algebraic closure. Let L be the language of ordered fields and let OF be the set of ordered field axioms. Let RCOF be the set of real closed ordered field axioms, i.e., RCOF consists of OF plus the axioms

$$\forall x_0 \cdots \forall x_n \, \forall u \, \forall v \, ((u < v \wedge x_n \cdot u^n + \cdots + x_0 < 0 < x_n \cdot v^n + \cdots + x_0)$$
$$\to \exists w \, (u < w < v \wedge x_n \cdot w^n + \cdots + x_0 = 0))$$

for all $n \in \mathbb{N}$.

LEMMA II.9.6. *The following facts are provable in* RCA$_0$. (i) RCOF *admits elimination of quantifiers.* (ii) *For any quantifier-free formula* φ, *if* RCOF *proves* φ *then* OF *proves* φ.

PROOF. The well known syntactical proofs of these results can be carried out in RCA$_0$. (See the notes at end of this section.) $\qquad \square$

THEOREM II.9.7 (existence and uniqueness of real closure). *The following is provable in* RCA$_0$. *Every countable ordered field K has a real closure. The real closure is unique in the sense that, if $h_1 \colon K \to \overline{K}_1$ and $h_2 \colon K \to \overline{K}_2$ are two real closures of K, there exists a unique isomorphism $h \colon \overline{K}_1 \to \overline{K}_2$ of $\overline{K}_1$ onto $\overline{K}_2$ such that $h(h_1(a)) = h_2(a)$ for all $a \in K$.*

PROOF. The proof of the existence of a real closure is similar to the proof of theorem II.9.4 relying now on lemma II.9.6 instead of lemma II.9.3. The uniqueness follows from the fact that for each $b_1 \in \overline{K}_1$ there exists an ordered pair (f, i) such that $f \in K[x]$ and b_1 is the unique $b \in \overline{K}_1$ such that $h_1(f)(b) = 0$ and there are exactly i elements $a \in \overline{K}_1$ such that $a < b$ and $h_1(f)(a) = 0$. By quantifier elimination there is a unique corresponding element $b_2 \in \overline{K}_2$ and this gives the isomorphism. $\qquad \square$

REMARKS II.9.8. (1) There is no analogous uniqueness result for algebraic closure. We shall see later (§IV.5) that RCA$_0$ does not prove that the algebraic closure of a countable field is unique. (2) A countable field K is said to be *formally real* if the equations $x_1^2 + \cdots + x_n^2 = -1$, $n \in \mathbb{N}$, have no solution in K. There is a well known theorem due to Artin and Schreier which states that every formally real field is orderable. We shall see later (§IV.4) that this theorem for countable formally real fields is not provable in RCA$_0$.

Notes for §II.9. For a somewhat different treatment of the material in this section, see Friedman/Simpson/Smith [78]. In proving lemmas II.9.3 and II.9.6, we used Tarski's syntactical quantifier elimination methods as presented, e.g., in Kreisel/Krivine [152]. If we specialize theorem II.9.4 to the ω-model REC, we obtain a result which is originally due to Rabin:

every recursive field has a recursive algebraic closure. This is one of the first theorems of a subject known as *recursive algebra*. For a recent survey of recursive algebra, see [53].

II.10. Separable Banach Spaces

In this section we show that some rudimentary portions of the theory of separable Banach spaces can be developed within RCA_0. The techniques of this section are based on those of §§II.5 and II.6.

Let K be a countable field. Within RCA_0, a *countable vector space A over K* consists of a set $|A| \subseteq \mathbb{N}$ together with operations $+ \colon |A| \times |A| \to |A|$ and $\cdot \colon |K| \times |A| \to |A|$ and a distinguished element $0 \in |A|$, such that $|A|, +, \cdot, 0$ satisfy the usual axioms for a vector space over K.

DEFINITION II.10.1 (separable Banach spaces). Within RCA_0, we define a (code for a) *separable Banach space $\widehat{A}$* to consist of a countable vector space A over the rational field $\mathbb{Q}$ together with a sequence of real numbers $\| \ \| \colon A \to \mathbb{R}$ satisfying

   (i) $\|q \cdot a\| = |q| \cdot \|a\|$ for all $q \in \mathbb{Q}$ and $a \in A$;
   (ii) $\|a + b\| \leq \|a\| + \|b\|$ for all $a, b \in A$.

A *point of $\widehat{A}$* is defined to be a sequence $\langle a_k \colon k \in \mathbb{N} \rangle$ of elements of A such that $\|a_k - a_{k+1}\| \leq 2^{-k-1}$ for all $k \in \mathbb{N}$.

Thus a code for a separable Banach space is simply a countable pseudo-normed vector space over the rationals. As usual we define a pseudometric on A by $d(a, b) = \|a - b\|$, for all $a, b \in A$. Thus $\widehat{A}$ is the complete separable metric space which is the completion of A under d, as in §II.5.

If $x = \langle a_k \colon k \in \mathbb{N} \rangle$ and $y = \langle b_k \colon k \in \mathbb{N} \rangle$ are points of $\widehat{A}$ and $\alpha = \langle q_k \colon k \in \mathbb{N} \rangle$ is a real number, we define $\|x\| = \lim_k \|a_k\|$, $x + y = \lim_k (a_k + b_k)$, and $\alpha \cdot x = \lim_k (q_k \cdot a_k)$. It is easy to show within RCA_0 that these limits exist and that $\| \ \| \colon \widehat{A} \to \mathbb{R}$, $+ \colon \widehat{A} \times \widehat{A} \to \widehat{A}$ and $\cdot \colon \mathbb{R} \times \widehat{A} \to \widehat{A}$ are continuous, etc. Thus $\widehat{A}$ enjoys the usual properties of a normed vector space over $\mathbb{R}$. In addition $\widehat{A}$ is separable in the sense of definition II.5.1 and complete in the sense of exercise II.5.2. Thus we are justified in referring to $\widehat{A}$ within RCA_0 as a separable Banach space.

We shall now present three examples of separable Banach spaces within RCA_0: ℓ_p, $C[0, 1]$, and $L_p[0, 1]$.

For all three examples we shall use the same underlying countable vector space A over $\mathbb{Q}$. Within RCA_0, we define $|A| \subseteq \mathbb{N}$ to be the set of (codes for) nonempty finite sequences of rational numbers $\langle r_0, \ldots, r_m \rangle$ such that either $m = 0$ or $r_m \neq 0$. Addition on $|A|$ is defined by putting $\langle r_0, \ldots, r_m \rangle + \langle s_0, \ldots, s_n \rangle = \langle r_0 + s_0, \ldots, r_k + s_k \rangle$ where $r_i, s_i = 0$ for $i > m, n$ respectively, and $k = \max\{i \colon i = 0 \vee r_i + s_i \neq 0\}$. For scalar multiplication on $|A|$, we put $q \cdot \langle r_0, \ldots, r_m \rangle = \langle 0 \rangle$ if $q = 0$, $\langle q \cdot r_0, \ldots, q \cdot r_m \rangle$ if $0 \neq q \in \mathbb{Q}$. It is then easily verified that A is a vector space over $\mathbb{Q}$. (This is the same as the vector space V_0 of §III.4.)

EXAMPLE II.10.2 (the Banach spaces ℓ_p, $1 \leq p < \infty$). We define an RCA$_0$ version of the ℓ_p spaces. Fix a real number p such that $1 \leq p < \infty$. Let A be as above. For all $\langle r_0, \ldots, r_m \rangle \in |A|$, put

$$\| \langle r_0, \ldots, r_m \rangle \| = \left(\sum_{i=0}^{m} |r_i|^p \right)^{1/p}.$$

Thus A becomes a code for a separable Banach space $\widehat{A}$ and we define $\ell_p = \widehat{A}$.

It can be shown in RCA$_0$ that the points of ℓ_p are in canonical one-to-one correspondence with the sequences $\langle x_i : i \in \mathbb{N} \rangle$, $x_i \in \mathbb{R}$, such that $\sum_{i=0}^{\infty} |x_i|^p$ converges. This correspondence is norm-preserving, so our $\ell_p = \widehat{A}$ can be identified with the usual ℓ_p sequence space as defined in Banach space textbooks.

EXAMPLE II.10.3 (the Banach space C[0, 1]). We define an RCA$_0$ version of the space of continuous real-valued functions C[0, 1]. Let $A, +, \cdot$ be as in the previous example. For $\langle r_0, \ldots, r_m \rangle \in A$, define

$$\| \langle r_0, \ldots, r_m \rangle \| = \sup_{0 \leq x \leq 1} |r_m x^m + r_{m-1} x^{m-1} + \cdots + r_1 x + r_0|.$$

We define C[0, 1] $= \widehat{A}$, the completion of A under the metric induced by this norm. Thus C[0, 1] is a separable Banach space.

We would like to be able to assert that the points of our C[0, 1] are in canonical one-to-one correspondence with the continuous real-valued functions on the closed unit interval [0, 1]. Unfortunately, the axioms of RCA$_0$ are not strong enough to prove this. This situation will be clarified in §IV.2 when we discuss the Weierstraß approximation theorem. There we shall see that points of our C[0, 1] are in canonical one-to-one correspondence with continuous real-valued functions on [0, 1] having a modulus of uniform continuity. See also the generalization to compact metric spaces in exercise IV.2.13.

EXAMPLE II.10.4 (the Banach spaces L$_p$[0, 1], $1 \leq p < \infty$). We define an RCA$_0$ version of the familiar spaces L$_p$[0, 1], $1 \leq p < \infty$. Again let A be as in example II.10.2. For $\langle r_0, \ldots, r_m \rangle \in A$ define

$$\| \langle r_0, \ldots, r_m \rangle \| = \left(\int_0^1 |r_m x^m + r_{m-1} x^{m-1} + \cdots + r_1 x + r_0|^p \, dx \right)^{1/p}.$$

Our use of the Riemann integral here will be justified in §IV.2. Under the above norm A again becomes a code for a separable Banach space $\widehat{A}$, and we define $L_p[0, 1] = \widehat{A}$. See also the generalization to compact metric spaces in exercise IV.2.15.

Unfortunately, RCA_0 is not strong enough to prove that the points of our $L_p[0, 1]$ are in canonical one-to-one correspondence with pth power absolutely integrable measurable functions on $[0, 1]$. Stronger axioms are needed in order to prove this. See also remark X.1.11 and the notes at the end of §IV.2.

We now discuss bounded linear operators.

DEFINITION II.10.5 (bounded linear operators). The following definition is made in RCA_0. Let $\widehat{A}$ and $\widehat{B}$ be separable Banach spaces. A (code for a) *bounded linear operator* from $\widehat{A}$ to $\widehat{B}$ is a sequence $F : A \to \widehat{B}$ of points of $\widehat{B}$, indexed by elements of A, such that (i) $F(q_1 a_1 + q_2 a_2) = q_1 F(a_1) + q_2 F(a_2)$ for all $q_1, q_2 \in \mathbb{Q}$ and $a_1, a_2 \in A$, (ii) there exists a real number α such that $\|F(a)\| \leq \alpha \cdot \|a\|$ for all $a \in A$.

For F and α as above and $x = \langle a_k : k \in \mathbb{N} \rangle \in \widehat{A}$, we define $F(x) = \lim_k F(a_k)$. Thus $\|F(x)\| \leq \alpha \cdot \|x\|$ for all $x \in \widehat{A}$. We write $F : \widehat{A} \to \widehat{B}$ to denote this state of affairs. If $\alpha \in \mathbb{R}$ is such that $\|F(x)\| \leq \alpha \cdot \|x\|$ for all $x \in \widehat{A}$, we write $\|F\| \leq \alpha$.

We now proceed to show within RCA_0 that bounded linear operators are the same thing as continuous linear operators.

DEFINITION II.10.6 (continuous linear operators). The following definition is made in RCA_0. Let $\widehat{A}$ and $\widehat{B}$ be separable Banach spaces. A *continuous linear operator* from $\widehat{A}$ to $\widehat{B}$ is a totally defined continuous function $\phi : \widehat{A} \to \widehat{B}$ (in the sense of §II.6) such that

$$\phi(\alpha_1 x_1 + \alpha_2 x_2) = \alpha_1 \phi(x_1) + \alpha_2 \phi(x_2)$$

for all $\alpha_1, \alpha_2 \in \mathbb{R}$ and $x_1, x_2 \in \widehat{A}$.

THEOREM II.10.7. *The following is provable in* RCA_0. *Given a continuous linear operator* $\phi : \widehat{A} \to \widehat{B}$, *there exists a bounded linear operator* $F : \widehat{A} \to \widehat{B}$ *such that*

$$F(x) = \phi(x) \quad \text{for all } x \in \widehat{A}. \tag{5}$$

Conversely, given a bounded linear operator $F : \widehat{A} \to \widehat{B}$, *there exists a continuous linear operator* $\phi : \widehat{A} \to \widehat{B}$ *such that* (5) *holds.*

PROOF. Given a continuous linear operator $\phi : \widehat{A} \to \widehat{B}$, let Φ be a code for ϕ and define $F : A \to \widehat{B}$ by $F(a) = \phi(a)$, for all $a \in A$. Clearly F is $\mathbb{Q}$-linear, i.e., satisfies condition II.10.5(i). To see that F is bounded, note that $\phi(0) = 0$, hence $(0, r)\Phi(0, 1)$ for some $r \in \mathbb{Q}^+$. Thus, for any $x \in \widehat{A}$, $\|x\| < r$ implies $\|\phi(x)\| \leq 1$. Therefore $\|F(a)\| \leq \|a\|/r$ for all $a \in A$, so F satisfies II.10.5(ii) with $\alpha = 1/r$. Thus $F : \widehat{A} \to \widehat{B}$ is a bounded linear operator, and it is easy to check that (5) holds.

For the converse, assume that $F : A \to \widehat{B}$ is the code of a bounded linear operator $F : \widehat{A} \to \widehat{B}$ with $\|F\| \leq \alpha$. Let $\varphi(a, r, b, s)$ be a Σ_1^0 formula saying that $a \in A$, $b \in B$, $r \in \mathbb{Q}^+$, $s \in \mathbb{Q}^+$, and $(F(a), r\alpha) < (b, s)$, i.e., $\|F(a) - b\| < s - r\alpha$. Write $\varphi(a, r, b, s) \equiv \exists n\, \theta(n, a, r, b, s)$ where θ is Σ_0^0. By Σ_0^0 comprehension let Φ be the set of all $(n, a, r, b, s) \in \mathbb{N} \times A \times \mathbb{Q}^+ \times B \times \mathbb{Q}^+$ such that $\theta(n, a, r, b, s)$ holds. It is straightforward to verify that Φ is a code of a totally defined continuous function $\phi : \widehat{A} \to \widehat{B}$ and that (5) holds.

This completes the proof. $\square$

We now prove an RCA$_0$ version of one of the most famous theorems in Banach space theory, known as the *Banach/Steinhaus theorem* or the *uniform boundedness principle*. The proof uses our RCA$_0$ version of the Baire category theorem, which was proved in §II.5.

THEOREM II.10.8 (Banach/Steinhaus theorem). *The following is provable in* RCA$_0$. *Let $\widehat{A}$ and $\widehat{B}$ be separable Banach spaces. Let $\langle F_n : n \in \mathbb{N} \rangle$ be a sequence of (codes for) bounded linear operators $F_n : \widehat{A} \to \widehat{B}$. Assume that for all $x \in \widehat{A}$ there exists M such that $\|F_n(x)\| < M$ for all $n \in \mathbb{N}$. Then there exists α such that, for all $x \in \widehat{A}$ and $n \in \mathbb{N}$, $\|F_n(x)\| \leq \alpha \cdot \|x\|$.*

PROOF. We reason in RCA$_0$. By lemma II.5.7 there exists a sequence of closed sets $\langle C_m : m \in \mathbb{N} \rangle$ in $\widehat{A}$ such that for all $x \in \widehat{A}$ and $m \in \mathbb{N}$, $x \in C_m$ if and only if $\|F_n(x)\| \leq m$ for all $n \in \mathbb{N}$. The hypothesis of the theorem implies $\widehat{A} = \bigcup_{m \in \mathbb{N}} C_m$. Hence by the Baire category theorem II.5.8 there exists $m \in \mathbb{N}$ such that C_m includes a nonempty open set. Let m_0 be such an m and let $a_0 \in A$ and $r_0 \in \mathbb{Q}^+$ be such that, for all $x \in \widehat{A}$, $\|x - a_0\| < r_0$ implies $x \in C_{m_0}$.

We claim that, for all $x \in \widehat{A}$ and $n \in \mathbb{N}$, $\|F_n(x)\| \leq 4m_0\|x\|/r_0$. If $x = 0$ this is trivial so assume $x \neq 0$. Then we have

$$\left\| a_0 - \left(a_0 + \frac{r_0 x}{2\|x\|} \right) \right\| = \left\| \frac{r_0 x}{2\|x\|} \right\| = \frac{r_0}{2} < r_0.$$

Thus

$$a_0 + \frac{r_0 x}{2\|x\|}$$

belongs to C_{m_0}, as does a_0, so for any $n \in \mathbb{N}$ we have

$$\frac{r_0}{2\|x\|} \|F_n(x)\| = \left\| F_n \left(\frac{r_0 x}{2\|x\|} \right) \right\|$$

$$\leq \left\| F_n \left(a_0 + \frac{r_0 x}{2\|x\|} \right) \right\| + \|F_n(a_0)\| \leq 2m_0.$$

From this our claim follows immediately. Thus we have the conclusion of the theorem with

$$\alpha = \frac{4m_0}{r_0}.$$

This completes the proof. $\square$

Notes for §II.10. A good reference for Banach space theory is Dunford/Schwartz [49]. The material of this section is from Brown/Simpson [27] and Brown [24]. For more on separable Banach spaces in subsystems of Z_2, see §§IV.2, IV.7, IV.9, and X.2.

II.11. Conclusions

In this chapter we have defined the formal system RCA_0 and developed a substantial part of ordinary mathematics within it. We have shown that many basic concepts concerning the real number system, complete separable metric spaces, continuous functions, mathematical logic, countable algebra, and separable Banach spaces can be adequately defined within RCA_0. Using primitive recursion (§II.3) and Σ_1^0 induction, we have shown that some nontrivial mathematical theorems are provable in RCA_0, including: nested interval completeness and the intermediate value property of the real line (§II.4); the Baire category theorem, paracompactness, and a version of the Tietze extension theorem for complete separable metric spaces (§§II.5–II.7); a strong version of the soundness theorem in mathematical logic (§II.8); existence of the algebraic closure of a countable field, and of the real closure of a countable ordered field (§II.9); the Banach/Steinhaus theorem (§II.10).

We conclude that RCA_0 may be viewed as a formal version of computable or constructive mathematics.

Chapter III

ARITHMETICAL COMPREHENSION

III.1. The Formal System ACA_0

The purpose of this chapter is to study a certain subsystem of second order arithmetic known as ACA_0. The acronym ACA stands for "arithmetical comprehension axiom." The axioms of ACA_0 assert the existence of subsets of $\mathbb{N}$ which are definable from given sets by formulas with no set quantifiers. This set existence principle is strong enough to permit a convenient development of large portions of ordinary mathematics which cannot be developed within the confines of RCA_0.

DEFINITION III.1.1 (arithmetical formulas). Let φ be a formula of the language L_2 of second order arithmetic. We say that φ is *arithmetical* if φ contains no set quantifiers. Note that an arithmetical formula may contain free set variables.

DEFINITION III.1.2 (definition of ACA_0). The axioms of ACA_0 are the basic axioms and the induction axiom (see definition I.2.4) together with comprehension axioms

$$\exists X \, \forall n \, (n \in X \leftrightarrow \varphi(n))$$

where $\varphi(n)$ is any arithmetical formula in which X does not occur freely.

The following lemma will be useful in showing that arithmetical comprehension is needed in order to prove various theorems of ordinary mathematics.

LEMMA III.1.3. *The following are pairwise equivalent over RCA_0.*

1. ACA_0.
2. Σ_1^0 comprehension, *i.e.,* $\exists X \, \forall n \, (n \in X \leftrightarrow \varphi(n))$ *restricted to Σ_1^0 formulas $\varphi(n)$ in which X does not occur freely.*
3. *For all one-to-one functions $f : \mathbb{N} \to \mathbb{N}$ there exists a set $X \subseteq \mathbb{N}$ such that $\forall n \, (n \in X \leftrightarrow \exists m \, (f(m) = n))$, i.e., X is the range of f.*

PROOF. The implications $1 \to 2$ and $2 \to 3$ are trivial. The implication $3 \to 2$ is immediate from lemma II.3.7. It remains to prove that $2 \to 1$, i.e., Σ_1^0 comprehension implies arithmetical comprehension. Since

105

each arithmetical formula is equivalent to a Σ_k^0 formula for some $k \in \omega$ (definition II.1.2.), it suffices to prove that Σ_1^0 comprehension implies Σ_k^0 comprehension. We prove this by induction on $k \in \omega$. For $k \leq 1$ the assertion is trivial. Let $\varphi(n)$ be Σ_{k+1}^0, $k \geq 1$. Write $\varphi(n)$ as $\exists j \, \psi(n, j)$ where $\psi(n, j)$ is Π_k^0. By Σ_k^0 comprehension let Y be the set of all (n, j) such that $\neg \psi(n, j)$ holds. Then by Σ_1^0 comprehension let X be the set of all n such that $\exists j \, ((n, j) \notin Y)$. Clearly $n \in X$ if and only if $\exists j \, \psi(n, j)$, i.e., $\varphi(n)$. This completes the proof. $\qquad\square$

We conclude this section with some remarks on models of ACA_0. It is not hard to show that ACA_0 has a minimum ω-model, ARITH, consisting of all subsets of ω which are first order definable over $(\omega, +, \cdot, 0, 1, <)$. Equivalently,

$$\mathrm{ARITH} = \{X \subseteq \omega : \exists n \, X \leq_{\mathrm{T}} \emptyset^{(n)}\}$$

where $\leq_{\mathrm{T}}$ denotes Turing reducibility and $\emptyset^{(n)}$ is the nth Turing jump of the empty set. For a proof of these results and other results about ω-models of ACA_0, see §VIII.1. For a discussion of non-ω-models and conservation results related to ACA_0, see chapter IX.

III.2. Sequential Compactness

In this section we show that the set existence axioms of ACA_0 are just strong enough to provide a good theory of sequential compactness and completeness. We begin with sequences of real numbers (the Bolzano/Weierstraß theorem). We then generalize to sequences of points in a compact metric space. Finally we consider sequences of continuous functions (the Ascoli lemma).

This section includes our first illustrations of the theme of Reverse Mathematics, which was mentioned in chapter I.

LEMMA III.2.1. *The following is provable in* ACA_0. *Let* $\langle x_n : n \in \mathbb{N} \rangle$ *be a bounded sequence of real numbers. Then* $x = \limsup_n x_n$ *exists. Moreover, there exists a subsequence* $\langle x_{n_k} : k \in \mathbb{N} \rangle$, $n_0 < \cdots < n_k < \cdots$, *which converges to* x.

PROOF. We reason within ACA_0. By a linear transformation we may assume that $0 \leq x_n \leq 1$ for all $n \in \mathbb{N}$. Define $f : \mathbb{N} \to \mathbb{N}$ by $f(k) = $ the largest $i < 2^k$ such that $i \cdot 2^{-k} \leq x_n \leq (i + 1) \cdot 2^{-k}$ for infinitely many $n \in \mathbb{N}$. This function f exists by arithmetical comprehension. Put $x = \langle q_k : n \in \mathbb{N} \rangle$ where $q_k = f(k) \cdot 2^{-k}$. It is straightforward to verify that x is a real number and that $\forall \epsilon > 0 \, \exists m \, \forall n \, (m < n \to x_n \leq x + \epsilon)$ and $\forall \epsilon > 0 \, \forall m \, \exists n \, (m < n \wedge |x - x_n| < \epsilon)$. In other words, $x = \limsup_n x_n$. Define the subsequence $\langle x_n : k \in \mathbb{N} \rangle$ by $n_0 = 0$, $n_{k+1} = $ least $n > n_k$ such that $|x - x_n| \leq 2^{-k}$. Clearly $x = \lim_k x_{n_k}$. This completes the proof. $\qquad\square$

THEOREM III.2.2. *The following assertions are pairwise equivalent over* RCA_0.

1. ACA_0.
2. *The* Bolzano/Weierstraß *theorem: Every bounded sequence of real numbers contains a convergent subsequence.*
3. *Every Cauchy sequence of real numbers is convergent. (A sequence $\langle x_n : n \in \mathbb{N} \rangle$ is called* Cauchy *if $\forall \epsilon > 0\, \exists m\, \forall n\, (m < n \to |x_m - x_n| < \epsilon))$.)*
4. *Every bounded sequence of real numbers has a least upper bound.*
5. *The* monotone convergence theorem: *Every bounded increasing sequence of real numbers is convergent.*

PROOF. The implication $1 \to 2$ is lemma III.2.1, and the implications $2 \to 3$ and $3 \to 5$ are obvious since every bounded increasing sequence is Cauchy. Also $4 \to 5$ is trivial, so it remains to prove $1 \to 4$ and $5 \to 1$.

We first prove $1 \to 4$. Assume 1 and let $\langle x_n : n \in \mathbb{N} \rangle$ be a bounded sequence of real numbers. We may safely assume that $0 \le x_n \le 1$ for all n. Define $f : \mathbb{N} \to \mathbb{N}$ by $f(k) =$ the largest $i < 2^k$ such that $\exists n\, (i \cdot 2^{-k} \le x_n)$. This f exists by arithmetical comprehension. Put $x = \langle q_k : k \in \mathbb{N} \rangle$ where $q_k = f(k) \cdot 2^{-k}$. It is straightforward to verify that x is a real number and that $x = \sup_n x_n$, i.e., $\forall n\, (x_n \le x)$ and $\forall y\, (y < x \to \exists n\, (y < x_n))$.

It remains to prove $5 \to 1$. Assume 5 and let $f : \mathbb{N} \to \mathbb{N}$ be a given one-to-one function. Put $c_n = \sum_{i=0}^{n} 2^{-f(i)}$. Clearly $c_0 < c_1 < \cdots < c_n < \cdots < 2$ for all $n \in \mathbb{N}$. Hence by the monotone convergence theorem 5 we have the existence of

$$c = \lim_n c_n = \sum_{i=0}^{\infty} 2^{-f(i)}.$$

It is easy to see that, for all k,

$$(\exists i\, (f(i) = k)) \leftrightarrow \forall n\, (|c_n - c| < 2^{-k} \to \exists i \le n\, (f(i) = k)).$$

The left hand side of this equivalence is Σ_1^0 while the right hand side is Π_1^0. Hence by Δ_1^0 comprehension (with parameters c and f) we obtain $\exists X\, \forall k\, (k \in X \leftrightarrow \exists i\, (f(i) = k))$. We have now proved from 5 that for all one-to-one functions $f : \mathbb{N} \to \mathbb{N}$ the range of f exists. Hence by lemma III.1.3 we have ACA_0. This completes the proof of theorem III.2.2. $\square$

REMARK. The implication $2 \to 1$ above is our first illustration of Reverse Mathematics. The point here is that the Bolzano/Weierstraß theorem (an ordinary mathematical statement) implies arithmetical comprehension (a set existence axiom). Thus no set existence axiom weaker than arithmetical comprehension will suffice to prove the Bolzano/Weierstraß theorem. See also the discussion in §I.9.

We shall now generalize part of the previous theorem to the context of complete separable metric spaces (as defined in §II.5).

DEFINITION III.2.3 (compactness). The following definition is made in RCA$_0$. A *compact metric space* is a complete separable metric space $\widehat{A}$ such that there exists an infinite sequence of finite sequences

$$\langle\langle x_{ij} : i \leq n_j \rangle : j \in \mathbb{N}\rangle, \qquad x_{ij} \in \widehat{A},$$

such that for all $z \in \widehat{A}$ and $j \in \mathbb{N}$ there exists $i \leq n_j$ such that $d(x_{ij}, z) < 2^{-j}$.

EXAMPLE III.2.4. The sequence $\langle\langle i \cdot 2^{-j} : i \leq 2^j \rangle : j \in \mathbb{N}\rangle$ shows that the closed unit interval $[0, 1] = \{x : 0 \leq x \leq 1\}$ is compact. More generally, any closed bounded interval in $\mathbb{R}$ is compact. These facts are provable in RCA$_0$.

LEMMA III.2.5 (compact product spaces). *The following is provable in* RCA$_0$. *Let* $\widehat{A}_k$, $k \in \mathbb{N}$, *be a countably infinite sequence of compact metric spaces. Assume that there exists a doubly infinite sequence of finite sequences*

$$\langle\langle x_{ijk} : i \leq n_{jk} \rangle : j, k \in \mathbb{N}\rangle, \qquad x_{ijk} \in \widehat{A}_k,$$

such that for all $j, k \in \mathbb{N}$ *and* $x \in \widehat{A}_k$ *there exists* $i \leq n_{jk}$ *such that* $d(x_{ijk}, x) \leq 2^{-j}$. *Then the infinite product space* $\widehat{A} = \prod_{k \in \mathbb{N}} \widehat{A}_k$ *is compact. A similar statement holds for finite products.*

PROOF. We first consider the case of a finite product $\widehat{A} = \prod_{i=1}^{m} \widehat{A}_k$. In this case, for each $j \in \mathbb{N}$, let l_j = the smallest l such that $m \cdot 2^{-l} \leq 2^{-j}$. Put $n_j = \prod_{k=1}^{m}(n_{l_j k} + 1) - 1$ and let $\langle x_{ij} : i \leq n_j \rangle$ be an enumeration of $\prod_{k=1}^{m}\{x_{il_j k} : i \leq n_{l_j k}\}$. Then $\langle\langle x_{ij} : i \leq n_j \rangle : j \in \mathbb{N}\rangle$ attests to the compactness of $\widehat{A}$.

In the case of a countably infinite product $\widehat{A} = \prod_{k \in \mathbb{N}} \widehat{A}_k$, for each $j \in \mathbb{N}$ let l_j = smallest l such that $(j + 2) \cdot 2^{-l} \leq 2^{-j-1}$. Put $n_j = \prod_{k=0}^{j+1}(n_{l_j k} + 1) - 1$ and let $\langle x_{ij} : i \leq n_j \rangle$ be an enumeration of $\prod_{k=0}^{j+1}\{x_{il_j k} : i \leq n_{l_j k}\}$. Again $\langle\langle x_{ij} : i \leq n_j \rangle : j \in \mathbb{N}\rangle$ attests to the compactness of $\widehat{A}$. This completes the proof of the lemma. $\qquad \square$

EXAMPLES III.2.6. Within RCA$_0$, we have:

1. Any closed bounded rectangle in $\mathbb{R}^m$ is compact.
2. The Cantor space $2^{\mathbb{N}} = \{0, 1\}^{\mathbb{N}}$ is compact.
3. The Hilbert cube $[0, 1]^{\mathbb{N}}$ is compact.
4. For any compact metric space $\widehat{A}$, the infinite product space $\widehat{A}^{\mathbb{N}} = \prod_{k \in \mathbb{N}} \widehat{A}$ is compact.

Our generalization of theorem III.2.2 to complete separable metric spaces is as follows.

THEOREM III.2.7. *The following assertions are pairwise equivalent over* RCA$_0$.

1. ACA$_0$.
2. *In any compact metric space, every sequence of points has a convergent subsequence.*
3. *In any complete separable metric space, every Cauchy sequence is convergent.* (*A sequence* $\langle x_n : n \in \mathbb{N} \rangle$, $x_n \in \widehat{A}$, *is said to be* Cauchy *if* $\forall \epsilon > 0 \, \exists m \, \forall n \, (m < n \to d(x_m, x_n) \leq \epsilon)$.)

PROOF. The proof of $1 \to 2$ is a straightforward generalization of the proof of lemma III.2.1. The proof of $1 \to 3$ is left as an exercise for the reader. (Compare exercise II.5.2.) The implications $2 \to 1$ and $3 \to 1$ are immediate from theorem III.2.2 since 2 and 3 are generalizations of III.2.2.2 and III.2.2.3 respectively.   □

We end this section by showing that ACA$_0$ is just strong enough to prove the Ascoli lemma for compact metric spaces. First we give the relevant definitions. Let $\widehat{A}$ and $\widehat{B}$ be complete separable metric spaces. Let $f_n : \widehat{A} \to \widehat{B}$, $n \in \mathbb{N}$, be a sequence of continuous functions. The sequence is said to be *equicontinuous* if for all $\epsilon > 0$ there exists $\delta > 0$ such that, for all $x, x' \in \widehat{A}$, $d(x, x') < \delta$ implies $d(f_n(x), f_n(x')) \leq \epsilon$ for all $n \in \mathbb{N}$. The sequence is said to be *uniformly convergent* if there exists a continuous function $f : \widehat{A} \to \widehat{B}$ such that for all $\epsilon > 0$ there exists m such that for all $n > m$ and $x \in \widehat{A}$, $d(f_n(x), f(x)) < \epsilon$.

THEOREM III.2.8 (Ascoli lemma). *The following is provable in* ACA$_0$. *Let* $\widehat{A}$ *and* $\widehat{B}$ *be compact metric spaces. Let* $\langle f_n : n \in \mathbb{N} \rangle$ *be an equicontinuous sequence of continuous functions* $f_n : \widehat{A} \to \widehat{B}$. *Then there exists a uniformly convergent subsequence* $\langle f_{n_k} : k \in \mathbb{N} \rangle$, $n_0 < n_1 < \cdots < n_k < \cdots$.

PROOF. We reason in ACA$_0$. Since $\widehat{A}$ is compact, let

$$\langle \langle x_{ij} : 1 \leq n_j \rangle : j \in \mathbb{N} \rangle, \qquad x_{ij} \in \widehat{A},$$

be as in III.2.3. Let I be the set of all $(i, j) \in \mathbb{N} \times \mathbb{N}$ such that $i \leq n_j$. For each $m \in \mathbb{N}$ put $z_m = \langle f_m(x_{ij}) : (i, j) \in I \rangle$. Thus $\langle z_m : n \in \mathbb{N} \rangle$ is a sequence of points in the infinite product space $\widehat{B}^I$. By III.2.6.4 this space is compact. Hence by III.2.7 there exists a convergent subsequence $\langle z_{m_k} : k \in \mathbb{N} \rangle$. It follows that $\langle f_{m_k}(x_{ij}) : k \in \mathbb{N} \rangle$ is convergent for each $(i, j) \in I$. Using arithmetical comprehension as in the last part of the proof of III.2.1, we may if necessary refine our subsequence so that

$$\forall i \, \forall j \, \forall k \, ((i \leq n_j \wedge j \leq k) \to d(f_{m_k}(x_{ij}), f_{m_{k+1}}(x_{ij})) \leq 2^{-k-1}).$$

By yet another application of arithmetical comprehension, define $h : \mathbb{N} \to \mathbb{N}$ by $h(l) =$ smallest n such that $d(a, a') < 2^{-n}$ implies $d(f_m(a), f_m(a')) \leq 2^{-l}$ for all $m \in \mathbb{N}$ and $a, a' \in A$. It follows that h is a *modulus of equicontinuity*, i.e., $d(x, x') < 2^{-h(l)}$ implies $d(f_m(x), f_m(x')) \le 2^{-l}$ for all $l, m \in \mathbb{N}$ and $x, x' \in \widehat{A}$.

Define a code F for a continuous function f from $\widehat{A}$ to $\widehat{B}$ by putting $(a, r)F(b, s)$ if and only if $a \in A$, $r \in \mathbb{Q}^+$, $b \in B$, $s \in \mathbb{Q}^+$, and there exist i, j, k and l such that $i \le n_j$, $(a, r) < (x_{ij}, 2^{-j})$, $h(l) \le j \le k$, $l \le k$, and $(f_{m_k}(x_{ij}), 2^{-l+1}) < (b, s)$. It is straightforward to check that f is totally defined on $\widehat{A}$, that

$$\forall i \, \forall j \, \forall k \, ((i \le n_j \wedge j \le k) \to d(f_{m_k}(x_{ij}), f(x_{ij})) \le 2^{-k}),$$

and that $d(x, x') < 2^{-h(l)}$ implies $d(f(x), f(x')) \le 2^{-l}$ for all $l \in \mathbb{N}$ and $x, x' \in \widehat{A}$. From these facts it follows that $d(f_m(x), f(x)) \le 3 \cdot 2^{-l}$ for all $l \in \mathbb{N}$ and $k \ge \max(l, h(l))$. Thus $\langle f_{m_k} : k \in \mathbb{N} \rangle$ converges uniformly to f. This proves the theorem.   □

As another instance of Reverse Mathematics, we have:

THEOREM III.2.9. *The following are pairwise equivalent over* RCA$_0$:

1. *arithmetical comprehension*;
2. *the Ascoli lemma*;
3. *The Bolzano/Weierstraß theorem*.

PROOF. The Bolzano/Weierstraß theorem is the special case of the Ascoli lemma in which $\widehat{A}$ and $\widehat{B}$ are closed bounded intervals and the f_n's are constant functions. This proves $2 \to 3$. The implications $1 \to 2$ and $3 \to 1$ have already been proved in III.2.8 and III.2.2 respectively.   □

Notes for §III.2. Theorem III.2.2 was stated without proof by Friedman [69]. The definition of compact metric spaces within RCA$_0$, as well as lemma III.2.5 and the examples in III.2.6, are due to Brown/Simpson [24, 27, 28]. For more on compact metric spaces, see §§IV.1 and IV.2 below. Theorems III.2.8 and III.2.9 are due to Simpson, previously unpublished. For a somewhat different treatment of the Ascoli lemma within ACA$_0$, see Simpson [236].

III.3. Strong Algebraic Closure

We saw in §II.9 that RCA$_0$ proves that every countable field has an algebraic closure. One might ask whether RCA$_0$ proves the stronger statement that every countable field is isomorphic to a subfield of its algebraic closure. We now show that ACA$_0$ is needed to prove this stronger statement.

DEFINITION III.3.1 (strong algebraic closure). The following definitions are made in RCA$_0$. Let K be a countable field. A *strong algebraic closure* of K is an algebraic closure $h \colon K \to \widetilde{K}$ (see definition II.9.2) with the further property that h is an isomorphism of K onto a subfield of $\widetilde{K}$.

The notion of *strong real closure* is defined similarly (compare definition II.9.5).

THEOREM III.3.2. *The following assertions are pairwise equivalent over* RCA$_0$.

1. ACA$_0$.
2. *Every countable field has a strong algebraic closure.*
3. *Every countable field is isomorphic to a subfield of a countable algebraically closed field.*
4. *Every countable ordered field has a strong real closure.*
5. *Every countable ordered field is isomorphic to a subfield of a countable real closed ordered field.*

PROOF. First assume ACA$_0$ and let K be a countable field. By theorem II.9.4 let $h: K \to \widetilde{K}$ be an algebraic closure of K. By Σ_1^0 comprehension let L be the set of all $b \in \widetilde{K}$ such that $\exists a\,(h(a) = b)$. Then L is a subfield of $\widetilde{K}$ and h is an isomorphism of K onto L. Hence $h: K \to \widetilde{K}$ is a strong algebraic closure of K. This proves that 1 implies 2.

The implication from 2 to 3 is trivial. We shall now prove that 3 implies 1. We reason in RCA$_0$. Assume 3. Instead of proving ACA$_0$ we shall prove the equivalent statement III.1.3.3. Let $f: \mathbb{N} \to \mathbb{N}$ be given. By theorem II.9.7 let $\overline{\mathbb{Q}}$ be the real closure of $\mathbb{Q}$. Let $\langle p_j : j \in \mathbb{N} \rangle$ be the enumeration of the rational primes in increasing order, i.e., $p_0 = 2$, $p_1 = 3$, $p_2 = 5$, ... For each $n \in \mathbb{N}$ let K_n be the subfield of $\overline{\mathbb{Q}}$ generated by $\{\sqrt{p_{f(i)}} : i < n\}$. Because we lack Σ_1^0 comprehension, we cannot form the subfield $\bigcup_{n \in \mathbb{N}} K_n$. However, we can apply lemma II.3.7 to find a field K and a monomorphism $g: K \to \overline{\mathbb{Q}}$ such that $\forall b\,(\exists n\,(b \in K_n) \leftrightarrow \exists a\,(g(a) = b))$. Intuitively, $K = \mathbb{Q}(\{\sqrt{p_{f(i)}} : i \in \mathbb{N}\})$. Now by 3 let $h: K \to L \subseteq M$ be an isomorphism of K onto a subfield L of a countable algebraically closed field M. Then for all $j \in \mathbb{N}$ we have $\exists i\,(f(i) = j)$ if and only if $\forall b\,((b \in M \wedge b^2 = p_j) \to b \in L)$. It follows by Δ_1^0 comprehension that $\exists X \forall j\,(j \in X \leftrightarrow \exists i\,(f(i) = j))$. By lemma III.1.3 this implies ACA$_0$.

We have now established the pairwise equivalence of 1, 2, and 3. An obvious modification establishes the pairwise equivalence of 1, 4, and 5. This completes the proof of theorem III.3.2. □

REMARK III.3.3 (formally real fields). Theorem III.3.2 remains true if the ordered fields are replaced by formally real fields. Again the same proof applies. For more on formally real fields see §IV.4.

Notes for §III.3. Theorem III.3.2 is from Friedman/Simpson/Smith [78]. The idea of the proof goes back to Fröhlich/Shepherdson [82].

III.4. Countable Vector Spaces

In this section we shall show that ACA_0 is just strong enough to prove that every countable vector space has a basis. We shall also obtain some strengthenings of this result.

Definition III.4.1 (countable vector spaces). The following definitions are made in RCA_0. Let K be a countable field (as defined in §II.9). A *countable vector space* V over K consists of a countable Abelian group $|V|, +_V, -_V, 0_V$ (see definition III.6.1 below) together with a function $\cdot_V : |K| \times |V| \to |V|$ which obeys the usual axioms for scalar multiplication, e.g., $a \cdot (u + v) = a \cdot u + a \cdot v$. For notational convenience we shall sometimes write $|V|$ as V.

A *basis* of V over K is a set $E \subseteq |V|$ such that each $v \in V$ can be expressed uniquely in the form $v = \sum_{e \in E_0} a_e \cdot e$ where E_0 is a finite subset of E and, for each $e \in E_0$, $0 \neq a_e \in K$.

Lemma III.4.2. ACA_0 *proves that every countable vector space over a countable field has a basis.*

Proof. We reason in ACA_0. Let V be a countable vector space over a countable field K. By arithmetical comprehension, there exists a set S consisting of all finite sequences $\langle v_0, \ldots, v_{n-1}, v_n \rangle$, $n \in \mathbb{N}$, such that $v_n = \sum_{i<n} a_i \cdot v_i$ for some $a_0, \ldots, a_{n-1} \in K$. Using S as a parameter, we can apply primitive recursion (§II.3) to define a sequence of vectors $e_0, e_1, \ldots, e_n, \ldots$ where $e_n =$ the least $v \in V$ such that $\langle e_0, \ldots, e_{n-1}, v \rangle \notin S$. (The recursion may end after finitely many steps.) Here $V = |V| \subseteq \mathbb{N}$ and "least" refers to the usual ordering of $\mathbb{N}$. The set $E = \{e_0, e_1, \ldots\}$ is easily shown to be a basis for V. This proves the lemma. $\square$

Theorem III.4.3. *The following assertions are pairwise equivalent over* RCA_0.

1. ACA_0.
2. *Every countable vector space over a countable field has a basis.*
3. *Every countable vector space over the rational field* $\mathbb{Q}$ *has a basis.*

Proof. Lemma III.4.2 gives the implication $1 \to 2$, and 2 implies 3 trivially. It remains to prove $3 \to 1$.

We reason within RCA_0. Assume 3. Our goal is to prove arithmetical comprehension. Let $f : \mathbb{N} \to \mathbb{N}$ be a one-to-one function. By lemma III.1.3, it suffices to prove that the range of f exists.

Let V_0 be the set of formal sums $\sum_{i \in I} q_i \cdot x_i$ where $I \subseteq \mathbb{N}$, I is finite, and $0 \neq q_i \in \mathbb{Q}$. Thus V_0 is a vector space over $\mathbb{Q}$ and $X = \{x_n : n \in \mathbb{N}\}$ is a basis of V_0. For each $m \in \mathbb{N}$ put

$$x'_m = x_{2f(m)} + m \cdot x_{2f(m)+1}$$

and let U be the subspace of V_0 generated by $X' = \{x'_m : m \in \mathbb{N}\}$. U exists by Δ^0_1 comprehension since $\sum_{i \in I} q_i \cdot x_i$ belongs to U if and only if

$\forall n \, (q_{2n} \neq 0 \rightarrow f(q_{2n+1}/q_{2n}) = n)$ and $\forall n \, (q_{2n} = 0 \rightarrow q_{2n+1} = 0)$. Note that X' is a basis of U.

Since U is a subspace of V_0, we may form the quotient space $V = V_0/U$ as follows. The elements of V are those $v \in V_0$ such that $\forall w \, ((w < v \land w \in V_0) \rightarrow v - w \notin U)$, i.e., v is the minimal representative of an equivalence class under the equivalence relation $v - w \in U$. The vector space operations on V are defined accordingly. For instance, for all $u, v \in V$ we put $u +_V v = $ the unique $w \in V$ such that w is equivalent to $u +_{V_0} v$ under the mentioned equivalence relation. Thus V is a vector space over $\mathbb{Q}$.

By our assumption 3, $V = V_0/U$ has a basis, call it X''. It follows that $X' \cup X''$ is a basis of V_0. Now for any $n \in \mathbb{N}$, we have $\exists m \, (f(m) = n)$ if and only if at least one of the unique expressions for x_{2n} and x_{2n+1} in terms of the basis $X' \cup X''$ involves an element x'_m from X' such that $f(m) = n$. Hence, by Δ^0_1 comprehension, the range of f exists.

This completes the proof of theorem III.4.3. □

REMARK. The point of the above theorem is that a fairly innocuous looking mathematical assertion ("every countable vector space over $\mathbb{Q}$ has a basis") is in fact equivalent to arithmetical comprehension. This is an instance of Reverse Mathematics. We shall now strengthen the theorem by showing that an even weaker looking assertion ("every countable vector space over $\mathbb{Q}$ either is finite dimensional or contains an infinite linearly independent set") is also equivalent to arithmetical comprehension.

A countable vector space V over K is said to be *finite dimensional* if it has a finite basis. A set $Y \subseteq |V|$ is said to be *linearly independent* if there is no equation $\sum_{i=0}^{k} a_i \cdot y_i = 0$ where $0 \neq a_i \in K$ and $y_0, \ldots, y_k$ are distinct elements of Y.

THEOREM III.4.4. *The following assertions are pairwise equivalent over* RCA$_0$.

1. ACA$_0$.
2. *Every countable vector space (over $\mathbb{Q}$) has a basis.*
3. *Every countable vector space over $\mathbb{Q}$ either is finite dimensional or contains an infinite linearly independent set. (Instead of the rational field $\mathbb{Q}$ we could use any infinite countable field.)*

PROOF. Lemma III.4.2 gives the implication from 1 to 2, and 2 implies 3 trivially. It remains to prove that 3 implies 1. We reason in RCA$_0$.

As in the proof of theorem III.4.3, let V_0 be an infinite dimensional vector space over $\mathbb{Q}$ which has a basis. For any finite set of vectors $v_0, \ldots, v_{n-1} \in V_0$, let $(v_0, \ldots, v_{n-1})$ be the set of $v \in V_0$ such that $v = \sum_{i<n} q_i \cdot v_i$ for some $q_0, \ldots, q_{n-1} \in \mathbb{Q}$. Note that, for V_0, the set S of finite sequences $\langle v_0, \ldots, v_{n-1}, v_n \rangle$ such that $v_n \in (v_0, \ldots, v_{n-1})$ exists in RCA$_0$

since we can use the basis plus determinants to test for linear independence of finite sets. Hence we can proceed as in the proof of lemma III.4.2 to define a sequence of vectors $e_n = $ least $v \in V_0$ such that $v \notin (e_0, \dots, e_{n-1})$. Note that $m < n$ implies $e_m < e_n$. Hence by Δ_1^0 comprehension the set of all e_n, $n \in \mathbb{N}$, exists and is therefore a basis of V_0.

Assume 3. As in the proof of theorem III.4.3, let $f : \mathbb{N} \to \mathbb{N}$ be a one-to-one function. We want to show that the range of f exists. Let us say that $m \in \mathbb{N}$ is *true* if $f(n) > f(m)$ for all $n > m$ and *false* otherwise. We may safely assume that 0 is false. Since the property of being false is Σ_1^0, lemma II.3.7 provides a one-to-one function $g : \mathbb{N} \to \mathbb{N}$ such that $\forall m\, (m$ is false $\leftrightarrow \exists k\, (g(k) = m))$. We may safely assume that $g(0) = 0$, hence $g(k) > 0$ for all $k > 0$.

By primitive recursion define vectors $u_k = e_0 + a_k \cdot e_{g(k)}$, $k > 0$, where the scalar $a_k \in \mathbb{Q}$, $a_k \neq 0$ is chosen so that for all vectors $v \leq k$, if $v \notin (u_1, \dots, u_{k-1})$ then $v \notin (u_1, \dots, u_{k-1}, u_k)$. To see that such a_k exists, note that since $e_0, e_{g(1)}, \dots, e_{g(k-1)}, e_{g(k)}$ are linearly independent, so are $e_0, u_1, \dots, u_{k-1}, e_{g(k)}$. Hence for any $v \notin (u_1, \dots, u_{k-1})$ there is at most one scalar $b_v \in \mathbb{Q}$ such that $v \in (u_1, \dots, u_{k-1}, e_0 + b_v \cdot e_{g(k)})$. Thus we need only choose a_k outside the finite set $\{0\} \cup \{b_v : v \in |V_0| \wedge v \leq k\}$. Now let U be the subspace of V_0 generated by $\{u_k : k > 0\}$. Here U exists by Δ_1^0 comprehension since $v \in U$ if and only if $v \in (u_1, \dots, u_v)$.

U is a subspace of V_0 so, as in the proof of theorem III.4.3, we may form the quotient space $V = V_0/U$. Since the e_m's for all true $m \in \mathbb{N}$ are linearly independent modulo U, we see that V is not finite dimensional. Hence by our assumption 3 there exists an infinite linearly independent set $Y \subseteq V$. Viewing Y as a subset of V_0, we see that Y is linearly independent modulo the subspace U. In particular, there is at most one way to express e_0 as an element of U plus a linear combination of elements of Y. Since $e_0 \notin U$, the linear combination of elements of Y must be nontrivial. Hence by deleting at most one element from Y, we may assume that Y is linearly independent modulo $\{e_0\} \cup U$. Hence Y is linearly independent modulo all the $e_{g(k)}$, $k \in \mathbb{N}$, i.e., all the e_m such that m is false.

Let $\langle y_j : j \in \mathbb{N} \rangle$ be the enumeration of the elements of Y in increasing order (lemma II.3.6). We claim that for each j, the number of true m with $e_m < y_j$ is at least j. To see this, suppose not and let n be the least integer such that $e_n \geq y_j$. Then the dimension of $(e_0, \dots, e_{n-1})$ modulo the e_m with m false is less than j. But the dimension of $(y_0, \dots y_{j-1})$ modulo the e_m with m false is j. Hence there is at least one $i < j$ such that $y_i \notin (e_0, \dots, e_{n-1})$. Hence $y_i \geq e_n$ since we defined $e_n = $ least v such that $v \notin (e_0, \dots, e_{n-1})$. Hence $e_n \leq y_i < y_j$, a contradiction.

From the previous claim it follows that if $e_m \geq y_j$ then $f(m) \geq j$. Hence for all j we have $\exists m\, (f(m) = j)$ if and only if $\exists m\, (e_m < y_{j+1} \wedge f(m) = j)$. Hence by Δ_1^0 comprehension the set of all j such that

$\exists m \, (f \, (m) \, = \, j)$ exists. This gives ACA_0 in view of lemma III.1.3. The proof of theorem III.4.4 is complete.   □

We end this section by mentioning a result on algebraic independence which is analogous to theorem III.4.3 on linear independence.

DEFINITION III.4.5 (algebraic independence). The following definitions are made in RCA_0. Let K and L be countable fields with $K \subseteq L$. A set $Y \subseteq L$ is said to be *algebraically independent over* K if there is no non-trivial polynomial equation $f \, (b_1, \dots, b_k) \, = \, 0$, $b_i \in Y$, $f \, (x_1, \dots, x_k) \in K[x_1, \dots, x_k]$. A *transcendence base* for L over K is a maximal algebraically independent set.

THEOREM III.4.6. *The following assertions are pairwise equivalent over* RCA_0.

1. ACA_0.
2. *For every pair of countable fields* $K \subseteq L$ *there exists a transcendence base for* L *over* K.
3. *Let* L *be any countable field of characteristic zero with no finite transcendence base. Then* L *contains an infinite algebraically independent set.*

PROOF. The ideas underlying the proof are the same as for theorem III.4.4. For details see Friedman/Simpson/Smith [78].   □

Notes for §III.4. The results of this section are due to Friedman/Simpson/ Smith [78]. The ideas in the proof of theorem III.4.4 are closely related to ideas of Dekker (see Rogers [208, §9.5]) and Metakides/Nerode [187]. In the recursion-theoretic Metakides/Nerode setting, the proof of theorem III.4.4 would amount to constructing a recursive, infinite dimensional vector space V over $\mathbb{Q}$ such that the complete recursively enumerable set is Turing reducible to any infinite linearly independent subset of V. Metakides/Nerode [187] contains a result which is somewhat weaker than this. (In the same recursion-theoretic setting, the proof of theorem III.4.3 would amount to constructing a recursive vector space V over $\mathbb{Q}$ such that the complete recursively enumerable set is Turing reducible to any basis of V.)

The proof of theorem III.4.6 is obtained by combining the proof of theorem III.4.4 with methods of Fröhlich/Shepherdson [82]. For details see Friedman/Simpson/Smith [78].

III.5. Maximal Ideals in Countable Commutative Rings

In this section we show that the axioms of ACA_0 are just strong enough to prove that every countable commutative ring has a maximal ideal. Prime ideals will be considered in §IV.6.

DEFINITION III.5.1 (countable commutative rings). The following definitions are made in RCA$_0$. A *countable commutative ring* R consists of a set $|R| \subseteq \mathbb{N}$ together with binary operations $+_R, \cdot_R \colon |R| \times |R| \to |R|$ and a unary operation $-_R \colon |R| \to |R|$ and distinguished elements $0_R, 1_R \in |R|$ satisfying the usual commutative ring axioms, including $\forall x \, \forall y \, (x \cdot y = y \cdot x)$ and $0 \neq 1$. For notational convenience we write $|R|$ as R. A *countable integral domain* is a countable commutative ring R satisfying $\forall x \, \forall y \, (x \cdot y = 0 \to (x = 0 \lor y = 0))$.

DEFINITION III.5.2 (ideals). The following definitions are made in RCA$_0$. Let R be a countable commutative ring. An *ideal* of R is a set $I \subseteq R$ such that $0 \in I$ and $1 \notin I$ and $\forall a \, \forall b \, ((a \in I \land b \in I) \to a + b \in I)$ and $\forall r \, \forall a \, ((r \in R \land a \in I) \to r \cdot a \in I)$. Given an ideal I we can form the *quotient ring* R/I. The elements of R/I are defined to be just those $r \in R$ such that $\forall s \, ((s < r \land s \in R) \to r - s \notin I))$, i.e., minimal elements of equivalence classes under the equivalence relation $r - s \in I$. Ring operations on R/I are defined accordingly.

DEFINITION III.5.3 (prime and maximal ideals). The following definitions are made in RCA$_0$. Let R be a countable commutative ring. A *prime ideal* of R is an ideal P such that $\forall r \, \forall s \, ((r \in R \land s \in R \land r \cdot s \in P) \to (r \in P \lor s \in P))$. This is equivalent to saying that R/P is an integral domain. A *maximal ideal* of R is an ideal M such that $\forall r \, ((r \in R \land r \notin M) \to \exists s \, (s \in R \land r \cdot s - 1 \in M))$. This is equivalent to saying that R/M is a field. Obviously every maximal ideal is prime.

LEMMA III.5.4. *ACA$_0$ proves that every countable commutative ring has a maximal ideal.*

PROOF. We reason in ACA$_0$. For any $X \subseteq R$ say that X is *good* if X does not generate R as an R-module, i.e., 1 is not of the form $\sum_{i=1}^{n} s_i \cdot a_i$ where $s_i \in R$, $a_i \in X$. Let $\langle r_n : n \in \mathbb{N} \rangle$ be an enumeration of the elements of R. Define $f \colon \mathbb{N} \to \{0, 1\}$ by $f(n) = 0$ if $\{r_m : m < n \land f(m) = 0\} \cup \{r_n\}$ is good, $f(n) = 1$ otherwise. Let M be the set of all r_m such that $f(m) = 0$. Clearly M is a maximal ideal of R. $\qquad \square$

THEOREM III.5.5. *The following assertions are pairwise equivalent over* RCA$_0$.

1. ACA$_0$.
2. *Every countable commutative ring has a maximal ideal.*
3. *Every countable integral domain has a maximal ideal.*

PROOF. Lemma III.5.4 gives the implication from 1 to 2, and the implication from 2 to 3 is trivial. It remains to prove that 3 implies 1.

Assume 3. Instead of proving ACA$_0$ we shall prove the equivalent statement III.1.3.3. Let $f \colon \mathbb{N} \to \mathbb{N}$ be given. We want to construct a countable integral domain R which, in a suitable sense, encodes the range of f. We proceed as follows. Let $R_0 = \mathbb{Q}[\langle x_n : n \in \mathbb{N} \rangle]$ be the polynomial ring over the rational field $\mathbb{Q}$ with countably many indeterminates. Let $K_0 = \mathbb{Q}(\langle x_n : n \in \mathbb{N}\rangle)$ be the field of fractions of R_0, i.e., K_0 is the field consisting of all fractions r/s where $r \in R_0$, $s \in R_0$, $s \neq 0$. Let $\varphi(b)$ be a Σ_1^0 formula asserting that $b \in K_0$ and b is of the form r/s where $r \in R_0$, $s \in R_0$, and s contains at least one monomial of the form $q x_{f(m_1)}^{e_1} x_{f(m_2)}^{e_2} \cdots x_{f(m_k)}^{e_k}$ with $q \in \mathbb{Q}$, $q \neq 0$, $k \geq 0$. By lemma II.3.7 let R be a countable integral domain and $h : R \to K_0$ a monomorphism such that $\forall b\, (\varphi(b) \leftrightarrow \exists a\, (h(a) = b))$. By 3 let M be a maximal ideal of R.

We claim that, for all $n \in \mathbb{N}$, $\exists m\, (f(m) = n)$ if and only if $h^{-1}(x_n) \notin M$. If $n = f(m)$ then $\varphi(1/x_n)$ holds, hence $h^{-1}(x_n)$ has an inverse $h^{-1}(1/x_n)$ in R, hence $h^{-1}(x_n) \notin M$ since M is an ideal of R. Conversely, if $h^{-1}(x_n) \notin M$, let $a \in R$ and $b \in M$ be such that $a \cdot h^{-1}(x_n) - 1 = b$. Put $h(b) = r/s$ where $r \in R_0$, $s \in R_0$, $s \neq 0$. Since $b \in M$ it follows that b is not invertible in R, hence r cannot contain any monomial of the form $q x_{f(m_1)}^{e_1} x_{f(m_2)}^{e_2} \cdots x_{f(m_k)}^{e_k}$, while of course s does contain at least one such monomial. But $h(a) \cdot x_n - 1 = h(b) = r/s$, hence $h(a) \cdot x_n \cdot s = r + s$. We conclude that $n = f(m)$ for some m. This proves our claim.

By Δ_1^0 comprehension let X be the set of all n such that $h^{-1}(x_n) \notin M$. Then $\forall n\, (n \in X \leftrightarrow \exists m\, (f(m) = n))$. This gives ACA_0 in view of lemma III.1.3. The proof of theorem III.5.5 is complete.   □

REMARK III.5.6 (localization). Roughly speaking, the idea of the above proof is that $R = R_0(R_0 \setminus P)^{-1}$ where P is a (carefully chosen) prime ideal of R_0. One describes this situation by saying that R is the *local ring* obtained from R_0 by *localizing* at the prime ideal P. It follows that R has a unique maximal ideal M, namely $M = P(R_0 \setminus P)^{-1}$. The prime ideal P is taken to be generated by the indeterminates x_n such that $n \notin$ range of f. Thus $x_n \in M$ if and only if $n \notin$ range of f.

REMARK III.5.7. Theorem III.5.5 provides yet another illustration of Reverse Mathematics. Namely, ACA_0 is both necessary and sufficient to prove the existence of maximal ideals in countable commutative rings. In §IV.6 we shall obtain an analogous result with maximal ideals replaced by prime ideals, and ACA_0 replaced by WKL_0.

REMARK III.5.8. Hatzikiriakou [108, 109] has shown that ACA_0 is equivalent over RCA_0 to the assertion that every countable commutative ring has a *minimal* prime ideal.

Notes for §III.5. The main results of this section are from Friedman/Simpson/Smith [78]. For general information about local rings and localization, see any textbook of commutative algebra, e.g., Zariski/Samuel [282, §IV.11].

III.6. Countable Abelian Groups

In this section we show that the axioms of ACA_0 are just strong enough to prove several basic results in the theory of countable Abelian groups. Later, in §§V.7 and VI.4, we shall return to this topic and show that the deeper theory of countable Abelian groups requires set existence axioms which are stronger than those of ACA_0.

We begin by discussing torsion subgroups.

DEFINITION III.6.1 (countable Abelian groups). The following definitions are made in RCA_0. A *countable Abelian group* A consists of a set $|A| \subseteq \mathbb{N}$ together with a binary operation $+_A : |A| \times |A| \to A$ and a unary operation $-_A : |A| \to |A|$ and a distinguished element $0_A \in |A|$ such that the system $|A|, +_A, -_A, 0_A$ obeys the usual Abelian group axioms, e.g., $\forall x \, (x + (-x) = 0)$ and $\forall x \, \forall y \, (x + y = y + x)$. For notational convenience we write $|A|$ as A. By primitive recursion define $f : \mathbb{N} \times A \to A$ by $f(0, a) = 0$, $f(n + 1, a) = f(n, a) + a$, and put $na = f(n, a)$. Thus $na = a + \cdots + a$ where the summation is repeated n times. A *torsion element* of A is an element $a \in A$ such that $\exists n \, (n \geq 1 \wedge na = 0)$.

THEOREM III.6.2 (torsion subgroup). ACA_0 *is equivalent over* RCA_0 *to the assertion that every countable Abelian group has a subgroup consisting of the torsion elements.*

PROOF. If A is a countable Abelian group, we can use arithmetical comprehension to form the set T consisting of all $a \in A$ such that $\exists n \, (n \geq 1 \wedge na = 0)$. It is then easy to see that T is a subgroup of A.

For the converse, assume that every countable Abelian group has a subgroup consisting of the torsion elements. Instead of proving ACA_0 we shall prove the equivalent assertion III.1.3.3. Let $f : \mathbb{N} \to \mathbb{N}$ be a given one-to-one function. Form a countable Abelian group A given by generators x_i, $i \in \mathbb{N}$, and relations $(2m + 1)x_{f(m)} = 0$, $m \in \mathbb{N}$. The elements of A are finite formal sums $\sum n_i x_i$ where $n_i \in \mathbb{Z}$ and $\forall m \, (m < |n_i| \to i \neq f(m))$. This set of formal sums exists by Σ_0^0 comprehension. Now by assumption let T be the subgroup of A consisting of the torsion elements. By Δ_1^0 comprehension let X be the set of all i such that $x_i \in T$. Then clearly $\forall i \, (i \in X \leftrightarrow \exists m \, (f(m) = i))$. By lemma III.1.3 this gives ACA_0. The proof of theorem III.6.2 is complete.   □

Next we turn to a discussion of divisible closures.

DEFINITION III.6.3 (divisible closure). The following definitions are made in RCA_0. Let D be a countable Abelian group. We say that D is *divisible* if for all $d \in D$ and all $n \geq 1$ there exists $c \in D$ such that $nc = d$. Given a countable Abelian group A, a *divisible closure* of A is a countable divisible Abelian group D together with a monomorphism

$h \colon A \to D$ such that for all nonzero $d \in D$ there exists $n \in \mathbb{N}$ such that $nd = h(a)$ for some nonzero $a \in A$.

We shall show that the existence of divisible closures is provable in RCA_0 but the uniqueness requires ACA_0.

THEOREM III.6.4 (existence of divisible closure). *It is provable in RCA_0 that every countable Abelian group A has a divisible closure.*

PROOF. Let A be a countable Abelian group. We may assume that $A = C/K$ where C is a countable free Abelian group and K is a subgroup of C. Since C is a direct sum of countably many copies of $\mathbb{Z}$, we may assume that $C \subseteq D$ where D is a direct sum of countably many copies of $\mathbb{Q}$. Thus D is a divisible closure of C.

Let us say that a finite set $X \subseteq D$ is *good* if $(X) \cap C \subseteq K$, where (X) is the subgroup of D generated by X. We claim that the set of all (codes for) good finite subsets of D exists. To see this, let $X = \{b_i : i < k\}$ and let m_i be the least $m \geq 1$ such that $mb_i \in C$. For X to be good it is necessary that $m_i b_i \in K$ and in this case $\sum n_i b_i \in K$ if and only if $\sum r_i b_i \in K$, where r_i is the residue of n_i modulo m_i. Thus to determine whether X is good we need only examine finitely many elements of (X), namely the elements $\sum r_i b_i$ where $0 \leq r_i < m_i$, $i < k$. Our claim follows by Δ_1^0 comprehension.

Let $\langle d_i : i \in \mathbb{N} \rangle$ be a one-to-one enumeration of the elements of D. Define $f \colon \mathbb{N} \to \{0, 1\}$ by primitive recursion putting $f(j) = 1$ if and only if $\{d_i : i < j \wedge f(i) = 1\} \cup \{d_j\}$ is good, $f(j) = 0$ otherwise. Let L be the set of all $d_i \in D$ such that $f(i) = 1$. Thus L is a subgroup of D and $L \cap C = K$. Putting $B = D/L$ we see that B is divisible and there is a canonical monomorphism of $A = C/K$ into B. Also, by the construction of L, for any $b \in D \setminus L$ there exists $n \in \mathbb{N}$ such that $nb + d \in C \setminus K$ for some $d \in L$. Hence $B = D/L$ is a divisible closure of A. This completes the proof of theorem III.6.4.   □

Before discussing uniqueness of divisible closure, let us mention one more concept. A countable Abelian group D is said to be *injective* if, for any homomorphism $h \colon A \to D$ and monomorphism $f \colon A \to B$, where A and B are countable Abelian groups, there exists a homomorphism $h' \colon B \to D$ such that $h'(f(a)) = h(a)$ for all $a \in A$. In RCA_0 we can easily prove that injectivity of D implies divisibility of D. (Consider homomorphisms from $\mathbb{Z}$ into D and their extensions to $\mathbb{Q}$.) The proof that divisibility implies injectivity requires ACA_0, as we shall now show.

THEOREM III.6.5 (uniqueness of divisible closure). *The following statements are pairwise equivalent over RCA_0.*

1. ACA_0.
2. *Every countable divisible Abelian group is injective.*
3. *The divisible closure of a countable Abelian group is unique.*

PROOF. We begin by proving that 1 implies 2. Reasoning in ACA_0, let D be a countable divisible Abelian group and let $h \colon A \to D$ be given where A is a subgroup of a countable Abelian group B. For any $b \in B$ let (b) be the subgroup of B generated by b. Let $\langle b_n \colon n \in \mathbb{N} \rangle$ be an enumeration of the elements of B and for each n let A_n be the subgroup of B generated by $A \cup \{b_0, \ldots, b_{n-1}\}$. We extend h to B by stages. Assume that we have already extended h to A_n. If $(b_n) \cap A_n = (0)$ define $h(b_n) = 0$. If $(b_n) \cap A_n \neq (0)$ let k_n be the least $k \geq 1$ such that $k b_n \in A_n$. Select $d \in D$ such that $k_n d = h(k_n b_n)$ and define $h(b_n) = d$. This gives a homomorphism of A_{n+1} into D since each element of A_{n+1} can be written uniquely in the form $a + j b_n$, $a \in A_n$, $0 \leq j < k_n$. Finally we extend h to all of B. Thus D is injective. This proves the implication from 1 to 2.

Next we prove that 2 implies 3. Reasoning in RCA_0, assume 2 and let $h_i \colon A \to D_i$, $i = 1, 2$, be divisible closures of a countable Abelian group A. By injectivity of D_2 let $h \colon D_1 \to D_2$ be such that $h(h_1(a)) = h_2(a)$ for all $a \in A$. Given $d_1 \in D_1$, $d_1 \neq 0$, let $a \in A$ be such that $a \neq 0$ and $h_1(a) = n d_1$ for some $n \in \mathbb{N}$. Then $n h(d_1) = h(n d_1) = h(h_1(a)) = h_2(a) \neq 0$ so $h(d_1) \neq 0$. Thus $h \colon D_1 \to D_2$ is a monomorphism. By injectivity of D_1 let $g \colon D_2 \to D_1$ be such that $g(h(d)) = d$ for all $d \in D_1$. Clearly $g \colon D_2 \to D_1$ is an epimorphism. Given $d_2 \in D_2$, $d_2 \neq 0$, let $a \in A$ be such that $a \neq 0$ and $h_2(a) = n d_2$ for some $n \in \mathbb{N}$. Then $n g(d_2) = g(n d_2) = g(h_2(a)) = g(h(h_1(a))) = h_1(a) \neq 0$ so $g(d_2) \neq 0$. Thus $g \colon D_2 \to D_1$ is a monomorphism and therefore an isomorphism of D_2 onto D_1. Moreover $g(h_2(a)) = g(h(h_1(a))) = h_1(a)$ for all $a \in A$. Thus the two divisible closures $h_i \colon A \to D_i$, $i = 1, 2$, are isomorphic over A. This proves the implication from 2 to 3.

It remains to prove that 3 implies 1. We reason in RCA_0. Assume 3. Instead of proving ACA_0 directly we shall prove the equivalent statement III.1.3.3. Let $f \colon \mathbb{N} \to \mathbb{N}$ be a given one-to-one function. Let $\langle p_k \colon k \in \mathbb{N} \rangle$ be the enumeration of the rational primes in increasing order, i.e., $p_0 = 2$, $p_1 = 3$, $p_2 = 5$, $\ldots$. Let A be the countable Abelian group given by generators x, y_{ij}, z_{ij}, $i \in \mathbb{N}$, $j \in \mathbb{N}$, and relations $x = y_{i0} = z_{i0}$, $y_{ij} = p_{f(i)} y_{i,j+1}$, $z_{ij} = p_{f(i)} z_{i,j+1}$. The elements of A may be described as finite formal sums

$$kx + \sum m_{ij} y_{ij} + \sum n_{ij} z_{ij} \tag{6}$$

where $k \in \mathbb{Z}$, $0 \leq m_{ij} < p_{f(i)}$, $0 \leq n_{ij} < p_{f(i)}$, $j \geq 1$. Let D_0 be the subgroup of A generated by the elements $d_{ij} = y_{ij} - z_{ij}$. D_0 exists by Δ_1^0 comprehension since (6) belongs to D_0 if and only if $kx + \sum (m_{ij} + n_{ij}) y_{ij} = 0$. Note also that $A = A_1 \oplus D_0 = A_2 \oplus D_0$ where A_1 and A_2 are the subgroups of A generated by the elements y_{ij} and z_{ij} respectively.

We claim that D_0 is divisible. To see this, let p be a prime. If $p = p_{f(i)}$ then $d_{ij} = p d_{i,j+1}$. If $p \neq p_{f(i)}$ let $m, n \in \mathbb{Z}$ be such that $mp + n p_{f(i)}^j = 1$.

Then $d_{ij} = (mp + np^j_{f(i)})d_{ij} = mpd_{ij}$. So d_{ij} is divisible by p for all primes p. By an easy application of Σ^0_1 induction it follows that D_0 is divisible.

Put $D = \mathbb{Q} \oplus D_0$ and define monomorphisms $h_1, h_2\colon A \to D$ by $h_1(y_{ij}) = h_2(z_{ij}) = (p^{-j}_{f(i)}, 0)$, $h_1(z_{ij}) = h_2(y_{ij}) = (p^{-j}_{f(i)}, d_{ij})$. By the previous claim, $h_i\colon A \to D$, $i = 1, 2$ are divisible closures of A. By 3 let $h\colon D \to D$ be an automorphism of D such that $h(h_1(a)) = h_2(a)$ for all $a \in A$. By Δ^0_1 comprehension let X be the set of all k such that $h((p_k^{-1}, 0)) \neq (p_k^{-1}, 0)$. We claim that $\forall k \, (k \in X \leftrightarrow \exists i \, (f(i) = k))$. If $k = f(i)$ then we have $h((p_k^{-1}, 0)) = h(h_1(y_{i1})) = h_2(y_{i1}) = (p_k^{-1}, d_{i1}) \neq (p_k^{-1}, 0)$ so $k \in X$. If $k \neq f(i)$ for all i, then we have $p_k h((p_k^{-1}, 0)) = h((1, 0)) = h(h_1(y_{i0})) = h_2(y_{i0}) = (1, 0)$ so $h((p_k^{-1}, 0)) = (p_k^{-1}, 0)$ since D_0 has no p_k-torsion. Thus $k \notin X$ in this case. Our claim is proved. By lemma III.1.3 this gives ACA_0. The proof of the theorem is complete. $\square$

REMARK III.6.6 (strong divisible closure). Let A be a countable Abelian group. A *strong divisible closure* of A is a divisible closure $h\colon A \to D$ such that h is an isomorphism of A onto a subgroup of D. Solomon [251, theorem 6.21] has shown that ACA_0 is equivalent over RCA_0 to the statement that every countable Abelian group has a strong divisible closure.

Notes for §III.6. The main results of this section are from Friedman/Simpson/Smith [78]. For general information on Abelian groups, see Fuchs [83] or Kaplansky [136].

III.7. König's Lemma and Ramsey's Theorem

In this section we consider two basic results of infinitary combinatorics, König's lemma and Ramsey's theorem. We show that these results are provable in ACA_0. We also obtain reversals by showing that each of the two results is equivalent to ACA_0 over RCA_0.

We first discuss König's lemma. It is important to distinguish between König's lemma and what we shall later call weak König's lemma. *König's lemma* says that every infinite, finitely branching tree has a path. *Weak König's lemma* makes this assertion only for trees of sequences of 0's and 1's. Weak König's lemma is very important and will be discussed throughly in the next chapter. The discussion here in chapter III refers only to the full König's lemma.

DEFINITION III.7.1 (König's lemma). The following definitions are made in RCA_0. A *tree* is a set $T \subseteq \mathbb{N}^{<\mathbb{N}}$ which is closed under initial segment, i.e., $\forall \sigma \, \forall \tau \, ((\sigma \in \mathbb{N}^{<\mathbb{N}} \wedge \sigma \subseteq \tau \wedge \tau \in T) \to \sigma \in T)$. We say that T is *finitely branching* if each element of T has only finitely many

immediate successors, i.e., $\forall \sigma\, (\sigma \in T \to \exists n\, \forall m\, (\sigma^\frown \langle m \rangle \in T \to m < n))$.
A *path* through T is a function $g \colon \mathbb{N} \to \mathbb{N}$ such that $g[n] \in T$ for all
$n \in \mathbb{N}$. Here we are using the initial sequence notation

$$g[n] = \langle g(0), g(1), \ldots, g(n-1) \rangle.$$

König's lemma is the assertion that every infinite, finitely branching tree
T has at least one path.

Theorem III.7.2. *The following assertions are pairwise equivalent over*
RCA_0.

1. ACA_0.
2. *König's lemma.*
3. *König's lemma restricted to trees $T \subseteq \mathbb{N}^{<\mathbb{N}}$ such that $\forall \sigma\, (\sigma \in T \to \sigma$*
   has only at most two immediate successors in T).

Proof. We first prove König's lemma from ACA_0. Let $T \subseteq \mathbb{N}^{<\mathbb{N}}$ be an
infinite, finitely branching tree. By arithmetical comprehension let T^* be
the set of all $\tau \in T$ such that there exist infinitely many $\sigma \in T$ such that
$\sigma \supseteq \tau$. Since T is infinite, the empty sequence $\langle\rangle$ belongs to T^*. Since
T is finitely branching, each $\tau \in T^*$ has at least one immediate successor
$\tau^\frown \langle n \rangle \in T^*$. Thus we may use primitive recursion to define a path g by
$g(k) =$ least n such that $g[k]^\frown \langle n \rangle \in T^*$. Thus $g[k]$ is an initial sequence
of g of length k. This proves that 1 implies 2.

The implication from 2 to 3 is trivial, so it remains to prove that 3 im-
plies 1. We reason in RCA_0. Assume 3 and let $f \colon \mathbb{N} \to \mathbb{N}$ be one-to-
one. By lemma III.1.3 it suffices to prove that the range of f exists, i.e.,
$\exists X\, \forall n\, (n \in X \leftrightarrow \exists m\, (f(m) = n))$. Define a tree $T \subseteq \mathbb{N}^{<\mathbb{N}}$ by putting
$\tau \in T$ if and only if

$$(\forall m < \mathrm{lh}(\tau))\, (\forall n < \mathrm{lh}(\tau))\, (f(m) = n \leftrightarrow \tau(n) = m+1) \qquad (7)$$

and

$$(\forall n < \mathrm{lh}(\tau))\, (\tau(n) > 0 \to f(\tau(n) - 1) = n). \qquad (8)$$

Clearly T exists by Σ^0_0 comprehension. If $\sigma \in T$ then σ has at most two
immediate successors in T, since by (8) the only possiblities are $\sigma^\frown \langle 0 \rangle$
and $\sigma^\frown \langle m+1 \rangle$ where $f(m) = \mathrm{lh}(\sigma)$. To see that T is infinite, let $k \in \mathbb{N}$
be given. By bounded Σ^0_1 comprehension (theorem II.3.9), let Y be the
set of all $n < k$ such that $\exists m\, f(m) = n$. Define $\sigma \in \mathbb{N}^{<\mathbb{N}}$, $\mathrm{lh}(\sigma) = k$ by
putting

$$\sigma(n) = \begin{cases} 0 & \text{if } n \notin Y \\ m+1 & \text{if } n \in Y \wedge f(m) = n \end{cases}$$

for all $n < k$. It is easy to check that $\sigma \in T$. This shows that T is
infinite. Hence by 3 there exists a path g though T. From (7) it is clear
that $\forall m\, \forall n\, (f(m) = n \leftrightarrow g(n) = m+1)$. By Δ^0_1 comprehension let X be

the set of all n such that $g(n) > 0$. Then $\forall n \, (\exists m \, (f(m) = n) \leftrightarrow n \in X)$. This completes the proof of theorem III.7.2.   □

We now turn to Ramsey's theorem.

DEFINITION III.7.3 (Ramsey's theorem). The following definitions are made in RCA$_0$. For any $X \subseteq \mathbb{N}$ and $k \in \mathbb{N}$, let $[X]^k$ be the set of all increasing sequences of length k of elements of X. In symbols, $s \in [X]^k$ if and only if $s \in \mathbb{N}^k$ and $(\forall j < k) \, (s(j) \in X \wedge (\forall i < j) \, (s(i) < s(j)))$. By RT$(k)$, i.e., *Ramsey's theorem for exponent k*, we mean the assertion that for all $l \in \mathbb{N}$ and all $f : [\mathbb{N}]^k \to \{0, 1, \ldots, l - 1\}$, there exist $i < l$ and an infinite set $X \subseteq \mathbb{N}$ such that $f(m_1, \ldots, m_k) = i$ for all $\langle m_1, \ldots, m_k \rangle \in [X]^k$.

The following lemma implies that for each $k \in \omega$, RT(k) is provable in ACA$_0$.

LEMMA III.7.4. ACA$_0$ *proves* RT(0) *and*

$$\forall k \, (\text{RT}(k) \to \text{RT}(k + 1)).$$

PROOF. We reason in ACA$_0$ and imitate a popular proof of Ramsey's theorem based on König's lemma. (Ramsey's original proof is simpler but apparently cannot be carried out in ACA$_0$.)

RT(0) is trivial. Assume RT(k) and let

$$f : [\mathbb{N}]^{k+1} \to \{0, 1, \ldots, l - 1\}$$

be given. Define a tree $T \subseteq \mathbb{N}^{<\mathbb{N}}$ by putting $t \in T$ if and only if, for all $n < \text{lh}(t)$, $t(n)$ is the least j such that

    (i) $t(m) < j$ for all $m < n$,

and

    (ii) $f(t(m_1), \ldots, t(m_k), j) = f(t(m_1), \ldots, t(m_k), t(m))$
       for all $m_1 < \cdots < m_k < m \le n$.

Clearly T is a tree and T exists by Σ^0_0 comprehension. Also T is finitely branching since, given $t \in T$ of length n, there are $\le l^{n^k}$ distinct j such that $t^\frown \langle j \rangle \in T$. Among combinatorists T is known as the *Erdős/Radó tree*.

We claim that for each j there exists $s \in T$ such that $s(n) = j$ for some $n < \text{lh}(s)$. To see this, fix j and let $t \in T$ be maximal such that $t(m) < j$ for all $m < \text{lh}(t)$, and $f(t(m_1), \ldots, t(m_k), t(m)) = f(t(m_1), \ldots, t(m_k), j)$ for all $m_1 < \cdots < m_k < m < \text{lh}(t)$. (There is at least one such t, namely $t = \langle \rangle$, the empty sequence.) Then clearly $t^\frown \langle j \rangle \in T$. This proves the claim.

The previous claim implies that T is infinite. Hence, by König's lemma in ACA$_0$ (theorem III.7.1), T has a path, call it g. From (i) we have that $m < n$ implies $g(m) < g(n)$. Define $f' : [\mathbb{N}]^k \to \{0, 1, \ldots, l - 1\}$ by $f'(m_1, \ldots, m_k) = f(g(m_1), \ldots, g(m_k), g(m))$ where $m_1 < \cdots < m_k < m$; by (ii) this does not depend on the choice of m. Using RT(k) let

$i < l$ and $X' \subseteq \mathbb{N}$ be such that X' is infinite and $f'(m_1, \ldots, m_k) = i$ for all $\langle m_1, \ldots, m_k \rangle \in [X']^k$. Then clearly $f(m_1, \ldots, m_k, m) = i$ for all $\langle m_1, \ldots, m_k, m \rangle \in [X]^{k+1}$, where X is the set of all $g(m)$, $m \in X'$. This proves $\mathrm{RT}(k+1)$. The proof of the lemma is complete. $\qquad \square$

LEMMA III.7.5. *It is provable in* RCA_0 *that* $\mathrm{RT}(3)$ *implies* ACA_0.

PROOF. Assume $\mathrm{RT}(3)$. By theorem III.1.3 it suffices to prove Σ_1^0 comprehension. Given a Σ_1^0 formula $\varphi(m)$ we want to prove $\exists Z \, \forall m \, (m \in Z \leftrightarrow \varphi(m))$. Let $\varphi(m) \equiv \exists n \, \theta(m, n)$ where $\theta(m, n)$ is Σ_0^0. Define $f : [\mathbb{N}]^3 \to \{0, 1\}$ by putting $f(a, b, c) = 1$ if

$$(\forall m < a) \, ((\exists n < b) \, \theta(m, n) \leftrightarrow (\exists n < c) \, \theta(m, n)), \qquad (9)$$

$f(a, b, c) = 0$ otherwise. By $\mathrm{RT}(3)$ let $i < 2$ and $X \subseteq \mathbb{N}$ be such that X is infinite and $f(a, b, c) = i$ for all $\langle a, b, c \rangle \in [X]^3$.

We claim that $i = 1$. It suffices to show that $f(a, b, c) = 1$ for at least one 3-tuple $\langle a, b, c \rangle \in [X]^3$. Let a be any element of X. By bounded Σ_1^0 comprehension (theorem II.3.9), let Y be the set of all $m < a$ such that $\exists n \, \theta(m, n)$. By Σ_1^0 induction we can prove that $\forall j \, \exists k \, (\forall m < j) \, (m \in Y \to (\exists n < k)\theta(m, n))$. In particular, taking $j = a$, we find that there exists k such that $\forall m \, (m \in Y \to (\exists n < k) \, \theta(m, n))$. Since X is infinite there exist $b \in X$, $c \in X$ such that $a < b < c$ and $k \le b$. Thus $\langle a, b, c \rangle \in [X]^3$ and (9) holds. Hence $f(a, b, c) = 1$. This proves the claim.

Since $i = 1$ and X is infinite, we have $\exists n \, \theta(m, n)$ if and only if $\forall a \, \forall b \, ((a \in X \wedge b \in X \wedge m < a < b) \to (\exists n < b) \theta(m, n))$. Hence by Δ_1^0 comprehension there exist Z such that $\forall m \, (m \in Z \leftrightarrow \exists n \, \theta(m, n))$. This completes the proof. $\qquad \square$

THEOREM III.7.6. *Over* RCA_0, ACA_0 *is equivalent to* $\mathrm{RT}(3)$. (*Here we could replace* $\mathrm{RT}(3)$ *by any* $\mathrm{RT}(k)$, $k \ge 3$, $k \in \omega$.)

PROOF. Immediate from lemmas III.7.4 and III.7.5. $\qquad \square$

REMARKS III.7.7. (1) The case $k = 2$ is anomalous. On the one hand, it is known that WKL_0 does not prove $\mathrm{RT}(2)$. In fact, there exists an ω-model of WKL_0 in which $\mathrm{RT}(2)$ fails. On the other hand, there exists an ω-model M of $\mathrm{WKL}_0 + \mathrm{RT}(2)$ which does not contain $\emptyset^{(1)}$, hence ACA_0 fails in M. For bibliographical references, see the notes at the end of this section. (2) It is known that $\forall k \, \mathrm{RT}(k)$ is not provable in ACA_0. However, by lemma III.7.4, $\forall k \, \mathrm{RT}(k)$ is provable from ACA_0 plus Π_2^1 induction.

We have now completed our discussion of König's lemma and Ramsey's theorem. We end this section with a brief discussion of the Radó selection lemma.

The *Radó selection lemma* is a well known combinatorial principle which plays an important role in transversal theory. Its general statement is as follows. Let X be an arbitrary set and let $\mathcal{F}$ be a family of functions such that $(\forall f \in \mathcal{F})\,(\mathrm{dom}(f) \subseteq X)$ and $(\forall$ finite $X_0 \subseteq X)\,(\exists f \in \mathcal{F})\,(X_0 \subseteq \mathrm{dom}(f))$. Assume that $\forall x \in X\,(\{f(x)\colon f \in \mathcal{F} \wedge x \in \mathrm{dom}(f)\}$ is finite). Then there exists a function F such that $\mathrm{dom}(F) = X$ and $(\forall$ finite $X_0 \subseteq X)\,(\exists f \in \mathcal{F})\,(X_0 \subseteq \mathrm{dom}(f) \wedge f{\restriction}X_0 = F{\restriction}X_0)$.

We consider two versions of the special case of the Radó selection lemma in which the underlying set X is countable. For convenience we take $X = \mathbb{N}$.

THEOREM III.7.8 (Radó selection lemma). *The following assertions are pairwise equivalent over* RCA$_0$.

1. ACA$_0$.
2. (countable Radó lemma, strong version) *Let $\langle f_i \colon i \in \mathbb{N}\rangle$ be a sequence of partially defined functions from $\mathbb{N}$ into $\mathbb{N}$. Assume that $(\forall$ finite $X \subseteq \mathbb{N})\,\exists i\,(X \subseteq \mathrm{dom}(f_i))$ and that $\forall m\, \exists n\, \forall i\,(m \in \mathrm{dom}(f_i) \to f_i(m) < n)$. Then there exists $f\colon \mathbb{N} \to \mathbb{N}$ such that $(\forall$ finite $X \subseteq \mathbb{N})\,\exists i\,(X \subseteq \mathrm{dom}(f_i) \wedge f{\restriction}X = f_i{\restriction}X)$.*
3. (countable Radó lemma, weak version) *Given a sequence of finite functions $\langle f_n \colon n \in \mathbb{N}\rangle$, $f_n \colon \{0, 1, \ldots, n\} \to \{0, 1\}$, there exists $f\colon \mathbb{N} \to \{0, 1\}$ such that $\forall m\, \exists n\,(m \le n \wedge f{\restriction}\{0, 1, \ldots, m\} = f_n{\restriction}\{0, 1, \ldots, m\})$.*

PROOF. The proof is left as an exercise for the reader. $\square$

Notes for §III.7. The original source for König's lemma is König [147]. Theorem III.7.2 has been stated without proof by Friedman [68, 69]. For a thorough discussion of Ramsey's theorem, including a facsimile of Ramsey's original proof, see Graham/Rothschild/Spencer [98]. Theorem III.7.6 is due to Simpson (unpublished) and is closely related to earlier results of Jockusch [133] and Paris [200]. The existence of an ω-model of WKL$_0$ in which RT(2) fails is due to Hirst [117, theorem 6.10] using a result of Jockusch [133, theorem 3.1]. The existence of an ω-model of WKL$_0$ + RT(2) in which ACA$_0$ fails is due to Seetapun; see Hummel [125]. Optimal results on the strength of RCA$_0$ + RT(2) are in Cholak/Jockusch/Slaman [36]. For more information on the Radó selection lemma and its role in transversal theory, see Mirsky [190]. Theorem III.7.8 is due jointly to Feng and Simpson; see Hirst [117, theorem 3.30].

III.8. Conclusions

We began this chapter by defining ACA$_0$ to consist of RCA$_0$ plus arithmetical comprehension. We then demonstrated that ACA$_0$ is considerably stronger than RCA$_0$ from the viewpoint of mathematical practice. Indeed, several mathematical theorems are equivalent over RCA$_0$ to ACA$_0$. Among these are: the least upper bound principle for sequences of real numbers (§III.2); sequential compactness of the closed unit interval $[0, 1]$

and of compact metric spaces ($\S$III.2); existence of the strong algebraic closure of an arbitrary countable field ($\S$III.3); the fact that every countable vector space over $\mathbb{Q}$ has a basis ($\S$III.4); the fact that every countable commutative ring has a maximal ideal ($\S$III.5); uniqueness of the divisible closure of an arbitrary countable Abelian group ($\S$III.6); König's lemma for subtrees of $\mathbb{N}^{<\mathbb{N}}$, and Ramsey's theorem for colorings of $[\mathbb{N}]^3$ ($\S$III.7). These equivalences provide our first illustrations of the theme of Reverse Mathematics.

Chapter IV

WEAK KÖNIG'S LEMMA

IV.1. The Heine/Borel Covering Lemma

The purpose of this chapter is to study a certain subsystem of second order arithmetic known as $\mathsf{WKL_0}$. In order to define $\mathsf{WKL_0}$, let $2^{<\mathbb{N}}$ (also denoted $\{0,1\}^{<\mathbb{N}}$) be the set of all (codes for) finite sequences of 0's and 1's, i.e., the set of all $s \in \mathbb{N}^{<\mathbb{N}}$ such that $\forall i\,(i < \mathrm{lh}(s) \rightarrow s(i) < 2)$. *Weak König's lemma* is the statement that every infinite tree $T \subseteq 2^{<\mathbb{N}}$ has a path. The axioms of $\mathsf{WKL_0}$ are those of $\mathsf{RCA_0}$ plus weak König's lemma.

By theorem III.7.2, the theorems of $\mathsf{WKL_0}$ are included in those of $\mathsf{ACA_0}$. It will become clear in §VIII.2 that this inclusion is strict. Hence by theorem III.2.2 it follows that $\mathsf{WKL_0}$ is not strong enough to prove the Bolzano/Weierstraß theorem, i.e., sequential compactness of the closed unit interval $0 \leq x \leq 1$. However, we shall show in this section that $\mathsf{WKL_0}$ is strong enough to prove the *Heine/Borel theorem*: Every covering of the closed unit interval $0 \leq x \leq 1$ by a sequence of open intervals has a finite subcovering. We shall then generalize this to compact metric spaces.

Also in this section we shall obtain a reversal showing that the Heine/ Borel theorem is in fact equivalent to $\mathsf{WKL_0}$ over $\mathsf{RCA_0}$. In subsequent sections of this chapter, we shall show that $\mathsf{WKL_0}$ is equivalent over $\mathsf{RCA_0}$ to several other ordinary mathematical theorems. Among those theorems are: the Gödel completeness theorem (§IV.3); the theorem that every continuous function on the closed unit interval $0 \leq x \leq 1$ attains a maximum value (§IV.2); the uniqueness theorem for countable algebraic closures (§IV.5); a theorem of Artin and Schreier concerning orderability of (countable) fields (§IV.4); the theorem that every countable commutative ring has a prime ideal (§IV.6); Brouwer's fixed point theorem (§IV.7); Peano's existence theorem for solutions of ordinary differential equations (§IV.8); and the Hahn/Banach theorem for separable Banach spaces (§IV.9). These results provide further illustrations of the theme of Reverse Mathematics.

Lemma IV.1.1 (Heine/Borel theorem for $[0, 1]$). *The following is provable in* WKL$_0$. *Given sequences of real numbers* $c_i, d_i, i \in \mathbb{N}$, *if*

$$\forall x \, (0 \le x \le 1 \to \exists i \, (c_i < x < d_i)),$$

then

$$\exists n \, \forall x \, (0 \le x \le 1 \to \exists i \le n \, (c_i < x < d_i)).$$

Proof. Reasoning in WKL$_0$, we shall first prove the theorem under the assumption that $\langle c_i : i \in \mathbb{N} \rangle$ and $\langle d_i : i \in \mathbb{N} \rangle$ are sequences of rational numbers.

For each $s \in 2^{<\mathbb{N}}$ put

$$a_s = \sum_{i < \mathrm{lh}(s)} \frac{s(i)}{2^{i+1}}$$

and

$$b_s = a_s + \frac{1}{2^{\mathrm{lh}(s)}}.$$

Thus for each $n \in \mathbb{N}$ we have partitioned the unit interval $0 \le x \le 1$ into 2^n subintervals of length 2^{-n}, namely $a_s \le x \le b_s$, $s \in 2^{<\mathbb{N}}$, $\mathrm{lh}(s) = n$. Form a tree $T \subseteq 2^{<\mathbb{N}}$ by putting $s \in T$ if and only if $\neg \exists i \le \mathrm{lh}(s) \, (c_i < a_s < b_s < d_i)$. T exists by Σ_0^0 comprehension since $c_i, d_i, a_s, b_s \in \mathbb{Q}$.

Assuming that $\forall x \, (0 \le x \le 1 \to \exists i \, (c_i < x < d_i))$, we claim that T has no path. To see this let $f : \mathbb{N} \to \{0, 1\}$ be given and put

$$x = \sum_{j=0}^{\infty} \frac{f(j)}{2^{j+1}},$$

i.e., the unique x such that $a_{f[n]} \le x \le b_{f[n]}$ for all $n \in \mathbb{N}$. Let i be such that $c_i < x < d_i$ and let n be so large that $n \ge i$ and $c_i < a_{f[n]} < b_{f[n]} < d_i$. Then $f[n] \notin T$ which proves the claim.

By weak König's lemma it follows that T is finite. Let n be such that $\forall s \, (s \in T \to \mathrm{lh}(s) < n)$. Then $\forall s \, (\mathrm{lh}(s) = n \to \exists i \le n \, (c_i < a_s < b_s < d_i))$. Hence $\forall x \, (0 \le x \le 1 \to \exists i \le n \, (c_i < x < d_i))$.

This proves the theorem under the assumption that $c_i, d_i \in \mathbb{Q}$. In general, consider the Σ_1^0 formula $\varphi(q, r)$ which says that $q \in \mathbb{Q} \wedge r \in \mathbb{Q} \wedge \exists i \, (c_i < q < r < d_i)$. By lemma II.3.7 there exists a function $f : \mathbb{N} \to \mathbb{Q} \times \mathbb{Q}$ such that $\forall q \, \forall r \, (\varphi(q, r) \leftrightarrow \exists j \, (f(j) = (q, r)))$. Thus we may replace the sequence $\langle (c_i, d_i) : i \in \mathbb{N} \rangle$ by the sequence $\langle (q_j, r_j) : j \in \mathbb{N} \rangle$ where $(q_j, r_j) = f(j)$. This reduces the theorem to the special case which has already been proved. □

Theorem IV.1.2. WKL$_0$ *is equivalent over* RCA$_0$ *to the Heine/Borel theorem for* $[0, 1]$.

PROOF. The previous lemma shows that WKL_0 proves Heine/Borel for $[0, 1]$. Reasoning in RCA_0, assume Heine/Borel for $[0, 1]$. We shall make use of the Cantor middle-third set $C \subseteq [0, 1]$ consisting of all real numbers of the form

$$\sum_{i=0}^{\infty} \frac{2f(i)}{3^{i+1}}, \qquad f \in 2^{\mathbb{N}}.$$

The idea of the proof will be that paths through $2^{<\mathbb{N}}$ can be identified with elements of C, so Heine/Borel compactness of $2^{\mathbb{N}}$ follows from Heine/Borel compactness of the closed unit interval $0 \leq x \leq 1$.

For each $s \in 2^{<\mathbb{N}}$ put

$$a_s = \sum_{i<\text{lh}(s)} \frac{2s(i)}{3^{i+1}}$$

and

$$b_s = a_s + \frac{1}{3^{\text{lh}(s)}}.$$

Thus $a_{\langle\rangle} = 0$, $b_{\langle\rangle} = 1$, and the closed interval $a_{s^\frown\langle 0\rangle} \leq x \leq b_{s^\frown\langle 0\rangle}$ (respectively $a_{s^\frown\langle 1\rangle} \leq x \leq b_{s^\frown\langle 1\rangle}$) is the left third (respectively the right third) of the closed interval $a_s \leq x \leq b_s$. Thus for each $x \in C$ there is a unique $f : \mathbb{N} \to \{0, 1\}$ such that $a_{f[n]} \leq x \leq b_{f[n]}$ for all $n \in \mathbb{N}$. Also, if $0 \leq x \leq 1$ and $x \notin C$, then $b_{s^\frown\langle 0\rangle} < x < a_{s^\frown\langle 1\rangle}$ for a unique $s \in 2^{<\mathbb{N}}$.

We also put

$$a_s' = a_s - \frac{1}{3^{\text{lh}(s)+1}}$$

and

$$b_s' = b_s + \frac{1}{3^{\text{lh}(s)+1}}$$

Note that the open intervals $a_s' < x < b_s'$ and $a_t' < x < b_t'$ are disjoint unless $s \subseteq t$ or $t \subseteq s$.

Let $T \subseteq 2^{<\mathbb{N}}$ be a tree with no path. We shall use the Heine/Borel theorem to show that T is finite. Let $\widetilde{T}$ be the set of $u \in 2^{<\mathbb{N}}$ such that $u \notin T \wedge \forall t\,(t \subsetneqq u \to t \in T))$. Then the open intervals $a_u' < x < b_u'$, $u \in \widetilde{T}$, are pairwise disjoint and cover C. Hence the closed unit interval $0 \leq x \leq 1$ is covered by the open intervals

$$a_u' < x < b_u', \qquad u \in \widetilde{T}$$

and

$$b_{s^\frown\langle 0\rangle} < x < a_{s^\frown\langle 1\rangle}, \qquad s \in 2^{<\mathbb{N}}.$$

By the Heine/Borel theorem, this covering has a finite subcovering. But the intervals $b_{s^\frown\langle 0\rangle} < x < a_{s^\frown\langle 1\rangle}$ are disjoint from C and clearly C is not covered by any proper subset of the intervals $a'_u < x < b'_u$, $u \in \widetilde{T}$. Hence $\widetilde{T}$ is finite. This completes the proof. □

In the remainder of this section, we generalize lemma IV.1.1 to the case of compact metric spaces (as defined in §III.2). For this we need the following generalization of weak König's lemma.

DEFINITION IV.1.3 (bounded König's lemma). Within RCA_0, a tree $T \subseteq \mathbb{N}^{<\mathbb{N}}$ is said to be *bounded* if there exists a function $g: \mathbb{N} \to \mathbb{N}$ such that for all $\tau \in T$ and $m < \mathrm{lh}(\tau)$, $\tau(m) < g(m)$. *Bounded König's lemma* is the assertion that every bounded infinite tree $T \subseteq \mathbb{N}^{<\mathbb{N}}$ has a path.

LEMMA IV.1.4. *Weak König's lemma is equivalent over RCA_0 to bounded König's lemma.*

PROOF. Any subtree of $2^{<\mathbb{N}}$ is bounded by the constant function 2. Therefore bounded König's lemma trivially implies weak König's lemma. For the converse, let an infinite bounded tree $T \subseteq \mathbb{N}^{<\mathbb{N}}$ be given. Let $g: \mathbb{N} \to \mathbb{N}$ be a bounding function, i.e., $\tau(j) < g(j)$ for all $\tau \in T$, $j < \mathrm{lh}(\tau)$. Given $\tau \in T$, define $\tau^* \in 2^{<\mathbb{N}}$ by putting

$$\tau^*\left(\sum_{i=0}^{j-1} g(i) + k\right) = \begin{cases} 0 & \text{if } k < \tau(j), \\ 1 & \text{if } \tau(j) \leq k < g(j), \end{cases}$$

for all $j < \mathrm{lh}(\tau)$. Thus $\mathrm{lh}(\tau^*) = \sum_{i=0}^{m-1} g(i) \geq m$ where $m = \mathrm{lh}(\tau)$. By Δ_1^0 comprehension, let T^* be the set of all $\sigma \in 2^{<\mathbb{N}}$ such that $\sigma \subseteq \tau^*$ for some $\tau \in T$ such that $\tau \leq g[\mathrm{lh}(\sigma)]$. Thus T^* is an infinite subtree of $2^{<\mathbb{N}}$. By weak König's lemma, let $f^*: \mathbb{N} \to \{0, 1\}$ be a path through T^*. Define $f: \mathbb{N} \to \mathbb{N}$ by putting $f(j) = $ the least k such that $f^*(\sum_{i=0}^{j-1} g(i)+k) = 1$. Thus f is a path through T. This completes the proof. □

THEOREM IV.1.5 (Heine/Borel for compact metric spaces). *The following is provable in WKL_0. Let $\widehat{A}$ be a compact metric space. If $\langle U_j : j \in \mathbb{N} \rangle$ is a covering of $\widehat{A}$ by open sets, then there exists a finite subcovering $\langle U_j : j \leq l \rangle$, $l \in \mathbb{N}$.*

PROOF. We reason in WKL_0. Since $\widehat{A}$ is compact, let $\langle \langle x_{ik} : k \leq n_i \rangle : i \in \mathbb{N} \rangle$ be a sequence of finite sequences of points $x_{ik} \in \widehat{A}$ such that for all $y \in \widehat{A}$ and $i \in \mathbb{N}$ there exists $k \leq n_i$ such that $d(x_{ik}, y) < 2^{-i}$. We may safely assume that each U_j is a basic open ball $B(a_j, r_j)$ where $a_j \in A$, $r_j \in \mathbb{Q}^+$. For any two points $x, y \in \widehat{A}$, we use the notation $d(x, y) = \langle d(x, y)_k : k \in \mathbb{N} \rangle$. Here we are viewing the real number $d(x, y)$ as a sequence of rational numbers (definition II.4.4).

By Δ_1^0 comprehension, form a tree T consisting of all $\tau \in \mathbb{N}^{<\mathbb{N}}$ such that

$$\forall i < \mathrm{lh}(\tau)\,[\tau(i) \leq n_i] \quad \text{and}$$

$$\forall i, j, k < \mathrm{lh}(\tau)\,[d\,(x_{i,\tau(i)}, x_{j,\tau(j)})_k \leq 2^{-i} + 2^{-j} + 2^{-k}] \quad \text{and}$$

$$\forall i, j, k < \mathrm{lh}(\tau)\,[d\,(x_{i,\tau(i)}, a_j)_k \geq r_j - 2^{-i} - 2^{-k}].$$

Obviously T is a bounded tree, the bounding function g being given by $g(i) = n_i + 1$. If $f : \mathbb{N} \to \mathbb{N}$ were a path through T, there would be a point

$$x = \lim_i x_{i,f(i)} \in \widehat{A}$$

such that $d(x, a_j) \geq r_j$ for all $j \in \mathbb{N}$, i.e., x does not belong to $U_j = B(a_j, r_j)$ for any $j \in \mathbb{N}$. This shows that T has no path. If follows by bounded König's lemma in WKL_0 (lemma IV.1.4) that T is finite. Let l be the least integer such that T contains no sequence of length l. Then clearly $\langle B(a_j, r_j) : j < l \rangle$ covers $\widehat{A}$. This completes the proof. $\qquad\square$

For use in the next section, we mention the following generalization of theorem IV.1.5.

Theorem IV.1.6. *The following is provable in* WKL_0. *Let $\widehat{A}$ be a compact metric space. Let $\langle\langle U_{nj} : j \in \mathbb{N}\rangle : n \in \mathbb{N}\rangle$ be a sequence of coverings of $\widehat{A}$ by open sets. Then there exists a sequence of finite subcoverings $\langle\langle U_{nj} : j \leq l_n\rangle : n \in \mathbb{N}\rangle$.*

Proof. This is a straightforward adaptation of the proof of theorem IV.1.5. We define a sequence of bounded trees $\langle T_n : n \in \mathbb{N}\rangle$ and argue as before that each of the trees in the sequence is finite. $\qquad\square$

The following two theorems will be needed in §IV.7.

Theorem IV.1.7. *The following is provable in* WKL_0. *Let X be a compact metric space. If C denotes a (code for a) closed set in X, the assertion that $C \neq \emptyset$ (i.e., C is nonempty) is expressible by a Π_1^0 formula.*

Proof. We reason in WKL_0. Since X is compact, let $\langle\langle x_{ik} : k \leq n_i\rangle : i \in \mathbb{N}\rangle$ be a sequence of finite sequences of points in X such that for all $y \in X$ and $i \in \mathbb{N}$ there exists $k \leq n_i$ such that $d(x_{ik}, y) < 2^{-i}$. Thus for each $i \in \mathbb{N}$ we have

$$X = \bigcup_{k \leq n_i} B(x_{ik}, 2^{-i}).$$

Now let C be a closed set in X, and let U be a code for the open set $X \setminus C$. Recall from definition II.5.6 that U is actually a subset of $\mathbb{N} \times A \times \mathbb{Q}^+$. A point $x \in X$ belongs to (the open set coded by) U if and only if $d(a, x) < r$ for some $(m, a, r) \in U$. Thus we have

$$X \setminus C = \bigcup_{(m,a,r) \in U} B(a, r).$$

We claim that $C = \emptyset$ if and only if the following condition $(*)$ holds: there exist i and a finite sequence of triples $(m_k, a_k, r_k) \in U$, $k \leq n_i$, such that $d(x_{ik}, a_k) + 2^{-i} < r_k$ for all $k \leq n_i$. Since this condition is Σ^0_1, the claim will suffice to prove our theorem.

To prove the claim, assume first that $C = \emptyset$. Then we have

$$X = \bigcup_{(m,a,r) \in U} \bigcup_{0 < q < r} B(a, q)$$

where q ranges over $\mathbb{Q}^+$. Hence by the Heine/Borel property (theorem IV.1.5), there exists a finite sequence of triples $(m_l, a_l, r_l) \in U$, $l \in L$, and $q_l \in \mathbb{Q}^+$, $l \in L$, such that $0 < q_l < r_l$ for all $l \in L$, and

$$X = \bigcup_{l \in L} B(a_l, q_l).$$

Let $i \in \mathbb{N}$ be such that $2^{-i} < \min_{l \in L}(r_l - q_l)$. Then for each $k \leq n_i$ we have $x_{ik} \in B(a_{l_k}, q_{l_k})$ for some $l_k \in L$, hence $d(x_{ik}, a_{l_k}) < q_{l_k}$, hence $d(x_{ik}, a_{l_k}) + 2^{-i} < r_{l_k}$. Thus $(*)$ holds. Conversely, if $(*)$ holds, then for all $k \leq n_i$ we have $B(x_{ik}, 2^{-i}) \subseteq B(a_k, r_k)$, hence $X = \bigcup_{k \leq n_i} B(a_k, r_k) \subseteq U$, hence $C = \emptyset$. This completes the proof. $\square$

The following theorem is a kind of choice principle for points in compact sets.

THEOREM IV.1.8. *The following is provable in* WKL$_0$. *Let X be a compact metric space, and let C_j, $j \in \mathbb{N}$, be a sequence of (codes for) nonempty closed sets in X. Then there exists a sequence of points $x_j \in C_j$, $j \in \mathbb{N}$.*

PROOF. As in the proof of theorems IV.1.5 and IV.1.6, construct a sequence of infinite trees T_j, $j \in \mathbb{N}$, such that $\forall i < \text{lh}(\tau)\,[\tau(i) \leq n_i]$ for all $\tau \in T_j$, and such that any path g through T_j gives rise to a point

$$x = \lim_i x_{i,g(i)} \in C_j.$$

Let $T = \oplus_{j \in \mathbb{N}} T_j$ be the interleaved tree, defined by putting $\tau \in T$ if and only if $\forall j\,[\tau_j \in T_j]$, where $\tau_j(i) = \tau((i, j))$ for all $(i, j) < \text{lh}(\tau)$. We also require that $\tau(k) = 0$ for all $k < \text{lh}(\tau)$ not of the form (i, j). Then T is a bounded tree, the bounding function h being given by $h((i, j)) = n_i + 1$, $h(k) = 1$ for all k not of the form (i, j). In order to show that T is infinite, we prove that for all n there exists $\tau \in T$ of length n such that

$$\forall j\, \forall m\, (m \geq \text{lh}(\tau_j) \rightarrow \tau_j \text{ has an extension of length } m \text{ in } T_j).$$

This Π^0_1 statement is easily proved by Π^0_1 induction on n, using the fact that each of the T_j's is infinite.

Since T is infinite and bounded, it follows by bounded König's lemma in WKL$_0$ (lemma IV.1.4) that T has an infinite path f. Then for each j we have a path f_j through T_j given by $f_j(i) = f((i, j))$. Thus $x_j = \lim_i x_{i, f_j(i)}$ belongs to C_j. This completes the proof. $\square$

Notes for §IV.1. The formal system WKL_0 was first defined by Friedman [69]. Theorem IV.1.2 was announced by Friedman [69]. Theorem IV.1.5 and its proof are taken from Brown's thesis [24]. Theorem IV.1.7 is from Blass/Hirst/Simpson [21]. Theorem IV.1.8 is due to Simpson, unpublished.

IV.2. Properties of Continuous Functions

In this section we shall show that WKL_0 is just strong enough to prove several basic results concerning continuous functions of a real variable. We shall also generalize some of these results so as to apply to continuous functions on compact metric spaces.

DEFINITION IV.2.1 (modulus of uniform continuity). The following definition is made in RCA_0. Let $\widehat{A}$ and $\widehat{B}$ be complete separable metric spaces, and let F be a continuous function from $\widehat{A}$ into $\widehat{B}$ (see definition II.6.1). A *modulus of uniform continuity* for F is a function $h: \mathbb{N} \to \mathbb{N}$ such that for all $n \in \mathbb{N}$ and all x and y in $\widehat{A}$, if $F(x)$ and $F(y)$ are defined and $d(x, y) < 2^{-h(n)}$, then $d(F(x), F(y)) < 2^{-n}$.

THEOREM IV.2.2 (properties of continuous functions). *The following is provable in* WKL_0. *Let* $X = \widehat{A}$ *be a compact metric space. Let* C *be a closed set in* X, *and let* F *be a continuous function from* C *into a complete separable metric space* $Y = \widehat{B}$. *Then* F *has a modulus of uniform continuity on* C. *If in addition* $X = C$ *and* $Y = \mathbb{R}$, *then* F *attains a maximum value.*

PROOF. We reason in WKL_0. Let $\varphi(n, a, r)$ be a Σ^0_1 formula which says that $a \in A$, $r \in \mathbb{Q}^+$, and $\exists b \, \exists s \, ((a, 2r)F(b, s) \wedge s < 2^{-n-1})$. Since $F(x)$ is defined for all $x \in C$, we can easily show that for all $x \in C$ and $n \in \mathbb{N}$ there exist a, r such that $\varphi(n, a, r)$ holds and $d(x, a) < r$. By lemma II.3.7, let $\langle (a_{ni}, r_{ni}): i, n \in \mathbb{N} \rangle$ be such that

$$\forall n \, \forall a \, \forall r \, (\varphi(n, a, r) \leftrightarrow \exists i \, (a, r) = (a_{ni}, r_{ni})).$$

Thus $\langle \langle \text{B}(a_{ni}, r_{ni}): i \in \mathbb{N} \rangle: n \in \mathbb{N} \rangle$ is a sequence of open coverings of C. By theorem IV.1.6, let $\langle \langle \text{B}(a_{ni}, r_{ni}): i \leq k_n \rangle: n \in \mathbb{N} \rangle$ be a sequence of finite subcoverings. Define $h: \mathbb{N} \to \mathbb{N}$ by putting $h(n) = $ the least j such that $2^{-j} < \min\{r_{ni}: i \leq k_n\}$. If $x, y \in C$ and $d(x, y) < 2^{-h(n)}$, let $i \leq k_n$ be such that $x \in \text{B}(a_{ni}, r_{ni})$. Then $x, y \in \text{B}(a_{ni}, 2r_{ni})$, so $F(x)$, $F(y)$ both belong to the closure of $\text{B}(b, s)$ where $s < 2^{-n-1}$. Hence $d(F(x), F(y)) < 2^{-n}$. This proves that h is a modulus of uniform continuity for F on C.

Assume now that $X = C$ and $Y = \mathbb{R}$. It is straightforward to show that

$$\alpha = \lim_n \max\{F(a_{ni}): i \leq k_n\}$$

exists and is the least upper bound of all $F(x)$, $x \in X$. It remains to show that $F(x) = \alpha$ for some $x \in X$. Suppose not. Let $\varphi(a, r, b, s)$ be a Σ_1^0 formula saying that $(a, r)F(b, s)$ holds and $b + s < \alpha$. By lemma II.3.7, there is a sequence $\langle (a_i, r_i, b_i, s_i) : i \in \mathbb{N} \rangle$ such that

$$\forall a, r, b, s \; (\varphi(a, r, b, s) \leftrightarrow \exists i \; (a, r, b, s) = (a_i, r_i, b_i, s_i)).$$

Since $F(x) < \alpha$ for all $x \in X$, the sequence $\langle \mathrm{B}(a_i, r_i) : i \in \mathbb{N} \rangle$ is an open covering of X. By theorem IV.1.5, there is a finite subcovering $\langle \mathrm{B}(a_i, r_i) : i \leq k \rangle$. Put $\beta = \max\{b_i + s_i : i \leq k\}$. Then $\beta < \alpha$ and for all $x \in \widehat{A}$ we have $F(x) \leq \beta$, contradicting the definition of α.

This completes the proof of theorem IV.2.2.   □

We now turn to Reverse Mathematics. We show that weak König's lemma is needed to prove some basic properties of continuous functions. Among other things, the following theorem says that WKL$_0$ is equivalent (over RCA$_0$) to the assertion that every continuous, real-valued function on the closed unit interval attains a maximum value.

A continuous function F is said to be *uniformly continuous* if for all $\epsilon > 0$ there exists $\delta > 0$ such that if $d(x_1, x_2) < \delta$ and $F(x_1)$ and $F(x_2)$ are defined, then $d(F(x_1), F(x_2)) < \epsilon$.

THEOREM IV.2.3 (reversals). *The following assertions are pairwise equivalent over* RCA$_0$.

1. *Weak König's lemma.*
2. *Every continuous function on the closed interval $0 \leq x \leq 1$ is uniformly continuous.*
3. *Every continuous function on $0 \leq x \leq 1$ is bounded.*
4. *Every bounded, uniformly continuous function on $0 \leq x \leq 1$ has a supremum.*
5. *Every bounded, uniformly continuous function on $0 \leq x \leq 1$ which has a supremum, attains it.*

PROOF. The fact that WKL$_0$ proves assertions 2, 3, 4, and 5 follows immediately as a special case of theorem IV.2.2. (These are essentially the standard proofs based on the Heine/Borel theorem.)

It remains to show that $\neg$WKL$_0$ implies $\neg2$, $\neg3$, $\neg4$, $\neg5$. We reason in RCA$_0$. Assume $\neg$WKL$_0$ and let $T \subseteq 2^{<\mathbb{N}}$ be an infinite tree with no path. Let C, a_s, b_s, and $\widetilde{T}$ be as in the proof of theorem IV.1.2. Since T is a tree, the closed intervals

$$a_u \leq x \leq b_u, \qquad u \in \widetilde{T} \tag{10}$$

are pairwise disjoint, and since T has no paths, they cover C. Thus any element of $0 \leq x \leq 1$ which does not belong to (10) must lie in an open interval $b_v < x < a_w$, $v \in \widetilde{T}$, $w \in \widetilde{T}$, which is disjoint from (10).

We shall now construct a counterexample to 2, i.e., a continuous real-valued function $\phi(x)$, $0 \leq x \leq 1$, which is not uniformly continuous. If $a_u \leq x \leq b_u$ for some $u \in \widetilde{T}$, define $\phi(x) = \mathrm{lh}(u)$. Otherwise define $\phi(x)$ by piecewise linearity, i.e.,

$$\phi(x) = \phi(b_v) + \frac{x - b_v}{a_w - b_v}(\phi(a_w) - \phi(b_v))$$

on each open interval $b_v < x < a_w$, $v \in \widetilde{T}$, $w \in \widetilde{T}$ which is disjoint from (10). The corresponding continuous function code Φ can be constructed by the same method as in the proof of theorem II.6.5. Since $\widetilde{T}$ is infinite, $\phi(x)$ is unbounded on $0 \leq x \leq 1$ and hence not uniformly continuous there. Thus $\phi_2(x) = \phi(x)$ is a counterexample to both 2 and 3.

Our counterexamples to 4 and 5 will be similar. Since $\mathsf{WKL_0}$ fails, it follows by theorem III.7.2 that $\mathsf{ACA_0}$ fails. Hence by theorem III.2.2 there exists a bounded increasing sequence of rational numbers $c_0 < c_1 < \cdots < c_n < \cdots < 2$ which has no least upper bound. Define a continuous real-valued function $\phi_4(x)$, $0 \leq x \leq 1$, as follows. If $a_u \leq x \leq b_u$, $u \in \widetilde{T}$, put $\phi_4(x) = c_{\mathrm{lh}(u)}$, otherwise define $\phi_4(x)$ by piecewise linearity as before. Thus $\sup\{\phi_4(x) \colon 0 \leq x \leq 1\} = \sup\{c_n \colon n \in \mathbb{N}\}$ does not exist although $\phi_4(x)$ is uniformly continuous and $0 < \phi_4(x) < 2$ for all x, $0 \leq x \leq 1$. Thus ϕ_4 is a counterexample to 4.

Finally define $\phi_5(x)$, $0 \leq x \leq 1$, as follows. If $a_u \leq x \leq b_u$, $u \in \widetilde{T}$, put $\phi_5(x) = 1 - 2^{-\mathrm{lh}(u)}$, otherwise define $\phi_5(x)$ by piecewise linearity as before. Then ϕ_5 is uniformly continuous and, since $\widetilde{T}$ is infinite, $\sup\{\phi_5(x) \colon 0 \leq x \leq 1\} = 1$. However, this supremum is clearly never attained. Thus we have a counterexample to 5.

This completes the proof of theorem IV.2.3.   □

We shall now discuss the Weierstraß polynomial approximation theorem.

LEMMA IV.2.4 (Weierstraß approximation theorem). *The following is provable in* $\mathsf{RCA_0}$. *Let* $\phi(x)$ *be a continuous real-valued function defined on* $0 \leq x \leq 1$.

1. *If* $\phi(x)$ *is uniformly continuous, then for each* $\epsilon > 0$ *there exists a polynomial* $f(x) \in \mathbb{Q}[x]$ *such that* $|\phi(x) - f(x)| < \epsilon$ *for all* x, $0 \leq x \leq 1$.
2. *If* $\phi(x)$ *possesses a modulus of uniform continuity, then there exists a sequence of polynomials* $\langle f_n(x) \colon n \in \mathbb{N} \rangle$, $f_n(x) \in \mathbb{Q}[x]$, *such that* $|\phi(x) - f_n(x)| < 2^{-n}$ *for all* $n \in \mathbb{N}$ *and* $0 \leq x \leq 1$.

PROOF. Straightforward imitation of the usual "constructive" proof of the Weierstraß theorem. (For references, see the notes at the end of this section.)   □

THEOREM IV.2.5. *The following assertions are pairwise equivalent over* RCA$_0$.

1. *Weak König's lemma.*
2. *For every continuous real-valued function $\phi(x)$, $0 \leq x \leq 1$, there exists a polynomial $f(x)$ such that $|\phi(x) - f(x)| < 1$.*
3. *For every continuous real-valued function $\phi(x)$, $0 \leq x \leq 1$, there exists a sequence of polynomials $\langle f_n(x) : n \in \mathbb{N} \rangle$, $f_n(x) \in \mathbb{Q}[x]$, such that $|\phi(x) - f_n(x)| < 2^{-n}$ for all $n \in \mathbb{N}$ and $0 \leq x \leq 1$.*

PROOF. Immediate from theorem IV.2.3 and lemma IV.2.4, since every polynomial $f(x)$ is bounded over $0 \leq x \leq 1$.   □

Next we turn to the Riemann integral.

LEMMA IV.2.6 (Riemann integral). *The following is provable in* RCA$_0$. *Let $\phi(x)$ be a continuous real-valued function on the closed bounded interval $a \leq x \leq b$. Assume in addition that $\phi(x)$ possesses a modulus of uniform continuity. Then the Riemann integral*

$$\int_a^b \phi(x)\,dx = \lim \sum_{i=1}^n \phi(x_i)\Delta x_i$$

exists. (Here the limit is taken over all partitions $a = a_0 < a_1 < \cdots < a_n = b$ and $a_i \leq x_i \leq a_{i+1}$, $\Delta x_i = a_{i+1} - a_i$, as $\max \Delta x_i$ approaches 0.) Furthermore $\int_a^x \phi(\xi)\,d\xi$ is continuously differentiable on $a \leq x \leq b$ and its derivative is $\phi(x)$.

PROOF. Straightforward adaptation of the usual argument, which employs a modulus of uniform continuity.   □

THEOREM IV.2.7. *The following assertions are pairwise equivalent over* RCA$_0$.

1. *Weak König's lemma.*
2. *For every continuous function $\phi(x)$ on a closed bounded interval $a \leq x \leq b$, the Riemann integral $\int_a^b \phi(x)\,dx$ exists and is finite.*

PROOF. The implication from 1 to 2 is immediate from theorem IV.2.2 and lemma IV.2.6. For the converse, assume that WKL$_0$ fails and let $T \subseteq 2^{<\mathbb{N}}$ be an infinite tree with no path. Let a_s, b_s, and $\widetilde{T}$ be as in the proof of theorem IV.2.3. Define a continuous function $\phi(x)$, $0 \leq x \leq 1$ as follows. If $a_u \leq x \leq b_u$ for some $u \in \widetilde{T}$, define $\phi(x) = 3^{\mathrm{lh}(u)} = |a_u - b_u|^{-1}$. Otherwise define $\phi(x)$ by piecewise linearity as in the proof of theorem IV.2.3. Since $\widetilde{T}$ is infinite, the Riemann integral $\int_0^1 \phi(x)\,dx$ would have to be infinite. Thus $\phi(x)$ is a counterexample to 2. This completes the proof.   □

REMARK IV.2.8 (Bishop-style constructivism). In lemmas IV.2.4 and IV.2.6 we needed to assume a modulus of uniform continuity, because in general its existence is not provable in RCA$_0$. However, it is interesting to note that "any continuous function which arises in practice" can be proved in RCA_0 to have a modulus of uniform continuity on any closed bounded subset of its domain. For instance, theorems II.6.2 through II.6.5 can be extended in this way. Thus lemmas IV.2.4 and IV.2.6 apply to "any continuous function which arises in practice." (We speak only of partial continuous functions from $\mathbb{R}^k$ into $\mathbb{R}$.)

This situation has prompted some authors, for example Bishop/Bridges [20, page 38], to build a modulus of uniform continuity into their definitions of continuous function. Such a procedure may be appropriate for Bishop since his goal is to replace ordinary mathematical theorems by their "constructive" counterparts. However, as explained in chapter I, our goal is quite different. Namely, we seek to draw out the set existence assumptions which are implicit in the ordinary mathematical theorems *as they stand*. (In particular, we have examined from this viewpoint the theorem that every continuous real-valued function on $0 \leq x \leq 1$ is uniformly continuous. See theorem IV.2.3 above.) Thus Bishop's procedure would not be appropriate for us.

EXERCISE IV.2.9. Show that WKL_0 is equivalent over RCA_0 to the assertion that every uniformly continuous real-valued function on the closed unit interval $0 \leq x \leq 1$ has a modulus of uniform continuity.

EXERCISE IV.2.10. Show that WKL_0 is equivalent over RCA_0 to the following assertion. Let f be a continuous real-valued function on a nonempty closed set C in a compact metric space. If $\alpha = \sup_{x \in C} f(x)$ exists, then $f(x_0) = \alpha$ for some $x_0 \in C$.

EXERCISE IV.2.11. Show that ACA_0 is equivalent over RCA_0 to each of the following assertions.

1. Every continuous real-valued function on a nonempty closed set in a compact metric space attains a maximum value.
2. Let f be a continuous real-valued function on a nonempty closed set C in the unit interval $0 \leq x \leq 1$. Then $\sup_{x \in C} f(x)$ exists.
3. For each nonempty closed set C in the unit interval, sup C exists.

EXERCISE IV.2.12 (uniform intermediate value theorem). Show that WKL_0 is equivalent over RCA_0 to each of the following assertions.

1. If ϕ_n, $n \in \mathbb{N}$, is a sequence of continuous real-valued functions on the closed unit interval $0 \leq x \leq 1$, then there exists a sequence of real numbers x_n, $n \in \mathbb{N}$, $0 \leq x_n \leq 1$, such that $\forall n \, (\phi_n(0) \leq 0 \leq \phi_n(1) \rightarrow \phi_n(x_n) = 0)$.
2. If ϕ_n, $n \in \mathbb{N}$, is a sequence of continuous real-valued functions on $0 \leq x \leq 1$ such that $\forall n \, (\phi_n(0) \leq 0 \leq \phi_n(1) \wedge \phi_n$ is monotone increasing), then there exists a sequence of real numbers x_n, $n \in \mathbb{N}$, $0 \leq x_n \leq 1$, such that $\forall n \, \phi_n(x_n) = 0$.

(Compare theorem II.6.6.)

We end this section with some additional exercises indicating an RCA$_0$ rendition of some functional analysis and measure theory over compact metric spaces.

EXERCISE IV.2.13 (the Banach space C(X)). Within RCA$_0$, let X be a compact metric space. Show that there exists a separable Banach space C(X) whose points are in canonical one-to-one correspondence with continuous real-valued functions $\phi : X \to \mathbb{R}$ equipped with a modulus of uniform continuity. Moreover, the norm on C(X) corresponds to the sup norm $\|\phi\| = \sup_{x \in X} |\phi(x)|$. (Note: We are using the RCA$_0$ notions of separable Banach space theory, as introduced in §II.10. See also lemma IV.2.4 and example II.10.3.)

Hint: The construction of C(X) within RCA$_0$ is as follows. Let $X = \widehat{A}$ and let d be the metric on X. Put $B = A \times \mathbb{Q}^+ \times \mathbb{Q}^+$. For $b = (a, r, s) \in B$ define a continuous function

$$\phi_b : X \to \mathbb{R} \text{ by } \phi_b(x) = \max(0, \min(s, 2s(r - d(a, x))/r)).$$

Put

$$C = \mathbb{Q} \times \{F : F \text{ is a finite nonempty subset of } B\}.$$

For $c = (q, F) \in C$ define a continuous function $\phi_c : X \to \mathbb{R}$ by $\phi_c(x) = q + \max\{\phi_b(x) : b \in F\}$. Finally C($X$) $= \widehat{C}$ under the sup norm given by $\|c\| = \|\phi_c\| = \sup_{x \in X} |\phi_c(x)|$.

DEFINITION IV.2.14 (Borel measures). Within RCA$_0$, let X be a compact metric space. A *Borel measure* on X is defined to be a bounded linear functional $\mu : $ C(X) $\to \mathbb{R}$ such that $\mu(\phi) \geq 0$ for all $\phi \geq 0$ in C(X). By normalizing, we may assume that $\mu(1) = 1$.

EXERCISE IV.2.15 (the Banach spaces L$_p$(X, μ), $1 \leq p < \infty$). Within RCA$_0$, let X be a compact metric space. Show that any Borel measure μ on X gives rise to separable Banach spaces L$_p$(X, μ), $1 \leq p < \infty$. Namely, if C(X) $= \widehat{C}$ under the sup norm as in exercise IV.2.13, then L$_p$(X, μ) $= \widehat{C}$ under the L$_p$-norm, $\|\phi\|_p = \mu(|\phi|^p)^{1/p}$.

EXAMPLE IV.2.16. Examples illustrating exercises IV.2.13 and IV.2.15 are C[0, 1] and L$_p$[0, 1] $=$ L$_p$([0, 1], μ), $1 \leq p < \infty$, where $\mu : $ C[0, 1] $\to \mathbb{R}$ is the Riemann integral, $\mu(\phi) = \int_0^1 \phi(x)\, dx$. See also lemma IV.2.4 and examples II.10.3 and II.10.4. See also the notes at the end of this section.

DEFINITION IV.2.17 (located sets). Within RCA$_0$, let X be a complete separable metric space. A nonempty closed set $K \subseteq X$ is said to be *located* if the distance function

$$d(x, K) = \inf\{d(x, y) : y \in K\}$$

is a continuous real-valued function on X.

EXERCISE IV.2.18 (Hausdorff metric). Within RCA_0, let X be a compact metric space. Show that there exists a compact metric space $K(X) \subseteq C(X)$ whose points are in one-to-one correspondence with the nonempty closed located sets $K \subseteq X$. Furthermore, the metric on $K(X)$ corresponds to the Hausdorff metric

$$d_H(K_1, K_2) = \sup\{d(x_1, K_2), d(x_2, K_1): x_1 \in K_1, x_2 \in K_2\}$$

or equivalently

$$d_H(K_1, K_2) = \sup_{x \in X} |d(x, K_1) - d(x, K_2)|.$$

Hint: The construction of $K(X)$ within RCA_0 is as follows. Let $X = \widehat{A}$ and let d be the metric on X. Put

$$A^* = \{F: F \text{ is a finite nonempty subset of } A\}.$$

Then $K(X) = \widehat{A^*}$ under the metric d^* given by

$$d^*(F_1, F_2) = \sup_{x \in X} |d(x, F_1) - d(x, F_2)|.$$

Notes for §IV.2. Theorem IV.2.3 is due to Simpson (unpublished, but see [243]). Theorem IV.2.2 is taken from Brown's thesis [24]. The results on polynomial approximation and the Riemann integral within RCA_0 and WKL_0 are due to Simpson, unpublished. An RCA_0 version of $C(X)$ similar to that of exercise IV.2.13 has been given by Brown [24, §III.E], who also proved an RCA_0 version of the Stone/Weierstraß theorem. Bishop-style constructive versions of the Weierstraß polynomial approximation theorem and the Stone/Weierstraß theorem are in Bishop/Bridges [20, page 106]. Regarding exercise IV.2.15, note that measure theory in subsystems of Z_2 has been studied by Yu/Simpson [280] and Brown/Giusto/Simpson [26]; see also Yu [275, 276, 277, 278, 279], Simpson [248], and section X.1 below. The results of exercise IV.2.18 on located sets and $K(X)$ in RCA_0 are from Giusto/Simpson [93].

IV.3. The Gödel Completeness Theorem

In the previous section we showed that weak König's lemma is provably equivalent over RCA_0 to several basic theorems on continuous functions of a real variable. We now show that weak König's lemma is also provably equivalent to several basic theorems of mathematical logic. We build on the results of §II.8.

DEFINITION IV.3.1. The following definition is made in RCA_0. As in §II.8 we assume a fixed countable language L. Let X be a countable set of sentences. A *completion* of X is a countable set of sentences $X^* \supseteq X$ such that X^* is consistent, complete, and closed under logical consequence.

Lemma IV.3.2. *The following is provable in* RCA$_0$. *Let X be a countable set of sentences. There exists a tree $T = T_X \subseteq 2^{<\mathbb{N}}$ such that the paths through T_X are just the characteristic functions of completions of X. Furthermore T_X is infinite if and only if X is consistent.*

Proof. Put $t \in T$ if and only if $\forall \sigma < \mathrm{lh}(t) \, [t(\sigma) = 1 \to \sigma \in \mathrm{Snt}]$ and $\forall \sigma < \mathrm{lh}(t) \, [\sigma \in X \to t(\sigma) = 1]$ and $\forall p < \mathrm{lh}(t) \, [\text{if } p \text{ is a proof}$ and $\forall i < \mathrm{lh}(p) \, (p(i) \text{ is a nonlogical axiom of } p \to t(p(i)) = 1)$ then $\forall i < \mathrm{lh}(p) \, (p(i) \in \mathrm{Snt} \to t(p(i)) = 1)]$ and $\forall \sigma < \mathrm{lh}(t) \, \forall \tau < \mathrm{lh}(t) \, [(\sigma \in \mathrm{Snt} \wedge \tau = \neg \sigma) \to t(\sigma) = 1 - t(\tau)]$. T exists by Σ^0_0 comprehension and clearly T has the desired properties.   □

Theorem IV.3.3. *The following are pairwise equivalent over* RCA$_0$.

1. *Weak König's lemma.*
2. Lindenbaum's lemma: *every countable consistent set of sentences has a completion.*
3. Gödel's completeness theorem: *every countable consistent set X of sentences has a model, i.e., there exists a countable model M such that $\forall \sigma \, (\sigma \in X \to M(\sigma) = 1)$.*
4. Gödel's compactness theorem: *if each finite subset of X has a model then X has a model.*
5. *The completeness theorem for propositional logic with countably many atoms.*
6. *The compactness theorem for propositional logic with countably many atoms.*

Proof. We reason in RCA$_0$. The implication $1 \to 2$ is immediate from the previous lemma. The implications $3 \to 4$, $4 \to 6$, $3 \to 5$, $5 \to 6$ are straightforward. It remains to prove $2 \to 3$ and $6 \to 1$.

Let X be a countable consistent set of sentences. Let C be an infinite set of new constant symbols, and let $\langle c_n : n \in \mathbb{N} \rangle$ be a one-to-one enumeration of C. Let Φ be the set of all formulas $\varphi(x)$ with one free variable x in the expanded language $L_1 = L \cup C$, and let $\langle \varphi_n(x) : n \in \mathbb{N} \rangle$ be an enumeration of Φ. We may safely assume that c_n does not occur in $\varphi_i(x)$, $i \le n$. Form Henkin sentences

$$\eta_n \equiv (\exists x \, \varphi_n(x)) \to \varphi_n(c_n)$$

and let $X_1 = X \cup \{\eta_n : n \in \mathbb{N}\}$. The usual syntactic argument shows that X_1 is consistent, so by Lindenbaum's lemma let X_1^* be a completion of X_1. A countable model M of X_1 can be read off as in the proof of theorem II.8.4. This proves $2 \to 3$.

Now consider propositional logic with countably many atomic formulas $\langle a_n : n \in \mathbb{N} \rangle$. A set X of formulas in this language is said to be *satisfiable* if and only if there exists a *model* of X, i.e., a function $f : \mathbb{N} \to \{0, 1\}$ such that each formula of X is true under the truth assignment

$$a_n \mapsto \begin{cases} \text{true} & \text{if } f(n) = 1, \\ \text{false} & \text{if } f(n) = 0. \end{cases}$$

The *compactness theorem* for propositional logic asserts that if each finite subset of X is satisfiable then X is satisfiable. We want to prove weak König's lemma from the compactness theorem.

Let $T \subseteq 2^{<\mathbb{N}}$ be an infinite tree. For each $n \in \mathbb{N}$ form a propositional formula

$$\sigma_n = \bigvee \{ \bigwedge \{ a_i^{s(i)} : i < n \} : s \in T, \mathrm{lh}(s) = n \}$$

where $a_i^1 = a_i$, $a_i^0 = \neg a_i$. Since T contains sequences of length n, σ_n is satisfiable. Also $\sigma_{n+1} \to \sigma_n$ is a tautology. Hence for each n, $\{\sigma_0, \sigma_1, \ldots, \sigma_n\}$ is satisfiable. From the compactness theorem it follows that $\{\sigma_n : n \in \mathbb{N}\}$ is satisfiable. Let $f : \mathbb{N} \to \{0, 1\}$ be a model of $\{\sigma_n : n \in \mathbb{N}\}$. Then clearly f is a path through T. This completes the proof of $6 \to 1$ and of theorem IV.3.3. $\qquad\square$

Notes for §IV.3. The material in this section is due to Simpson, unpublished. Lemma IV.3.2 is inspired by Jockusch/Soare [134].

IV.4. Formally Real Fields

A famous result of Artin and Schreier (see van der Waerden [270]) asserts that every formally real field is orderable. (This result was an essential ingredient in the Artin/Schreier solution of Hilbert's 17th problem; see [23].) The purpose of this section is to show that WKL$_0$ is just strong enough to prove the Artin/Schreier result for countable fields.

DEFINITION IV.4.1 (formally real fields). Within RCA$_0$, let K be a countable field. We say that K is *formally real* if -1 is not a sum of squares in K. An equivalent condition is that K does not contain a sequence of elements $\langle c_0, c_1, \ldots, c_n \rangle$, $c_i \neq 0$, $n \in \mathbb{N}$, such that $\sum_{i=0}^n c_i^2 = 0$.

DEFINITION IV.4.2 (orderable fields). Within RCA$_0$, a countable field K is said to be *orderable* if there exists a binary relation $<_K$ on K which makes K into an ordered field. An equivalent condition is the existence of a "positive cone" $P \subseteq K$ such that $0 \notin P$ and $\forall a \, \forall b \, ((a \in P \land b \in P) \to (a + b \in P \land a \cdot b \in P))$ and $\forall a \, ((a \in K \land a \neq 0) \to (a \in P \leftrightarrow -a \notin P))$.

LEMMA IV.4.3. WKL$_0$ *proves that every countable, formally real field is orderable.*

PROOF. We reason in WKL$_0$. Let K be countable, formally real field. Let $\langle a_i : i \in \mathbb{N} \rangle$ be an enumeration of the nonzero elements of K. For each $t \in 2^{<\mathbb{N}}$ and $i < \mathrm{lh}(t)$ put

$$t_i = \begin{cases} 1 & \text{if } t(i) = 1, \\ -1 & \text{if } t(i) = 0. \end{cases}$$

Let T be the set of all $t \in 2^{<\mathbb{N}}$ such that for all $i, j, k < \mathrm{lh}(t)$,

(i) $a_i + a_j = a_k$ and $t_i = t_j = 1$ imply $t_k = 1$;
(ii) $a_i \cdot a_j = a_k$ and $t_i = t_j = 1$ imply $t_k = 1$;
(iii) $a_i = -a_j$ implies $t_i = -t_j$.

Clearly T is a tree. Assume for a contradiction that T is finite. Let $n \in \mathbb{N}$ be such that T contains no $t \in 2^{<\mathbb{N}}$ of length n. Then for each $t \in 2^{<\mathbb{N}}$ of length n there exist $i, j, k < n$ such that either $t_i a_i + t_j a_j + t_k a_k = 0$ or $t_i a_i t_j a_j + t_k a_k = 0$ or $t_i a_i + t_j a_j = 0$. Hence $f_t = 0$ where

$$f_t = \prod_{i,j,k<n} (t_i a_i + t_j a_j + t_k a_k)^2 (t_i a_i t_j a_j + t_k a_k)^2 (t_i a_i + t_j a_j)^2.$$

Now expand f_t as a sum of monomial terms of the form $\alpha_t = \prod_{i<n} t_i^{e_i} a_i^{e_i}$ where $e_i \in \mathbb{N}$. Note that if all of the e_i are even, then α_t is a nonzero square, and furthermore there is at least one monomial α_t of this type. On the other hand, if some e_i is odd, we have $\sum \{\alpha_t : \mathrm{lh}(t) = n\} = 0$ because each summand with $t_i = 1$ is cancelled by a corresponding summand with $t_i = -1$. Thus $\sum \{f_t : \mathrm{lh}(t) = n\} = 0$ leads to an expression of 0 as a nontrivial sum of squares, contradicting the assumption that K is formally real. This proves that T is infinite. By weak König's lemma let $g : \mathbb{N} \to \{0, 1\}$ be a path through T. Let P be the set of all $a_i \in K$ such that $g(i) = 1$. Clearly P is a positive cone for K so K is orderable. This completes the proof. □

In order to prove the converse of lemma IV.4.3, we shall need the following result which gives a useful equivalent characterization of weak König's lemma.

LEMMA IV.4.4 (WKL$_0$ and Σ_1^0 separation). *The following are pairwise equivalent over* RCA$_0$.

1. WKL$_0$.

2. (Σ_1^0 separation) *Let* $\varphi_i(n)$, $i = 0, 1$ *be* Σ_1^0 *formulas in which X does not occur freely. If* $\neg \exists n \, (\varphi_0(n) \wedge \varphi_1(n))$ *then*

$$\exists X \, \forall n \, ((\varphi_0(n) \to n \in X) \wedge (\varphi_1(n) \to n \notin X)).$$

3. *If* $f, g : \mathbb{N} \to \mathbb{N}$ *are one-to-one with* $\forall m \, \forall n \, f(m) \neq g(n)$, *then*

$$\exists X \, \forall m \, (f(m) \in X \wedge g(m) \notin X).$$

PROOF. First assume WKL$_0$ and let $\varphi_i(n)$, $i = 0, 1$, be Σ_1^0 with $\neg \exists n \, (\varphi_0(n) \wedge \varphi_1(n))$. Let $\varphi_i(n) \equiv \exists m \, \theta_i(m, n)$ where $\theta_i(m, n)$ is Σ_0^0. Let T be the set of all $t \in 2^{<\mathbb{N}}$ such that

$$(\forall i < 2) \, (\forall m < \mathrm{lh}(t)) \, (\forall n < \mathrm{lh}(t)) \, (\theta_i(m, n) \to t(n) = 1 - i).$$

T exists by Σ^0_0 comprehension. Clearly T is an infinite tree. By weak König's lemma let X be a set whose characteristic function is a path through T. Then clearly X satisfies the conclusion of 2. This proves that 1 implies 2. The equivalence of 2 and 3 is immediate from lemma II.3.7.

It remains to prove that 2 implies 1. Assume 2 and let $T \subseteq 2^{<\mathbb{N}}$ be an infinite tree. Let $\theta(n, \sigma)$ be the Σ^0_0 formula $\exists \tau\,(\mathrm{lh}(\tau) = n \wedge \tau \in T \wedge \tau \supseteq \sigma)$. Let $\varphi(\sigma, i)$ be the Σ^0_1 formula $\exists n\,(\theta(n, \sigma^\frown\langle i \rangle) \wedge \neg\theta(n, \sigma^\frown\langle 1 - i \rangle))$. Clearly $\neg\exists\sigma\,(\varphi(\sigma, 0) \wedge \varphi(\sigma, 1))$ so by the assumption 2 let X be such that $\forall\sigma\,((\varphi(\sigma, 0) \to \sigma \in X) \wedge (\varphi(\sigma, 1) \to \sigma \notin X))$. Now define a sequence of sequences $\sigma_0 \subseteq \sigma_1 \subseteq \cdots \subseteq \sigma_k \subseteq \cdots$ in $2^{<\mathbb{N}}$ by $\sigma_0 = $ empty sequence, $\sigma_{k+1} = \sigma_k^\frown\langle 0 \rangle$ if $\sigma_k \in X$, $\sigma_{k+1} = \sigma_k^\frown\langle 1 \rangle$ if $\sigma_k \notin X$. Clearly $\mathrm{lh}(\sigma_k) = k$ for all k. We claim that $\theta(n, \sigma_k)$ holds for all k and n with $k \le n$. Fix n. We prove the claim by induction on $k \le n$. Trivially $\theta(n, \sigma_0)$ since T is infinite. Assume inductively that $\theta(n, \sigma_k)$ holds for some $k < n$. Clearly either $\theta(n, \sigma_k^\frown\langle 0 \rangle)$ or $\theta(n, \sigma_k^\frown\langle 1 \rangle)$ must hold. If $\neg\theta(\sigma_k^\frown\langle 0 \rangle)$ then we have $\varphi(\sigma_k, 1)$, hence $\sigma_k \notin X$ so $\sigma_{k+1} = \sigma_k^\frown\langle 1 \rangle$ whence $\theta(n, \sigma_{k+1})$. If $\neg\theta(\sigma_k^\frown\langle 1 \rangle)$ then we have $\varphi(\sigma_k, 0)$, hence $\sigma_k \in X$ so $\sigma_{k+1} = \sigma_k^\frown\langle 0 \rangle$ whence $\theta(n, \sigma_{k+1})$. In any case $\theta(n, \sigma_{k+1})$ holds so our claim is proved. In particular we have $\theta(n, \sigma_n)$, i.e., $\sigma_n \in T$, for all n. so $f = \bigcup\{\sigma_n : n \in \mathbb{N}\}$ is a path through T. This proves weak König's lemma from 2. The proof of lemma IV.4.4 is complete. $\square$

We now show that weak König's lemma is needed to prove the orderability of countable, formally real fields.

THEOREM IV.4.5. *The following assertions are pairwise equivalent over* RCA$_0$.

1. WKL$_0$.
2. *Every countable, formally real field is orderable.*
3. *Every countable, formally real field has a real closure.*

PROOF. Assertions 2 and 3 are equivalent in view of theorem II.9.7. Lemma IV.4.3 shows that 1 implies 2. It remains to prove that 2 implies 1. Assume 2. Instead of proving weak König's lemma directly, we shall prove the equivalent statement IV.4.4.3. Let $f, g \colon \mathbb{N} \to \mathbb{N}$ be functions such that $\forall i\,\forall j\, f(i) \ne g(j)$. Let $\langle p_k : k \in \mathbb{N} \rangle$ be an enumeration of the rational primes, $p_0 = 2$, $p_1 = 3$, $p_2 = 5, \ldots$. By theorem II.9.7 let $\overline{\mathbb{Q}}$ be the real closure of the rational field $\mathbb{Q}$. For each $n \in \mathbb{N}$ let K_n be the subfield of $\overline{\mathbb{Q}}(\sqrt{-1})$ generated by

$$\left\{ \sqrt[4]{p_{f(i)}} : i < n \right\} \cup \left\{ \sqrt{-\sqrt{p_{g(j)}}} : j < n \right\} \cup \left\{ \sqrt{p_k} : k < n \right\}.$$

Because we lack Σ^0_1 comprehension we cannot form the subfield $\bigcup_{n \in \mathbb{N}} K_n$. However, we can apply lemma II.3.7 to find a field K and an embedding $h \colon K \to \mathbb{Q}(\sqrt{-1})$ such that $\forall b\,(\exists n\,(b \in K_n) \leftrightarrow \exists a\,(h(a) = b))$.

Note that each K_n is embeddable into $\overline{\mathbb{Q}}$ by taking $\sqrt{p_{f(i)}}$ to $\sqrt{p_{f(i)}}$ and $\sqrt{p_k}$ to $-\sqrt{p_k}$ whenever $k \neq f(i)$ for all $i < n$. Since $\overline{\mathbb{Q}}$ is an ordered field, it follows that each K_n is formally real. Hence K is formally real. Hence by 2, K is orderable. Fix an ordering of K. Since $h^{-1}\sqrt{p_{f(i)}}$ has a square root in K (namely $h^{-1}\sqrt[4]{p_{f(i)}}$), we must have $h^{-1}\sqrt{p_{f(i)}} > 0$. On the other hand, since $-h^{-1}\sqrt{p_{g(j)}}$ has a square root in K (namely $h^{-1}\sqrt{-\sqrt{p_{g(j)}}}$), we must have $h^{-1}\sqrt{p_{g(j)}} < 0$. By Δ_1^0 comprehension let X be the set of all $k \in \mathbb{N}$ such that $h^{-1}\sqrt{p_k} > 0$. Then $\forall i \, (f(i) \in X)$ and $\forall j \, (g(j) \notin X)$. By lemma IV.4.4 this gives weak König's lemma. The proof of the theorem is complete. □

REMARK IV.4.6. WKL$_0$ is also equivalent over RCA$_0$ to the assertion that every countable torsion-free Abelian group is orderable. This result of Hatzikiriakou/Simpson [113] is related to a recursive counterexample of Downey/Kurtz [47]. Solomon [251] has obtained additional Reverse Mathematics results concerning orderability of countable groups.

REMARK IV.4.7. WKL$_0$ is also equivalent over RCA$_0$ to the theorem on extension of valuations for countable fields: Given a monomorphism of countable fields $h \colon K_1 \to K_2$ and a valuation ring V_1 of K_1, there exists a valuation ring V_2 of K_2 such that $V_1 = h^{-1}(V_2)$. This result is due to Hatzikiriakou/Simpson [112].

EXERCISE IV.4.8. Show that RCA$_0$ proves Π_1^0 *separation*: For any Π_1^0 formulas $\psi_1(n)$ and $\psi_0(n)$ in which Z does not occur freely, $\neg \exists n \, (\psi_1(n) \wedge \psi_0(n)) \to \exists Z \, \forall n \, ((\psi_1(n) \to n \in Z) \wedge (\psi_0(n) \to n \notin Z))$. This is in contrast to lemma IV.4.4.

Notes for §IV.4. The main results of this section are from Friedman/Simpson/Smith [78]. A corollary of theorem IV.4.5 is that there exists a recursive, formally real field with no recursive ordering. This result is originally due to Ershov [54]. An improvement of Ershov's result due to Metakides/Nerode [187] states that for any recursive tree $T \subseteq 2^{<\mathbb{N}}$ there exists a recursive, formally real field K such that the space of all orderings of K is recursively homeomorphic to $[T]$, the closed set in $2^{\mathbb{N}}$ consisting of all paths through T.

IV.5. Uniqueness of Algebraic Closure

In §II.9 we showed that RCA$_0$ proves that every countable field has an algebraic closure. In this section we show that WKL$_0$ is needed to prove that these algebraic closures are unique.

Lemma IV.5.1 (uniqueness of algebraic closure). *It is provable in* WKL$_0$ *that every countable field K has a unique algebraic closure.* (*Uniqueness means that if $h_i \colon K \to \widetilde{K}_i$, $i = 1, 2$ are two algebraic closures of K then there exists an isomorphism $h \colon \widetilde{K}_1 \to \widetilde{K}_2$ of $\widetilde{K}_1$ onto $\widetilde{K}_2$ such that $h(h_1(a)) = h_2(a)$ for all $a \in K$.*)

Proof. Let K be a countable field. The existence of an algebraic closure $h \colon K \to \widetilde{K}$ is provable in RCA$_0$ (theorem II.9.4). For the uniqueness, let $h_i \colon K \to \widetilde{K}_i$, $i = 1, 2$ be two algebraic closures of K. Let $\langle a_i \colon i \in \mathbb{N} \rangle$ be an enumeration of the elements of K and let $\langle b_i \colon i \in \mathbb{N} \rangle$ be an enumeration of the elements of $\widetilde{K}_1$. Let $\langle p_i(x) \colon i \in \mathbb{N} \rangle$ be a sequence of nonconstant polynomials $p_i(x) \in K[x]$ such that $h_1(p_i)(b_i) = 0$. (We do not demand that $p_i(x)$ be irreducible.) Let T be the set of all $t \in \mathbb{N}^{<\mathbb{N}}$ such that $\forall i \, (i < \mathrm{lh}(t) \to t(i) \in \widetilde{K}_2)$ and for all $i, j, k < \mathrm{lh}(t)$

(i) $b_i + b_j = b_k$ implies $t(i) + t(j) = t(k)$;
(ii) $b_i \cdot b_j = b_k$ implies $t(i) \cdot t(j) = t(k)$;
(iii) $h_1(a_i) = b_j$ implies $h_2(a_i) = t(j)$;
(iv) $h_2(p_i)(t(i)) = 0$.

The idea is that $t \in T$ encodes a partial isomorphism of $\widetilde{K}_1$ onto $\widetilde{K}_2$ over K. Clearly T is a subtree of $\mathbb{N}^{<\mathbb{N}}$. By considering finitely generated algebraic extensions of K, we can show that T is infinite. (For details see Friedman/Simpson/Smith [78].) Also T is a bounded tree since by (iv) we have $t(i) \leq g(i) = \max\{c \colon c \in \widetilde{K}_2 \wedge h_2(p_i)(c) = 0\}$. Hence by bounded König's lemma in WKL$_0$ (lemma IV.1.4), there exists a path f through T. Define $h \colon \widetilde{K}_1 \to \widetilde{K}_2$ by $h(b_i) = f(i)$. Clearly h is an isomorphism of $\widetilde{K}_1$ onto $\widetilde{K}_2$ and by (iii) we have $h(h_1(a_i)) = h_2(a_i)$ for all $a_i \in K$. This completes the proof. $\square$

We now prove the converse.

Theorem IV.5.2. *The following are equivalent over* RCA$_0$.

1. WKL$_0$.
2. *Every countable field has a unique algebraic closure.*

Proof. Lemma IV.5.1 gives half of the theorem. For the other half, assume 2. Instead of proving weak König's lemma directly, we shall prove the equivalent statement IV.4.4.3. Let $f, g \colon \mathbb{N} \to \mathbb{N}$ be functions such that $\forall i \, \forall j \, f(i) \neq g(j)$. By theorem II.9.7 let $\overline{\mathbb{Q}}$ be the real closure of the rational field $\mathbb{Q}$. Let $\langle p_n \colon n \in \mathbb{N} \rangle$ be the enumeration of the rational primes in increasing order, i.e., $p_0 = 2$, $p_1 = 3$, $p_2 = 5$, ... For each $n \in \mathbb{N}$ let K_n be the subfield of $\overline{\mathbb{Q}}$ generated by

$$\{\sqrt{p_{f(i)}} \colon i < n\} \cup \{\sqrt{p_{g(j)}} \colon j < n\}.$$

By lemma II.3.7 let K be a countable field and $h_1 \colon K \to \overline{\mathbb{Q}}$ a monomorphism such that $\forall b \, (\exists n \, (b \in K_n) \leftrightarrow \exists a \, (h_1(a) = b))$. Define another monomorphism $h_2 \colon K \to \overline{\mathbb{Q}}$ by putting $h_2(h_1^{-1}(\sqrt{p_{f(i)}})) = \sqrt{p_{f(i)}}$ and

$h_2(h_1^{-1}(\sqrt{p_{g(j)}})) = -\sqrt{p_{g(j)}}$. Thus $h_1, h_2 \colon K \to \overline{\mathbb{Q}}(\sqrt{-1})$ are two algebraic closures of K. By 2 there exists an automorphism $h \colon \overline{\mathbb{Q}}(\sqrt{-1}) \to \overline{\mathbb{Q}}(\sqrt{-1})$ such that $h(h_1(a)) = h_2(a)$ for all $a \in K$. Let X be the set of all m such that $h(\sqrt{p_m}) = \sqrt{p_m}$. Then

$$h(\sqrt{p_{f(i)}}) = h(h_1(h_1^{-1}(p_{f(i)}))) = h_2(h_1^{-1}(\sqrt{p_{f(i)}})) = \sqrt{p_{f(i)}}$$

and

$$h(\sqrt{p_{g(j)}}) = h(h_1(h_1^{-1}(p_{g(j)}))) = h_2(h_1^{-1}(\sqrt{p_{g(j)}})) = -\sqrt{p_{g(j)}}$$

so $f(i) \in X$ and $g(j) \notin X$. By lemma IV.4.4 this implies weak König's lemma. The proof of the theorem is complete. $\qquad\square$

Notes for §IV.5. The results of this section are from Friedman/Simpson/Smith [78].

IV.6. Prime Ideals in Countable Commutative Rings

In this section we show that WKL_0 is just strong enough to accommodate the development of an important topic in commutative algebra.

DEFINITION IV.6.1 (prime ideals). Within RCA_0, let R be a countable commutative ring. A *prime ideal* of R is a set $P \subseteq R$ such that P is an ideal of R (definition III.5.2) and $\forall a \, \forall b \, (a \cdot b \in P \to (a \in P \vee b \in P))$.

A basic theorem of commutative algebra asserts that every commutative ring has a prime ideal. The usual way to prove this theorem is to obtain a maximal ideal (by Zorn's lemma) and then to observe that maximal ideals are prime. This method cannot work in WKL_0 since by theorem III.5.5 the existence of maximal ideals is not provable in WKL_0. Nevertheless, we have:

LEMMA IV.6.2 (existence of prime ideals). *It is provable in WKL_0 that every countable commutative ring possesses a prime ideal.*

PROOF. We reason in WKL_0. Let R be a countable commutative ring and let $\langle a_i \colon i \in \mathbb{N}\rangle$ be an enumeration of the elements of R. Use primitive recursion (theorem II.3.4) to define a sequence of (codes for) finite sets $X_s \subseteq R$, $s \in 2^{<\mathbb{N}}$, beginning with $X_{\langle\rangle} = \{0\}$. (Here $\langle\rangle$ denotes the empty sequence.) Let $s \in 2^{<\mathbb{N}}$ be given and suppose that X_s has already been defined. Let

$$\mathrm{lh}(s) = 4 \cdot \langle(i, j), m\rangle + k, \quad 0 \le k < 4, \tag{11}$$

where (i, j) denotes the pairing function (theorem II.2.2).

Case 1: $k = 0$. If $a_i \cdot a_j \in X_s$ put $X_{s^\frown\langle 0\rangle} = X_s \cup \{a_i\}$ and $X_{s^\frown\langle 1\rangle} = X_s \cup \{a_j\}$; otherwise put $X_{s^\frown\langle 0\rangle} = X_s$ and $X_{s^\frown\langle 1\rangle} = \emptyset = $ the empty set.

Case 2: $k = 1$. Put $X_{s^\frown\langle 0\rangle} = \emptyset$. If $a_i \in X_s$ and $a_j \in X_s$ put $X_{s^\frown\langle 1\rangle} = X_s \cup \{a_i + a_j\}$, otherwise $X_{s^\frown\langle 1\rangle} = X_s$.

Case 3: $k = 2$. Put $X_{s^\frown\langle 0\rangle} = \emptyset$. If $a_i \in X_s$ put $X_{s^\frown\langle 1\rangle} = X_s \cup \{a_i \cdot a_j\}$, otherwise $X_{s^\frown\langle 1\rangle} = X_s$.

Case 4: $k = 3$. Put $X_{s^\frown\langle 0\rangle} = \emptyset$. If $1 \in X_s$ put $X_{s^\frown\langle 1\rangle} = \emptyset$, otherwise $X_{s^\frown\langle 1\rangle} = X_s$.

If $\mathrm{lh}(s)$ is not as in (11), put $X_{s^\frown\langle 0\rangle} = \emptyset$ and $X_{s^\frown\langle 1\rangle} = X_s$. This completes the construction of X_s for all $s \in 2^{<\mathbb{N}}$.

Let S be the set of all $s \in 2^{<\mathbb{N}}$ such that $X_s \neq \emptyset$. Clearly S is a tree.

We claim that for each $n \in \mathbb{N}$ there exists $s \in S$ of length n such that X_s does not generate R as an R-module. For $n = 0$ the claim is trivial. If $n \equiv 1, 2,$ or 3 mod 4 and the claim holds for n then trivially it holds for $n + 1$. Suppose $n \equiv 0$ mod 4 and the claim holds for n. Let $s \in S$ be of length n such that X_s does not generate R as an R-module. Let $n = 4 \cdot ((i, j), m)$. If $a_i \cdot a_j \notin X_s$ then trivially our claim holds for $n + 1$. If $a_i \cdot a_j \in X_s$ then we make a subclaim that $X_{s^\frown\langle 0\rangle} = X_s \cup \{a_i\}$ and $X_{s^\frown\langle 1\rangle} = X_s \cup \{a_j\}$ do not both generate R as an R-module. If they did, we would have $1 = c + ra_i = d + sa_j$ where $r, s \in R$ and c, d are finite linear combinations of elements of X_s with coefficients from R. Then $1 = cd + csa_j + dra_i + rsa_ia_j$ so X_s generates R as a R-module, a contradiction. This proves the subclaim. Hence our claim holds for $n+1$. The claim for all $n \in \mathbb{N}$ now follows by Π_1^0 induction on n.

The above claim implies that S is infinite. Hence by weak König's lemma S has a path, call it f. If it were now possible to form the set of all $a \in R$ such that $\exists n\,(a \in X_{f[n]})$, then clearly this set would be a prime ideal of R and the proof of lemma IV.6.2 would be complete. (Here $f[n]$ denotes the initial sequence of f of length n.) Unfortunately, we cannot form this set because we lack Σ_1^0 comprehension. However, we can use bounded Σ_1^0 comprehension to finish the proof as follows.

We may safely assume that our enumeration $\langle a_i : i \in \mathbb{N}\rangle$ of R is such that $a_0 = 0$ and $a_1 = 1$. Let T be the set of all $t \in 2^{<\mathbb{N}}$ such that

(i) $0 < \mathrm{lh}(t)$ implies $t(0) = 0$;
(ii) $1 < \mathrm{lh}(t)$ implies $t(1) = 1$;
(iii) if $i, j, k < \mathrm{lh}(t)$ then
    (a) $t(i) = t(j) = 0$ and $a_i + a_j = a_k$ imply $t(k) = 0$;
    (b) $t(i) = 0$ and $a_i \cdot a_j = a_k$ imply $t(k) = 0$;
    (c) $t(i) = t(j) = 1$ and $a_i \cdot a_j = a_k$ imply $t(k) = 1$.

Clearly T is a tree. We claim that T is infinite. To see this, let $m \in \mathbb{N}$ be given. By bounded Σ_1^0 comprehension (theorem II.3.9) let Y be the set of all $i < m$ such that $\exists n\,(a_i \in X_{f[n]})$. Define $t \in 2^{<\mathbb{N}}$, $\mathrm{lh}(t) = m$ by putting $t(i) = 0$ if $i \in Y$, $t(i) = 1$ if $i \notin Y$. Then clearly $t \in T$ and $\mathrm{lh}(t) = m$. This proves that T is infinite. Hence by another application of weak König's lemma there exists a path g through T. Let P be the set of all $a_i \in R$ such that $g(i) = 0$. Then clearly P is a prime ideal of R.

This completes the proof of lemma IV.6.2.   □

We now turn to the reversal. First, a definition:

DEFINITION IV.6.3 (radical ideals). Within RCA_0, let R be a countable commutative ring. A *radical ideal* of R is an ideal $J \subseteq R$ (cf. definition III.5.2) such that $a^n \in J$ implies $a \in J$ for all $a \in R$, $n \in \mathbb{N}$.

Clearly every prime ideal of R is a radical ideal of R.

THEOREM IV.6.4 (reversal). *The following assertions are pairwise equivalent over* RCA_0.

1. WKL_0.
2. *Every countable commutative ring contains a prime ideal.*
3. *Every countable commutative ring contains a radical ideal.*

PROOF. The implication from 1 to 2 has already been proved as lemma IV.6.2. The implication from 2 to 3 is trivial. It remains to prove that 3 implies 1. We reason in RCA_0. Assume 3. Instead of proving weak König's lemma directly, we shall prove the equivalent IV.4.4.3.

Let $f, g \colon \mathbb{N} \to \mathbb{N}$ be given with $\forall i \, \forall j \, (f(i) \neq g(j))$. Let $R_0 = \mathbb{Q}[\langle x_n \colon n \in \mathbb{N}\rangle]$ be the polynomial ring over rational field $\mathbb{Q}$ with countably many indeterminates x_n, $n \in \mathbb{N}$. Let $I \subseteq R_0$ be the ideal generated by the polynomials $x_{f(m)}^{m+1}$ and $x_{g(m)}^{m+1} - 1$, $m \in \mathbb{N}$. To see that I exists, note that any given $f \in R_0$ can be put into a normal form $f^* \equiv f$ modulo I, where if x_n^k occurs in f^* then $n \neq f(m), g(m)$ for all $m < k$. Thus $f \in I$ if and only if $f^* = 0$, so I exists by Δ_1^0 comprehension. Form the quotient ring $R = R_0/I$. By our assumption 3, let J be a radical ideal in R. Let J_0 be the ideal in R_0 which corresponds to J. Then J_0 is a radical ideal in R_0 and $I \subseteq J_0$. It follows that $x_{f(m)} \in J_0$ and $x_{g(m)} \notin J_0$ for all $m \in \mathbb{N}$. Setting $X = \{n \colon x_n \in J_0\}$ we obtain $\forall m \, (f(m) \in X \wedge g(m) \notin X)$. Thus by IV.4.4 we have weak König's lemma. This completes the proof.   □

COROLLARY IV.6.5. RCA_0 *is not strong enough to prove that every countable commutative ring has a prime (or even radical) ideal.*

PROOF. Immediate from theorem IV.6.4 and the fact (to be proved in §VIII.2) that the theorems of RCA_0 are strictly included in those of WKL_0.
                                                                              □

EXERCISE IV.6.6. Show that the following is provable in WKL_0. Let R be a countable commutative ring. Let $\varphi(a)$ and $\psi(a)$ be Σ_1^0 such that

1. $\forall a \, \forall b \, ((\varphi(a) \wedge \psi(b)) \to (a \in R \wedge b \in R \wedge a \neq b))$,
2. $\varphi(0) \wedge \psi(1)$,
3. $\forall a \, \forall b \, ((\varphi(a) \wedge \varphi(b)) \to \varphi(a + b))$,
4. $\forall a \, \forall r \, ((\varphi(a) \wedge r \in R) \to \varphi(r \cdot a))$,
5. $\forall a \, \forall b \, ((\psi(a) \wedge \psi(b)) \to \psi(a \cdot b))$.

Then R has a prime ideal P such that $\forall a \ (\varphi(a) \to a \in P)$ and $\forall a \ (\psi(a) \to a \notin P)$.

Notes for §IV.6. The main results in this section are from Friedman/Simpson/Smith [78, 79].

IV.7. Fixed Point Theorems

In this and the next two sections, we resume the study of analysis in WKL_0, which was begun in §§IV.1 and IV.2.

A famous theorem of Brouwer states that any continuous mapping of a k-simplex into itself has a fixed point. The purpose of this section is to show that Brouwer's theorem and its generalization to infinite-dimensional spaces are provable in WKL_0. We shall also obtain a reversal showing that Brouwer's theorem is equivalent to weak König's lemma over RCA_0.

We begin by presenting one of the well known proofs of Brouwer's theorem, within WKL_0. We use the proof via Sperner's lemma.

DEFINITION IV.7.1 (k-simplices). The following definitions are made in RCA_0. For $k \in \mathbb{N}$, a k-*simplex* S is the convex hull of $k + 1$ affinely independent points $s_0, \dots, s_k$ in $\mathbb{R}^n$, called the *vertices* of S. We can coordinatize S by identifying each point $x \in S$ with the unique $(k+1)$-tuple $(x_0, \dots, x_k)$ such that $x = \sum_{i=0}^{k} x_i s_i$, $\sum_{i=0}^{k} x_i = 1$, and $x_i \geq 0$ for all $i \leq k$. Clearly S is a compact metric space.

If S is a simplex, a *face* of S is any simplex whose vertices are a subset of the vertices of S. For any point $x \in S$, the *carrier* of x is the smallest face of S which contains x.

DEFINITION IV.7.2 (simplicial subdivision). Within RCA_0, let S be a k-simplex. A *simplicial subdivision* of S is a finite set of k-simplices $S_0, \dots, S_m$ such that $S = S_0 \cup \dots \cup S_m$ and, for all $i < j \leq m$, $S_i \cap S_j$ is either empty or a common face of S_i and S_j.

DEFINITION IV.7.3 (admissible labeling). Within RCA_0, let S be a k-simplex, and let P be a finite set of points in S which includes the vertices of S. An *admissible labeling of* P is a mapping from P into $\{0, \dots, k\}$ such that (i) the vertices of S are mapped to the full set of labels $\{0, \dots, k\}$; and (ii) for every $x \in P$, the label of x is the same as the label of one of the vertices of the carrier of x.

LEMMA IV.7.4 (Sperner's lemma). *The following is provable in* RCA_0. *Let* S *be a* k-simplex, $k \in \mathbb{N}$, *and let* $S_0, \dots, S_m$ *be a simplicial subdivision of* S. *Suppose that the vertices of* $S_0, \dots, S_m$ *are admissibly labeled. Then for some* $i \leq m$, *the vertices of* S_i *are mapped to the full set of labels* $\{0, \dots, k\}$.

PROOF. The proof consists of elementary combinatorial reasoning which is straightforwardly formalized in RCA_0. We shall now present this proof.

We shall actually prove that the number of S_i's receiving a full set of labels is odd. The proof is by induction on k. For $k = 0$ the result is trivial. For $k = 1$, note that S is a line segment and $S_0, \ldots, S_m$ is a partition of S into subsegments. Since the endpoints of S are labeled 0 and 1, it is clear that there are an odd number of S_i's whose endpoints are labeled 0 and 1. This is the base of the induction.

Now suppose $k > 1$. Let T be the face of S with vertices labeled $\{0, \ldots, k-1\}$. Let $T_0, \ldots, T_n$ be the simplicial subdivision of T induced by $S_0, \ldots, S_m$. Clearly the induced labeling of the vertices of $T_0, \ldots, T_n$ is admissible. Hence by induction hypothesis the number of T_j's with vertices labeled $\{0, \ldots, k-1\}$ is odd. For $i \leq m$ let d_i be the number of faces of S_i with vertices labeled $\{0, \ldots, k-1\}$. By admissibility, each such face is either one of the T_j's or a common face of two S_i's. Since the number of such faces which are T_j's is odd, it follows that $d_0 + \cdots + d_m$ is odd. Hence there are an odd number of S_i's with d_i odd. But if d_i is odd, it is easy to see that $d_i = 1$, and this holds if and only if the vertices of S_i receive the full set of labels $\{0, \ldots, k\}$. This completes the proof. $\square$

LEMMA IV.7.5. *The following is provable in* WKL_0. *Let S be a k-simplex. Then every continuous function $f : S \to S$ has a fixed point, i.e., $f(x) = x$ for some $x \in S$.*

PROOF. We reason in WKL_0. Suppose the conclusion fails. Then $|f(x) - x| > 0$ for all $x \in S$. By a simple argument involving the Heine/Borel property (cf. exercise IV.2.10), we see that there exists $\epsilon > 0$ such that $|f(x) - x| > \epsilon$ for all $x \in S$. Put $\epsilon^* = \epsilon/(3k + 3)$. Let $\varphi(x, i)$ be a Σ_1^0 formula which says that $i \leq k$, $x_i > 0$ and $y_i < x_i + \epsilon^*$, where $x = (x_0, \ldots, x_k)$ and $f(x) = y = (y_0, \ldots, y_k)$. It is clear that, for each $x \in S$, $\varphi(x, i)$ holds for at least one $i \leq k$.

By theorem IV.2.2, f is uniformly continuous, so let $\delta > 0$ be such that $|x - x'| < \delta$ implies $|f(x) - f(x')| < \epsilon^*$. Let $S_0, \ldots, S_n$ be a subdivision of S into k-simplices of diameter less than the minimum of δ and ϵ^*. If x is any vertex of this simplicial subdivision, we define $\mathrm{label}(x) = i$ for some i such that $\varphi(x, i)$ holds. It is straightforward to verify that this labeling is admissible, so by Sperner's lemma in RCA_0 (lemma IV.7.4) there exists j such that the vertices of S_j receive a full set of labels. It is then easy to see that $|f(x') - x'| < \epsilon$ holds for any $x' \in S_j$. This contradiction completes the proof. $\square$

The following is our version of Brouwer's fixed point theorem.

THEOREM IV.7.6 (Brouwer fixed point theorem in WKL_0). *The following is provable in* WKL_0. *Let C be the convex hull of a nonempty finite set of points in $\mathbb{R}^n$, $n \in \mathbb{N}$. Then every continuous function $f : C \to C$ has a fixed point.*

PROOF. Let k be the dimension of C. We can find a k-simplex S in $\mathbb{R}^n$ such that $C \subseteq S$. Using elementary linear algebra in RCA$_0$ (cf. exercise II.4.11), we can show that C is a *retract* of S, i.e., there is a continuous function $r : S \to C$ such that $r(x) = x$ for all $x \in C$. Given a continuous function $f : C \to C$, consider $g : S \to C$ given by $g(x) = f(r(x))$. By theorem IV.7.5, let $x \in S$ be such that $g(x) = x$. Then $x \in C$, hence $r(x) = x$, hence $f(x) = g(x) = x$. This completes the proof. $\square$

We shall now obtain a reversal showing that weak König's lemma is needed to prove Brouwer's theorem, even for the unit square.

THEOREM IV.7.7 (reversal). *The following are pairwise equivalent over* RCA$_0$.

1. *Let C be the convex hull of a nonempty finite set of points in $\mathbb{R}^n$, $n \in \mathbb{N}$. Then every continuous function $f : C \to C$ has a fixed point.*
2. *Let C be the unit square, $[0, 1] \times [0, 1]$. Then every continuous function $f : C \to C$ has a fixed point.*
3. *Weak König's lemma.*

PROOF. The implication $1 \to 2$ is trivial, and $3 \to 1$ has already been proved as theorem IV.7.6. It remains to prove $2 \to 3$. Working within RCA$_0$, assume that weak König's lemma is false. Let $T \subseteq 2^{<\mathbb{N}}$ be an infinite tree with no infinite path. We shall use T to construct a continuous function $f : C \to C$ with no fixed point, where $C = [0, 1] \times [0, 1]$.

Let ∂C be the boundary of C, i.e., the four edges of the unit square. It suffices to show that ∂C is a retract of C. For, once we have a retraction map $r : C \to \partial C$, we can let $f : C \to \partial C$ consist of r followed by a 90° rotation of ∂C. Clearly such an f has no fixed point.

We claim that there exists a *singular covering* of $[0, 1]$, i.e., a covering of $[0, 1]$ by an infinite sequence of closed rational intervals $I_n = [a_n, b_n]$, $a_n, b_n \in \mathbb{Q}$, $a_n < b_n$, $n \in \mathbb{N}$, such that for all $m \neq n$, $I_m \cap I_n$ consists of at most one point. To see this, define intervals $[c_\sigma, d_\sigma]$, $\sigma \in 2^{<\mathbb{N}}$, by putting $c_{\langle\rangle} = 0$, $d_{\langle\rangle} = 1$, $c_{\sigma^\frown\langle 0\rangle} = c_\sigma$, $c_{\sigma^\frown\langle 1\rangle} = d_{\sigma^\frown\langle 0\rangle} = (c_\sigma + d_\sigma)/2$, and $d_{\sigma^\frown\langle 1\rangle} = d_\sigma$. Let $\langle \sigma_n : n \in \mathbb{N}\rangle$ be an enumeration of $\widetilde{T} = \{\sigma \in 2^{<\mathbb{N}} : \sigma \notin T \wedge \sigma[\text{lh}(\sigma) - 1] \in T\}$, and put $I_n = [a_n, b_n] = [c_{\sigma_n}, d_{\sigma_n}]$. Clearly $\langle I_n : n \in \mathbb{N}\rangle$ has the desired properties, so our claim is proved.

For each $n \in \mathbb{N}$, put

$$A_n = \left(\bigcup_{m \leq n} I_m \times I_n \right) \cup \left(\bigcup_{m \leq n} I_n \times I_m \right)$$

and

$$B_n = ([0, 1] \times I_n) \cup (I_n \times [0, 1]).$$

Note that $C = \bigcup_{n \in \mathbb{N}} A_n$. Note also that A_n is properly included in B_n.

Our retraction map $r: C \to \partial C$ will be defined in stages. We begin by defining r on ∂C to be the identity map. At stage n, we assume that r has already been defined on ∂C and on A_m for all $m < n$, and we define r on A_n. Let $P_{n0}, \ldots, P_{nk_n}$ be the connected components of A_n. Since A_n is properly included in B_n, it follows that ∂P_{ni} has at least one edge e_{ni} which, except for its endpoints, lies inside $B_n \setminus \partial C$, hence is disjoint from $\bigcup_{m<n} A_m$. Let e'_{ni} be e_{ni} minus its endpoints. We define r on P_{ni} to consist of a continuous retraction of P_{ni} onto $\partial P_{ni} \setminus e'_{ni}$, followed by a continuous mapping of $\partial P_{ni} \setminus e'_i$ into ∂C which is compatible with the part of r that has already been defined. This defines r on $A_n = \bigcup_{i \leq k_n} P_{ni}$.

It can be shown that the above construction gives rise to a continuous function r defined on all of $C = \bigcup_{n \in \mathbb{N}} A_n$. Clearly r is a retraction of C onto ∂C. This completes the proof.   □

We shall now obtain an infinite-dimensional generalization of Brouwer's theorem, within WKL_0. The theorem which we shall prove is closely related to the Schauder/Tychonoff fixed point theorem. First, we need the following technical lemma.

LEMMA IV.7.8. *The following is provable in* WKL_0. *Let C be a closed set in a compact metric space X. Given $\epsilon > 0$, there exists a finite set of points $c_1, \ldots, c_m \in C$ such that for all $x \in C$, $d(x, c_i) < \epsilon$ for some i, $1 \leq i \leq m$.*

PROOF. By compactness, there exists a finite set of points $x_1, \ldots, x_n \in X$ such that for all $x \in X$ there exists i such that $d(x, x_i) < \epsilon/2$. By theorem IV.1.7, we see that the formula

$$\varphi(i) \equiv \exists x \, (x \in C \text{ and } d(x, x_i) \leq \epsilon/2)$$

is equivalent to a Π^0_1 formula. By bounded Π^0_1 comprehension, let $I \subseteq \{1, \ldots, n\}$ be the set of i such that $\varphi(i)$ holds. By theorem IV.1.8, let $\langle c_i : i \in I \rangle$ be a sequence of points such that $c_i \in C$ and $d(c_i, x_i) \leq \epsilon/2$. Then for all $x \in C$ we have $d(x, x_i) < \epsilon/2$ for some i, hence $i \in I$, hence $d(c_i, x_i) \leq \epsilon/2$, hence $d(x, c_i) < \epsilon$. We can renumber the c_i's as $c_1, \ldots, c_m$ where $m = |I| \leq n$. This completes the proof.   □

The following is our version of the Schauder/Tychonoff fixed point theorem. Recall from examples II.5.5 and III.2.6 that the Hilbert cube $[0, 1]^{\mathbb{N}}$ is compact.

THEOREM IV.7.9 (Schauder fixed point theorem in WKL_0). *The following is provable in* WKL_0. *Let C be a nonempty closed convex set in $[-1, 1]^{\mathbb{N}}$. Then every continuous function $f: C \to C$ has a fixed point.*

PROOF. Suppose not. Let $f: C \to C$ be continuous such that $f(x) \neq x$ for all $x \in C$. For $m \geq 1$ and $x = \langle x_i : i \in \mathbb{N} \rangle \in \mathbb{R}^{\mathbb{N}}$, we put $\|x\|_m = \max_{i < m} |x_i|$. Let us write $B_m(x, \epsilon)$ (respectively $B_m^*(x, \epsilon)$) for the open (respectively closed) ball consisting of all $y \in \mathbb{R}^{\mathbb{N}}$ such that $\|y - x\|_m < \epsilon$ (respectively $\|y - x\|_m \leq \epsilon$). By the Heine/Borel covering principle in WKL_0 (theorem IV.1.5), there exist $m \in \mathbb{N}$ and $\epsilon > 0$ such that $f(x) \notin B_m(x, \epsilon)$ for all $x \in C$. We shall obtain a contradiction by finding a point $x \in C$ such that $f(x) \in B_m(x, \epsilon)$.

By the previous lemma, let $c_0, \ldots, c_k \in C$ be such that C is covered by the open sets $U_i = B_m(c_i, \epsilon)$, $i \leq k$. Let $D \subseteq C$ be the convex hull of $c_0, \ldots, c_k$. Recall from example II.5.5 that $\mathbb{R}^{\mathbb{N}} = \widehat{A}$ where A is the set of eventually 0 sequences of rational numbers. Let $\varphi(k, a, r, i)$ be a Σ_1^0 formula which says that $n \in \mathbb{N}$, $a \in A$, $r \in \mathbb{Q}^+$, $i \leq k$, and

$$B_n^*(a, r) \subseteq (\mathbb{R}^{\mathbb{N}} \setminus C) \cup f^{-1}(U_i).$$

It is straightforward to show that $\mathbb{R}^{\mathbb{N}}$ is covered by open sets $B_n(a, r)$ such that $\varphi(k, a, r, i)$ holds for some $i \leq k$. Hence by the Heine/Borel principle there exists a covering of C by finitely many open sets $B_{n_{ij}}(a_{ij}, r_{ij})$, $i \leq k$, $j < l_i$, with $\varphi(k_{ij}, a_{ij}, r_{ij}, i)$. Put $V_i = \bigcup_{j<l_i} B_{n_{ij}}(a_{ij}, r_{ij})$. Thus C is covered by the open sets V_i, $i \leq k$, and $f(V_i) \subseteq U_i$. Put

$$n = \max\{m\} \cup \{n_{ij} : i \leq k, j < l_i\}.$$

For any $x = \langle x_i : i \in \mathbb{N} \rangle \in \mathbb{R}^{\mathbb{N}}$, let us write $\overline{x} = \langle x_i : i < n \rangle \in \mathbb{R}^n$. Let $\overline{D}$ be the convex hull of $\overline{c}_0, \ldots, \overline{c}_k$ in $\mathbb{R}^k$, and let $\overline{V}_i = \bigcup_{j<l_i} B_{n_{ij}}(\overline{a}_{ij}, r_{ij}) \subseteq \mathbb{R}^n$. Thus $\overline{D}$ is covered by open sets $\overline{V}_i$, $i \leq k$, in $\mathbb{R}^n$. As in the proof of theorem II.7.2, let $g_i : \overline{D} \to [0, 1]$, $i \leq k$, be a sequence of continuous functions such that $\sum_{i=0}^k g_i(x) = 1$ for all $\overline{x} \in \overline{D}$, and $g_i(\overline{x}) > 0$ implies $\overline{x} \in \overline{V}_i$. Define $g : \overline{D} \to \overline{D}$ by $g(\overline{x}) = \sum_{i=0}^k g_i(\overline{x}) \overline{c}_i$.

By theorem IV.7.6, there is $\overline{x}' \in \overline{D}$ such that $g(\overline{x}') = \overline{x}'$. Put

$$x' = \sum_{i=0}^k g_i(\overline{x}') c_i$$

and note that $x' \in C$. By bounded Σ_1^0 comprehension, let I be the set of all $i \leq k$ such that $g_i(\overline{x}') > 0$. Then for all $i \in I$ we have $\overline{x}' \in \overline{V}_i$, hence $x' \in V_i$, hence $f(x') \in U_i = B_m(c_i, \epsilon)$, i.e., $\|f(x') - c_i\|_m < \epsilon$. Since $\sum g_i(\overline{x}') = 1$, it follows that

$$\|f(x') - x'\|_m = \left\| \sum_{i \in I} g_i(\overline{x}') f(x') - \sum_{i \in I} g_i(\overline{x}') c_i \right\|_m$$

$$\leq \sum_{i \in I} g_i(\overline{x}') \|f(x') - c_i\|_m < \epsilon.$$

Thus $f(x') \in B_m(x', \epsilon)$ and the proof is complete. $\square$

Notes for §IV.7. Shioji/Tanaka [219] proved versions of the Brouwer and Schauder fixed point theorems, within WKL_0. Our results in this section are variants of those of Shioji/Tanaka [219]. Our proof of Brouwer's theorem within WKL$_0$ is modeled after the well known proof of Brouwer's theorem via Sperner's lemma (cf. Tompkins [267]). The fact that Brouwer's theorem for $[0, 1] \times [0, 1]$ implies weak König's lemma is due to Shioji/Tanaka [219], based on a recursive counterexample due to Orevkov [199].

IV.8. Ordinary Differential Equations

In this section we discuss Peano's existence theorem for solutions of ordinary differential equations. Peano's theorem says that, if $f(x, y)$ is continuous in some neighborhood of $(0, 0)$, then the initial value problem

$$y' = f(x, y), \qquad y(0) = 0 \qquad (12)$$

has a solution $y = \phi(x)$ which is continuously differentiable in some neighborhood of $x = 0$. Here y' denotes the derivative of the unknown function $y = y(x)$. We shall show that Peano's theorem is provable in WKL$_0$. We shall also show that Peano's theorem is equivalent to weak König's lemma over RCA$_0$.

We begin by proving Peano's theorem in WKL$_0$. The proof will be based on theorem IV.7.9, our WKL$_0$ version of the Schauder fixed point theorem.

THEOREM IV.8.1 (Peano's theorem in WKL$_0$). *The following is provable in* WKL$_0$. *Let $f(x, y)$ be a continuous real-valued function on the rectangle $-a \leq x \leq a, -b \leq y \leq b$ where $a, b > 0$. Then the initial value problem*

$$\frac{dy}{dx} = f(x, y), \qquad y(0) = 0$$

has a continuously differentiable solution $y = \phi(x)$ on the interval $-\alpha \leq x \leq \alpha, \alpha = \min(a, b/M)$, where

$$M = \max\{|f(x, y)|: \, -a \leq x \leq a, -b \leq y \leq b\}.$$

PROOF. We reason in WKL$_0$.

Note first that M exists by theorem IV.2.2.

Let $A = \{q_i : i \in \mathbb{N}\}$ be an enumeration of the rational numbers in the closed interval $[-\alpha, \alpha]$. Thus $\widehat{A} = [-\alpha, \alpha]$. We may safely assume that $q_0 = 0$. Let C be the closed convex set in $\mathbb{R}^{\mathbb{N}}$ consisting of all sequences $\langle y_i : i \in \mathbb{N} \rangle$ such that $y_0 = 0$ and $|y_i - y_j| \leq M \cdot |q_i - q_j|$ for all $i, j \in \mathbb{N}$. C is included in the compact product space $\prod_{i \in \mathbb{N}}[-M\alpha, M\alpha]$ (cf. lemma III.2.5).

To each $\langle y_i : i \in \mathbb{N} \rangle \in C$ is associated a continuous function

$$\phi: [-\alpha, \alpha] \to \mathbb{R}$$

such that $\phi(q_i) = y_i$ for all $i \in \mathbb{N}$. Namely, the code Φ of ϕ is given by putting $(q_i, r)\Phi(b, s)$ if and only if $M \cdot r + |b - y_i| < s$. Thus we shall identify points of C with continuous functions $\phi \colon [-\alpha, \alpha] \to \mathbb{R}$ satisfying $\phi(0) = 0$ and

$$|\phi(x) - \phi(x')| \leq M \cdot |x - x'|$$

for $|x|, |x'| \leq \alpha$.

We define a continuous function $F \colon C \to C$ as follows. For $\langle y_i : i \in \mathbb{N} \rangle \in C$, we put

$$F(\langle y_i : i \in \mathbb{N} \rangle) = \left\langle \int_0^{q_i} f(x, \phi(x)) \, dx : i \in \mathbb{N} \right\rangle$$

where $\phi \colon [-\alpha, \alpha] \to \mathbb{R}$ is the continuous function associated to $\langle y_i : i \in \mathbb{N} \rangle$ as in the previous paragraph. For all $i, j \in \mathbb{N}$ we have

$$\left| \int_0^{q_i} f(x, \phi(x)) \, dx - \int_0^{q_j} f(x, \phi(x)) \, dx \right| = \left| \int_{q_i}^{q_j} f(x, \phi(x)) \, dx \right|$$

$$\leq M \cdot |q_i - q_j|$$

so $F(\langle y_i : i \in \mathbb{N} \rangle) \in C$. Using a modulus of uniform continuity for f (cf. theorem IV.2.2), it is straightforward to construct a code for F.

Now by theorem IV.7.9 let $\langle y_i : i \in \mathbb{N} \rangle \in C$ be a fixed point of F, i.e.,

$$F(\langle y_i : i \in \mathbb{N} \rangle) = \langle y_i : i \in \mathbb{N} \rangle.$$

Let $\phi \colon [-\alpha, \alpha] \to \mathbb{R}$ be the continuous function associated to $\langle y_i : i \in \mathbb{N} \rangle$. Then for all $i \in \mathbb{N}$ we have

$$\phi(q_i) = \int_0^{q_i} f(x, \phi(x)) dx.$$

It follows easily that

$$\frac{d\phi(x)}{dx} = f(x, \phi(x)) \text{ and } \phi(0) = 0$$

for all x in $[-\alpha, \alpha]$. This proves our theorem. $\qquad\square$

We remark that it is straightforward to extend the previous theorem so as to apply to initial value problems of the form

$$y_1' = f_1(x, y_1, \ldots, y_n), \qquad y_1(0) = 0$$
$$y_2' = f_2(x, y_1, \ldots, y_n), \qquad y_2(0) = 0$$

$$\vdots$$

$$y_n' = f_n(x, y_1, \ldots, y_n), \qquad y_n(0) = 0$$

where $n \in \mathbb{N}$.

We now turn to the reversal of theorem IV.8.1. The following theorem says that Peano's theorem is equivalent over RCA_0 to weak König's lemma.

THEOREM IV.8.2 (reversal). *The following assertions are pairwise equiv-alent over* RCA$_0$.

1. WKL$_0$.
2. *Peano's theorem, as stated in* IV.8.1.
3. *If $f(x, y)$ is continuous and has a modulus of uniform continuity in some neighborhood of $x = 0$, $y = 0$, then the initial value problem* (12) *has a continuously differentiable solution $y = \phi(x)$ in some interval containing $x = 0$.*

PROOF. The implication from 1 to 2 is given by theorem IV.8.1, and the implication from 2 to 3 is trivial. It remains to prove that 3 implies 1. Assume 3. Instead of proving weak König's lemma directly, we shall prove Σ_1^0 separation (lemma IV.4.4.2).

Let $\varphi(n, i)$ be a Σ_1^0 formula such that $\neg \exists n\, (\varphi(n, 0) \wedge \varphi(n, 1))$. Working in RCA$_0$, we shall construct a a continuous function $f(x, y)$ on the rectangle $|x| \leq 1$, $|y| \leq 1$, such that $|f(x, y)| \leq 1$, $f(-x, y) = -f(x, y)$, and for each $n \geq 1$, if $y = \phi(x)$ is any solution of $y' = f(x, y)$ on the interval $-2^{-n+1} \leq x \leq -2^{-n}$, then

$$\phi(-2^{-n+1}) = \phi(-2^{-n}); \tag{13}$$

$$\varphi(n, 0) \text{ and } \phi(-2^{-n+1}) = 0 \text{ imply } \phi(-2^{-n} - 2^{-n-1}) > 2^{-3(n+2)}; \tag{14}$$

$$\varphi(n, 1) \text{ and } \phi(-2^{-n+1}) = 0 \text{ imply } \phi(-2^{-n} - 2^{-n-1}) < 2^{-3(n+2)}. \tag{15}$$

Moreover, $f(x, y)$ will have a modulus of uniform continuity on the rectangle $|x| \leq 1$, $|y| \leq 1$.

Once we obtain $f(x, y)$ as above, we can apply IV.8.2.3 to get a contin-uously differentiable function $\phi(x)$ which is a solution of the initial value problem (12) on some interval containing $x = 0$. Using the property $f(-x, y) = -f(x, y)$, we may assume that $\phi(x)$ is a solution of (12) on some interval of the form

$$-2^{-N} \leq x \leq 0,$$

where $N \in \mathbb{N}$. By (13) we have $\phi(-2^{-n+1}) = \phi(-2^{-n})$ for all $n > N$. Since $\phi(0) = 0$ and ϕ is continuous, it follows by Σ_1^0 induction that $\phi(-2^{-n+1}) = 0$ for all $n > N$. Hence by (14) we have that $\varphi(n, 0)$ implies $\phi(-2^{-n} - 2^{-n-1}) > -2^{-3(n+2)}$, while by (15) we have that $\varphi(n, 1)$ implies $\phi(-2^{-n} - 2^{-n-1}) < -2^{-3(n+2)}$, for all $n > N$. Let A be the set of rational numbers in the interval $|x| \leq 2^{-N}$, and let Φ be a code for ϕ. By minimization (theorem II.3.5), there exists

$$g : \mathbb{N} \setminus \{0, 1, \ldots, N\} \to \mathbb{N} \times A \times \mathbb{Q}^+ \times A \times \mathbb{Q}^+$$

defined by $g(n) = $ the least $(k, a, r, b, s) \in \Phi$ such that

$$\left| a - (-2^{-n} - 2^{-n-1}) \right| < r \text{ and } s < 2^{-3(n+2)}.$$

Writing $g(n) = (k_n, a_n, r_n, b_n, s_n)$, we have

$$\left| b_n - \phi(-2^{-n} - 2^{-n-1}) \right| < 2^{-3(n+2)}.$$

By Δ_1^0 comprehension, let X be the set of $n > N$ such that $b_n > 0$. Thus for $n > N$ we have $\varphi(n, 0)$ implies $n \in X$, while $\varphi(n, 1)$ implies $n \notin X$. This together with bounded Σ_1^0 comprehension (theorem II.3.9) gives Σ_1^0 separation. Hence by lemma IV.4.4 we have weak König's lemma.

It remains to construct $f(x, y)$ as above. We shall need certain auxiliary functions $h_n(x)$ and $j_n(x, y)$, $n \in \mathbb{N}$.

Write $\varphi(n, i)$ as $\exists m\, \theta(m, n, i)$ where θ is Σ_0^0. Define

$$q(x) = \max(1 - |x|, 0).$$

For $n \in \mathbb{N}$ define

$$h_n(x) = \begin{cases} 2^{-k} \cdot q\left(2^k \left(x - \tfrac{1}{2}\right)\right) & \text{if } k = \text{least } m \text{ such that } \theta(m, n, 0), \\ -2^{-k} \cdot q\left(2^k \left(x - \tfrac{1}{2}\right)\right) & \text{if } k = \text{least } m \text{ such that } \theta(m, n, 1), \\ 0 & \text{otherwise.} \end{cases}$$

Thus $\varphi(n, 0)$ (respectively $\varphi(n, 1)$) implies that $h_n(x)$ is positive (respectively negative) on an interval of length 2^{-k+1} centered at $x = 1/2$, where $k = $ the least m such that $\theta(m, n, 0)$ (respectively $\theta(m, n, 1)$) holds.

We shall need information on the solutions of the equation

$$y' = s(x, y) = 9x(1 - x)y^{1/3}.$$

It is easy to verify that $y' = s(x, y)$ with initial condition $y(0) = y_0$ has on the interval $0 \le x \le 1$ the unique solution

$$y(x) = (\operatorname{sgn} y_0)\,[x^2(3 - 2x) + |y_0|^{2/3}]^{3/2}$$

for $y_0 \ne 0$. Here $\operatorname{sgn} t = 1$ if $t > 0$, -1 if $t < 0$. For $y_0 = 0$, there is a family of solutions

$$y(x) = \begin{cases} 0 & \text{for } 0 \le x \le c, \\ \pm\,[x^2(3 - 2x) - c^2(3 - 2c)]^{3/2} & \text{for } c \le x \le 1, \end{cases}$$

where $0 \le c \le 1$. Each possible real value for $y(1)$ is assumed by exactly one of these solutions. Also, if a solution $y(x)$ has $y(x_0) \ne 0$ where $0 \le x_0 < 1$, then $|y(x)|$ must increase throughout the interval $x_0 \le x \le 1$. Thus the indicated solutions are the only ones on the interval $0 \le x \le 1$.

Now we define the functions $j_n(x, y)$, $n \in \mathbb{N}$, by

$$j_n(x, y) = \begin{cases} h_n(x) & \text{for } 0 \le x \le 1, \\ s(x - 1, y) & \text{for } 1 \le x \le 2, \\ -s(x - 2, y) & \text{for } 2 \le x \le 4, \\ -h_n(x - 3) & \text{for } 3 \le x \le 4. \end{cases}$$

If $y(x)$ is a solution of $y' = j_n(x, y)$ over $0 \leq x \leq 4$, then $y(2)$ determines $y(x)$ throughout $1 \leq x \leq 2$ and hence also $0 \leq x \leq 1$. Using the identities $h_n(x) = h_n(1 - x)$ and $s(x, y) = s(1 - x, y)$, we have $j_n(x, y) = -j_n(4 - x, y)$. This implies that $y_1(x) = y(4 - x)$ is also a solution over $0 \leq x \leq 4$. Since $y_1(2) = y(2)$, we have $y_1(x) = y(x)$ in the interval $0 \leq x \leq 2$. It now follows that $y(x) = y(4 - x)$ for $0 \leq x \leq 4$. Thus, if $y(x)$ is any solution of $y' = j_n(x, y)$ on $0 \leq x \leq 4$, we have $y(0) = y(4)$. If in addition $y(0) = 0$, then we have $y(2) > 1$ if $\varphi(n, 0)$, $y(2) < -1$ if $\varphi(n, 1)$, and $-1 \leq y(2) \leq 1$ if $\neg\varphi(n, 0) \wedge \neg\varphi(n, 1)$.

Finally, define $f(x, y)$ for $x \leq 0$ by

$$f(x, y) = \sum_{n=1}^{\infty} 2^{-2(n+2)} j_n(2^{n+2}(x + 2^{-n+1}), 2^{3(n+2)} y),$$

and for $x \geq 0$ by $f(x, y) = -f(-x, y)$. Note that under the transformation

$$\hat{x} = 2^{n+2}(x + 2^{-n+1}),$$
$$\hat{y} = 2^{3(n+2)} \cdot y,$$

a solution of $y' = j_n(x, y)$ on the interval $0 \leq x \leq 4$ becomes a solution of

$$y' = 2^{-2(n+2)} \cdot j_n(2^{n+2}(x + 2^{-n+1}), 2^{3(n+2)} y)$$

on the interval $-2^{-n+1} \leq x \leq -2^{-n}$. The properties of $f(x, y)$ listed earlier are now easily verified.

This completes the proof.   □

A consequence of the previous theorem is that Peano's existence theorem for solutions of ordinary differential equations is not provable in RCA_0. In view of this fact, it is interesting to note that a version of Picard's existence and uniqueness theorem is provable in RCA_0. We formalize this as follows.

THEOREM IV.8.3 (Picard's theorem in RCA_0). *The following is provable in* RCA_0. *Assume that* $f(x, y)$ *has a modulus of uniform continuity* $h : \mathbb{N} \to \mathbb{N}$ *and satisfies a Lipschitz condition*

$$|f(x, y_1) - f(x, y_2)| \leq L \cdot |y_1 - y_2|$$

and $|f(x, y)| \leq M$ *throughout the rectangle* $-a \leq x \leq a$, $-b \leq y \leq b$, *where* L, M, a, *and* b *are positive real numbers. Then the initial value problem* (12) *has a unique solution* $y = \phi(x)$ *on the interval* $-\alpha \leq x \leq \alpha$, $\alpha = \min(a, b/M)$. *Moreover* $\phi(x)$ *has a modulus of uniform continuity on this interval.*

PROOF. We reason in RCA_0.

As in the proof of theorem IV.8.1, let C be the compact convex set consisting of all continuous real-valued functions $\phi(x)$, $|x| \leq \alpha$, satisfying

$\phi(0) = 0$ and with modulus of uniform continuity given by

$$|\phi(x_1) - \phi(x_2)| \le M \cdot |x_1 - x_2|$$

for $|x_1|, |x_2| \le \alpha$. Also as in that proof, let $F : C \to C$ be given by

$$F(\phi)(x) = \int_0^x f(\xi, \phi(\xi)) \, d\xi.$$

Define a sequence of functions $\phi_n \in C$, $n \in \mathbb{N}$, by putting $\phi_0(x) = 0$ for all $|x| \le \alpha$, and $\phi_{n+1} = F(\phi_n)$ for all $n \in \mathbb{N}$. We claim that, for all $n \in \mathbb{N}$,

$$|\phi_{n+1}(x) - \phi_n(x)| \le \frac{L^n M |x|^{n+1}}{(n+1)!}. \tag{16}$$

In proving this claim, the base step $n = 0$ is given by

$$|\phi_1(x) - \phi_0(x)| = \left| \int_0^x f(\xi, 0) \, d\xi \right| \le M |x|.$$

For the inductive step, note that (16) implies

$$\begin{aligned}
|\phi_{n+2}(x) - \phi_{n+1}(x)| &\le \int_0^x |f(\xi, \phi_{n+1}(\xi)) - f(\xi, \phi_n(\xi))| \, d\xi \\
&\le L \cdot \int_0^x |\phi_{n+1}(\xi) - \phi_n(\xi)| \, d\xi \\
&\le \frac{L^{n+1} M}{(n+1)!} \int_0^x |\xi|^{n+1} \, d\xi \\
&= \frac{L^{n+1} M |x|^{n+2}}{(n+2)!}.
\end{aligned}$$

The inequalities (16) together with lemma II.6.5 imply that $\phi_n(x)$ converges uniformly to a function $\phi(x)$ in C. It is straightforward to verify that $\phi(x)$ is a fixed point of F, i.e.,

$$\phi(x) = \int_0^x f(\xi, \phi(\xi)) \, d\xi.$$

Thus $y = \phi(x)$ is a solution to the initial value problem (12).

To prove uniqueness, suppose that $y = \widehat{\phi}(x)$ is another solution, and consider the function

$$\psi(x) = (\phi(x) - \widehat{\phi}(x))^2 e^{-2Lx}.$$

Then we have $\psi(0) = 0$ and, for $x > 0$,

$$\psi'(x) + 2L\psi(x) = 2(\phi(x) - \widehat{\phi}(x))(\phi'(x) - \widehat{\phi}'(x)) e^{-2Lx}.$$

The absolute value of the right hand side is

$$2 \cdot |\phi(x) - \widehat{\phi}(x)| \cdot |f(x, \phi(x)) - f(x, \widehat{\phi}(x))| \cdot e^{-2Lx}$$
$$\leq 2 \cdot |\phi(x) - \widehat{\phi}(x)| \cdot L \cdot |\phi(x) - \widehat{\phi}(x)| \cdot e^{-2Lx}$$
$$= 2L\psi(x)$$

so $\psi'(x) \leq 0$, hence $\psi(x) \leq \psi(0) = 0$ for $x > 0$. This implies $\psi(x) = 0$ since obviously $\psi(x) \geq 0$ by definition. Thus $\phi(x) = \widehat{\phi}(x)$ for $x \geq 0$. Similarly, by considering

$$\psi(x) = (\phi(x) - \widehat{\phi}(x))^2 e^{2Lx}$$

we obtain $\phi(x) = \widehat{\phi}(x)$ for $x \leq 0$. This completes the proof.   □

Once again, we remark that the result of theorem IV.8.3 extends straightforwardly to the case of an initial value problem involving n unknown functions, $n \in \mathbb{N}$.

Notes for §IV.8. For a somewhat different treatment of the material in this section, see Simpson [236]. The results of this section are due to Simpson [236]. The proof of Peano's theorem in WKL$_0$ given here (IV.8.1), based on a WKL$_0$ version of Schauder's fixed point theorem, is essentially due to Shioji/Tanaka [219]. The fact that Peano's theorem implies weak König's lemma (theorem IV.8.2) is due to Simpson [236], based on a recursive counterexample due to Aberth [2]. See also Pour-El/Richards [203]. Our successive approximation proof of the Picard existence and uniqueness theorem (theorem IV.8.3) follows Aberth [2]. See also Birkhoff/Rota [19, pages 99–115].

IV.9. The Separable Hahn/Banach Theorem

In §II.10 we developed some of the rudimentary theory of separable Banach spaces, within RCA$_0$. We shall now show that a version of the Hahn/Banach theorem for separable Banach spaces can be proved in WKL$_0$. Indeed, this theorem is equivalent to weak König's lemma over RCA$_0$.

For our WKL$_0$ proof of the separable Hahn-Banach theorem, we shall use an idea of Kakutani. The following lemma is a WKL$_0$ version of a famous theorem of functional analysis, known as the Markov/Kakutani fixed point theorem.

Given a closed convex set $C \subseteq \mathbb{R}^{\mathbb{N}}$, a continuous function $f : C \to C$ is called *affine* if

$$f\left(\sum_{i=0}^{k} \alpha_i x_i\right) = \sum_{i=0}^{k} \alpha_i f(x_i)$$

for all $k \in \mathbb{N}$, $x_0, \ldots, x_k \in C$, and $\alpha_0, \ldots, \alpha_k \geq 0$ with $\sum_{i=0}^{k} \alpha_i = 1$. A sequence of continuous functions $f_n \colon C \to C$, $n \in \mathbb{N}$ is said to be *commutative* if $f_m f_n(x) = f_n f_m(x)$ for all $m, n \in \mathbb{N}$ and $x \in C$.

LEMMA IV.9.1 (Markov/Kakutani theorem in WKL$_0$). *The following is provable in* WKL$_0$. *Let C be a closed convex set in $[-1, 1]^{\mathbb{N}}$. Let $f_n \colon C \to C$, $n \in \mathbb{N}$, be a commutative sequence of continuous affine maps. Then these maps have a common fixed point, i.e., there exists $x \in C$ such that $f_n(x) = x$ for all $n \in \mathbb{N}$.*

PROOF. We reason in WKL$_0$.

For each $n \in \mathbb{N}$, let C_n be the set of fixed points of f_n, i.e., the set of $x \in C$ such that $f_n(x) = x$. Since f_n is continuous and affine, it follows easily that C_n is closed and convex. For all $m, n \in \mathbb{N}$ and $x \in C_m$, we have $f_m f_n(x) = f_n f_m(x) = f_n(x)$, so $f_n(x) \in C_m$. Thus $f_n(C_m) \subseteq C_m$. For each $n \in \mathbb{N}$, put $C_n^* = \bigcap_{m<n} C_m$. Thus C_n^* is also closed and convex, and we have $f_n(C_n^*) \subseteq C_n^*$.

We claim that C_n^* is nonempty for all $n \in \mathbb{N}$. Since C_n^* is a closed set in a compact metric space, the statement $C_n^* \neq \emptyset$ is Π_1^0 (theorem IV.1.7). Thus we can prove our claim by Π_1^0 induction on $n \in \mathbb{N}$. By assumption, $C_0^* = C$ is nonempty. If C_n^* is nonempty, then by applying Schauder's fixed point theorem (IV.7.9) to $f_n \colon C_n^* \to C_n^*$, we see that f_n has a fixed point in C_n^*, i.e., $C_{n+1}^* = C_n^* \cap C_n$ is nonempty. This gives the inductive step. Our claim is proved.

By Heine/Borel compactness of C (theorem IV.1.5), we conclude that $\bigcap_{n \in \mathbb{N}} C_n$ is nonempty, i.e., there exists $x \in C$ such that $f_n(x) = x$ for all $n \in \mathbb{N}$. This proves the lemma. $\square$

We need the following definition.

DEFINITION IV.9.2 (subspaces, extensions). The following definitions are made in RCA$_0$. Given a separable Banach space $\widehat{A}$, a *subspace* of $\widehat{A}$ consists of a separable Banach space $\widehat{S}$ together with a bounded linear mapping $\psi \colon \widehat{S} \to \widehat{A}$ such that $\|x\| = \|\psi(x)\|$ for all $x \in \widehat{S}$. We identify $x \in \widehat{S}$ with $\psi(x) \in \widehat{A}$. If $\widehat{B}$ is another separable Banach space and $F \colon \widehat{S} \to \widehat{B}$ is a bounded linear operator, we say that $\widetilde{F} \colon \widehat{A} \to \widehat{B}$ *extends* F if $\widetilde{F}(x) = F(\psi(x))$ for all $x \in \widehat{S}$.

Given a separable Banach space $\widehat{A}$, a *bounded linear functional* on $\widehat{A}$ is a bounded linear operator $f \colon \widehat{A} \to \mathbb{R}$. The following is our WKL$_0$ version of the Hahn/Banach theorem for separable Banach spaces.

THEOREM IV.9.3 (Hahn/Banach theorem in WKL$_0$). *The following is provable in* WKL$_0$. *Let $\widehat{A}$ be a separable Banach space and let $\widehat{S}$ be a subspace of $\widehat{A}$. Let $f \colon \widehat{S} \to \mathbb{R}$ be a bounded linear functional such that $\|f\| \leq \alpha$, where α is a positive real number. Then there exists a bounded linear functional $\widetilde{f} \colon \widehat{A} \to \mathbb{R}$ extending f such that $\|\widetilde{f}\| \leq \alpha$.*

PROOF. We may safely assume that $\alpha = 1$.

Let $A = \{a_i : i \in \mathbb{N}\}$ and $S = \{s_i : i \in \mathbb{N}\}$. We may safely assume that $a_0 = s_0 = 0$. Let C_0 be the closed convex set in $\mathbb{R}^{\mathbb{N}}$ consisting of those sequences $\langle z_i : i \in \mathbb{N} \rangle$ such that $z_0 = 0$ and $|z_i - z_j| \leq \|a_i - a_j\|$ for all $i, j \in \mathbb{N}$. C_0 is included in the compact product space

$$\prod_{i \in \mathbb{N}} [-\|a_i\|, \|a_i\|]$$

(cf. lemma III.2.5).

To each $z = \langle z_i : i \in \mathbb{N} \rangle \in C_0$ is associated a continuous function

$$g = g_z : \widehat{A} \to \mathbb{R}$$

such that $g(a_i) = z_i$ for all $i \in \mathbb{N}$. Namely, the code G of g is given by putting $(a_i, r) G(b, s)$ if and only if $r + |b - z_i| < s$. Thus we shall identify points of C_0 with continuous functions $g : \widehat{A} \to \mathbb{R}$ satisfying

$$|g(x) - g(x')| \leq \|x - x'\|$$

for all $x, x' \in \widehat{A}$.

Let $C_1 = \{g \in C_0 : g(\psi(s)) = f(s) \text{ for all } s \in \widehat{S}\}$. Clearly C_1 is a compact convex subset of C_0. We claim that C_1 is nonempty. To see this, note that $C_1 = \bigcap_{k \in \mathbb{N}} C_{1,k}$ where

$$C_{1,k} = \{g \in C_0 : g(\psi(s_j)) = f(s_j) \text{ for all } j \leq k\}.$$

Thus, by Heine/Borel compactness (theorem IV.1.5), it suffices to show $C_{1,k} \neq \emptyset$ for all $k \in \mathbb{N}$. Putting $g(x) = \min_{j \leq k} (f(s_j) + \|x - \psi(s_j)\|)$, it is straightforward to check that $g \in C_{1,k}$. This proves our claim.

Next let

$$C_2 = \{g \in C_1 : g(x + \psi(s)) = g(x) + g(\psi(s)) \text{ for all } x \in \widehat{A} \text{ and } s \in \widehat{S}\}.$$

Clearly C_2 is a compact convex subset of C_1. Note that C_2 is the set of common fixed points of the maps $T_j : C_1 \to C_1$, $j \in \mathbb{N}$, given by

$$(T_j g)(x) = g(x + \psi(s_j)) - g(\psi(s_j)).$$

It is also straightforward to verify that T_j, $j \in \mathbb{N}$, is a commuting sequence of continuous affine maps from C_1 to C_1. Hence by lemma IV.9.1 we have $C_2 \neq \emptyset$.

Finally let

$$C_3 = \{g \in C_2 : g(x + y) = g(x) + g(y) \text{ for all } x, y \in \widehat{A}\}.$$

Clearly C_3 is a compact convex subset of C_2. Note that C_3 is the set of common fixed points of the maps $U_j : C_2 \to C_2$, $j \in \mathbb{N}$, given by

$$(U_j g)(x) = g(x + a_j) - g(a_j).$$

It is straightforward to verify that U_j, $j \in \mathbb{N}$, is a commuting sequence of continuous affine maps from C_2 to C_2. Hence by lemma IV.9.1 we have $C_3 \neq \emptyset$.

For any $g \in C_3$, we have $g(nx) = ng(x)$ for all $n \in \mathbb{N}$ and $x \in \widehat{A}$. Hence

$$ng\left(\frac{m}{n}x\right) = g(mx) = mg(x)$$

for all $m, n \geq 1$. Hence $g(qx) = qg(x)$ for all $q \in \mathbb{Q}$. From this it follows that $g(\alpha x) = \alpha g(x)$ for all $\alpha \in \mathbb{R}$ and $x \in \widehat{A}$. Thus any $g \in C_3$ is a bounded linear functional on $\widehat{A}$ extending f. This completes the proof.   □

We now turn to the reversal of the previous theorem. We shall show that the separable Hahn/Banach theorem is equivalent to WKL$_0$ over RCA$_0$.

THEOREM IV.9.4 (reversal). *The separable Hahn-Banach theorem* (*as stated in theorem* IV.9.3) *is equivalent over* RCA$_0$ *to weak König's lemma.*

PROOF. Theorem IV.9.3 tells us that weak König's lemma implies the separable Hahn/Banach theorem. For the converse, we reason in RCA$_0$ and assume the separable Hahn/Banach theorem. Instead of proving weak König's lemma directly, we shall prove Σ_1^0 separation (lemma IV.4.4.2).

Let $\varphi(n, i)$ be a Σ_1^0 formula such that $\neg\exists n(\varphi(n, 0) \wedge \varphi(n, 1))$. Write $\varphi(n, i)$ as $\exists m \, \theta(m, n, i)$ where θ is Σ_0^0. Define

$$\delta_{mn} = \begin{cases} 2^{-k} & \text{if } k = (\text{least } j \leq m) \, \theta(j, n, 0), \\ -2^{-k} & \text{if } k = (\text{least } j \leq m) \, \theta(j, n, 1), \\ 0 & \text{otherwise}, \end{cases}$$

and let $\delta_n = \langle \delta_{mn} : m \in \mathbb{N} \rangle$. Note that δ_n is a real number. For $(p, q) \in \mathbb{Q} \times \mathbb{Q}$, let

$$\|(p, q)\|_{mn} = \begin{cases} \max\left(\left|\frac{1 - \delta_{mn}}{1 + \delta_{mn}} p + q\right|, |p - q|\right) & \text{if } \delta_{mn} > 0, \\ \max\left(\left|\frac{1 + \delta_{mn}}{1 - \delta_{mn}} p - q\right|, |p + q|\right) & \text{if } \delta_{mn} < 0, \\ \max(|p + q|, |p - q|) & \text{if } \delta_{mn} = 0. \end{cases}$$

Let $\|(p, q)\|_n = \langle \|(p, q)\|_{mn} : m \in \mathbb{N} \rangle$ and note that $\|(p, q)\|_n$ is a real number.

Let A be the set of (codes for) finite nonempty sequences of elements of $\mathbb{Q} \times \mathbb{Q}$. Define addition and scalar multiplication on A in the obvious way so as to make A a vector space over $\mathbb{Q}$ (cf. §II.10). For $\langle (p_i, q_i) : i \leq n \rangle \in A$, define

$$\|\langle (p_i, q_i) : i \leq n \rangle\| = \sum_{i=0}^{n} 2^{-i-1} \cdot \|(p_i, q_i)\|_i.$$

Let $\widehat{A}$ be the separable Banach space coded by A with this norm.

Intuitively, $\widehat{A}$ is the ℓ_1-sum of separable Banach spaces $\widehat{A}_n$ where, for each $n \in \mathbb{N}$, $\widehat{A}_n$ is the 2-dimensional Banach space $\mathbb{R} \times \mathbb{R}$ with unit ball determined by δ_n. The various cases are:

1. $\delta_n > 0$. Here the unit ball is the parallelogram with vertices $(0, 1)$, $(0, -1)$, $(-1 - \delta_n, -\delta_n)$ and $(1 + \delta_n, \delta_n)$.
2. $\delta_n < 0$. Here the unit ball is the parallelogram with vertices $(0, 1)$, $(0, -1)$, $(-1 + \delta_n, -\delta_n)$ and $(1 - \delta_n, \delta_n)$.
3. $\delta_n = 0$. Here the unit ball is the parallelogram with vertices $(0, 1)$, $(0, -1)$, $(-1, 0)$ and $(1, 0)$.

Let S be the set of (codes for) finite nonempty sequences of pairs of rational numbers of the form $(p, 0)$. S is a subset of A and so inherits the addition, scalar multiplication, and norm described above. Let $\widehat{S}$ be the separable Banach space coded by S and note that $\widehat{S}$ is a subspace of $\widehat{A}$. Intuitively, $\widehat{S}$ is the ℓ_1-sum of the x-axes of the 2-dimensional spaces $\widehat{A}_n$, $n \in \mathbb{N}$.

Let $f : S \to \mathbb{R}$ be defined by

$$f(\langle (p_0, 0), \dots, (p_n, 0) \rangle) = \sum_{i=0}^{n} 2^{-i-1} \cdot p_i.$$

Note that f is linear on S, and

$$
\begin{aligned}
|f(\langle (p_0, 0), \dots, (p_n, 0) \rangle)| &= \left| \sum_{i=0}^{n} 2^{-i-1} \cdot p_i \right| \\
&\leq \sum_{i=0}^{n} 2^{-i-1} \cdot |p_i| \\
&\leq \sum_{i=0}^{n} 2^{-i-1} \cdot \|(p_i, 0)\|_i \\
&= \|\langle (p_0, 0), \dots, (p_n, 0) \rangle\|.
\end{aligned}
$$

Thus f encodes a bounded linear functional on $\widehat{S}$ with $\|f\| \leq 1$ (cf. definition II.10.5).

Now apply the separable Hahn/Banach theorem to obtain an extension $\widetilde{f}$ of f to $\widehat{A}$ with $\|\widetilde{f}\| \leq 1$.

For $n \in \mathbb{N}$ let $z_n \in A$ be the sequence of length $n + 1$ of the form $\langle (0, 0), \dots, (0, 0), (0, 1) \rangle$. Note that

$$|\widetilde{f}(z_n)| \leq \|z_n\| = 2^{-n-1}.$$

Moreover, if $\delta_n > 0$, then we have

$$
\begin{aligned}
|2^{-n-1}(1 + \delta_n) + \delta_n \widetilde{f}(z_n)| &= |\widetilde{f}(\langle (0, 0), \dots, (0, 0), (1 + \delta_n, \delta_n) \rangle)| \\
&\leq \|\langle (0, 0), \dots, (0, 0), (1 + \delta_n, \delta_n) \rangle\| \\
&= 2^{-n-1},
\end{aligned}
$$

which implies $\widetilde{f}(z_n) = -2^{-n-1}$. Similarly, if $\delta_n < 0$, then $\widetilde{f}(z_n) = 2^{-n-1}$.

With this in mind, let $\widetilde{f}(z_n) = \langle \widetilde{f}(z_n)_k : k \in \mathbb{N} \rangle$ (viewed as a sequence of rational numbers; cf. the definition of real numbers in §II.4). By Δ_1^0 comprehension, let $X = \{n \in \mathbb{N} : \widetilde{f}(z_n)_{n+2} \leq 0\}$. Suppose that $\varphi(n, 0)$ holds. Then $\delta_n > 0$ and so $\widetilde{f}(z_n) = -2^{-n-1}$. Since

$$|\widetilde{f}(z_n) - \widetilde{f}(z_n)_{n+2}| \leq 2^{-n-2},$$

it follows that $\widetilde{f}(z_n)_{n+2} < 0$, hence $n \in X$. Similarly, if $\varphi(n, 1)$ holds, then $\delta_n < 0$, hence $\widetilde{f}(z_n) = 2^{-n-1}$, hence $\widetilde{f}(z_n)_{n+2} > 0$, hence $n \notin X$. Thus we have Σ_1^0 separation. By lemma IV.4.4, we have weak König's lemma. This completes the proof of the theorem.   □

Notes for §IV.9. Theorems IV.9.3 and IV.9.4 are due to Brown/Simpson [27]. The proof of theorem IV.9.3 given here (using ideas of Kakutani) is essentially due to Shioji/Tanaka [219]. Lemma IV.9.1 is a variant of Shioji/Tanaka [219, theorem 7.1]. An even more elegant proof of theorem IV.9.3 is given in Humphreys/Simpson [128]. The fact that the separable Hahn/Banach theorem implies weak König's lemma (over RCA₀) is due to Brown/Simpson [27], based on a recursive counterexample due to Bishop and Metakides/Nerode/Shore [188].

For more information on functional analysis in RCA₀ and WKL₀, see Brown [24], Brown/Simpson [27, 28], Humphreys [126], Humphreys/Simpson [127, 128], and §X.2 below.

IV.10. Conclusions

In this chapter we have seen that several key mathematical theorems are provable in WKL₀ and indeed equivalent to weak König's lemma over RCA₀ in the sense of Reverse Mathematics. Among them are: the Heine/Borel theorem for $[0, 1]$ and for compact metric spaces (§IV.1); various properties of continuous real-valued functions on $[0, 1]$ and on compact metric spaces, including uniform continuity, the maximum principle, Riemann integrability, and Weierstraß approximation (§IV.2); the completeness and compactness theorems in mathematical logic (§IV.3), existence of real closure for countable formally real fields (§IV.4), uniqueness of algebraic closure of countable fields (§IV.5), existence of prime ideals in countable commutative rings (§IV.6), the Brouwer and Schauder fixed point theorems (§IV.7), the Peano existence theorem for solutions of ordinary differential equations (§IV.8), and the separable Hahn/Banach theorem (§IV.9).

Our principal technique for proving mathematical theorems in WKL₀ has been to use compactness arguments of various kinds. For the reversals, we have made extensive use of Σ_1^0 separation (see lemma IV.4.4).

Chapter V

ARITHMETICAL TRANSFINITE RECURSION

In §I.11 we introduced the formal system ATR_0 of arithmetical transfinite recursion. We explained that ATR_0 is much stronger than ACA_0 from the viewpoint of mathematical practice and is of great importance with respect to Reverse Mathematics.

The purpose of this chapter is to present some details of results concerning mathematics and Reverse Mathematics in ATR_0, which were merely outlined in §I.11. Models of ATR_0 will be considered in later chapters; see especially §§VII.2–VII.3 and VIII.3–VIII.5.

V.1. Countable Well Orderings; Analytic Sets

The purpose of this preliminary section is to present some basic definitions and results concerning countable well orderings. Our discussion of countable well orderings will be continued in §V.2 and concluded in §V.6.

In this section we shall introduce and use the notion of *analytic set*. Analytic sets (sometimes known in the literature as Σ_1^1 sets) are of fundamental importance in the branch of ordinary mathematics known as *classical descriptive set theory*. We shall investigate in §§V.3, V.4, and V.5 and in chapter VI the extent to which classical descriptive set theory can be developed formally within subsystems of second order arithmetic.

All of the results in this preliminary section will be proved within the relatively weak formal system ACA_0, which was studied in chapter III. The stronger system ATR_0, which is the main concern of the present chapter, will be introduced in §V.2.

Recall (from §II.3) that $\mathbb{N} \times \mathbb{N}$ is identified with a subset of $\mathbb{N}$ via the pairing function $(i, j) = (i + j)^2 + i$. We can use this identification to discuss binary relations on $\mathbb{N}$. A set $X \subseteq \mathbb{N} \times \mathbb{N} \subseteq \mathbb{N}$ is said to be *reflexive* if $\forall i \, \forall j \, ((i, j) \in X \to ((i, i) \in X \land (j, j) \in X))$. If X is reflexive we write $\mathrm{field}(X) = \{i : (i, i) \in X\}$ and

$$i \leq_X j \leftrightarrow (i, j) \in X,$$
$$i <_X j \leftrightarrow ((i, j) \in X \land (j, i) \notin X).$$

DEFINITION V.1.1 (countable well orderings). The following definitions are made within RCA$_0$. Let $X \subseteq \mathbb{N}$ be reflexive. We say that X is *well founded* if it has no infinite descending sequence, i.e., there is no $f : \mathbb{N} \to \text{field}(X)$ such that $f(n+1) <_X f(n)$ for all $n \in \mathbb{N}$. We say that X is a *countable linear ordering* if it is a reflexive linear ordering of its field, i.e.,

$$\forall i \, \forall j \, \forall k \, ((i \leq_X j \wedge j \leq_X k) \to i \leq_X k),$$

$$\forall i \, \forall j \, ((i \leq_X j \wedge j \leq_X i) \to i = j),$$

$$\forall i \, \forall j \, (i, j \in \text{field}(X) \to (i \leq_X j \vee j \leq_X i)).$$

We say that X is a *countable well ordering* if it is both well founded and a countable linear ordering.

Let $\text{WF}(X)$, $\text{LO}(X)$, and $\text{WO}(X)$ be formulas saying that X is respectively well founded, a countable linear ordering, and a countable well ordering. Clearly $\text{WO}(X)$ is a Π_1^1 formula with a single free variable, X. The main result of this section is theorem V.1.9 which says that the Π_1^1 formula $\text{WO}(X)$ is not equivalent to any Σ_1^1 formula.

An important tool is the *Kleene/Brouwer ordering*. Recall that Seq is the set of codes for finite sequences of natural numbers. We define KB to be the set of all pairs $(\sigma, \tau) \in \text{Seq} \times \text{Seq}$ such that either $\sigma \supseteq \tau$ (i.e., $\text{lh}(\sigma) \geq \text{lh}(\tau) \wedge \forall i \, (i < \text{lh}(\tau) \to \sigma(i) = \tau(i)))$ or

$$\exists j < \min(\text{lh}(\sigma), \text{lh}(\tau)) \, [\sigma(j) < \tau(j) \wedge \forall i < j \, (\sigma(i) = \tau(i))].$$

Thus $\leq_{\text{KB}}$ is a binary relation whose field is Seq. It is straightforward to verify (in RCA$_0$ for instance) that $\leq_{\text{KB}}$ is a dense liner ordering with no left endpoint and with the empty sequence $\langle \rangle$ as its right endpoint.

DEFINITION V.1.2 (the Kleene/Brouwer ordering). The following definition is made in RCA$_0$. Recall that a *tree* is a set $T \subseteq \text{Seq}$ such that $\forall \sigma \, \forall \tau \, ((\sigma \in \text{Seq} \wedge \sigma \subseteq \tau \wedge \tau \in T) \to \sigma \in T)$. We write $\text{KB}(T) = \text{KB} \cap (T \times T) = $ the restriction of $\leq_{\text{KB}}$ to T, i.e.,

$$\text{KB}(T) = \{(\sigma, \tau): \sigma, \tau \in T \wedge \sigma \leq_{\text{KB}} \tau\}.$$

Thus $\text{KB}(T)$ is a linear ordering. We refer to $\text{KB}(T)$ as the *Kleene/Brouwer ordering of* T.

Recall that a *path* through a tree T is a function $f : \mathbb{N} \to \mathbb{N}$ such that $\forall n \, (f[n] \in T)$, where $f[n] = \langle f(0), f(1), \ldots, f[n-1] \rangle$. The following lemma says that T has a path if and only if $\text{KB}(T)$ is not a well ordering.

LEMMA V.1.3. *The following is provable in* ACA$_0$. *Let* $T \subseteq \text{Seq}$ *be a tree. Then*

$$\text{WO}(\text{KB}(T)) \leftrightarrow \forall f \, \exists n \, (f[n] \notin T).$$

PROOF. If f is a path through T, we have $f[n + 1] \nsupseteq f[n]$ hence $f[n + 1] <_{\mathrm{KB}} f[n]$ for all n, so $\langle f[n] : n \in \mathbb{N} \rangle$ is a descending sequence witnessing that $\mathrm{KB}(T)$ is not a well ordering.

Conversely, suppose that T is not well ordered under $\leq_{\mathrm{KB}}$. Let $\langle \sigma_m : m \in \mathbb{N} \rangle$ be a descending sequence, i.e., $\sigma_{m+1} <_{\mathrm{KB}} \sigma_m$ and $\sigma_m \in T$ for all $m \in \mathbb{N}$. Put $S = \{\sigma \in T : \exists m\, (\sigma \subseteq \sigma_m)\}$. The existence of S is assured by arithmetical comprehension, and clearly S is a subtree of T.

We claim that S is finitely branching. Suppose not. Let $\sigma \in S$ be such that $\sigma^\frown\langle i \rangle \in S$ for infinitely many $i \in \mathbb{N}$. If $\sigma^\frown\langle i \rangle \in S$ let $f(i)$ be the least m such that $\sigma^\frown\langle i \rangle \subseteq \sigma_m$. Then $i < j$ implies $\sigma_{f(i)} <_{\mathrm{KB}} \sigma_{f(j)}$ so $\{\sigma_{f(i)} : \sigma^\frown\langle i \rangle \in S\}$ is an infinite ascending sequence under $\leq_{\mathrm{KB}}$. This contradicts the fact that $\{\sigma_m : m \in \mathbb{N}\}$ is an infinite descending sequence. Our claim is proved.

Clearly S is infinite so by König's lemma (a consequence of ACA_0; see section III.7), S has a path. Hence T has a path. This completes the proof of lemma V.1.3. □

The next lemma is a formal version of the well known *Kleene normal form theorem* for Σ^1_1 relations.

LEMMA V.1.4 (normal form theorem). *Let $\varphi(X)$ be a Σ^1_1 formula. Then we can find an arithmetical (in fact Σ^0_0) formula $\theta(\sigma, \tau)$ such that ACA_0 proves*

$$\forall X\, (\varphi(X) \leftrightarrow \exists f\, \forall m\, \theta(X[m], f[m])).$$

(Here f ranges over total functions from $\mathbb{N}$ into $\mathbb{N}$. Also

$$X[m] = \langle \xi_0, \xi_1, \ldots, \xi_{m-1} \rangle$$

where $\xi_i = 1$ if $i \in X$, 0 if $i \notin X$. Note that $\varphi(X)$ may contain free variables other than X. If this is the case, then $\theta(\sigma, \tau)$ will also contain those free variables.)

PROOF. Let us first prove the result under the assumption that φ is arithmetical. In this special case we can write φ in prenex normal form as

$$\forall m_1 \exists n_1 \cdots \forall m_k \exists n_k\, \chi(X, m_1, n_1, \ldots, m_k, n_k)$$

where χ is quantifier-free and does not mention X except in atomic formulas of the form $m_i \in X$, $n_i \in X$, $i = 1, \ldots, k$. (We can accomplish this by treating $+$ and $\cdot$ as ternary relation symbols instead of binary function symbols.) Given $X \subseteq \mathbb{N}$ we say that $g_i : \mathbb{N}^i \to \mathbb{N}$, $i = 1, \ldots, k$ are *Skolem functions for X* if

$$\forall m_1 \cdots \forall m_k\, \chi(X, m_1, g_1(m_1), \ldots, m_k, g_k(m_1, \ldots, m_k)).$$

From arithmetical comprehension it follows that $\varphi(X)$ holds if and only if there exist Skolem functions for X. Thus $\varphi(X)$ holds if and only if $\exists f\, \forall m\, \theta(X[m], f[m])$ where $\theta(X[m], f[m])$ is the following arithmetical (in fact Σ_0^0) assertion: for all $m_1, \ldots, m_k$ less than m, if $\langle 1, m_1 \rangle$, $f(\langle 1, m_1 \rangle)$, $\ldots$, $\langle k, m_1, \ldots, m_k \rangle$, $f(\langle k, m_1, \ldots, m_k \rangle)$ are all less than m, then

$$\chi(X, m_1, f(\langle 1, m_1 \rangle), \ldots, m_k, f(\langle k, m_1, \ldots, m_k \rangle))$$

holds. This proves lemma V.1.4 in the special case when φ is arithmetical.

Suppose now that φ is Σ_1^1. Let $\varphi(X) \equiv \exists Y \, \varphi'(X, Y)$ where φ' is arithmetical. By the special case which was already proved, we have

$$\forall X \, \forall Y \, [\varphi'(X, Y) \leftrightarrow \exists f \, \forall m \, \theta'((X \oplus Y)[m], f[m])]$$

where θ' is arithmetical (in fact Σ_0^0) and

$$X \oplus Y = \{2n : n \in X\} \cup \{2n + 1 : n \in Y\}.$$

By a straightforward use of the pairing function we can convert θ' to another arithmetical (in fact Σ_0^0) formula θ such that

$$\forall X \, (\exists Y \, \exists f \, \forall m \, \theta'((X \oplus Y)[m], f[m]) \leftrightarrow \exists h \, \forall m \, \theta(X[m], h[m])).$$

This completes the proof of lemma V.1.4.   □

One of the purposes of this book is to study the formalization of ordinary mathematics within subsystems of second order arithmetic. Accordingly, we shall now relate the previous lemma to the branch of ordinary mathematics known as *classical descriptive set theory*. An excellent textbook for this theory is Kechris [138]. The notion which we now require from classical descriptive set theory is that of *analytic set*. Of course we face the usual difficulty that L_2 (the language of second order arithmetic) is not powerful enough to discuss analytic sets directly. But, also as usual, there is no real loss since we can instead discuss *codes* for analytic sets. The appropriate codes are given by definitions V.1.5 and V.1.6 below.

As our underlying space for descriptive set theory, we choose the *Cantor space*. (Our reasons for this choice are explained in remark V.5.8, below.) When formalizing descriptive set theory in L_2, we shall often identify a set $X \subseteq \mathbb{N}$ with its characteristic function $X : \mathbb{N} \to \{0, 1\}$ given by $X(n) = 1$ if $n \in X$, 0 if $n \notin X$. Such a characteristic function will be called a *point of the Cantor space*. Thus each $X \subseteq \mathbb{N}$ is a point of the Cantor space, and conversely. We shall use $2^{\mathbb{N}}$ informally to denote the Cantor space (just as in §II.4 we used $\mathbb{R}$ informally to denote the space of all real numbers).

The following definitions are made in RCA_0.

DEFINITION V.1.5 (analytic codes). An *analytic code* (i.e., a code for an analytic subset of the Cantor space $2^{\mathbb{N}}$) is a set $A \subseteq \mathrm{Seq}$ such that A is a tree and each finite sequence $\sigma \in A$ is of the form $\sigma = \langle (\xi_0, m_0), \ldots, (\xi_{k-1}, m_{k-1}) \rangle$ where $\forall j < k \, (\xi_j \in \{0, 1\} \wedge m_j \in \mathbb{N})$. In other words, $\langle \xi_0, \ldots, \xi_{k-1} \rangle \in 2^{<\mathbb{N}}$ and $\langle m_0, \ldots, m_{k-1} \rangle \in \mathrm{Seq}$.

DEFINITION V.1.6 (analytic codes, continued). If $X \in 2^{\mathbb{N}}$ and A is an analytic code, we say that X *is a point of* A (abbreviated $X \in A$) if

$\exists f \,\forall k\, A(X[k], f[k])$. Here f ranges over total functions from $\mathbb{N}$ into $\mathbb{N}$, and we write $A(X[k], f[k])$ to mean that

$$\langle (X(0), f(0)), \ldots, (X(k-1), f(k-1)) \rangle \in A.$$

(There is a conflict here between the new notation $X \in A$ and the old notation $\sigma \in A$ of definition V.1.5. However, this conflict should cause no confusion since X is a point of $2^{\mathbb{N}}$ while $\sigma \in$ Seq.) We abbreviate $\neg(X \in A)$ as $X \notin A$.

The following theorem (which is nothing but a reformulation of lemma V.1.4) says that analytic sets are in a sense the same thing as Σ_1^1 formulas. This theorem will be applied in §§V.3, V.5, and V.6.

THEOREM V.1.7 (analytic codes and Σ_1^1 formulas). *For an analytic code A, the formula $X \in A$ is Σ_1^1. Conversely, for any Σ_1^1 formula $\varphi(X)$, ACA_0 proves*

$$(\exists \text{ analytic code } A)\, \forall X\, (X \in A \leftrightarrow \varphi(X)).$$

PROOF. It is obvious from definition V.1.6 that the formula $X \in A$ is Σ_1^1. For the converse, given a Σ_1^1 formula $\varphi(X)$, let $\theta(\sigma, \tau)$ be an arithmetical formula as provided by lemma V.1.4. Thus ACA_0 proves $\forall X\, (\varphi(X) \leftrightarrow \exists f \,\forall j\, \theta(X[j], f[j]))$. By arithmetical comprehension, let A be the set of all $\sigma \in$ Seq of the form $\sigma = \langle (\xi_0, m_0), \ldots, (\xi_{k-1}, m_{k-1}) \rangle$ such that $\forall j < k\, (\xi_j < 2)$ and

$$\forall j \leq k\, \theta(\langle \xi_0, \ldots, \xi_{j-1} \rangle, \langle m_0, \ldots, m_{j-1} \rangle).$$

Clearly A is a tree and has the other desired properties. □

The following uniform variant of theorem V.1.7 will sometimes be useful.

THEOREM V.1.7'. *For any Σ_1^1 formula $\varphi(n, X)$, ACA_0 proves the existence of a sequence of analytic codes $\langle A_n : n \in \mathbb{N} \rangle$ such that $\forall n \,\forall X\, (\varphi(n, X) \leftrightarrow X \in A_n)$.*

PROOF. Lemma V.1.4 provides an arithmetical formula $\theta(n, \sigma, \tau)$ such that ACA_0 proves $\forall n \,\forall X\, (\varphi(n, X) \leftrightarrow \exists f \,\forall j\, \theta(n, X[j], f[j]))$. Let A be the set of all ordered pairs $(n, \langle (\xi_0, m_0), \ldots, (\xi_{k-1}, m_{k-1}) \rangle)$ such that $\forall j < k\, (\xi_j < 2)$ and

$$\forall j \leq k\, \theta(n, \langle \xi_0, \ldots, \xi_{j-1} \rangle, \langle m_0, \ldots, m_{j-1} \rangle).$$

Then A encodes the sequence $\langle A_n : n \in \mathbb{N} \rangle$ where $A_n = \{\sigma : (n, \sigma) \in A\}$. This sequence has the desired properties. □

We now relate the notion of analytic code to the Kleene/Brouwer ordering. Given an analytic code A and a point $X \in 2^{\mathbb{N}}$, put

$$\mathsf{T}_A(X) = \{\tau \in \text{Seq} : A(X[\mathrm{lh}(\tau)], \tau)\}.$$

Thus $\mathsf{T}_A(X)$ is a tree, and $X \in A$ holds if and only if $\mathsf{T}_A(X)$ has a path. Combining the previous results, we obtain:

LEMMA V.1.8. *For any Π_1^1 formula $\psi(X)$, ACA_0 proves the existence of an analytic code A such that*

$$\forall X \, (\psi(X) \leftrightarrow \mathrm{WO}(\mathrm{KB}(\mathrm{T}_A(X)))).$$

PROOF. Let $\varphi(X)$ be the Σ_1^1 formula $\neg\psi(X)$. By theorem V.1.7 we get an analytic code A such that $\forall X \, (\varphi(X) \leftrightarrow X \in A)$. Thus $\forall X$ $(\psi(X) \leftrightarrow \mathrm{T}_A(X)$ has no path). By lemma V.1.3 we get $\forall X \, (\psi(X) \leftrightarrow \mathrm{WO}(\mathrm{KB}(\mathrm{T}_A(X))))$. This completes the proof.   □

The above lemma may be interpreted as saying that the Π_1^1 formula $\mathrm{WO}(X)$ is in a sense "universal" Π_1^1, provably in ACA_0. We are now ready to prove the next theorem, which says that $\mathrm{WO}(X)$ is not equivalent to any Σ_1^1 formula, again provably in ACA_0.

THEOREM V.1.9. *For any Σ_1^1 formula $\varphi(X)$, ACA_0 proves*

$$\neg \forall X \, (\varphi(X) \leftrightarrow \mathrm{WO}(X)).$$

PROOF. We reason in ACA_0. Suppose by way of contradiction that $\forall X \, (\varphi(X) \leftrightarrow \mathrm{WO}(X))$ where $\varphi(X)$ is Σ_1^1. We diagonalize by putting $\psi(X) \equiv (X$ is an analytic code and $\neg\varphi(\mathrm{KB}(\mathrm{T}_X(X))))$. Since $\psi(X)$ is Π_1^1, lemma V.1.8 provides an analytic code A such that $\forall X \, (\psi(X) \leftrightarrow \mathrm{WO}(\mathrm{KB}(\mathrm{T}_A(X)))$. Thus $\psi(A)$ if and only if $\neg\psi(A)$. This contradiction completes the proof.   □

We shall now reformulate the previous theorem in the terminology of analytic sets. A well known theorem of classical descriptive set theory, due to Lusin and Sierpinski, says that the set of all countable well orderings is not analytic. We shall now show that this theorem is provable in ACA_0.

THEOREM V.1.10. *The following is provable in ACA_0. There is no analytic code A such that*

$$\forall X \, (X \in A \leftrightarrow \mathrm{WO}(X)).$$

PROOF. This is equivalent to theorem V.1.9 in view of theorem V.1.7.   □

There is a stronger theorem (also due to Lusin and Sierpinski) which reads as follows. *Let A be an analytic set of countable well orderings; then the order types of the well orderings in A are bounded by some countable ordinal.* This is known as the Σ_1^1 *bounding principle*. We shall see (in §V.6) that this theorem is not provable in ACA_0 but is provable in the stronger formal system ATR_0. We shall also see (in §§V.3, V.4, and V.5) that many other theorems of classical descriptive set theory are not provable in ACA_0 but are provable in ATR_0.

We end the section with some exercises.

EXERCISE V.1.11. Show that ACA_0 is equivalent over RCA_0 to the assertion that for all trees $T \subseteq \mathrm{Seq}$, $\mathrm{WO}(\mathrm{KB}(T)) \leftrightarrow \forall f \, \exists n \, f[n] \notin T$.

Hint: The forward direction is given by lemma V.1.3. For the reversal, use a tree as in the proof of theorem III.7.2.

EXERCISES V.1.12. Let A_n, $n \in \mathbb{N}$, be a sequence of analytic codes.

1. Prove in RCA$_0$ that there exists an analytic code A' such that

$$\forall X \, (X \in A' \leftrightarrow \exists n \, (X \in A_n)).$$

2. Prove in Σ^1_1-AC$_0$ that there exists an analytic code A'' such that

$$\forall X \, (X \in A'' \leftrightarrow \forall n \, (X \in A_n)).$$

3. Prove in RCA$_0$ that there exists an analytic code A^* such that

$$\forall n \, \forall X \, (\{n\} \cup \{n + m + 1 : m \in X\} \in A^* \leftrightarrow X \in A_n).$$

Note: These analytic codes are denoted $A' = \bigcup_{n\in\mathbb{N}} A_n$, $A'' = \bigcap_{n\in\mathbb{N}} A_n$, $A^* = \bigoplus_{n\in\mathbb{N}} A_n$ respectively.

Notes for §V.1. For background on descriptive set theory, including analytic sets and the Kleene/Brouwer ordering, see Kechris [138], Mansfield/Weitkamp [171], Moschovakis [191], and Rogers [208]. The result stated in exercise V.1.11 is due to Hirst [121].

V.2. The Formal System ATR$_0$

The purpose of this section is to introduce the formal system ATR$_0$ and to illustrate some of the proof techniques which are available in it. (Another important proof technique, the method of pseudohierarchies, will be introduced in §V.4.)

The acronym ATR stands for *arithmetical transfinite recursion*. Before discussing arithmetical transfinite recursion, we shall first discuss a related but much weaker principle known as *arithmetical transfinite induction*.

In ordinary mathematics, a fundamental property of countable well orderings is that proofs by transfinite induction may be carried out along them. In other words, if we have a countable well ordering X and we are trying to prove that some property $\varphi(j)$ holds for each $j \in \text{field}(X)$, we may legitimately assume that $\varphi(i)$ holds for all $i <_X j$. We now point out that this procedure is formally valid in ACA$_0$ provided $\varphi(j)$ is arithmetical. In other words:

LEMMA V.2.1 (arithmetical transfinite induction). *For any arithmetical formula $\varphi(j)$, ACA$_0$ proves*

$$(\text{WO}(X) \wedge \forall j \, (\forall i \, (i <_X j \rightarrow \varphi(i)) \rightarrow \varphi(j))) \rightarrow \forall j \, \varphi(j).$$

PROOF. By arithmetical comprehension, let Y be the set of all j such that $\neg\varphi(j)$. By hypothesis we have that for all $j \in Y$ there exists $i \in Y$ such that $i <_X j$. If Y is nonempty, define $f : \mathbb{N} \to Y$ by $f(0) = \text{least}$ $j \in Y$; $f(n+1) = \text{least } i \in Y$ such that $i <_X f(n)$. Thus f is a descending sequence through X, contradicting the assumption WO(X). Hence Y is empty, i.e., $\forall j \, \varphi(j)$. □

The above lemma says that arithmetical transfinite induction is provable in ACA$_0$. Having made this preliminary remark, we now turn to the discussion of arithmetical transfinite recursion. It will become clear that arithmetical transfinite recursion (unlike arithmetical transfinite induction) is very much stronger than ACA$_0$.

The idea of arithmetical transfinite recursion is as follows. Suppose we are given a countable well ordering X and an arithmetical formula $\theta(n, Y)$. To each $j \in \text{field}(X)$ we wish to associate a set Y_j. We define the Y_j's by transfinite recursion along X. Assume that Y_i has already been defined for each $i <_X j$. Then we define

$$Y^j = \{(m, i): i <_X j \wedge m \in Y_i\}$$

and

$$Y_j = \{n: \theta(n, Y^j)\}.$$

Intuitively, Y^j is the cumulative result of comprehension by θ applied repeatedly along X up to (but not including) j. Then Y_j is the result of applying θ one more time.

In accordance with the above informal description, we make the following formal definition.

DEFINITION V.2.2. Let $\theta(n, Y)$ be any formula. Define $H_\theta(X, Y)$ to be the formula which says that $\text{LO}(X)$ and that Y is equal to the set of all pairs (n, j) such that $j \in \text{field}(X)$ and $\theta(n, Y^j)$ where $Y^j = \{(m, i): i <_X j \wedge (m, i) \in Y\}$. Intuitively $H_\theta(X, Y)$ says that Y is the result of iterating θ along X. We also define $H_\theta(k, X, Y)$ to be the formula which says that $\text{LO}(X)$ and $k \in \text{field}(X)$ and Y is equal to the set of all pairs (n, j) as above such that in addition $j <_X k$. Intuitively $H_\theta(k, X, Y)$ says that $Y = Y^k = $ the result of iterating θ along X up to k. Thus $H_\theta(X, Y)$ and $k \in \text{field}(X)$ imply $H_\theta(k, X, Y^k)$.

(Note that $\theta(n, Y)$ may contain free variables other than those displayed. If this is the case, then $H_\theta(X, Y)$ and $H_\theta(k, X, Y)$ will also contain those free variables. Note also that if $\theta(n, Y)$ is arithmetical, then so is $H_\theta(X, Y)$.)

LEMMA V.2.3. *The following is provable in* ACA$_0$. *Let* WO(X) *be assumed. Then there is at most one Y such that $H_\theta(X, Y)$. Also, for each k, there is at most one Y such that $H_\theta(k, X, Y)$.*

PROOF. Suppose WO(X) and $H_\theta(X, Y)$ and $H_\theta(X, Z)$. We shall show that $Y^j = Z^j$ for all j, by arithmetical transfinite induction (lemma V.2.1). By the induction hypothesis we may assume that $Y^i = Z^i$ for all $i <_X j$. Then $Y_i = \{m: \theta(m, Y^i)\} = \{m: \theta(m, Z^i)\} = Z_i$. Hence $Y^j = \{(m, i): i <_X j \wedge m \in Y_i\} = \{(m, i): i <_X j \wedge m \in Z_i\} = Z^j$. By arithmetical transfinite induction we have $Y^j = Z^j$ for all j. It follows easily that $Y = Z$. This completes the proof of the first part. The proof of the second part is similar. $\square$

We now define the formal system of arithmetical transfinite recursion, ATR$_0$.

DEFINITION V.2.4 (definition of ATR$_0$). ATR$_0$ is the formal system in the language of second order arithmetic whose axioms consist of ACA$_0$ plus all instances of

$$\forall X \, (\mathrm{WO}(X) \rightarrow \exists Y \, H_\theta(X, Y))$$

where θ is arithmetical.

The system ATR$_0$ is properly stronger than ACA$_0$. To see this, consider the minimum ω-model

$$\mathrm{ARITH} = \{Z \subseteq \omega \colon Z \text{ is arithmetical}\}$$

of ACA$_0$ (§§I.3, III.1, VIII.1).

PROPOSITION V.2.5. *The ω-model ARITH is not a model of ATR$_0$.*

PROOF. Let $\theta(n, Y)$ be the arithmetical formula which says that $n \in \mathrm{TJ}(Y)$, i.e., n is an element of the Turing jump of Y. Let X be the canonical reflexive well ordering of $\mathbb{N}$, i.e., $X = \{(i, j) \colon i \leq j\}$. Then $\mathrm{WO}(X)$ holds and there exists a unique set Y such that $H_\theta(X, Y)$ holds. Namely $Y = \{(n, j) \colon n \in Y_j\}$ where Y_j is the Turing jump of $Y^j = \{(m, i) \colon i < j \wedge m \in Y_i\}$. Thus Y^j is essentially $\emptyset^{(j)}$, the jth Turing jump of the empty set. Thus $\mathrm{ARITH} = \{Z \subseteq \omega \colon \exists j \, (Z \text{ is recursive in } Y^j)\}$. Since Y_j is the Turing jump of Y^j and hence is not recursive in Y^j, it follows that $Y \notin \mathrm{ARITH}$. (Another way to see this is to observe that $Y = \emptyset^{(\omega)} =$ essentially the truth set for first order arithmetic. Hence $Y \notin \mathrm{ARITH}$ by Tarski's theorem on the undefinability of truth.) Thus for this particular X and θ we have $\mathrm{ARITH} \models (\mathrm{WO}(X) \wedge \neg \exists Y \, H_\theta(X, Y))$. So ARITH is not a model of ATR$_0$. $\square$

For those readers who happen to be familiar with hyperarithmetical sets (see also §VIII.3), we point out the following:

PROPOSITION V.2.6. *The ω-model*

$$\mathrm{HYP} = \{Z \subseteq \omega \colon Z \text{ is hyperarithmetical}\}$$

is not a model of ATR$_0$.

PROOF. Let $\theta(n, Y)$ say that n belongs to the Turing jump of Y. Let X be a *recursive pseudowellordering*, i.e., a recursive linear ordering which has infinite descending sequences but no hyperarithmetical infinite descending sequences. Thus $H_\theta(X, Y)$ says that Y is what is sometimes known as a pseudohierarchy on X (compare §V.4). By lemma VIII.3.23 (see also Harrison [106]), there is no hyperarithmetical Y such that $H_\theta(X, Y)$. Thus for this particular X and θ we have $\mathrm{HYP} \models (\mathrm{WO}(X) \wedge \neg \exists Y \, H_\theta(X, Y))$. $\square$

For more information on models of ATR_0, see chapters VII and VIII of the present work, and also Simpson [234].

It will become clear in this chapter that the formal system ATR_0 is much more powerful than ACA_0 from the standpoint of ordinary mathematical practice. We shall see that many theorems of ordinary mathematics which are not provable in ACA_0 are provable in ATR_0. Among these theorems are: Lusin's theorem on Borel sets (§V.3), the perfect set theorem (every uncountable analytic set contains a perfect set, §V.4), determinacy of open games in Baire space (§V.8), the open Ramsey theorem (§V.9), and the Ulm structure theorem for countable reduced Abelian p-groups (§V.7). Furthermore, in accordance with our theme of Reverse Mathematics (§I.9), we shall obtain reversals showing that (special cases of) all of these theorems are in fact equivalent to ATR_0 over a weak base theory. For example, the fact that every uncountable closed subset of the Cantor space contains a perfect set is equivalent to ATR_0 over ACA_0. Thus the axioms of ATR_0 are necessary to prove the perfect set theorem, in the sense that no weaker axioms could possibly suffice. The same remark applies to each of the other theorems just mentioned.

Thus ATR_0 plays a significant role with respect to the formalization of ordinary mathematics. A partial explanation for this phenomenon has to do with *countable ordinals*. Countable ordinals arise in a variety of contexts in ordinary mathematics. Sometimes they appear explicitly in the statement of a theorem (e.g., Ulm's theorem, or various properties of Borel sets). At other times they are involved overtly or covertly in the proof of a theorem. (This is the case with the open Ramsey theorem, for example.) It will turn out that ATR_0 is the weakest set of axioms which permits the development of a decent theory of countable ordinals.

A countable ordinal is essentially an equivalence class of countable well orderings under the equivalence relation of isomorphism. The fundamental fact that the countable ordinals are linearly ordered depends on having sufficiently many *comparison maps*, i.e., isomorphisms, between countable well orderings. We shall now show that ATR_0 proves the existence of the needed comparison maps. In §V.6 it will turn out that ATR_0 is actually equivalent to the existence of these comparison maps.

DEFINITION V.2.7 (comparison maps). The following definitions are made in RCA_0. If $LO(X)$ and $LO(Y)$, we say that X is *isomorphic to Y* if there exists an isomorphism between them, i.e., a function $f : \text{field}(X) \to \text{field}(Y)$ such that $\forall i \, \forall j \, (i \leq_X j \leftrightarrow f(i) \leq_Y f(j))$ and $(\forall k \in \text{field}(Y))$ $(\exists i \in \text{field}(X))\,(f(i) = k)$. We write $|X| = |Y|$ to mean that X is isomorphic to Y. We write $f : |X| = |Y|$ to mean that f is an isomorphism of X onto Y.

We say that X is an *initial section* of Y if there exists $k \in \text{field}(Y)$ such that $\forall i \, \forall j \, (i \leq_X j \leftrightarrow (i \leq_Y j \wedge j <_Y k))$. In this case we call X the *initial section of Y determined by k*.

We write $f: |X| < |Y|$ to mean that f is an isomorphism of X onto some initial section of Y. We write $f: |X| > |Y|$ to mean that f is an isomorphism of some initial section of X onto Y. The notations $|X| < |Y|$, $|X| > |Y|$, $f: |X| \leq |Y|$, $f: |X| \geq |Y|$, $|X| \leq |Y|$, and $|X| \geq |Y|$ are defined in the obvious way.

We say that f is a *comparison map from X to Y* if $f: |X| \leq |Y|$ or $f: |X| \geq |Y|$. We say that X and Y are *comparable* if there exists a comparison map from X to Y.

LEMMA V.2.8 (uniqueness of comparison maps). *The following is provable in* RCA$_0$. *If* WO(X) *and* LO(Y) *and* X *and* Y *are comparable, then the comparison map is unique.*

PROOF. We may restrict ourselves to the special case when X and Y are isomorphic. Given two isomorphisms $f: |X| = |Y|$ and $g: |X| = |Y|$, by Δ_1^0 comprehension let Z be the set of $m \in \text{field}(X)$ such that $f(m) \neq g(m)$. Clearly for all $m \in Z$ there exists $n \in Z$ such that $n <_X m$. Thus, if Z is nonempty, we can use primitive recursion (§II.3) to define $h: \mathbb{N} \to Z$ by $h(0) = $ any element of Z, $h(i + 1) = $ least $n \in Z$ such that $n <_X h(i)$. Then h is a descending sequence through X. This contradicts WO(X). Hence Z is empty, i.e., $f = g$. $\qquad\square$

LEMMA V.2.9 (comparability of countable well orderings). *It is provable in* ATR$_0$ *that any two countable well orderings are comparable. In other words,* ATR$_0$ *proves*

$$\forall W \, \forall X \, ((\text{WO}(W) \wedge \text{WO}(X)) \to (|W| \leq |X| \vee |W| \geq |X|)).$$

PROOF. Assume WO(W) and WO(X). Let $\theta(n, Y)$ say that $n \in \text{field}(W)$ and Y is an isomorphism of the initial section of W determined by n onto some initial section of X. Clearly θ is arithmetical, so by arithmetical transfinite recursion let Y be such that $H_\theta(X, Y)$ holds. Thus $(n, j) \in Y$ if and only if Y^j is an isomorphism of the initial section of W determined by n onto some initial section of X. By arithmetical transfinite induction (lemma V.2.1), it follows straightforwardly that Y is a comparison map between W and X. $\qquad\square$

DEFINITION V.2.10 (countable ordinals). Within RCA$_0$ we define a *countable ordinal code* to be a countable well ordering (in the sense of definition V.1.1). Two countable ordinal codes X and Y are said to be *equal* (as countable ordinals) if $|X| = |Y|$. We use $\alpha, \beta, \gamma, \ldots$ as special variables ranging over countable ordinals. Thus $\alpha = |X|$ means that X is a code for the countable ordinal α. If $\alpha = |X|$ and $\beta = |Y|$ we write $\alpha < \beta$ to mean that $|X| < |Y|$, etc.

Lemma V.2.9 says that the countable ordinals (as just defined) form a linear ordering. In §V.6 we shall see that ATR$_0$ is the weakest natural theory in which this can be proved. Thus we shall have a partial explanation of why ATR_0 is needed for the proofs of many ordinary mathematical theorems which depend (explicitly or implicitly) on countable ordinals.

Notes for §V.2. The system ATR_0 was introduced by Friedman [68, 69] (see also Friedman [62, chapter II]) and Steel [256, chapter I]. Other key references on ATR_0 are Friedman/McAloon/Simpson [76] and Simpson [234, 235, 247].

V.3. Borel Sets

In this section and the next, we shall show that several basic theorems of classical descriptive set theory are provable in ATR_0. The theorems in question concern Borel and analytic sets.

As our basic space for descriptive set theory we take the Cantor space, $2^{\mathbb{N}}$. As explained in §V.1, a point of the Cantor space is any set $X \subseteq \mathbb{N}$. Such a set is identified with its characteristic function $X: \mathbb{N} \to \{0,1\}$ where $X(n) = 1$ if $n \in X$, 0 if $n \notin X$.

In §V.1 we introduced the appropriate codes for analytic sets (definitions V.1.5 and V.1.6). We now introduce codes for Borel sets.

DEFINITION V.3.1 (Borel codes). Within RCA_0 we define a *Borel code* (i.e., a code for a Borel subset of $2^{\mathbb{N}}$) to be a set $B \subseteq$ Seq such that B is a tree, B has no path, and there is exactly one $m \in \mathbb{N}$ such that $\langle m \rangle \in B$.

Let $\sigma \in B$ where B is a Borel code. We say that σ is an *interior node* of B if $\exists n \, (\sigma^\frown \langle n \rangle \in B)$. Otherwise σ is called an *end node* of B.

DEFINITION V.3.2 (evaluation maps). Given a Borel code B and a point $X \in 2^{\mathbb{N}}$, an *evaluation map for B at X* is defined in RCA_0 to be a function $f: B \to \{0, 1\}$ such that:

(i) if σ is an end node of B, then

$$f(\sigma) = \begin{cases} 1 & \text{if } \sigma(\text{lh}(\sigma) - 1) = 2n + 2 + X(n), \\ 0 & \text{if } \sigma(\text{lh}(\sigma) - 1) = 2n + 3 - X(n), \\ 1 & \text{if } \sigma(\text{lh}(\sigma) - 1) = 1, \\ 0 & \text{if } \sigma(\text{lh}(\sigma) - 1) = 0; \end{cases}$$

(ii) if σ is an interior node of B and $\sigma \neq \langle \rangle$, then

$$f(\sigma) = \begin{cases} 1 & \text{if } \sigma(\text{lh}(\sigma) - 1) \text{ is odd and } \forall n \, (\sigma^\frown \langle n \rangle \in B \to f(\sigma^\frown \langle n \rangle) = 1), \\ 1 & \text{if } \sigma(\text{lh}(\sigma) - 1) \text{ is even and } \exists n \, (\sigma^\frown \langle n \rangle \in B \wedge f(\sigma^\frown \langle n \rangle) = 1), \\ 0 & \text{otherwise}; \end{cases}$$

(iii) $f(\langle \rangle) = f(\langle m \rangle)$ for the unique m such that $\langle m \rangle \in B$.

In order to motivate the above definition, note that: (i) an end node corresponds to a subbasic open set $\{X \in 2^{\mathbb{N}}: X(n) = 1\}$, $\{X \in 2^{\mathbb{N}}: X(n) = 0\}$, $2^{\mathbb{N}}$, or $\emptyset$; (ii) an interior node other than $\langle\rangle$ corresponds to an operation of countable intersection or union. Intuitively, the class of Borel sets is the smallest class containing the subbasic neighborhoods (i) and closed under the operations (ii). This will become clearer in definition V.3.4 and lemma V.3.5.

LEMMA V.3.3 (existence of evaluation maps). *The following is provable in* ATR$_0$. *Given* $X \in 2^{\mathbb{N}}$ *and a Borel code* B, *there exists an evaluation map for* B *at* X. *This evaluation map is unique.*

PROOF. We reason in ATR$_0$. Since B has no path, the Kleene/Brouwer ordering KB(B) is a well ordering (definition V.1.2, lemma V.1.3). We define the desired evaluation map $f : B \to \{0, 1\}$ by means of arithmetical transfinite recursion (definition V.2.4) along KB(B). Uniqueness of f is proved by arithmetical transfinite induction (lemma V.2.1) along KB(B).

The details of the recursion are as follows. We first write down an arithmetical formula $\theta(n, Y)$ which is virtually a transcription of definition V.3.2. Thus $\theta(n, Y)$ says: (i) if σ is an end node of B and $\sigma(\mathrm{lh}(\sigma) - 1) = 2m + 2 + X(m)$, then $n = 1$, etc.; (ii) if $\sigma \neq \langle\rangle$ is an interior node of B and $\sigma(\mathrm{lh}(\sigma) - 1)$ is odd and $\forall m \, (\sigma^\frown\langle m\rangle \in B \to (1, \sigma^\frown\langle m\rangle) \in Y)$, then $n = 1$, etc.; and (iii) if $\sigma = \langle\rangle$ and $(1, \langle m\rangle) \in Y$ for some m, then $n = 1$, etc. Then, by arithmetical transfinite recursion along KB(B), there exists Y such that $\mathrm{H}_\theta(\mathrm{KB}(B), Y)$. We set $f = \{(\sigma, n): (n, \sigma) \in Y\}$. For each $\sigma \in B$ set $f^\sigma = \{(\tau, n): \tau \leq_{\mathrm{KB}} \sigma \wedge (\tau, n) \in f\}$. By arithmetical transfinite induction along KB(B) it is straightforward to verify that f^σ is a function from $\{\tau: \tau \leq_{\mathrm{KB}(B)} \sigma\}$ into $\{0, 1\}$ and that this function satisfies the clauses of definition V.3.2 up to σ. (Recall that $\sigma^\frown\langle m\rangle$ is strictly below σ in Kleene/Brouwer ordering.) Thus f is the desired evaluation map. Uniqueness of f follows by lemma V.2.3 or can be proved directly by arithmetical transfinite induction along KB(B). $\square$

DEFINITION V.3.4. Within ATR$_0$, given a point X and a Borel code B, we write E(f, X, B) to mean that f is an evaluation map for B at X. Note that the formula E(f, X, B) is arithmetical (in the parameter B). We say that X is a point of B (abbreviated $X \in B$) if $\exists f \, (\mathrm{E}(f, X, B) \wedge f(\langle\rangle) = 1)$.

(This new notation $X \in B$ conflicts with the notation $\sigma \in B$ of definition V.3.2. However, no confusion should result, since $X \in 2^{\mathbb{N}}$ while $\sigma \in \mathrm{Seq}$.)

We say that $X \notin B$ if $\exists f \, (\mathrm{E}(f, X, B) \wedge f(\langle\rangle) = 0)$. By lemma V.3.3 we have $\forall X \, (X \notin B \leftrightarrow \neg(X \in B))$, provided of course that B is a Borel code.

We now list some simple closure properties of the class of Borel subsets of the Cantor space $2^{\mathbb{N}}$. In the statement of the following lemma, X ranges over points of $2^{\mathbb{N}}$.

LEMMA V.3.5. *The following facts are provable in* ATR$_0$.

1. *There exist Borel codes B^0 and B^1 such that $\forall X (X \notin B^0)$ and $\forall X (X \in B^1)$.*
2. *For each $n \in \mathbb{N}$ and $\xi \in \{0, 1\}$ there exists a Borel code B_n^ξ such that $\forall X (X \in B_n^\xi \leftrightarrow X(n) = \xi)$.*
3. *Given a Borel code B, there exists a Borel code $\overline{B}$ such that $\forall X (X \in \overline{B} \leftrightarrow X \notin B)$.*
4. *Given a sequence of Borel codes $\langle B_n : n \in \mathbb{N} \rangle$, there exist Borel codes $\bigcup_{n \in \mathbb{N}} B_n$ and $\bigcap_{n \in \mathbb{N}} B_n$ such that*

$$\forall X \left(X \in \bigcup_{n \in \mathbb{N}} B_n \leftrightarrow \exists n (X \in B_n) \right)$$

and

$$\forall X \left(X \in \bigcap_{n \in \mathbb{N}} B_n \leftrightarrow \forall n (X \in B_n) \right).$$

5. *Given a Borel code B and a sequence of Borel codes $\langle B_n : n \in \mathbb{N} \rangle$, there exists a Borel code B' such that $\forall X (X \in B' \leftrightarrow X' \in B)$, where*

$$X'(n) = \begin{cases} 1 & \text{if } X \in B_n, \\ 0 & \text{if } X \notin B_n. \end{cases}$$

PROOF.

1. $B^0 = \{\langle\rangle, \langle 0 \rangle\}$; $B^1 = \{\langle\rangle, \langle 1 \rangle\}$.
2. $B_n^\xi = \{\langle\rangle, \langle 2n + 2 + \xi \rangle\}$.
3. $\overline{B} = \{\overline{\sigma} : \sigma \in B\}$ where

$$\overline{\sigma}(i) = \begin{cases} \sigma(i) + 1 & \text{if } \sigma(i) \text{ is even,} \\ \sigma(i) - 1 & \text{if } \sigma(i) \text{ is odd.} \end{cases}$$

4. $$\bigcup_{n \in \mathbb{N}} B_n = \{\langle\rangle, \langle 0 \rangle\} \cup \{\langle 0, n \rangle ^\frown \tau : n \in \mathbb{N} \wedge \tau \in B_n\};$$

$$\bigcap_{n \in \mathbb{N}} B_n = \{\langle\rangle, \langle 1 \rangle\} \cup \{\langle 1, n \rangle ^\frown \tau : n \in \mathbb{N} \wedge \tau \in B_n\}.$$

5. $B' = B \cup \{\sigma ^\frown \tau : \sigma \text{ is an end node of } B$

and $\exists n (\sigma(\mathrm{lh}(\sigma) - 1) = 2n + 3 \wedge \tau \in B_n)\}$

$\cup \{\sigma ^\frown \tau : \sigma \text{ is an end node of } B$

and $\exists n (\sigma(\mathrm{lh}(\sigma) - 1) = 2n + 2 \wedge \tau \in \overline{B_n})\}.$

It is straightforward to verify that these trees are Borel codes and have the desired properties.   □

REMARK V.3.6 (properties of Borel sets). Intuitively, lemma V.3.5 says that: (1) $\emptyset$ and $2^{\mathbb{N}}$ are Borel sets; (2) the subbasic open sets $\{X \in 2^{\mathbb{N}} : X(n) = 0\}$ and $\{X \in 2^{\mathbb{N}} : X(n) = 1\}$ are Borel sets; the class of Borel sets is closed under (3) complementation and (4) countable union and countable intersection; (5) for any Borel function $F: 2^{\mathbb{N}} \to 2^{\mathbb{N}}$ and Borel set $B \subseteq 2^{\mathbb{N}}$, the inverse image $F^{-1}(B)$ is Borel. (Here the function F is given by $F(X) = X'$.) Whenever possible we shall identify these Borel sets with their codes as constructed in the proof of lemma V.3.5. In particular we shall denote by $\emptyset$, $2^{\mathbb{N}}$, and $\{X: X(n) = \xi\}$ the corresponding Borel codes B^0, B^1, and B_n^{ξ}.

The following lemma will be useful.

LEMMA V.3.7. *The following is provable in* ATR$_0$. *Let* $\langle X_n: n \in \mathbb{N} \rangle$, $X_n \in 2^{\mathbb{N}}$, *be a sequence of points, and let* $\langle B_n: n \in \mathbb{N} \rangle$ *be a sequence of Borel codes. Then there exists a set* $Z \subseteq \mathbb{N}$ *such that* $\forall n\, (n \in Z \leftrightarrow X_n \in B_n)$.

PROOF. The proof of this lemma is similar to that of lemma V.3.3. Given any sequence of countable well orderings $\langle W_n: n \in \mathbb{N} \rangle$, we can form the sum

$$\sum_{n \in \mathbb{N}} W_n = \{((i,n),(j,n)): (i,j) \in W_n\}$$
$$\cup \{((i,m),(j,n)): (i,i) \in W_m \wedge (j,j) \in W_n \wedge m < n\}.$$

Intuitively $\sum_{n \in \mathbb{N}} W_n$ consists of W_0 followed by W_1 followed by $\ldots$. Clearly $\sum_{n \in \mathbb{N}} W_n$ is a countable well ordering. In particular, taking $W_n = \text{KB}(B_n)$, we see that $\sum_{n \in \mathbb{N}} \text{KB}(B_n)$ is a countable well ordering. Using arithmetical transfinite recursion along $\sum_{n \in \mathbb{N}} \text{KB}(B_n)$ we define a sequence of functions $\langle f_n: n \in \mathbb{N} \rangle$ and prove that $\forall n\, (f_n$ is an evaluation map for B_n at $X_n)$. The details of this recursion are as for lemma V.3.3, so we omit them. Now by arithmetical comprehension let $Z = \{n: f_n(\langle\rangle) = 1\}$. Thus $Z = \{n: X_n \in B_n\}$. This completes the proof. $\square$

A classical theorem of Souslin asserts that $\Delta_1^1 = $ Borel, i.e., every Borel set is Δ_1^1 (i.e., both analytic and coanalytic) and conversely. There is a generalization known as *Lusin's separation theorem*, which reads as follows. Let A_1 and A_0 be disjoint Σ_1^1 (i.e., analytic) sets. Then there exists a Borel set B such that $A_1 \subseteq B$ and $A_0 \cap B = \emptyset$.

We shall now show that these theorems of Souslin and Lusin are provable in ATR$_0$. We begin with the "easy half" of Souslin's theorem.

THEOREM V.3.8. *The following is provable in* ATR$_0$. *Given a Borel code* B, *there exist analytic codes* A_1 *and* A_0 *such that* $\forall X\, (X \in A_1 \leftrightarrow X \in B)$ *and* $\forall X\, (X \in A_0 \leftrightarrow X \notin B)$.

PROOF. By definition V.3.4 the formulas $X \in B$ and $X \notin B$ are Σ_1^1 (with parameter B). Hence by theorem V.1.7 there exist analytic codes A_1 and A_0 as desired. $\square$

THEOREM V.3.9 (Lusin's theorem in ATR$_0$). *Let* A_1 *and* A_0 *be analytic codes. If* $\neg \exists X\, (X \in A_1 \wedge X \in A_0)$ *then there exists a Borel code* B *such that* $\forall X\, (X \in A_1 \to X \in B)$ *and* $\forall X\, (X \in A_0 \to X \notin B)$.

PROOF. Recall the definition of analytic codes (definition V.1.5). Without loss of generality, assume that $\langle\rangle \in A_0$ and $\langle\rangle \in A_1$. Let $T = A_1 * A_0 \subseteq$ Seq be the set of all finite sequences of the form

$$\tau = \langle(\xi_0, m_0, n_0), \ldots, (\xi_{k-1}, m_{k-1}, n_{k-1})\rangle \tag{$*$}$$

such that $\tau_1 \in A_1$ and $\tau_0 \in A_0$, where $\tau_1 = \langle(\xi_0, m_0), \ldots, (\xi_{k-1}, m_{k-1})\rangle$ and $\tau_0 = \langle(\xi_0, n_0), \ldots, (\xi_{k-1}, n_{k-1})\rangle$.

Clearly T is a tree and $\langle\rangle \in T$. From the assumption $\neg\exists X\,(X \in A_1 \wedge X \in A_0)$ it follows that T has no path. Hence the Kleene/Brouwer ordering $\mathrm{KB}(T)$ is a well ordering.

We use arithmetical transfinite recursion along $\mathrm{KB}(T)$ to define for each $\tau \in T$ a tree $B_\tau \subseteq$ Seq. These trees will turn out to be Borel codes. Assume that $\tau \in T$ and that $B_{\tau^\frown\langle(\xi,m,n)\rangle}$ has already been defined for each (ξ, m, n) such that $\tau^\frown\langle(\xi, m, n)\rangle \in T$. We define B_τ as follows:

$$B_\tau = \bigcup_{\xi<2}\bigcup_{m\in\mathbb{N}}\bigcap_{\eta<2}\bigcap_{n\in\mathbb{N}} C_\tau^{\xi,m,\eta,n}$$

where

$$C_\tau^{\xi,m,\eta,n} = \begin{cases} B_{\mathrm{lh}(\tau)}^\xi(=\{X: X(\mathrm{lh}(\tau))=\xi\}) & \text{if } \xi\neq\eta, \\ B_{\tau^\frown\langle(\xi,m,n)\rangle} & \text{if } \xi=\eta \text{ and } \tau^\frown\langle(\xi,m,n)\rangle \in T, \\ B^0(=\emptyset) & \text{if } \xi=\eta \text{ and } \tau_1^\frown\langle(\xi,m)\rangle \notin A_1, \\ B^1(=2^\mathbb{N}) & \text{if } \xi=\eta \text{ and } \tau_1^\frown\langle(\xi,m)\rangle \in A_1 \\ & \qquad \text{and } \tau_0^\frown\langle(\xi,n)\rangle \notin A_0. \end{cases}$$

At each stage of the recursion we are applying the operations of countable union and countable intersection as defined in the proof of lemma V.3.5. (See lemma V.3.5 and remark V.3.6.)

We claim that for each $\tau \in T$, B_τ is a Borel code. The proof of this claim is by arithmetical transfinite induction along $\mathrm{KB}(T)$. Unfortunately the statement which is to be proved, "B_τ is a Borel code," is Π_1^1 rather than arithmetical. Thus there is a difficulty in showing that the tree B_τ has no path. We overcome this difficulty as follows. First, we note that B_τ consists of sequences of the form $\rho^\frown\sigma$ where $\mathrm{lh}(\rho) \leq 10$ and $\sigma \in B_{\tau^\frown\langle(\xi,m,n)\rangle}$ for some $\tau^\frown\langle(\xi, m, n)\rangle \in T$. Thus, by arithmetical transfinite recursion along $\mathrm{KB}(T)$, we can define for each $\tau \in T$ a function $g_\tau: B_\tau \to T$ such that $g_\tau(\langle\rangle) = \tau$, $g_\tau(\sigma_1) \subseteq g_\tau(\sigma_2)$ whenever $\sigma_1 \subseteq \sigma_2 \in B_\tau$, and $g_\tau(\sigma_1) \neq g_\tau(\sigma_2)$ whenever $\sigma_1 \subseteq \sigma_2 \in B_\tau$ with $\mathrm{lh}(\sigma_1) + 10 \leq \mathrm{lh}(\sigma_2)$. (We omit the details of this recursion.) Thus any path through B_τ would be mapped by g_τ to a path through T. Since T has no path, it follows that B_τ has no path. Hence B_τ is a Borel code.

In particular we have a Borel code $B = B_{\langle\rangle}$. We shall now show that B satisfies the conclusion of the theorem.

Let X be given such that $X \in A_1$. By definition V.1.6 let $f : \mathbb{N} \to \mathbb{N}$ be such that $\forall k \, (\langle (X(j), f(j)) : j < k \rangle \in A_1)$. Let S be the set of all $\tau \in T$ of the form $(*)$ such that $\forall j < k \, (\xi_j = X(j) \wedge m_j = f(j))$. Thus S is a subtree of T. Hence $\mathrm{KB}(S)$ is a well ordering. We claim that $X \in B_\tau$ for each $\tau \in S$. The proof of this claim is by arithmetical transfinite induction along $\mathrm{KB}(S)$. (Unfortunately, the statement to be proved, "$X \in B_\tau$," is Δ_1^1 rather than arithmetical. We overcome this difficulty as follows. By lemma V.3.7 let Z be the set of $\tau \in T$ such that $X \in B_\tau$. Instead of proving that $X \in B_\tau$ for all $\tau \in S$, we shall prove an equivalent arithmetical assertion: $\tau \in Z$ for all $\tau \in S$. The proof is by arithmetical transfinite induction along $\mathrm{KB}(S)$.) Given $\tau \in S$, put $\xi = X(\mathrm{lh}(\tau))$ and $m = f(\mathrm{lh}(\tau))$. Let $\eta < 2$ and $n \in \mathbb{N}$ be arbitrary. If $\eta = 1 - \xi$ we have $X \in B_{\mathrm{lh}(\tau)}^\xi = C_\tau^{\xi,m,\eta,n}$. If $\eta = \xi$ and $\tau^\frown \langle (\xi, m, n) \rangle \in S$, we have $X \in B_{\tau^\frown \langle (\xi,m,n) \rangle} = C_\tau^{\xi,m,\eta,n}$ by induction hypothesis. If $\eta = \xi$ and $\tau^\frown \langle (\xi, m, n) \rangle \notin S$, we must have $\tau^\frown \langle (\xi, m, n) \rangle \notin T$. But clearly $\tau_1^\frown \langle (\xi, m) \rangle \in A_1$, so we must have $\tau_0^\frown \langle (\xi, n) \rangle \notin A_0$. Hence $X \in 2^\mathbb{N} = C_\tau^{\xi,m,\eta,n}$ in this case also. Thus $X \in \bigcap_{\eta<2} \bigcap_{n \in \mathbb{N}} C_\tau^{\xi,m,\eta,n}$. Hence $X \in B_\tau$. This completes the proof of our claim. In particular, taking $\tau = \langle \rangle$, we obtain $X \in B_{\langle \rangle} = B$.

The previous paragraph shows that $\forall X \, (X \in A_1 \to X \in B)$. A similar argument, which we omit, shows that $\forall X \, (X \in A_0 \to X \notin B)$.

This completes the proof of theorem V.3.9. □

As a corollary of theorem V.3.9 we obtain:

THEOREM V.3.10 (Souslin's theorem in ATR$_0$). *If A_1 and A_0 are analytic codes such that $\forall X \, (X \in A_1 \leftrightarrow X \notin A_0)$, then there exists a Borel code B such that $\forall X \, (X \in B \leftrightarrow X \in A_1)$. Conversely, given any Borel code B, there exist analytic codes A_1 and A_0 with these properties.*

PROOF. Immediate from theorems V.3.8 and V.3.9. □

The following uniform version of theorem V.3.9 will be used in the proof of theorem V.3.11.

THEOREM V.3.9'. *The following is provable in ATR$_0$. Let $\langle A_n^1 : n \in \mathbb{N} \rangle$ and $\langle A_n^0 : n \in \mathbb{N} \rangle$ be sequences of analytic codes such that $\neg \exists n \, \exists X \, (X \in A_n^1 \wedge X \in A_n^0)$. Then there exists a sequence of Borel codes $\langle B_n : n \in \mathbb{N} \rangle$ such that $\forall n \, \forall X \, ((X \in A_n^1 \to X \in B_n) \wedge (X \in A_n^0 \to X \notin B_n))$.*

PROOF. For each n let $T_n = A_n^1 * A_n^0$ be as in the proof of theorem V.3.9. For each n, $\mathrm{KB}(T_n)$ is a well ordering. Hence the sum $\sum_{n \in \mathbb{N}} \mathrm{KB}(T_n)$ is a well ordering (see the proof of lemma V.3.7). By arithmetical transfinite recursion along $\sum_{n \in \mathbb{N}} \mathrm{KB}(T_n)$ define for each (τ, n) with $\tau \in T_n$ a Borel code B_n^τ as in the proof of theorem V.3.9. Setting $B_n = B_n^{\langle \rangle}$ we obtain a sequence of Borel codes $\langle B_n : n \in \mathbb{N} \rangle$ which has the desired properties. The details are as for theorem V.3.9. □

We end this section by pointing out that an interesting consequence of Lusin's separation theorem V.3.9 is also provable in ATR_0. This consequence concerns Borel sets in the plane. Recall that $2^{\mathbb{N}} \times 2^{\mathbb{N}}$ is homeomorphic to $2^{\mathbb{N}}$ via the pairing function $(X, Y) \mapsto X \oplus Y$ where $(X \oplus Y)(2n) = X(n)$, $(X \oplus Y)(2n + 1) = Y(n)$. Thus any Borel set $C \subseteq 2^{\mathbb{N}}$ may be regarded as a binary relation $C \subseteq 2^{\mathbb{N}} \times 2^{\mathbb{N}}$. Formally, if C is a Borel code, we write $C(X, Y)$ to mean that $X \oplus Y \in C$.

THEOREM V.3.11 (Borel domain theorem in ATR_0). *The following is provable in* ATR_0. *The domain of any single-valued Borel relation is Borel. In other words, let C be a Borel code such that*

$$\forall X \, (\exists \text{ at most one } Y) \, C(X, Y).$$

Then there exists a Borel code B such that

$$\forall X \, (X \in B \leftrightarrow \exists Y \, C(X, Y)).$$

PROOF. By definition V.3.4 the formula $\exists Y \, C(X, Y)$ is Σ_1^1. By theorem V.1.7' let $\langle A_n^1 : n \in \mathbb{N} \rangle$ and $\langle A_n^0 : n \in \mathbb{N} \rangle$ be sequences of analytic codes such that $\forall \xi \, \forall n \, \forall X \, (X \in A_n^\xi \leftrightarrow \exists Y \, (C(X, Y) \wedge Y(n) = \xi))$. From the hypothesis $\forall X \, (\exists \text{ at most one } Y) \, C(X, Y)$ it follows that $\neg \exists n \, \exists X \, (X \in A_n^1 \wedge X \in A_n^0)$. Hence, by theorem V.3.9', there exists a sequence of Borel codes $\langle B_n : n \in \mathbb{N} \rangle$ such that $\forall n \, \forall X \, (X \in A_n^1 \rightarrow X \in B_n)$ and $\forall n \, \forall X \, (X \in A_n^0 \rightarrow X \notin B_n)$. For each X define X' by $X'(n) = 1$ if $X \in B_n$, 0 if $X \notin B_n$ (lemma V.3.7). From the hypothesis $\forall X \, (\exists \text{ at most one } Y) \, C(X, Y)$ it follows that $\forall X \, \forall Y \, (C(X, Y) \rightarrow Y = X')$. By lemma V.3.5.5, let B be a Borel code such that $\forall X \, (X \in B \leftrightarrow C(X, X'))$. Then clearly $\forall X \, (X \in B \leftrightarrow \exists Y \, C(X, Y))$. This completes the proof. □

REMARK V.3.12. Our proof of Lusin's theorem in ATR_0 made heavy use of arithmetical transfinite recursion. In §V.5 we shall obtain reversals showing that the use of arithmetical transfinite recursion (or of some equivalent set existence axiom) was essential here. Namely, both Lusin's theorem V.3.9 and its consequence, theorem V.3.11, are in an appropriate sense equivalent to ATR_0. (Note: It can be shown that Souslin's theorem V.3.10 holds in the ω-model HYP. Hence by proposition V.2.6 Souslin's theorem is not equivalent to ATR_0.)

We end this section with some exercises.

EXERCISE V.3.13. A *coanalytic set* is defined to be the complement of an analytic set; see definition VI.2.3. Show that ATR_0 proves the existence of two disjoint coanalytic sets which cannot be separated by a Borel set.

EXERCISE V.3.14. Show that the following strong converse of theorem V.3.11 is provable in ATR_0. Any Borel set $B \subseteq 2^{\mathbb{N}}$ is the domain of a single-valued closed set $C \subseteq 2^{\mathbb{N}} \times \mathbb{N}^{\mathbb{N}}$.
Hint: Use lemma V.3.3.

EXERCISE V.3.15. Show that the following generalization of theorem V.3.11 is provable in ATR_0. If $C \subseteq 2^{\mathbb{N}} \times 2^{\mathbb{N}}$ is Borel and if $\forall X \,(\exists$ at most countably many $Y)\, C(X, Y)$, then there exists a Borel set $B \subseteq 2^{\mathbb{N}}$ such that $\forall X \,(X \in B \leftrightarrow \exists Y\, C(X, Y))$.

EXERCISES V.3.16 (Borel uniformization). Let $B \subseteq 2^{\mathbb{N}} \times 2^{\mathbb{N}}$ be Borel. We say that B is *Borel uniformizable* if there exists a Borel set $C \subseteq B$ such that $\forall X \,(\exists Y\, B(X, Y) \;\leftrightarrow\; \exists Y\, C(X, Y))$ and $\forall X \,(\exists$ at most one $Y)\, C(X, Y)$. Show that the following results are provable in ATR_0.

1. If $\forall X \,(\{\, Y : B(X, Y)\,\}$ is countable), then B is Borel uniformizable.
2. Same as 1 with "countable" replaced by "K_σ". A K_σ set is the union of countably many compact sets.
3. Same as 1 with "countable" replaced by "non-meager".
4. Same as 1 with "countable" replaced by "of positive measure". Here we are referring to the fair coin measure, as in X.1.3.

Notes for §V.3. The results of this section are due to Simpson (previously unpublished).

V.4. Perfect Sets; Pseudohierarchies

In this section we continue our investigation of the extent to which classical descriptive set theory can be formalized within ATR_0. This investigation was begun in §§V.1 and V.3.

DEFINITION V.4.1 (perfect trees). Within RCA_0, a finite sequence $\tau \in \mathbb{N}^{<\mathbb{N}}$ is said to be an *extension* of $\sigma \in \mathbb{N}^{<\mathbb{N}}$ if $\sigma \subseteq \tau$, i.e., if $\text{lh}(\sigma) \leq \text{lh}(\tau) \wedge \forall i \,(i < \text{lh}(\sigma) \to \sigma(i) = \tau(i))$. Two finite sequences $\tau_1, \tau_2 \in \mathbb{N}^{<\mathbb{N}}$ are said to be *incompatible* if neither is an extension of the other, i.e., if $\exists i \,(i < \min(\text{lh}(\tau_1), \text{lh}(\tau_2)) \wedge \tau_1(i) \neq \tau_2(i))$. A tree $T \subseteq \mathbb{N}^{<\mathbb{N}}$ is said to be *perfect* if each element of T has a pair of incompatible extensions in T, i.e., if $(\forall \sigma \in T)\,(\exists \tau_1, \tau_2 \in T)\,(\sigma \subseteq \tau_1 \wedge \sigma \subseteq \tau_2 \wedge \tau_1, \tau_2$ are incompatible).

In this section, we shall be mainly concerned with perfect trees $P \subseteq 2^{<\mathbb{N}}$. Such trees may be regarded as codes for perfect closed subsets of $2^{\mathbb{N}}$.

DEFINITION V.4.2. Within RCA_0, let A be an analytic set (given by an analytic code, definitions V.1.5 and V.1.6). We say that A *is countable* if there exists a sequence $\langle X_n : n \in \mathbb{N} \rangle$ such that $\forall X \,(X \in A \to \exists n \,(X = X_n))$. We say that A *contains a nonempty perfect set* if there exists a nonempty perfect tree $P \subseteq 2^{<\mathbb{N}}$ such that $\forall X \,(X$ is a path through $P \to X \in A)$.

The purpose of this section is to prove within ATR_0 the following theorem, which is known as the *perfect set theorem*.

THEOREM V.4.3 (perfect set theorem in ATR_0). *The following is provable in* ATR_0. *Let A be an analytic code. If A is not countable, then A contains a nonempty perfect set.*

REMARKS V.4.4 (the continuum hypothesis). The perfect set theorem may be regarded as a form of the continuum hypothesis (applied to analytic sets). The paths of a nonempty perfect tree $P \subseteq 2^{<\mathbb{N}}$ are clearly in one-to-one correspondence with the points of $2^{\mathbb{N}}$. Thus the perfect set theorem says that A is either countable or of cardinality $2^{\aleph_0}$.

In §VI.3 we shall study another theorem of classical descriptive set theory which may also be regarded as a form of the continuum hypothesis. This is *Silver's theorem* to the effect that the set of equivalence classes of a coanalytic equivalence relation on $2^{\mathbb{N}}$ is either countable or of cardinality $2^{\aleph_0}$.

The proof of theorem V.4.3 will be based on the following definition and lemma.

DEFINITION V.4.5. Within ACA_0, let A be an analytic code. For any finite sequence $\tau = \langle (\xi_0, m_0), \ldots, (\xi_{k-1}, m_{k-1}) \rangle \in A$ we write $\tau' = \langle \xi_0, \ldots, \xi_{k-1} \rangle$. Note that $\tau' \in 2^{<\mathbb{N}}$ and $\text{lh}(\tau') = \text{lh}(\tau)$. Two finite sequences $\tau_1, \tau_2 \in A$ are said to be *strongly incompatible* if τ_1' and τ_2' are incompatible. Let A' be the set of all $\sigma \in A$ such that σ has a pair of strongly incompatible extensions in A. Note that A' is again an analytic code, and $A' \subseteq A$.

LEMMA V.4.6. *The following is provable in* ATR_0. *Let A be an analytic code. For any countable well ordering X, there exists a sequence of analytic codes $\langle A_j : j \in \text{field}(X) \rangle$ such that for all $j \in \text{field}(X)$ and $\sigma \in \text{Seq}$,*

$$\sigma \in A_j \leftrightarrow (\sigma \in A \wedge \forall i \, (i <_X j \rightarrow \sigma \in A_i')).$$

PROOF. This is a straightforward instance of arithmetical transfinite recursion. Let $\theta(\sigma, j, Y)$ be the following arithmetical formula: $j \in \text{field}(X) \wedge \sigma \in A \wedge \forall i \, (i <_X j \rightarrow \exists$ strongly incompatible $\tau_1, \tau_2 \supseteq \sigma$ such that $(\tau_1, i), (\tau_2, i) \in Y)$. Given a countable well ordering X, let Y be the result of iterating θ along X, i.e., let Y be such that $H_\theta(X, Y)$ holds. For each $j \in \text{field}(X)$ set $A_j = Y_j = \{\sigma : (\sigma, j) \in Y\}$. Then $\langle A_j : j \in \text{field}(X) \rangle$ has the desired properties. $\square$

REMARK V.4.7. The thought behind lemma V.4.6 is that we wish to define, for each countable ordinal α, an analytic code A_α, where $A_0 = A$, $A_{\alpha+1} = A_\alpha'$, and $A_\delta = \bigcap_{\alpha<\delta} A_\alpha$ for limit ordinals δ. Lemma V.4.6 says that this definition can be carried out up to any given countable ordinal $\alpha = |X|$. (See also definition V.2.10.)

REMARK V.4.8 (the method of pseudohierarchies). In order to finish the proof of theorem V.4.3, we shall introduce a technique which has not previously appeared in this book. The new technique is known as *the method of pseudohierarchies*. In the present context, the method of pseudohierarchies takes the form of a generalization of lemma V.4.6 in which the countable well ordering X is replaced by a countable linear ordering which is not a well ordering. The sequence $\langle A_j : j \in \text{field}(X) \rangle$ is then called a pseudohierarchy.

Rather than obtain theorem V.4.3 as an application of an abstract result on the existence of pseudohierarchies, we shall simply present the proof of theorem V.4.3 in the simplest possible way. After that, we shall comment on pseudohierarchies in general (see lemma V.4.12 below).

PROOF OF THEOREM V.4.3. We reason within ATR_0. Let A be a given analytic code. The proof splits into two cases.

Case 1. Assume that there exists a countable well ordering X and a sequence $\langle A_j : j \in \text{field}(X) \rangle$ as in lemma V.4.6 such that in addition $A_j = \emptyset$ for some $j \in \text{field}(X)$. (Here $\emptyset$ denotes the empty set.)

Fix j such that $A_j = \emptyset$. Then for each $\sigma \in A$ there is a unique i such that $i <_X j$ and $\sigma \in A_i$ and $\sigma \notin A_i'$. With this i we define a function $Y_\sigma : \mathbb{N} \to \{0, 1\}$ by: $Y_\sigma(n) = 1$ if there exists $\tau \in A_i$ such that $\tau \supseteq \sigma$ and $\text{lh}(\tau) > n$ and $\tau'(n) = 1$; $Y_\sigma(n) = 0$ otherwise. (Here τ' is as in definition V.4.5.) Thus $\langle Y_\sigma : \sigma \in A \rangle$ is a sequence of points in the Cantor space $2^{\mathbb{N}}$; the sequence exists by arithmetical comprehension. We claim that $\forall Y (Y \in A \to \exists \sigma (\sigma \in A \wedge Y = Y_\sigma))$. To see this, suppose $Y \in A$. By definition V.1.6 let $f : \mathbb{N} \to \mathbb{N}$ be such that $\forall k \, A(Y[k], f[k])$. Let $i <_X j$ be such that $\forall k \, A_i(Y[k], f[k])$ but $\neg \forall k \, A_i'(Y[k], f[k])$. Let k be such that $\neg A_i'(Y[k], f[k])$. Put $\sigma = \langle (Y(0), f(0)), \ldots, (Y(k-1), f(k-1)) \rangle$. Clearly $\sigma \in A_i$, $\sigma \notin A_i'$, and $Y = Y_\sigma$. This proves our claim. Thus A is countable.

Theorem V.4.3 has now been proved under the hypothesis of case 1.

Case 2. Assume that the hypothesis of case 1 does not hold (for the given analytic code A).

Let $\varphi(X)$ be the following Σ_1^1 formula: $\text{LO}(X)$ and there exists a sequence of analytic codes $\langle A_j : j \in \text{field}(X) \rangle$ such that $\forall j \, \forall \sigma (\sigma \in A_j \leftrightarrow (j \in \text{field}(X) \wedge \sigma \in A \wedge \forall i (i <_X j \to \sigma \in A_i')))$ and $\forall j (j \in \text{field}(X) \to A_j \neq \emptyset)$. By lemma V.4.6 and our assumption, we have $\forall X (\text{WO}(X) \to \varphi(X))$. But by theorem V.1.9 we have $\neg \forall X (\text{WO}(X) \leftrightarrow \varphi(X))$. Hence $\exists X (\varphi(X) \wedge \neg \text{WO}(X))$. In other words, there exists an X and a sequence $\langle A_j : j \in \text{field}(X) \rangle$ such that X is a countable linear ordering and

$$\forall j \, \forall \sigma (\sigma \in A_j \leftrightarrow (j \in \text{field}(X) \wedge \sigma \in A \wedge \forall i (i <_X j \to \sigma \in A_i')))$$

and $\forall j (j \in \text{field}(X) \to A_j \neq \emptyset)$ and X is not a well ordering.

Fix an X and a sequence $\langle A_j : j \in \text{field}(X) \rangle$ as above. In particular $\forall i \, \forall j (i <_X j \to A_j \subseteq A_i')$. Since X is not a well ordering, let $f : \mathbb{N} \to \text{field}(X)$ be a descending sequence through X, i.e., for all n, $f(n+1) <_X f(n)$. Hence $A_{f(n)} \subseteq A_{f(n+1)}'$, i.e., each $\sigma \in A_{f(n)}$ has a pair of strongly incompatible extensions in $A_{f(n+1)}$. Since also $A_{f(0)} \neq \emptyset$, we can define by recursion a function $g \colon 2^{<\mathbb{N}} \to A$ such that for all $\rho \in 2^{<\mathbb{N}}$, $g(\rho) \in A_{f(\mathrm{lh}(\rho))}$ and moreover $g(\rho^\frown\langle 0\rangle)$ and $g(\rho^\frown\langle 1\rangle)$ are strongly incompatible extensions of $g(\rho)$. Let P be the set of all $\sigma \in 2^{<\mathbb{N}}$ such that $\exists \rho\,(\rho \in 2^{<\mathbb{N}} \wedge \sigma \subseteq g(\rho)')$. (The notation τ' for $\tau \in A$ was defined in V.4.5.) Clearly P is a nonempty perfect subtree of $2^{<\mathbb{N}}$ and $\forall Y\,(Y$ is a path through $P \to Y \in A)$.

This completes the proof of theorem V.4.3.

Let B be a Borel set (given by a Borel code). We say that B *is countable* if there exists a sequence $\langle X_n : n \in \mathbb{N}\rangle$ such that $\forall X\,(X \in B \to \exists n\,(X = X_n))$. We say that B *contains a nonempty perfect set* if there exists a nonempty perfect tree $P \subseteq 2^{<\mathbb{N}}$ such that $\forall X\,(X$ is a path through $P \to X \in B)$.

COROLLARY V.4.9. *The following is provable in* ATR_0. *Let B be a Borel code. Either B is countable or B contains a nonempty perfect set.*

PROOF. This is an immediate consequence theorem V.4.3 in view of theorem V.3.8.   □

In the next section we shall see that both the perfect set theorem V.4.3 and its corollary, V.4.9 (or even the special case of V.4.9 in which B is a closed subset of $2^{\mathbb{N}}$), are provably equivalent to ATR_0 over the weak base theory ACA_0. Thus ATR_0 is the weakest subsystem of second order arithmetic in which these results can be proved.

A further important result on perfect sets is the Cantor-Bendixson theorem. We shall see in chapter VI that this theorem is not provable in ATR_0 but is provable in the stronger system $\Pi_1^1\text{-}\mathrm{CA}_0$.

EXERCISE V.4.10. Show that the following is provable in ATR_0. Let A_n, $n \in \mathbb{N}$, be a sequence of analytic codes. If $\forall n\,(A_n$ is countable), then $\bigcup_{n\in\mathbb{N}} A_n$ (as defined in exercise V.1.12) is countable. Hint: Use theorem V.4.3.

EXERCISE V.4.11. Show that the following is provable in ATR_0. If A_n, $n \in \mathbb{N}$, is a sequence of analytic codes, then there exists a sequence of points X_m, $m \in \mathbb{N}$, such that $\forall n \,\forall X\,((X \in A_n \wedge A_n$ countable$) \to \exists m\, X = X_m)$.

We end this section with an abstract formulation of the method of pseudohierarchies.

Let θ be a given arithmetical formula as in definition V.2.4. By a *hierarchy* for θ we mean a set Y such that $\mathrm{H}_\theta(X, Y)$ holds for some X such that $\mathrm{WO}(X)$. Thus, the principal axiom of ATR_0 asserts the existence of "sufficiently many" hierarchies. By a *pseudohierarchy* for θ we mean a set Y such that $\mathrm{H}_\theta(X, Y)$ holds for some X such that $\mathrm{LO}(X) \wedge \neg\mathrm{WO}(X)$. The following lemma asserts the existence of "sufficiently many" pseudohierarchies.

LEMMA V.4.12 (existence of pseudohierarchies). *The following is provable in* ACA$_0$. *Let* $\theta(n, Y)$ *be an arithmetical formula as in definition* V.2.4. *Let* $\varphi(X, Y)$ *be a* Σ^1_1 *formula. If*

$$\forall X \, (\mathrm{WO}(X) \to \exists Y \, (\mathrm{H}_\theta(X, Y) \wedge \varphi(X, Y)))$$

then

$$\exists X \, \exists Y \, (\mathrm{LO}(X) \wedge \neg\mathrm{WO}(X) \wedge \mathrm{H}_\theta(X, Y) \wedge \varphi(X, Y)).$$

PROOF. Let $\varphi'(X)$ be the following Σ^1_1 formula:

$$\mathrm{LO}(X) \wedge \exists Y \, (\mathrm{H}_\theta(X, Y) \wedge \varphi(X, Y)).$$

By hypothesis we have $\forall X \, (\mathrm{WO}(X) \to \varphi'(X))$. But by theorem V.1.9 we have $\neg\forall X \, (\mathrm{WO}(X) \leftrightarrow \varphi'(X))$. Hence $\exists X \, (\varphi'(X) \wedge \neg\mathrm{WO}(X))$, Q.E.D.

□

These pseudohierarchies provide a powerful and apparently indispensable proof technique within ATR$_0$. The idea of lemma V.4.12 has already been applied in the proof of theorem V.4.3, case 2, above. Other applications of the same idea are in §§V.7 and V.8.

Notes for §V.4. Pseudohierarchies were introduced by Spector [254] and Gandy [88] in the context of hyperarithmetical theory; see §VIII.3 below. Further work on pseudohierarchies is in Harrison [106], Friedman [62, chapters II and III], Steel [256, chapter I], and Friedman/McAloon/Simpson [76].

V.5. Reversals

In §§V.3 and V.4 we have seen that several theorems of classical descriptive set theory are provable in the formal system ATR$_0$. We shall now show that each of these theorems is, in a suitable sense, equivalent to ATR$_0$.

We begin with the reversal of Lusin's separation theorem (theorem V.3.9). We shall essentially show that Lusin's theorem implies arithmetical transfinite recursion. There is a slight conceptual difficulty here since the statement of Lusin's theorem mentions Borel sets. Our concept of Borel set (definitions V.3.1, V.3.2, V.3.4) depends on arithmetical transfinite recursion in order to prove the existence of the needed evaluation maps (lemma V.3.3). Therefore, in the absence of arithmetical transfinite recursion, it is not even clear how to state Lusin's theorem in a meaningful way. In order to circumvent this difficulty, we adopt the following procedure. We first deduce from Lusin's theorem a simple consequence, the so-called Σ^1_1 separation principle, which does not mention Borel sets. We then show that the Σ^1_1 separation principle implies arithmetical transfinite recursion.

THEOREM V.5.1 (ATR$_0$ and Σ^1_1 separation). *The following are equivalent over* RCA$_0$:

1. *Arithmetical transfinite recursion.*
2. *The* Σ^1_1 separation principle: *For any* Σ^1_1 *formulas* $\varphi_1(n)$ *and* $\varphi_0(n)$ *in which* Z *does not occur freely, we have*

$$\neg\exists n\, (\varphi_1(n) \wedge \varphi_0(n)) \rightarrow \exists Z\, \forall n\, ((\varphi_1(n) \rightarrow n \in Z) \wedge (\varphi_0(n) \rightarrow n \notin Z)).$$

Here Z *ranges over subsets of* $\mathbb{N}$.

PROOF. We first show how to prove the Σ^1_1 separation principle in ATR$_0$, via Lusin's theorem. Reasoning within ATR$_0$, assume $\neg\exists n\, (\varphi_1(n) \wedge \varphi_0(n))$. Let $\langle X_n : n \in \mathbb{N}\rangle$ be a fixed sequence of distinct points in the Cantor space $2^{\mathbb{N}}$. (E.g., we may take $X_n(m) = 1$ if $m < n$, 0 otherwise.) By theorem V.1.7 there exist analytic codes A_i, $i < 2$, such that $\forall X\, (X \in A_i \leftrightarrow \exists n\, (X = X_n \wedge \varphi_i(n)))$. By Lusin's theorem in ATR$_0$ (theorem V.3.9), let B be a Borel code such that $\forall X\, ((X \in A_1 \rightarrow X \in B) \wedge (X \in A_0 \rightarrow X \notin B))$. By lemma V.3.7 let $Z \subseteq \mathbb{N}$ be such that $\forall n\, (n \in Z \leftrightarrow X_n \in B)$. Then $\varphi_1(n)$ implies $X_n \in A_1$ which implies $X_n \in B$, i.e., $n \in Z$, and similarly $\varphi_0(n)$ implies $n \notin Z$. This proves the implication $1 \rightarrow 2$.

For the converse implication, assume the Σ^1_1 separation principle. In particular we have arithmetical (in fact Δ^1_1) comprehension. Let X be a given countable well ordering and let $\theta(n, Y)$ be a given arithmetical formula. We wish to prove the existence of a Z such that $H_\theta(X, Z)$ holds (cf. definition V.2.4). Define Σ^1_1 formulas

$$\varphi_1(j, n) \equiv \exists Y\, (H_\theta(j, X, Y) \wedge \theta(n, j, Y))$$

and

$$\varphi_0(j, n) \equiv \exists Y\, (H_\theta(j, X, Y) \wedge \neg\theta(n, j, Y)).$$

Then by lemma V.2.3 we have $\neg\exists j\, \exists n\, (\varphi_1(j, n) \wedge \varphi_0(j, n))$. Hence by Σ^1_1 separation there exists $W \subseteq \mathbb{N}$ such that $\forall j\, \forall n\, ((\varphi_1(j, n) \rightarrow (n, j) \in W) \wedge (\varphi_0(j, n) \rightarrow (n, j) \notin W))$. For each j put

$$W^j = \{(m, i) : i <_X j \wedge (m, i) \in W\}.$$

We claim that $H_\theta(j, X, W^j)$ and $\forall n\, ((n, j) \in W \leftrightarrow \theta(n, j, W^j))$ hold for all $j \in \mathrm{field}(X)$. Assume inductively that the claim holds for all $i <_X j$. By definition V.2.2 it follows that $H_\theta(j, X, W^j)$ holds. By lemma V.2.3 we have $\forall n\, ((\varphi_1(j, n) \leftrightarrow \theta(n, j, W^j)) \wedge (\varphi_0(j, n) \leftrightarrow \neg\theta(n, j, W^j)))$. Hence by the choice of W we have $\forall n\, ((n, j) \in W \leftrightarrow \theta(n, j, W^j))$. Our claim now follows by arithmetical transfinite induction (lemma V.2.1) along X.

From the claim and definition V.2.2 it follows that $H_\theta(X, Z)$ holds if we define $Z = \{(n, j) : (n, j) \in W \wedge j \in \mathrm{field}(X)\}$. This completes the proof. $\square$

We now turn to the reversal of theorem V.3.11. Theorem V.3.11 says that the domain of a single-valued Borel relation is Borel. As in the case of Lusin's theorem, our procedure for the the reversal will be to formulate a consequence of theorem V.3.11 which does not mention Borel sets, and then to prove ATR_0 from this consequence.

THEOREM V.5.2 (ATR_0 and unique paths). *The following are pairwise equivalent over RCA_0.*

1. *Arithmetical transfinite recursion.*
2. *The scheme*

$$\forall i \, (\exists \text{ at most one } X) \, \varphi(i, X) \rightarrow \exists Z \, \forall i \, (i \in Z \leftrightarrow \exists X \, \varphi(i, X)),$$

   where $\varphi(i, X)$ is any arithmetical formula in which Z does not occur.
3. *For any sequence of trees $\langle T_i : i \in \mathbb{N} \rangle$, if $\forall i \, (T_i$ has at most one path) then $\exists Z \, \forall i \, (i \in Z \leftrightarrow T_i$ has a path).*

PROOF. We first show how to prove 2 within ATR_0, via theorem V.3.11. Assume $\forall i \, (\exists \text{ at most one } Y) \, \varphi(i, Y)$ where φ is arithmetical. Let $\langle X_i : i \in \mathbb{N} \rangle$ be any fixed sequence of distinct points in $2^{\mathbb{N}}$. By theorems V.1.7 and V.3.10 let C be a Borel code such that $\forall X \, \forall Y \, (C(X, Y) \leftrightarrow \exists i \, (X = X_i \wedge \varphi(i, Y)))$. Then $\forall X \, (\exists \text{ at most one } Y) \, C(X, Y)$ so by theorem V.3.11 let B be a Borel code such that $\forall X \, (X \in B \leftrightarrow \exists Y \, C(X, Y))$. By lemma V.3.7 let $Z \subseteq \mathbb{N}$ be such that $\forall i \, (i \in Z \leftrightarrow X_i \in B)$. Then clearly $\forall i \, (i \in Z \leftrightarrow \exists Y \varphi(i, Y))$. This proves the implication $1 \rightarrow 2$.

Next we prove the converse, $2 \rightarrow 1$. Assume 2. In particular we have arithmetical comprehension. We wish to prove arithmetical transfinite recursion. Let X be a given countable well ordering, and let $\theta(n, j, Y)$ be a given arithmetical formula. Let $\varphi(i, Y)$ be the following arithmetical formula: $\exists n \, \exists j \, (i = (n, j) \wedge H_\theta(j, X, Y) \wedge \theta(n, j, Y))$. By lemma V.2.3 we have $\forall i \, (\exists \text{ at most one } Y) \, \varphi(i, Y)$. Hence by 2 let $Z \subseteq \mathbb{N}$ be such that $\forall i \, (i \in Z \leftrightarrow \exists Y \, \varphi(i, Y))$. For each k set $Z^k = \{(n, j) : j <_X k \wedge (n, j) \in Z\}$. By arithmetical transfinite induction along X (lemma V.2.1) we see that $H_\theta(j, X, Z^j)$ and $\forall n \, ((n, j) \in Z \leftrightarrow \theta(n, j, Z^j))$ for all $j \in \text{field}(X)$. From this it follows easily that $H_\theta(X, Z)$ holds. Thus we have arithmetical transfinite recursion.

It remains to prove that V.5.2.2 is equivalent to V.5.2.3. The statement V.5.2.3 is the special case of V.5.2.2 with $\varphi(i, X) \equiv (X$ encodes a path through T_i). So the implication from V.5.2.2 to V.5.2.3 is trivial. For the converse, we prove two lemmas.

LEMMA V.5.3. *It is provable in RCA_0 that* V.5.2.3 *implies arithmetical comprehension.*

PROOF. Assume V.5.2.3. Instead of arithmetical comprehension we shall prove the equivalent statement III.1.3.3. Let $f : \mathbb{N} \rightarrow \mathbb{N}$ be given. Define a sequence of trees $\langle T_i : i \in \mathbb{N} \rangle$ by putting $\tau \in T_i$ if and only if $\forall m \, (m < \text{lh}(\tau) \rightarrow (\tau(m) = 0 \wedge f(m) \neq i))$. Clearly $\forall i \, (T_i$ has at most one path) and $\forall i$ (T_i has a path $\leftrightarrow \forall m\,(f(m) \neq i)$). So by the assumption V.5.2.3 there exists Z such that $\forall i\,(i \in Z \leftrightarrow \forall m\,(f(m) \neq i))$. Hence by lemma III.1.3 we have arithmetical comprehension.   □

The next lemma is an improvement of lemma V.1.4, our formal version of the Kleene normal form theorem.

LEMMA V.5.4. *For any arithmetical formula $\varphi(X)$ we can find an arithmetical (in fact Σ_0^0) formula $\theta(\sigma, \tau)$ such that ACA_0 proves*

$$\forall X\,(\varphi(X) \leftrightarrow \exists f\,\forall m\,\theta(X[m], f[m]))$$

and

$$\forall X\,(\exists \text{ at most one } f)\,\forall m\,\theta(X[m], f[m])).$$

(Here X ranges over subsets of $\mathbb{N}$ and f ranges over total functions from $\mathbb{N}$ into $\mathbb{N}$. Also $X[m] = \langle \xi_0, \xi_1, \ldots, \xi_{m-1} \rangle$ where $\xi_i = 1$ if $i \in X$, 0 if $i \notin X$. Note that $\varphi(X)$ may contain free variables other than X. If this is the case, then $\theta(\sigma, \tau)$ will also contain those free variables.)

PROOF. Replacing $\forall n$ by $\neg\exists n\,\neg$, we may safely assume that the given arithmetical formula φ contains no universal quantifiers. Let $\langle \exists n\,\psi_i : i < k \rangle$ be a list of all subformulas of φ of the form $\exists n\,\psi$ where n is any number variable. For each $i < k$ let $m_{i1}, \ldots, m_{ik_i}$ be a list of the free number variables occurring in $\exists n\,\psi_i$. Functions $g_i \colon \mathbb{N}^{k_i} \to \mathbb{N}$, $i < k$ are called *minimal Skolem functions* if for all $i < k$ and all $m_{i1}, \ldots, m_{ik_i} \in \mathbb{N}$,

$$g_i(m_{i1}, \ldots, m_{ik_i}) = \begin{cases} 0 & \text{if } \neg\exists n\,\psi_i(m_{i1}, \ldots, m_{ik_i}, n), \\ n_i + 1 & \text{if } n_i = \text{least } n \text{ such that } \psi_i(m_{i1}, \ldots, m_{ik_i}, n). \end{cases}$$

By arithmetical comprehension there is for any given X a unique set of minimal Skolem functions. Given any functions $g_i \colon \mathbb{N}^{k_i} \to \mathbb{N}$, $i < k$, we associate to each subformula ψ of φ a formula $\overline{\psi}$ in terms of the given g_i, $i < k$, as follows: $\psi \equiv \overline{\psi}$ if ψ is atomic, $\overline{\psi_1 \wedge \psi_2} \equiv \overline{\psi_1} \wedge \overline{\psi_2}$, $\overline{\neg\psi} \equiv \neg\overline{\psi}$, and $\overline{\exists n\,\psi_i} \equiv (g_i(m_{i1}, \ldots, m_{ik_i}) > 0)$. Thus, for any given X, we see that $\varphi(X)$ holds if and only if there exist functions $g_i \colon \mathbb{N}^{k_i} \to \mathbb{N}$, $i < k$, such that $\overline{\varphi}$ holds and, for all $i < k$ and all $m_{i1}, \ldots, m_{ik_i}, n_i, n \in \mathbb{N}$,

$$\begin{cases} g_i(m_{i1}, \ldots, m_{ik_i}) = 0 \to \neg\overline{\psi_i}(m_{i1}, \ldots, m_{ik_i}, n), \\ g_i(m_{i1}, \ldots, m_{ik_i}) = n_i + 1 \to \overline{\psi_i}(m_{i1}, \ldots, m_{ik_i}, n_i), \quad \text{and} \qquad (*) \\ (g_i(m_{i1}, \ldots, m_{ik_i}) = n_i + 1 \wedge n < n_i) \to \neg\overline{\psi_i}(m_{i1}, \ldots, m_{ik_i}, n). \end{cases}$$

Furthermore, for a given X, these functions g_i, $i < k$, are unique if they exist. Now let $\forall m\,\theta(X[m], f[m])$ say that $f(n) = 0$ for all n not of the form $\langle i, m_{i1}, \ldots, m_{ik_i} \rangle$, $i < k$, and furthermore that $\overline{\varphi}$ and $(*)$ hold when the g_i, $i < k$ are defined by $g_i(m_{i1}, \ldots, m_{ik_i}) = f(\langle i, m_{i1}, \ldots, m_{ik_i} \rangle)$. Clearly this θ has the desired properties. Lemma V.5.4 is proved.   □

We are now ready to finish the proof of theorem V.5.2. Assume V.5.2.3. Hence by lemma V.5.3 we have arithmetical comprehension. We wish to prove V.5.2.2. Assume $\forall i$ ($\exists$ at most one X) $\varphi(i, X)$ where φ is arithmetical. By lemma V.5.4 there is an arithmetical formula θ such that

$$\forall i \, (\forall X \, \varphi(i, X) \leftrightarrow \exists f \, \forall k \, \theta(i, X[k], f[k]))$$

and $\forall i$ ($\exists$ at most one pair (X, f) such that $\forall k \, \theta(i, X[k], f[k])$). Define a sequence of trees $\langle T_i : i \in \mathbb{N} \rangle$ by putting $\tau \in T_i$ if and only if τ is of the form $\langle (\xi_0, n_0), \ldots, (\xi_{k-1}, n_{k-1}) \rangle$ and $\forall j < k \, (\xi_j \in \{0, 1\} \wedge n_j \in \mathbb{N})$ and $\forall j \leq k \, (\theta(i, \langle \xi_0, \ldots, \xi_{j-1} \rangle, \langle n_0, \ldots, n_{j-1} \rangle))$. ($T_i$ is in fact an analytic code. Compare the proof of theorem V.1.7'.) Clearly $\forall i$ (T_i has at most one path) and $\forall i$ (T_i has a path $\leftrightarrow \exists X \, \varphi(i, X)$). By the assumption V.5.2.3 there exists $Z \subseteq \mathbb{N}$ such that $\forall i \, (i \in Z \leftrightarrow T_i$ has a path). Hence $\forall i \, (i \in Z \leftrightarrow \exists X \, \varphi(i, X))$. This completes the proof of theorem V.5.2. $\quad\square$

We now turn to the reversal of the perfect set theorem V.4.3 and of its corollary, V.4.9.

THEOREM V.5.5 (ATR$_0$ and the perfect set theorem). *The following are pairwise equivalent over* ACA$_0$.

1. *Arithmetical transfinite recursion.*
2. *The* perfect set theorem: *For every analytic code A, if A is uncountable, then A has a nonempty perfect subset.*
3. *For every tree $T \subseteq 2^{<\mathbb{N}}$, if T has uncountably many paths, then T has a nonempty perfect subtree.*
4. *For every tree $T \subseteq \mathbb{N}^{<\mathbb{N}}$, if T has uncountably many paths, then T has a nonempty perfect subtree.*

(A tree T is said to have *uncountably many paths* if for all sequences of functions $\langle f_n : n \in \mathbb{N} \rangle$ there exists a function f such that f is a path through T and $\forall n \, (f \neq f_n)$.)

PROOF. That 1 implies 2 has already been proved as theorem V.4.3. To show that 2 implies 3, let T be a given subtree of $2^{<\mathbb{N}}$. Let A be the set of all finite sequences of the form $\langle (\xi_0, 0), \ldots, (\xi_{k-1}, 0) \rangle$ such that $\langle \xi_0, \ldots, \xi_{k-1} \rangle \in T$. Then A is an analytic code. Thus 2 contains 3 as a special case.

The proof that 3 implies 4 will be based on a canonical homeomorphism of the Baire space $\mathbb{N}^{\mathbb{N}}$ into the Cantor space $2^{\mathbb{N}}$. Given $f : \mathbb{N} \to \mathbb{N}$, define $f^* : \mathbb{N} \to \{0, 1\}$ by

$$f^*(n) = \begin{cases} 1 & \text{if } \exists k \, (n = k + \sum_{i=0}^{k} f(i)), \\ 0 & \text{otherwise.} \end{cases}$$

LEMMA V.5.6. *The following is provable in* RCA$_0$. *For any tree $T \subseteq \mathbb{N}^{<\mathbb{N}}$ there exists a tree $T^* \subseteq 2^{<\mathbb{N}}$ such that $\forall f \, (f$ is a path through $T \leftrightarrow f^*$ is a path through T^*).*

PROOF. Let T^* be the set of all $\tau \in 2^{<\mathbb{N}}$ of the form

$$\underbrace{\langle 0, \ldots, 0 \rangle}_{m_0} {}^\frown \langle 1 \rangle {}^\frown \underbrace{\langle 0, \ldots, 0 \rangle}_{m_1} {}^\frown \langle 1 \rangle {}^\frown \cdots {}^\frown \underbrace{\langle 0, \ldots, 0 \rangle}_{m_{k-1}} {}^\frown \langle 1 \rangle {}^\frown \underbrace{\langle 0, \ldots, 0 \rangle}_{n}$$

where $\langle m_0, m_1, \ldots, m_{k-1} \rangle \in T$ and $n \in \mathbb{N}$. Clearly T^* has the desired property.   □

In particular, if T has uncountably many paths, then so does T^*. On the other hand, if T^* has a nonempty perfect subtree P, then by recursion we can define a nonempty perfect subtree $Q \subseteq P$ such that any path g through Q has $g(n) = 1$ for infinitely many n; hence $g = f^*$ for some $f: \mathbb{N} \to \mathbb{N}$. Let R be the set of all $\langle m_0, \ldots, m_{k-1} \rangle \in \mathbb{N}^{<\mathbb{N}}$ such that

$$\underbrace{\langle 0, \ldots, 0 \rangle}_{m_0} {}^\frown \langle 1 \rangle {}^\frown \underbrace{\langle 0, \ldots, 0 \rangle}_{m_1} {}^\frown \langle 1 \rangle {}^\frown \cdots {}^\frown \underbrace{\langle 0, \ldots, 0 \rangle}_{m_{k-1}} {}^\frown \langle 1 \rangle$$

belongs to Q. Then clearly R is a nonempty perfect subtree of T. In sum, V.5.5.4 for T follows from V.5.5.3 applied to T^*.

It remains to prove that V.5.5.4 implies arithmetical transfinite recursion. Instead of proving arithmetical transfinite recursion directly, we shall prove the equivalent statement V.5.2.3. Let $\langle T_i : i \in \mathbb{N} \rangle$ be a given sequence of trees such that $\forall i$ (T_i has at most one path). Form a tree $T \subseteq \mathbb{N}^{<\mathbb{N}}$ by

$$T = \{\langle\rangle\} \cup \{\langle i \rangle {}^\frown \tau : i \in \mathbb{N} \wedge \tau \in T_i\}.$$

Clearly T has no nonempty perfect subtree. Therefore, by V.5.5.4, T has only countably many paths, i.e., there exists a sequence $\langle f_n : n \in \mathbb{N} \rangle$ such that $\forall f$ (f is a path through $T \to \exists n$ ($f = f_n$)). By arithmetical comprehension let Z be the set of all $i \in \mathbb{N}$ such that $\exists n$ ($f_n(0) = i \wedge f_n$ is a path through T). Then clearly $\forall i$ ($i \in Z \leftrightarrow T_i$ has a path). Hence by theorem V.5.2 we have arithmetical transfinite recursion.

This completes the proof of theorem V.5.5.   □

The results of this section, especially theorem V.5.1, will be applied in later sections to show that other theorems of ordinary mathematics are equivalent to ATR$_0$.

EXERCISE V.5.7. Show that Σ^1_1-AC$_0$ proves Π^1_1 *separation*: For any Π^1_1 formulas $\psi_1(n)$ and $\psi_0(n)$ in which Z does not occur freely, $\neg \exists n$ ($\psi_1(n) \wedge \psi_0(n)$) $\to$ $\exists Z \, \forall n$ (($\psi_1(n) \to n \in Z$) $\wedge$ ($\psi_0(n) \to n \notin Z$)). This is in contrast to theorem V.5.1.

REMARK V.5.8 (Cantor space versus Baire space). In our treatment of classical descriptive set theory in §V.1 and §§V.3–V.5, we have chosen to work with the Cantor space $2^{\mathbb{N}}$. Since it is customary to work with the Baire space $\mathbb{N}^{\mathbb{N}}$, we are obliged to explain our choice. We adduce the following considerations. (1) There is no real loss of generality, since the Baire space, or any uncountable complete separable metric space, is Borel-isomorphic to $2^{\mathbb{N}}$. (2) In this book, the second order variables of the language of Z_2 range over points of the Cantor space (i.e., subsets of $\mathbb{N}$) rather than points of the Baire space (i.e., functions from $\mathbb{N}$ to $\mathbb{N}$). It is therefore natural for us here to work with $2^{\mathbb{N}}$ rather than $\mathbb{N}^{\mathbb{N}}$. (3) Results concerning $2^{\mathbb{N}}$ are easily compared to chapter IV, which is also concerned with closed subsets of $2^{\mathbb{N}}$ (coded by trees $T \subseteq 2^{<\mathbb{N}}$). The same would not hold for $\mathbb{N}^{\mathbb{N}}$. (4) Our results concerning closed subsets of Cantor space are sometimes sharper than the corresponding results for Baire space. Consider for instance the reversal of the perfect set theorem for closed subsets of $2^{\mathbb{N}}$, i.e., the implication $3 \to 1$ in theorem V.5.5. The corresponding result for $\mathbb{N}^{\mathbb{N}}$ follows trivially from this, but the converse requires a further trick, lemma V.5.6. Thus the reversal of the perfect set theorem for Cantor space is more definitive than the corresponding result for Baire space. A similar remark will also apply to the reversal of the Cantor/Bendixson theorem, in §VI.1.

Notes for §V.5. The equivalence $1 \leftrightarrow 4$ of theorem V.5.5 has been announced by Friedman [68, 69]. Theorem V.5.1 has been announced by Simpson [243]. The other results of this section are due to Simpson (previously unpublished).

V.6. Comparability of Countable Well Orderings

In this section we complete the discussion of countable well orderings which was begun in §V.2 (definitions V.2.7 and V.2.10, lemmas V.2.8 and V.2.9). We show that the set existence axioms of ATR_0 are indispensable for a decent theory of countable ordinals. Clearly a minimum requirement for a decent theory of countable ordinals is that any two countable well orderings are comparable (definition V.2.7). We show that this assertion is equivalent to arithmetical transfinite recursion.

Let CWO be the assertion that any two countable well orderings are comparable, i.e.,

$$\forall X \, \forall Y \, ((\mathrm{WO}(X) \wedge \mathrm{WO}(Y)) \to (|X| \leq |Y| \vee |Y| \leq |X|)).$$

We begin by proving:

LEMMA V.6.1. *Over* RCA_0, CWO *implies arithmetical comprehension.*

PROOF. Reason in RCA_0 and assume CWO. Instead of proving arithmetical comprehension directly, we shall prove the equivalent assertion III.1.3.3. Let a one-to-one function $f : \mathbb{N} \to \mathbb{N}$ be given. By Δ^0_1 comprehension let $X = \{(m, n): f(m) \leq f(n)\}$. Clearly $\mathrm{LO}(X)$, and by bounded Σ^0_1 comprehension each initial section of X is finite; hence $\mathrm{WO}(X)$. Comparing X with the standard well ordering of $\mathbb{N}$, we get a bijection $g: \mathbb{N} \to \mathbb{N}$ such that $\forall m \, \forall n \, (m \leq n \leftrightarrow g(m) \leq_X g(n))$, i.e., $\forall m \, \forall n \, (m \leq n \leftrightarrow f(g(m)) \leq f(g(n)))$. Hence for all k we have

$$\exists m \, (f(m) = k) \leftrightarrow \exists m \, (m \leq k \wedge f(g(m)) = k).$$

Hence by Δ_1^0 comprehension $\exists Y \, \forall k \, (k \in Y \leftrightarrow \exists m \, (f(m) = k))$, i.e., $\mathrm{rng}(f)$ exists. By lemma III.1.3 this gives arithmetical comprehension. □

An important consequence of CWO is the so-called Σ_1^1 *bounding principle*:

LEMMA V.6.2 (Σ_1^1 bounding principle). *The following is provable in* RCA_0. *Assume* CWO. *Then for any* Σ_1^1 *formula* $\varphi(X)$ *we have*

$$\forall X \, (\varphi(X) \to \mathrm{WO}(X)) \to \exists Y \, (\mathrm{WO}(Y) \wedge \forall X \, (\varphi(X) \to |X| < |Y|)).$$

PROOF. Assume CWO. By lemma V.6.1 we have arithmetical comprehension. Assume the hypothesis $\forall X \, (\varphi(X) \to \mathrm{WO}(X))$. If the conclusion fails, then by CWO we have $\forall Y \, (\mathrm{WO}(Y) \to \exists X \, (\varphi(X) \wedge |X| \geq |Y|))$. Hence $\forall Y \, (\mathrm{WO}(Y) \leftrightarrow \varphi'(Y))$ where $\varphi'(Y)$ is the following Σ_1^1 formula: $\mathrm{LO}(Y) \wedge \exists X \, (\varphi(X) \wedge |X| \geq |Y|)$. This contradicts theorem V.1.9. □

LEMMA V.6.3. *The following is provable in* RCA_0. *Assume* CWO. *If* $\mathrm{WO}(X)$ *and* $\mathrm{WO}(Y)$ *and* X *is isomorphic to a subordering of* Y, *then* $|X| \leq |Y|$.

PROOF. If not, then by CWO we would have $|Y| < |X|$, hence Y would be isomorphic to a subordering of an initial section of Y. Thus there would be $f: \mathrm{field}(Y) \to \mathrm{field}(Y)$ and $k \in \mathrm{field}(Y)$ such that $\forall m \, \forall n \, (m \leq_Y n \leftrightarrow f(m) \leq_Y f(n) <_Y k)$. By arithmetical transfinite induction along Y (lemmas V.6.1 and V.2.1) it is straightforward to prove that $m \leq_Y f(m)$ for all $m \in \mathrm{field}(Y)$. In particular $k \leq_Y f(k)$, a contradiction. □

The key to the proof that CWO implies arithmetical transfinite recursion is the next definition.

DEFINITION V.6.4 (double descent tree). The following definition is made in RCA_0. If X and Y are countable linear orderings, the *double descent tree* $\mathrm{T}(X, Y)$ is the set of all finite sequences of the form

$$\langle (m_0, n_0), (m_1, n_1), \ldots, (m_{k-1}, n_{k-1}) \rangle$$

such that

$$m_0 >_X m_1 >_X \cdots >_X m_{k-1}$$

and

$$n_0 >_Y n_1 >_Y \cdots >_Y n_{k-1}.$$

We write $X * Y = \mathrm{KB}(\mathrm{T}(X, Y)) =$ the Kleene/Brouwer ordering of $\mathrm{T}(X, Y)$.

Lemma V.6.5. *The following is provable in* RCA_0. *Assume* CWO.

(i) *If* $\mathrm{WO}(X) \wedge \mathrm{LO}(Y)$ *then* $\mathrm{WO}(X * Y)$.

(ii) *If* $\mathrm{WO}(X) \wedge \mathrm{LO}(Y) \wedge \neg\mathrm{WO}(Y)$ *then* $|X| \leq |X * Y|$.

Proof. Assume CWO. By lemma V.6.1 we have arithmetical comprehension. If $\neg\mathrm{WO}(X * Y)$ then by lemma V.1.3 there is a path through $T(X, Y)$. Let $\langle (m_i, n_i) : i \in \mathbb{N} \rangle$ be such a path. Then $\langle m_i : i \in \mathbb{N} \rangle$ is a descending sequence through X. This proves part (i).

For part (ii), assume that $\mathrm{WO}(X) \wedge \mathrm{LO}(Y) \wedge \neg\mathrm{WO}(Y)$. Let $T(X)$ be the *descent tree* of X, i.e., $T(X)$ is the set of all finite descending sequences $\langle m_0, m_1, \ldots, m_{k-1} \rangle$, $m_0 >_X m_1 >_X \cdots >_X m_{k-1}$. For each $m \in \mathrm{field}(X)$ let σ_m be the KB-least $\sigma \in T(X)$ such that $\sigma(\mathrm{lh}(\sigma) - 1) = m$. Then $m <_X n$ implies $\sigma_m \leq_{\mathrm{KB}} \sigma_n{}^\frown\langle m \rangle <_{\mathrm{KB}} \sigma_n$. Thus $\langle \sigma_m : m \in \mathrm{field}(X) \rangle$ is an isomorphism of X onto a suborldering of $\mathrm{KB}(T(X))$. Hence by lemmas V.1.3 and V.6.3 we have $|X| \leq |\mathrm{KB}(T(X))|$. Now let $\langle n_i : i \in \mathbb{N} \rangle$ be a fixed descending sequence through Y. Define $f : T(X) \to T(X, Y)$ by

$$f(\langle m_0, m_1, \ldots, m_{k-1} \rangle) = \langle (m_0, n_0), (m_1, n_1), \ldots, (m_{k-1}, n_{k-1}) \rangle.$$

Clearly $\sigma <_{\mathrm{KB}(T(X))} \tau$ implies $f(\sigma) <_{\mathrm{KB}} f(\tau)$, so f is an isomorphism of $\mathrm{KB}(T(X))$ onto a suborldering of $\mathrm{KB}(T(X, Y))$. Hence by part (i) and lemma V.6.3 we have

$$|X| \leq |\mathrm{KB}(T(X))| \leq |\mathrm{KB}(T(X, Y))| = |X * Y|.$$

This completes the proof. $\qquad\qquad\qquad\qquad\qquad\qquad\qquad\qquad\qquad\qquad\square$

Definition V.6.6 (sum of two linear orderings). Within RCA_0, if $\mathrm{LO}(X)$ and $\mathrm{LO}(Y)$, we define

$$X + Y = \{(2m, 2n) : (m, n) \in X\} \cup \{(2m + 1, 2n + 1) : (m, n) \in Y\}$$
$$\cup \{(2m, 2n + 1) : (m, m) \in X \wedge (n, n) \in Y\}.$$

Clearly $\mathrm{LO}(X + Y)$ and $|X| \leq |X + Y|$. Intuitively $X + Y$ consists of X followed by Y. Also $\mathrm{WO}(X + Y)$ if and only if $\mathrm{WO}(X) \wedge \mathrm{WO}(Y)$.

Definition V.6.7 (sum of a sequence of linear orderings). In RCA_0, if $\langle X_i : i \in \mathbb{N} \rangle$ is a sequence of countable linear orderings, we define

$$\sum_{i \in \mathbb{N}} X_i = \{((m, i), (n, i)) : (m, n) \in X_i\}$$

$$\cup \{((m, i), (n, j)) : (m, m) \in X_i \wedge (n, n) \in X_j \wedge i < j\}.$$

Clearly $\mathrm{LO}(\sum_{i \in \mathbb{N}} X_i)$. Intuitively $\sum_{i \in \mathbb{N}} X_i$ is the countable linear ordering $X_0 + X_1 + \cdots + X_i + \cdots$. Also $\mathrm{WO}(\sum_{i \in \mathbb{N}} X_i)$ if and only if $\forall i\, \mathrm{WO}(X_i)$.

If $\mathrm{LO}(X)$ we write

$$X \cdot \mathbb{N} = \sum_{i \in \mathbb{N}} X = X + X + \cdots + X + \cdots.$$

We are now ready to prove the main theorem of this section:

Theorem V.6.8 (ATR$_0$ and CWO). *The following are equivalent over* RCA$_0$.

1. *Arithmetical transfinite recursion*.
2. CWO, *i.e., comparability of countable well orderings, i.e., the statement*

$$\forall X \, \forall Y \, ((\mathrm{WO}(X) \wedge \mathrm{WO}(Y)) \rightarrow (|X| \leq |Y| \vee |Y| \leq |X|)).$$

Proof. That ATR$_0$ implies CWO has already been proved as lemma V.2.9. For the converse, assume CWO. By lemma V.6.1 we have arithmetical comprehension. We wish to prove arithmetical transfinite recursion.

Instead of proving arithmetical transfinite recursion directly, we shall prove the Σ_1^1 separation principle V.5.1.2. Assume that $\neg \exists n \, (\varphi_1(n) \wedge \varphi_0(n))$ where $\varphi_1(n)$ and $\varphi_0(n)$ are Σ_1^1. By theorem 1.7' and lemma V.1.3 there exist sequences of countable linear orderings $\langle X_n : n \in \mathbb{N} \rangle$ and $\langle Y_n : n \in \mathbb{N} \rangle$ such that $\forall n \, (\varphi_1(n) \leftrightarrow \neg \mathrm{WO}(X_n))$ and $\forall n \, (\varphi_0(n) \leftrightarrow \neg \mathrm{WO}(Y_n))$. Our assumption $\neg \exists n \, (\varphi_1(n) \wedge \varphi_0(n))$ implies that

$$\forall n \, (\mathrm{WO}(X_n) \vee \mathrm{WO}(Y_n)).$$

By lemma V.6.5(i) and the Σ_1^1 bounding principle V.6.2, there exists a countable well ordering U such that

$$\forall X \, \forall n \, (\mathrm{LO}(X) \wedge \neg \mathrm{WO}(X)) \rightarrow |X * Y_n| < |U|).$$

Put $Z_n = (U + X_n) * Y_n$. The choice of U and lemma V.6.5(ii) imply that

$$\forall n \, ((\neg \mathrm{WO}(X_n) \rightarrow |Z_n| < |U|) \wedge (\neg \mathrm{WO}(Y_n) \rightarrow |U| \leq |Z_n|)).$$

By lemmas V.6.5(i) and V.6.2 there exists a countable well ordering V such that $|U| < |V|$ and $\forall n \, (|Z_n| < |V|)$. By arithmetical comprehension we may safely assume that U is an initial section of V. Note that $|Z_n + V \cdot \mathbb{N}| = |V + V \cdot \mathbb{N}|$ for all n. Put

$$Z = \sum_{n \in \mathbb{N}} (Z_n + V \cdot \mathbb{N})$$

and

$$W = (V + V \cdot \mathbb{N}) \cdot \mathbb{N} = \sum_{n \in \mathbb{N}} (V + V \cdot N).$$

By CWO and lemma V.6.3 there exists an isomorphism f of Z onto W. For each $n \in \mathbb{N}$ let f_n be the induced isomorphism of $Z_n + V \cdot \mathbb{N}$ onto $V + V \cdot \mathbb{N}$. Thus $|Z_n| < |U|$ if and only if the image of Z_n under f_n is an initial section of U. Hence by arithmetical comprehension, there exists $S \subseteq \mathbb{N}$ such that

$$\forall n \, (n \in S \leftrightarrow |Z_n| < |U|).$$

In particular $\forall n \, ((\varphi_1(n) \rightarrow n \in S) \wedge (\varphi_0(n) \rightarrow n \notin S))$. Thus we have Σ_1^1 separation. By theorem V.5.1 we have arithmetical transfinite recursion. This completes the proof of theorem V.6.8. □

We can also show that ATR_0 is equivalent to the Σ_1^1 bounding principle:

THEOREM V.6.9 (ATR_0 and Σ_1^1 bounding). *The following are equivalent over* RCA_0.

1. *Arithmetical transfinite recursion.*
2. *For any analytic code* A, *if* $\forall X \, (X \in A \to \mathrm{WO}(X))$ *then*

$$\exists Y \, (\mathrm{WO}(Y) \wedge \forall X \, (X \in A \to |X| \leq |Y|)).$$

PROOF. By lemmas V.2.9 and V.6.2, ATR_0 proves the Σ_1^1 bounding principle. Hence ATR_0 proves assertion 2.

For the converse, assume 2. Given $\mathrm{WO}(X_0)$ and $\mathrm{WO}(X_1)$, consider the Σ_1^1 formula $\varphi(X) \equiv (X = X_0 \vee X = X_1)$. By theorem V.1.7 let A be an analytic code such that $\forall X \, (X \in A \leftrightarrow (X = X_0 \vee X = X_1))$. By 2 there exists Y such that $\mathrm{WO}(Y) \wedge |X_0| \leq |Y| \wedge |X_1| \leq |Y|$. It follows that X_0 and X_1 are comparable. Thus we have CWO. By V.6.8 we have ATR_0. This completes the proof. $\square$

REMARK V.6.10. Girard [90] (see Hirst [121, theorem 2.6]) has shown that ACA_0 is equivalent over RCA_0 to the statement that $\mathrm{WO}(X)$ implies $\mathrm{WO}(Y)$, $|Y| = 2^{|X|}$. See also Hirst [122].

Notes for §V.6. Early versions of theorem V.6.8 are due to Steel [256, chapter I] and Friedman [68, 69] (see also [62, chapter II]). Recent refinements are due to Friedman/Hirst [74, 75] and Shore [223]. See also Hirst [119, 120, 121].

V.7. Countable Abelian Groups

In this section we shall show that ATR_0 is equivalent over RCA_0 to some well known theorems concerning countable reduced Abelian groups.

Let G be a countable Abelian group, and let p be a prime number. G is a *p-group* if for every $a \in G$ we have $p^n a = 0$ for some $n \in \mathbb{N}$. A key theorem in the classification of countable Abelian p-groups is Ulm's theorem, which is based on the following arithmetical transfinite recursion: $G_0 = G$, $G_{\alpha+1} = pG_\alpha$, and $G_\delta = \bigcap_{\alpha<\delta} G_\alpha$ for limit ordinals δ. This recursion ends with the least β such that $G_\beta = G_{\beta+1}$. If $G_\beta = 0$, the sequence $\langle G_\alpha : \alpha \leq \beta \rangle$ is called an *Ulm resolution* of G. In this case, the *Ulm invariants* of G, defined here in a form due to Kaplansky, are the numbers $\dim(P_\alpha/P_{\alpha+1})$, where $P_\alpha = \{a \in G_\alpha : pa = 0\}$ and the dimension is computed as a vector space over the field of integers modulo p. Each Ulm invariant is either a natural number or ∞, and the sequence of Ulm invariants can be written as $U_G(\alpha) = \dim(P_\alpha/P_{\alpha+1})$, $\alpha < \beta$. *Ulm's theorem* states that two countable reduced Abelian p-groups are isomorphic if and only if they have the same Ulm invariants.

In one formulation, Ulm's theorem does not even require arithmetical comprehension:

Theorem V.7.1. *The following is provable in* RCA$_0$. *If G and H are countable reduced Abelian p-groups with Ulm resolutions $\langle G_\alpha : \alpha \leq \beta \rangle$ and $\langle H_\alpha : \alpha \leq \beta \rangle$ respectively, and if $U_G(\alpha) = U_H(\alpha)$ for all $\alpha < \beta$, then $G \cong H$, i.e., G and H are isomorphic.*

Proof. Richman's constructive proof of Ulm's theorem [205] goes through in RCA$_0$.   □

From V.7.1 it may appear that Ulm's theorem does not require strong set existence axioms. However, the hypothesis of this particular formulation of Ulm's theorem is already very strong, since it implies that the given group G has an Ulm resolution. We shall see that the existence of Ulm resolutions is equivalent to arithmetical transfinite recursion over RCA$_0$. We shall also see that a certain weak sounding consequence of Ulm's theorem is likewise equivalent to ATR$_0$ over RCA$_0$.

We first prove a lemma concerning uniqueness of Ulm resolutions.

Lemma V.7.2. *The following is provable in* ACA$_0$. *If $\langle G_\alpha : \alpha \leq \beta \rangle$ and $\langle G'_\alpha : \alpha \leq \beta' \rangle$ are two Ulm resolutions of a countable reduced Abelian p-group G, then there is an isomorphism of countable well orderings $f : \beta \cong \beta'$ such that $G_\alpha = G'_{f(\alpha)}$ for all $\alpha \leq \beta$.*

Proof. By arithmetical induction along β, we easily prove

$$\forall \alpha \leq \beta \, \exists \gamma \leq \beta' \, G_\alpha = G'_\gamma.$$

Symmetrically we also have

$$\forall \gamma \leq \beta' \, \exists \alpha \leq \beta \, G'_\gamma = G_\alpha.$$

Define f by $f(\alpha) = \gamma$ if and only if $G_\alpha = G'_\gamma$. It is easy to see that this works.   □

An Abelian group H is said to be a *direct summand* of an Abelian group G if there exists an Abelian group K such that $G \cong H \oplus K$. Let us define a countable Abelian p-group G to be *fat* if it has an Ulm resolution $\langle G_\alpha : \alpha \leq \beta \rangle$ such that $U_G(\alpha) = \infty$ for all $\alpha < \beta$.

The main result of this section is:

Theorem V.7.3 (ATR$_0$ and Ulm resolutions). *The following statements are pairwise equivalent over* RCA$_0$.

1. *Arithmetical transfinite recursion.*
2. *Every countable reduced Abelian p-group has an Ulm resolution.*
3. *If G and H are fat countable Abelian p-groups, then either G is a direct summand of H or H is a direct summand of G.*

Proof. We first prove 1 → 2, using the method of pseudohierarchies (§V.4). Assume ATR$_0$ and let G be a countable reduced Abelian p-group. We claim that there exists an Ulm resolution of G. Suppose otherwise, and define $\langle G_\alpha : \alpha \le \beta \rangle$ to be a *pseudoresolution* if β is a linear ordering and $pG_{\alpha_1} \supseteq G_{\alpha_2}$ for all $\alpha_1 < \alpha_2 \le \beta$, and $G_\beta \ne 0$. By arithmetical transfinite recursion, every countable well ordering carries a pseudoresolution. The property of being a linear ordering which carries a pseudoresolution is Σ_1^1; hence there exists a linear ordering β which is not a well ordering but carries a pseudoresolution. Let $\beta > \alpha_0 > \alpha_1 > \ldots$ be a descending sequence through β. Define $H_n = G_{\alpha_n}$ where $\langle G_\alpha : \alpha \le \gamma \rangle$ is the pseudoresolution. Thus $H_n \subseteq pH_{n+1}$ for all $n \in \mathbb{N}$. Hence $H = \bigcup_{n \in \mathbb{N}} H_n$ is a divisible subgroup of G, and clearly $H \ne 0$. Hence G is not reduced. This contradiction implies that G has an Ulm resolution. Thus we have proved $1 \to 2$.

Next we prove $2 \to 3$. Our first claim is that 2 implies arithmetical comprehension. Reasoning in RCA$_0$, let $f : \mathbb{N} \to \mathbb{N}$ be one-to-one. Let G be an Abelian group with generators $x_n, y_n, n \in \mathbb{N}$, and relations $px_n = 0$ and $py_n = x_{f(n)}$. The elements of G can be written in normal form as $\sum_{i \in I} k_i x_i + \sum_{j \in J} l_j y_j$ where $0 < k_i < p$, $0 < l_j < p$, and I and J are finite subsets of $\mathbb{N}$. Thus G exists in RCA$_0$. Clearly any Ulm resolution of G is of length 2. By V.7.3.2, let $\langle G_0, G_1, G_2 \rangle$ be an Ulm resolution of G. We have $G_0 = G$, $G_1 = pG$, and $G_2 = 0$. It is easy to check that $n \in \mathrm{rng}(f)$ if and only if $x_n \in G_1$. Thus $\mathrm{rng}(f)$ exists. By lemma III.1.3 this gives arithmetical comprehension.

Now suppose G and H are fat Abelian p-groups with Ulm resolutions $\langle G_\alpha : \alpha \le \beta_1 \rangle$ and $\langle H_\alpha : \alpha \le \beta_2 \rangle$, respectively. Define

$$K = G \oplus H = \{(a, b) : a \in G, b \in H\}.$$

Clearly K is reduced, so by our assumption V.7.3.2, K has an Ulm resolution $\langle K_\alpha : \alpha \le \beta_3 \rangle$. Define $\pi_1(K_\alpha) = \{a : \exists b (a, b) \in K_\alpha\}$. Then $\langle \pi_1(K_\alpha) : \alpha \le \beta_3 \rangle$ exists by arithmetical comprehension, and clearly $\langle \pi_1(K_\alpha) : \alpha \le \gamma_1 \rangle$ is an Ulm resolution of G for some $\gamma_1 \le \beta_3$. By lemma V.7.2 there exists an isomorphism $f : \beta_1 \cong \gamma_1$ such that $G_\alpha = \pi_1(K_{f(\alpha)})$ for all $\alpha \le \beta_1$. Similarly, there exists an isomorphism $g : \beta_2 \cong \gamma_2 \le \beta_3$ such that $H_\alpha \cong \pi_2(K_{g(\alpha)})$ for all $\alpha \le \beta_2$. Thus $\langle G_\alpha \oplus H_\alpha : \alpha \le \beta_3 \rangle$ is an Ulm resolution of K. By lemma V.7.2 it follows that $K_\alpha = G_\alpha \oplus H_\alpha$ and hence $U_K(\alpha) = \infty$ for all $\alpha \le \beta_3$. Moreover $\beta_3 = \max(\gamma_1, \gamma_2)$, say $\beta_3 = \gamma_1$, so by theorem V.7.1 it follows that $G \cong K = G \oplus H$, i.e., H is a direct summand of G. This completes the proof of $2 \to 3$.

It remains to prove $3 \to 1$. Again, we shall first prove that 3 implies arithmetical comprehension. We shall then complete the argument by showing that 3 implies CWO.

Reasoning in RCA$_0$, let β be any countable ordinal. We construct a reduced Abelian p-group $G(\beta)$ of Ulm rank β, as follows. Define an *unsecured sequence* to be a finite sequence $s = \langle \alpha_1, \ldots, \alpha_n \rangle$, $\beta > \alpha_1 > \cdots > \alpha_n$, $n \in \mathbb{N}$. The generators of $G(\beta)$ are x_s for all unsecured sequences s. The relations are $px_t = x_s$, $t = s^\frown \langle \alpha \rangle$, t unsecured, and

$x_{\langle\rangle} = 0$. The elements of $G(\beta)$ have the normal form $\sum_{s \in S} m_s x_s$ where $s \neq \langle\rangle$, $0 < m_s < p$, and S is finite. Put $\mu(x_s) = \alpha_n$ if $s = \langle \alpha_1, \ldots, \alpha_n \rangle$, $n \geq 1$, and for $a = \sum_s m_s x_s \in G(\beta)$ put $\mu(a) = \min_s \mu(x_s)$. Note that $\mu(0)$ is undefined. It is easy to see that $G(\beta)$ is a reduced Abelian p-group with canonical Ulm resolution $\langle G_\alpha : \alpha \leq \beta \rangle$ where $G_\alpha = \{a : \mu(a) \geq \alpha\} \cup \{0\}$. In particular $x_{\langle \alpha \rangle} \in G_\alpha \setminus G_{\alpha+1}$ and $U_{G(\beta)}(\alpha) \geq 1$.

Let $H(\beta)$ be the direct sum of countably many copies of $G(\beta)$. Then $H(\beta)$ inherits an Ulm resolution from $G(\beta)$. Moreover $U_{H(\beta)}(\alpha) = \infty$ for all $\alpha < \beta$, i.e., $H(\beta)$ is fat.

Lemma V.7.4. *It is provable in* RCA_0 *that* V.7.3.3 *implies arithmetical comprehension.*

Proof. Reasoning in RCA_0, let $f : \mathbb{N} \to \mathbb{N}$ be one-to-one. Define $m \prec n \equiv f(m) < f(n)$, and let ω_0 be (the ordinal encoded by) the countable well ordering $\mathbb{N}, \prec$. Note that for all $\alpha < \omega_0$, $\{\beta : \beta < \alpha\}$ is finite, by bounded Σ_1^0 comprehension. Let ω be (the ordinal encoded by) the countable well ordering $\mathbb{N}, <$, i.e., the standard ordering of $\mathbb{N}$. Consider the groups $H(\omega_0)$ and $H(\omega + 1)$. By our assumption V.7.3.3, one is a direct summand of the other. In $H(\omega + 1)$ there is a nonzero element, $x_{\langle \omega \rangle}$, which is divisible by p^n for all $n \in \mathbb{N}$. We claim that in $H(\omega_0)$ there is no such element. For if $p^n \sum_{t \in T} n_t x_t = \sum_{s \in S} m_s x_s$, then each $s \in S$ is an initial segment of some $t \in T$, and $\mathrm{lh}(t) = \mathrm{lh}(s) + n$. Since t is unsecured, it follows that $\mu(x_s)$ has at least n elements preceding it. This proves our claim. It follows that $H(\omega + 1)$ is not a direct summand of $H(\omega_0)$. Hence $H(\omega_0)$ is a direct summand of $H(\omega + 1)$. Define $g : \omega_0 \to \omega$ by $g(n) = $ the least m such that $x_{\langle n \rangle} \in H_m \setminus H_{m+1}$, where $\langle H_m : m \leq \omega + 1 \rangle$ is the canonical Ulm resolution of $H(\omega + 1)$. Then g is an isomorphism of ω_0 onto ω. Define $h : \mathbb{N} \to \mathbb{N}$ by $h(x) = $ least n such that $f(n) > x$. Then $x \in \mathrm{rng}(f)$ if and only if $\exists y \prec h(x)\, (f(y) = x)$, if and only if $\exists z < g(h(x))\, (f(g^{-1}(z)) = x)$. Thus $\mathrm{rng}(f)$ exists by Δ_1^0 comprehension. By lemma III.1.3 this gives arithmetical comprehension.   □

Now let β_1 and β_2 be two ordinals. Consider the canonical Ulm resolutions $\langle H_\alpha(\beta_1) : \alpha \leq \beta_1 \rangle$ and $\langle H_\alpha(\beta_2) : \alpha \leq \beta_2 \rangle$ of $H(\beta_1)$ and $H(\beta_2)$ respectively. $H(\beta_1)$ and $H(\beta_2)$ are fat, so by V.7.3.3 we have, say, $H(\beta_1)$ is a direct summand of $H(\beta_2)$. Put $K_\alpha = H(\beta_1) \cap H_\alpha(\beta_2)$, and let $\beta_0 \leq \beta_2$ be the least α such that $K_\alpha = 0$. Then $\langle K_\alpha : \alpha \leq \beta_0 \rangle$ is another Ulm resolution of $H(\beta_1)$. By lemma V.7.2 this gives an isomorphism $f : \beta_1 \cong \beta_0$. Thus we have comparability of countable well orderings, CWO. Hence by theorem V.6.8 we get arithmetical transfinite recursion. This completes the proof of $3 \to 1$ and of theorem V.7.3.   □

Remark V.7.5. In addition to statements such as V.7.3.2 and V.7.3.3, there are other statements equivalent to ATR_0 that are purely about countable Abelian groups, with no mention of ordinal numbers or Ulm invariants. The following exercise presents one such result.

EXERCISE V.7.6. Show that the following statement is equivalent over ACA_0 to ATR_0: For any countable reduced Abelian p-groups G and H, there is a common direct summand K such that every common direct summand is embeddable in K.

REMARK V.7.7 (a conjecture of Friedman). Consider the following statement S: If G and H are two countable reduced Abelian p-groups, and if each of G and H is a direct summand of the other, then G is isomorphic to H. Note that S is easily proved in ATR_0 as a consequence of Ulm's theorem. Note also that S is a simple statement about countable Abelian groups and does not explicitly mention ordinal numbers or Ulm invariants (compare remark V.7.5). Friedman has conjectured that S is equivalent over ACA_0 to ATR_0.

Notes for §V.7. A nice exposition of Ulm's theorem is in Kaplansky [136]. The main results of this section are due to Friedman/Simpson/Smith [78]. Exercise V.7.6 is due to Friedman (unpublished manuscript, May 4, 1986).

V.8. Σ_1^0 and Δ_1^0 Determinacy

In this section we show that ATR_0 is just strong enough to prove a certain special case of the so-called axiom of determinacy.

Recall that the *Baire space* $\mathbb{N}^{\mathbb{N}}$ is the space of all total functions $f : \mathbb{N} \to \mathbb{N}$. Recall the notation

$$f[n] = \langle f(0), f(1), \ldots, f(n-1) \rangle \in \mathrm{Seq}.$$

Define $\mathrm{Seq}_0 = \{\sigma \in \mathrm{Seq}: \mathrm{lh}(\sigma) \text{ is even}\}$ and $\mathrm{Seq}_1 = \{\sigma \in \mathrm{Seq}: \mathrm{lh}(\sigma)$ is odd$\}$. A 0-*strategy* is a function $S_0: \mathrm{Seq}_0 \to \mathbb{N}$. A 1-*strategy* is a function $S_1: \mathrm{Seq}_1 \to \mathbb{N}$. If S_0 is a 0-strategy and S_1 is a 1-strategy, let $S_0 \otimes S_1$ be the function $f: \mathbb{N} \to \mathbb{N}$ defined by $f(2k) = S_0(f[2k])$, $f(2k+1) = S_1(f[2k+1])$.

The *axiom of determinacy* is the assertion that for all $\mathcal{F} \subseteq \mathbb{N}^{\mathbb{N}}$ either $\exists S_0 \,\forall S_1 \,(S_0 \otimes S_1 \in \mathcal{F})$ or $\exists S_1 \,\forall S_0 \,(S_0 \otimes S_1 \notin \mathcal{F})$. The intuitive idea behind the axiom of determinacy is as follows. Consider an infinite game $G_{\mathcal{F}}$ between two players, 0 and 1. The rules of the game are that player 0 picks $f(0)$, then player 1 picks $f(1), \ldots$, then player 0 picks $f(2k)$, then player 1 picks $f(2k+1)$, then $\ldots$. Finally player 0 wins if $f \in \mathcal{F}$, and player 1 wins if $f \notin \mathcal{F}$. The axiom of determinacy asserts that one player or the other has a winning strategy.

The axiom of determinacy is generally regarded as false. Nevertheless, the axiom of determinacy is the basis of an intricate theory known as *modern descriptive set theory*. In this theory, some of the known results concerning Borel and analytic sets are generalized to projective and hyperprojective classes, assuming the axiom of determinacy.

Modern descriptive set theory is a generalization of classical descriptive set theory, and classical descriptive set theory is a branch of ordinary mathematics. Therefore, from the viewpoint of the Main Question of this book, it is of interest to investigate the extent to which special cases of the axiom of determinacy are provable in subsystems of second order arithmetic.

The main result of this section is that ATR_0 is just strong enough to prove all instances of the axiom of determinacy in which the set $\mathcal{F} \subseteq \mathbb{N}^{\mathbb{N}}$ is open or clopen, i.e., Σ_1^0 or Δ_1^0 definable. We formalize this as follows:

DEFINITION V.8.1 (Σ_1^0 and Δ_1^0 determinacy). By Σ_1^0 (respectively Π_1^0) *determinacy* we mean the scheme

$$\exists S_0 \, \forall S_1 \, \varphi(S_0 \otimes S_1) \vee \exists S_1 \, \forall S_0 \, \neg\varphi(S_0 \otimes S_1)$$

where φ is Σ_1^0 (respectively Π_1^0). By Δ_1^0 *determinacy* we mean the scheme

$$\forall f \, (\varphi(f) \leftrightarrow \psi(f)) \rightarrow (\exists S_0 \, \forall S_1 \, \varphi(S_0 \otimes S_1) \vee \exists S_1 \, \forall S_0 \, \neg\varphi(S_0 \otimes S_1))$$

where φ and ψ are Σ_1^0 and Π_1^0, respectively. The schemes of Σ_k^i, Π_k^i, and Δ_k^i determinacy, $i < 2$, $1 \leq k \leq \infty$, are defined similarly.

(In the above definition, S_0 and S_1 range over 0-strategies and 1-strategies respectively, and f ranges over total functions from $\mathbb{N}$ into $\mathbb{N}$.)

THEOREM V.8.2. ATR_0 *proves Σ_1^0 determinacy.*

PROOF. We reason in ATR_0. Let $\varphi(f)$ be a Σ_1^0 formula. We can write $\varphi(f)$ in normal form as $\varphi(f) \equiv \exists n \, \theta(f[n])$ where θ is arithmetical (see lemma V.1.4). By arithmetical comprehension let W be the set of all $\sigma \in \mathrm{Seq}_0$ such that $\exists n \, (n \leq \mathrm{lh}(\sigma) \wedge \theta(\sigma[n]))$. Thus $\forall f \, (\varphi(f) \leftrightarrow \exists k \, (f[2k] \in W))$. Intuitively, W is the set of positions at which player 0 has "already won". It is to be proved that

$$\exists S_0 \, \forall S_1 \, \exists k \, ((S_0 \otimes S_1)[2k] \in W) \vee \exists S_1 \, \forall S_0 \, \forall k \, ((S_0 \otimes S_1)[2k] \notin W).$$

We proceed much as in the proof of the perfect set theorem in ATR_0, theorem V.4.3. By arithmetical transfinite recursion, there exists for each countable well ordering X a sequence of sets $\langle W_j : j \in \mathrm{field}(X) \rangle$ such that

$$\forall j \, \forall \sigma \, (\sigma \in W_j \leftrightarrow (\sigma \in W \vee \exists m \, \forall n \, \exists i \, (i <_X j \wedge \sigma^\frown \langle m, n \rangle \in W_i))). \tag{$*$}$$

The intuitive idea here is that to each countable ordinal β we associate a set $W_\beta \subseteq \mathrm{Seq}_0$ by $W_0 = W$, $W_\beta = \{\sigma : \exists m \, \forall n \, (\sigma^\frown \langle m, n \rangle \in \bigcup_{\alpha < \beta} W_\alpha)\}$ for $\beta > 0$. Thus each $\sigma \in W_\beta$ is a "winning position of order β" for player 0. The proof splits into two cases.

Case 1. There exists a countable well ordering X and a sequence of sets $\langle W_j : j \in \mathrm{field}(X) \rangle$ such that $(*)$ holds and in addition $\langle \rangle \in W_l$ for some $l \in \mathrm{field}(X)$.

In this case, for each $\sigma \in W_l$, let $g(\sigma)$ be the unique $j \leq_X l$ such that $\sigma \in W_j \wedge \forall i\, (i <_X j \to \sigma \notin W_i)$. For each $\sigma \in \mathrm{Seq}_0$ define $S_0(\sigma) = 0$ if $\sigma \in W$ or if $\sigma \notin W_l$, otherwise $S_0(\sigma) =$ the least m such that $\forall n\, \exists i\, (i <_X g(\sigma) \wedge \sigma^\frown \langle m, n \rangle \in W_i)$. This m exists by $(*)$. Thus S_0 is a 0-strategy.

We claim that S_0 is a winning strategy for player 0, i.e.,

$$\forall S_1\, \exists k\, ((S_0 \otimes S_1)[2k] \in W).$$

To see this, consider any 1-strategy S_1 and put $f = S_0 \otimes S_1$. By case hypothesis, $\langle \rangle \in W_l$. By choice of S_0, $f[2k] \in W_l$ implies $f[2k+2] \in W_l$. Thus by induction $f[2k] \in W_l$ for all k. Also, by choice of S_0, $f[2k] \notin W$ implies $g(f[2k + 2]) <_X g(f[2k])$. Since X is a well ordering, there must exist k such that $f[2k] \in W$. This proves the claim.

Case 2. Assume that the hypothesis of case 1 does not hold.

In this case we use the method of pseudohierarchies. By lemma V.4.12 (or by a direct argument as in case 2 of the proof of theorem V.4.3), there exists a countable linear ordering X and a sequence of sets $\langle W_j : j \in \text{field}(X) \rangle$ such that $(*)$ holds and in addition $\forall j\, (j \in \text{field}(X) \to \langle \rangle \notin W_j)$ and X is not a well ordering. Let $g \colon \mathbb{N} \to \text{field}(X)$ be a fixed descending sequence through X, i.e., $g(k + 1) <_X g(k)$ for all k. For each $\sigma \in \mathrm{Seq}_1$ of length $2k + 1$ define $S_1(\sigma) =$ least n such that $\sigma^\frown \langle n \rangle \notin W_{g(k+1)}$ if such an n exists, otherwise $S_1(\sigma) = 0$. Thus S_1 is a 1-strategy.

We claim that S_1 is a winning strategy for player 1, i.e.,

$$\forall S_0\, \forall k\, ((S_0 \otimes S_1)[2k] \notin W).$$

To see this, consider a 0-strategy S_0 and put $f = S_0 \otimes S_1$. By case hypothesis, $\langle \rangle \notin W_{g(0)}$. If $f[2k] \notin W_{g(k)}$, then by $(*)$ we have $\forall m\, \exists n\, \forall i\, (i <_X g(k) \to f[2k]^\frown \langle m, n \rangle \notin W_i)$. Hence by choice of S_1 we have $f[2k+2] \notin W_{g(k+1)}$. Thus by induction $f[2k] \notin W_{g(k)}$ for all k. In particular $\forall k\, (f[2k] \notin W)$ proving our claim.

This completes the proof of theorem V.8.2. $\qquad \sqcap$

As a consequence of Σ_1^0 determinacy in ATR_0, we obtain a form of the axiom of choice in ATR_0:

THEOREM V.8.3 (Σ_1^1 axiom of choice). ATR_0 proves the Σ_1^1 axiom of choice, i.e., the scheme

$$\forall k\, \exists X\, \varphi(k, X) \to \exists Y\, \forall k\, \varphi(k, (Y)_k)$$

where φ is any Σ_1^1 formula and $(Y)_k = \{i : (i, k) \in Y\}$.

PROOF. We reason in ATR_0. By theorem V.1.7' it suffices to prove: For any sequence of trees $\langle T_k : k \in \mathbb{N} \rangle$ such that $\forall k\, (T_k$ has a path$)$, there exists a sequence $\langle g_k : k \in \mathbb{N} \rangle$ such that $\forall k\, (g_k$ is a path through $T_k)$. We shall obtain this as a consequence of Σ_1^0 determinacy.

In the intuitive game-theoretic terminology, consider the following Σ_1^0 game. Player 0 chooses k, then player 1 chooses $g(0), g(1), \ldots$. Player

1 wins if and only if g is a path through T_k. Since $\forall k$ (T_k has a path), player 0 cannot have a winning strategy. Hence by Σ_1^0 determinacy there exists a winning strategy for player 1. This strategy provides the desired choice function.

Formally, let $\varphi(f)$ be the following Σ_1^0 formula:

$$\exists j \,(\langle f(1), f(3), \ldots, f(2j-1)\rangle \notin T_{f(0)}).$$

We claim that $\forall S_0 \, \exists S_1 \, \neg\varphi(S_0 \otimes S_1)$. To see this, let S_0 be given. Put $k = S_0(\langle\rangle)$ and let g be any path through T_k. Define $S_1(\sigma) = g(j)$ for all σ of length $2j + 1$. Put $f = S_0 \otimes S_1$. Clearly $f(0) = k$ and $f(2j+1) = g(j)$ for all j. Since g is a path through T_k, we have $\forall j \,(\langle f(1), f(3), \ldots, f(2j-1)\rangle \in T_{f(0)})$, i.e., $\neg\varphi(f)$. This proves our claim. Hence by Σ_1^0 determinacy there exists S_1 such that $\forall S_0 \, \neg\varphi(S_0 \otimes S_1)$. Define a sequence of functions $\langle g_k : k \in \mathbb{N}\rangle$ by

$$g_k(j) = S_1(\langle k, \underbrace{0, \ldots, 0}_{2j}\rangle).$$

We claim that g_k is a path through T_k. To see this, define $S_0(\langle\rangle) = k$ and $S_0(\sigma) = 0$ for all other $\sigma \in \mathrm{Seq}_0$. Put $f = S_0 \otimes S_1$. Then $f(2j+1) = g_k(j)$ for all j. We have $\neg\varphi(S_0 \otimes S_1)$, i.e., $\neg\varphi(f)$, i.e.,

$$\neg\exists j \,(\langle g_k(0), g_k(1), \ldots, g_k(j-1)\rangle \notin T_k),$$

i.e., g_k is a path through T_k. This completes the proof.   □

Remark V.8.4 (Σ_1^1 choice versus ATR_0). The Σ_1^1 axiom of choice is not equivalent to ATR_0. For instance, the ω-model HYP (proposition V.2.6) satisfies the Σ_1^1 axiom of choice but does not satisfy ATR_0. It is also true that ATR_0 proves the existence of a countable ω-model of ACA_0 plus the Σ_1^1 axiom of choice. For more information on models of the Σ_1^1 axiom of choice, see §§VIII.3, VIII.4, VIII.5, and IX.4.

We now turn to the reversal of Σ_1^0 determinacy. We shall in fact show that Δ_1^0 determinacy implies arithmetical transfinite recursion. We begin with:

Lemma V.8.5. *It is provable in RCA_0 that Δ_1^0 determinacy implies arithmetical comprehension.*

Proof. Reasoning in RCA_0, assume Δ_1^0 determinacy. Instead of arithmetical comprehension we shall prove the equivalent statement III.1.3.3. Let $g : \mathbb{N} \to \mathbb{N}$ be given. We shall prove the existence of a set X such that $\forall k \, (k \in X \leftrightarrow \exists m \,(g(m) = k))$.

In the intuitive game-theoretic terminology, consider the following game of length 3. Player 0 chooses k, then player 1 chooses n, then player 0 chooses m. Player 0 wins if and only if $g(n) \neq k$ and $g(m) = k$. Clearly player 0 cannot have a winning strategy. Hence by Δ_1^0 determinacy player 1 has a winning strategy. The desired set X then exists by Δ^0_1 comprehension, using this strategy as a parameter.

Formally, let $\varphi(f)$ be the following Σ^0_1 or Π^0_1 formula:

$$g(f(1)) \neq f(0) \wedge g(f(2)) = f(0).$$

We claim that $\forall S_0 \exists S_1 \neg\varphi(S_0 \otimes S_1)$. Given S_0, put $k = S_0(\langle\rangle)$. Let n be such that $g(n) = k$ if such an n exists, otherwise let $n = 0$. Consider any S_1 such that $S_1(\langle k\rangle) = n$. Put $m = S_0(\langle k, n\rangle)$. Then $g(n) = k \vee g(m) \neq k$, i.e., $\neg\varphi(S_0 \otimes S_1)$. This proves our claim. Hence by Δ^0_1 determinacy there exists S_1 such that $\forall S_0 \neg\varphi(S_0 \otimes S_1)$. We claim that

$$\forall m \, \forall k \, (g(m) = k \rightarrow g(S_1(\langle k\rangle)) = k).$$

If not, let m and k be such that $g(m) = k$ and $g(S_1(\langle k\rangle)) \neq k$. Put $n = S_1(\langle k\rangle)$ and consider any S_0 such that $S_0(\langle\rangle) = k$ and $S_0(\langle k, n\rangle) = m$. Then $g(n) \neq k \wedge g(m) = k$, i.e., $\varphi(S_0 \otimes S_1)$. This contradiction proves our claim. Hence

$$\forall k \, (\exists m \, (g(m) = k) \leftrightarrow g(S_1(\langle k\rangle)) = k).$$

Applying Δ^0_1 comprehension we get $\exists X \, \forall k \, (k \in X \leftrightarrow \exists m \, (g(m) = k))$. Hence by lemma III.1.3 we have arithmetical comprehension. This completes the proof of lemma V.8.5. $\square$

In order to prove that Δ^0_1 determinacy implies arithmetical transfinite recursion, we consider the following family of Δ^0_1 games. Let X and Y be countable linear orderings. Assume that at least one of X and Y is a well ordering. Let $G(X, Y)$ be the game in which player 0 builds a descending sequence $f(0) >_X f(2) >_X \cdots$ through X and player 1 builds a descending sequence $f(1) >_Y f(3) >_Y \cdots$ through Y. The winner of $G(X, Y)$ is that player whose descending sequence keeps going the longest. Clearly 0 has a winning strategy for $G(X, Y)$ whenever $\neg\mathrm{WO}(X)$. Also, if 0 has a winning strategy S_0 for $G(X, Y)$, then 1 has a winning strategy for $G(Y, X)$, namely, he disregards 0's initial move $f(0)$ and thereafter plays S_0. We formalize this as follows:

LEMMA V.8.6. *The following is provable in* RCA$_0$. *Assume*

$$\mathrm{LO}(X) \wedge \mathrm{LO}(Y) \wedge (\mathrm{WO}(X) \vee \mathrm{WO}(Y)).$$

Let $\varphi(f, X, Y)$ *be the* Σ^0_1 *formula*

$$\exists j \, (f(2j+3) \not<_Y f(2j+1) \wedge \forall i \, (i \leq j \rightarrow f(2i+2) <_X f(2i))).$$

Let $\psi(f, X, Y)$ *be the* Π^0_1 *formula*

$$\neg\exists j \, (f(2j+2) \not<_X f(2j) \wedge \forall i \, (i < j \rightarrow f(2i+3) <_Y f(2i+1))).$$

Then:

1. $\forall f \, (\varphi(f, X, Y) \leftrightarrow \psi(f, X, Y))$.
2. $\neg\mathrm{WO}(X) \rightarrow \exists S_0 \, \forall S_1 \, \varphi(S_0 \otimes S_1, X, Y)$.
3. $\exists S_0 \, \forall S_1 \, \varphi(S_0 \otimes S_1, X, Y) \rightarrow \exists S_1 \, \forall S_0 \, \neg\varphi(S_0 \otimes S_1, Y, X)$.

PROOF. 1. Let $f : \mathbb{N} \to \mathbb{N}$ be given. Since $\mathrm{WO}(X) \vee \mathrm{WO}(Y)$ there must exist i such that $f(2i+2) \not<_X f(2i) \vee f(2i+3) \not<_Y f(2i+1)$. Let j be the least such i. If $f(2j+2) <_X f(2j)$ then we have $\varphi(f, X, Y) \wedge \psi(f, X, Y)$. If $f(2j+2) \not<_X f(2j)$ then we have $\neg\varphi(f, X, Y) \wedge \neg\psi(f, X, Y)$.

2. Let $g : \mathbb{N} \to \mathrm{field}(X)$ be a descending sequence through X. Define $S_0(\sigma) = g(j)$ for all σ of length $2j$. Given any 1-strategy S_1, put $f = S_0 \otimes S_1$. Then $f(2j+2) = g(j+1) <_X g(j) = f(2j)$ for all j. Hence $\psi(f, X, Y)$. Hence $\varphi(f, X, Y)$, i.e., $\varphi(S_0 \otimes S_1, X, Y)$, in view of part 1.

3. Let S_0 be such that $\forall S_1 \, \varphi(S_0 \otimes S_1, X, Y)$. Define

$$S_1'(\sigma) = S_0(\langle \sigma(1), \ldots, \sigma(2j) \rangle)$$

for all σ of length $2j + 1$. We claim that $\forall S_0' \, \neg\varphi(S_0' \otimes S_1', Y, X)$. To see this, let S_0' be given. Set $k = S_0'(\langle \rangle)$ and define $S_1(\sigma) = S_0'(\langle k \rangle^\frown \sigma)$ for all $\sigma \in \mathrm{Seq}_1$. Put $f = S_0 \otimes S_1$ and $f' = S_0' \otimes S_1'$. Beginning with $f'(0) = k$ we have inductively

$$f'(2j+1) = S_1'(f'[2j+1]) = S_0(\langle f'(1), \ldots, f'(2j) \rangle)$$
$$= S_0(\langle f(0), \ldots, f(2j-1) \rangle) = S_0(f[2j]) = f(2j)$$

and

$$f(2j+1) = S_1(f[2j+1]) = S_0'(\langle k \rangle^\frown f[2j+1])$$
$$= S_0'(\langle k, f(0), \ldots, f(2j) \rangle) = S_0'(f'[2j+2]) = f'(2j+2).$$

Thus $f(2j) = f'(2j+1)$ and $f(2j+1) = f'(2j+2)$ for all j. By assumption $\varphi(S_0 \otimes S_1, X, Y)$, i.e., $\varphi(f, X, Y)$. Let j be such that

$$f(2j+3) \not<_Y f(2j+1) \wedge \forall i \, (i \leq j \to f(2i+2) <_X f(2i)).$$

It follows that

$$f'(2j+4) \not<_Y f'(2j+2) \wedge \forall i \, (i \leq j \to f'(2i+3) <_X f'(2i+1)).$$

Thus $\neg\psi(f, Y, X)$, i.e., $\neg\varphi(f', Y, X)$, i.e., $\neg\varphi(S_0' \otimes S_1', Y, X)$. This proves our claim. The proof of lemma V.8.6 is complete. $\square$

With the above lemmas, we are now ready to prove:

THEOREM V.8.7 (ATR$_0$ and determinacy). *The following are pairwise equivalent over* RCA$_0$:

1. *arithmetical transfinite recursion;*
2. Σ_1^0 *determinacy;*
3. Δ_1^0 *determinacy.*

PROOF. The implication from 1 to 2 has already been proved as theorem V.8.2. The implication from 2 to 3 is trivial. It remains to prove that 3 implies 1. Assume Δ_1^0 determinacy. By lemma V.8.5 we have arithmetical comprehension. We wish to prove arithmetical transfinite recursion. Instead of proving arithmetical transfinite recursion directly, we shall prove the Σ_1^1 separation principle V.5.1.2. Assume that $\neg\exists k \, (\varphi_1(k) \wedge \varphi_0(k))$ where $\varphi_1(k)$ and $\varphi_0(k)$ are Σ_1^1. We seek a set $Z \subseteq \mathbb{N}$ such

that $\forall k\,((\varphi_1(k) \rightarrow k \in Z) \wedge (\varphi_0(k) \rightarrow k \notin Z))$. By theorem V.1.7'
and lemma V.1.3 there exist sequences of countable linear orderings
$\langle X_k : k \in \mathbb{N} \rangle$ and $\langle Y_k : k \in \mathbb{N} \rangle$ such that $\forall k\,(\varphi_1(k) \leftrightarrow \neg\mathrm{WO}(X_k))$ and
$\forall k\,(\varphi_0(k) \leftrightarrow \neg\mathrm{WO}(Y_k))$. Our assumption $\neg\exists k\,(\varphi_1(k) \wedge \varphi_0(k))$ implies
that $\forall k\,(\mathrm{WO}(X_k) \vee \mathrm{WO}(Y_k))$.

In the intuitive game-theoretic terminology, consider the following Δ^0_1
game G'. Player 0 chooses k, then player 1 chooses $i \in \{0, 1\}$, then players
0 and 1 play $G(X_k, Y_k)$ if $i = 0$, $G(Y_k, X_k)$ if $i = 1$. (The family of Δ^0_1
games $G(X, Y)$ was defined in the discussion preceding lemma V.8.6.) We
claim that 0 has no winning strategy for G'. To see this, suppose that 0
begins G' by choosing k. If 1 has a winning strategy for $G(X_k, Y_k)$, he
can win G' by playing $i = 0$ followed by that strategy. If 1 does not have a
winning strategy for $G(X_k, Y_k)$, then by Δ^0_1 determinacy, 0 has a winning
strategy for $G(X_k, Y_k)$. Hence 1 has a winning strategy for $G(Y_k, X_k)$,
so he can win G' by playing $i = 1$ followed by that strategy. In either
case, 1 can win G'. This proves our claim. Hence, by Δ^0_1 determinacy,
1 has a winning strategy for G'. Let Z be the set of all k such that if
0 begins G' by choosing k, then 1 responds with $i = 1$ according to
his winning strategy. It is easy to see that Z is the desired separating
set.

Formally, let φ and ψ be as in lemma V.8.6. For each $f : \mathbb{N} \rightarrow \mathbb{N}$ define
$f'(n) = f(n + 2)$ for all n. Let $\varphi'(f)$ be the Σ^0_1 formula

$$(f(1) = 0 \wedge \varphi(f', X_{f(0)}, Y_{f(0)})) \vee$$
$$(f(1) = 1 \wedge \varphi(f', Y_{f(0)}, X_{f(0)})) \vee f(1) \geq 2.$$

Let $\psi'(f)$ be the Π^0_1 formula

$$(f(1) = 0 \wedge \psi(f', X_{f(0)}, Y_{f(0)})) \vee$$
$$(f(1) = 1 \wedge \psi(f', Y_{f(0)}, X_{f(0)})) \vee f(1) \geq 2.$$

By lemma V.8.6.1 we have $\forall f\,(\varphi'(f) \leftrightarrow \psi'(f))$.

We claim that $\forall S_0 \exists S_1 \neg\varphi'(S_0 \otimes S_1)$. To see this, let S_0 be given and
set $k = S_0(\langle\rangle)$. Case 1: Assume that $\exists S'_1 \forall S'_0 \neg\varphi(S'_0 \otimes S'_1, X_k, Y_k)$. Then
choose such an S'_1 and set $i = 0$. Case 2: Assume that the hypothesis
of case 1 does not hold. Then by lemma V.8.6.1 and Δ^0_1 determinacy
we have $\exists S'_0 \forall S'_1 \varphi(S'_0 \otimes S'_1, X_k, Y_k)$. Hence by lemma V.8.6.3 we have
$\exists S'_1 \forall S'_0 \neg\varphi(S'_0 \otimes S'_1, Y_k, X_k)$. Choose such an S'_1 and set $i = 1$. In either
case define $S_1(\sigma) = i$ for σ of length 1, $S_1(\sigma) = S'_1(\langle \sigma(2), \ldots, \sigma(2j+2)\rangle)$
for σ of length $2j + 3$, and $S'_0(\sigma) = S_0(\langle k, i\rangle^\frown\sigma)$ for all $\sigma \in \mathrm{Seq}_0$. Thus
$S'_0 \otimes S'_1 = (S_0 \otimes S_1)'$ so by choice of S'_1 and i we have

$$(i = 0 \wedge \neg\varphi((S_0 \otimes S_1)', X_k, Y_k)) \vee (i = 1 \wedge \neg\varphi((S_0 \otimes S_1)', Y_k, X_k)).$$

Hence $\neg\varphi'(S_0 \otimes S_1)$. This proves our claim.

Hence by Δ_1^0 determinacy there exists S_1 such that $\forall S_0 \, \neg\varphi'(S_0 \otimes S_1)$. We claim that

$$\forall k \, ((\neg\mathrm{WO}(X_k) \to S_1(\langle k \rangle) = 1) \wedge (\neg\mathrm{WO}(Y_k) \to S_1(\langle k \rangle) = 0)).$$

To see this, let k be given. Define $i = S_1(\langle k \rangle)$ and $S_1'(\sigma) = S_1(\langle k, i \rangle {}^\frown \sigma)$ for all $\sigma \in \mathrm{Seq}_1$. If $\neg\mathrm{WO}(X_k)$, then by V.8.6.2 let S_0' be such that $\varphi(S_0' \otimes S_1', X_k, Y_k)$. Define $S_0(\langle \rangle) = k$ and $S_0(\sigma) = S_0'(\langle \sigma(2), \ldots, \sigma(2j+1) \rangle)$ for σ of length $2j+2$. Then $S_0' \otimes S_1' = (S_0 \otimes S_1)'$, hence $\varphi((S_0 \otimes S_1)', X_k, Y_k)$. Since $\neg\varphi'(S_0 \otimes S_1)$, we must have $(S_0 \otimes S_1)(1) \neq 0$. Hence $S_1(\langle k \rangle) = (S_0 \otimes S_1)(1) = 1$. On the other hand, if $\neg\mathrm{WO}(Y_k)$, then by V.8.6.2 let S_0' be such that $\varphi(S_0' \otimes S_1', Y_k, X_k)$. Defining S_0 as before, we have again $S_0' \otimes S_1' = (S_0 \otimes S_1)'$, hence this time $\varphi((S_0 \otimes S_1)', Y_k, X_k)$. Since $\neg\varphi'(S_0 \otimes S_1)$, we must have $(S_0 \otimes S_1)(1) \neq 1$. Hence $S_1(\langle k \rangle) = (S_0 \otimes S_1)(1) = 0$. This proves our claim.

By Δ_1^0 comprehension let $Z \subseteq \mathbb{N}$ be the set of all k such that $S_1(\langle k \rangle) = 1$. Then by the previous claim we have $\forall k \, ((\varphi_1(k) \to k \in Z) \wedge (\varphi_0(k) \to k \notin Z))$. This proves Σ_1^1 separation. Hence by theorem V.5.1 we have arithmetical transfinite recursion. This completes the proof of theorem V.8.7.   □

EXERCISE V.8.8. Assume $\mathrm{WO}(X) \wedge \mathrm{WO}(Y)$. Let $G(X, Y)$ be the game of lemma V.8.6, in which the players play descending sequences through X and Y respectively. Show that player 1 has a winning strategy for $G(X, Y)$ if and only if $|X| \leq |Y|$. Formally,

$$\forall X \, \forall Y \, ((\mathrm{WO}(X) \wedge \mathrm{WO}(Y)) \to$$
$$(\exists S_1 \, \forall S_0 \, \neg\varphi(S_0 \otimes S_1, X, Y) \leftrightarrow |X| \leq |Y|),$$

where φ is as in lemma V.8.6.

Notes for §V.8. For an exposition of modern descriptive set theory based on the axiom of determinacy, see Moschovakis [191]. The fact that ATR_0 proves Σ_1^1 choice (theorem V.8.3) is essentially due to Friedman [62, chapter II]. A version of theorem V.8.7 is in Steel [256]; this was one of the earliest results of Reverse Mathematics. Our proof of the reversal in theorem V.8.7 is new. Related results are in §§VI.5 and VI.7. See also Tanaka [263, 264].

V.9. The Σ_1^0 and Δ_1^0 Ramsey Theorems

In this section we shall discuss a certain "infinite exponent" generalization of Ramsey's theorem, namely the so-called *open Ramsey theorem*. We shall prove that the open Ramsey theorem is equivalent to ATR_0 over RCA_0.

The *Ramsey space* is the space $[\mathbb{N}]^{\mathbb{N}}$ of all total functions $f : \mathbb{N} \to \mathbb{N}$ such that f is strictly increasing, i.e., $f(m) < f(n)$ for all $m, n \in \mathbb{N}$ such that $m < n$. For $f \in [\mathbb{N}]^{\mathbb{N}}$ and $m \in \mathbb{N}$ we write as usual

$$f[m] = \langle f(0), f(1), \ldots, f(m-1) \rangle.$$

Thus $f[m] \in [\mathbb{N}]^m$ (see definition III.7.3).

Given $f, g \in [\mathbb{N}]^{\mathbb{N}}$ we define $f \cdot g \in [\mathbb{N}]^{\mathbb{N}}$ by $(f \cdot g)(n) = f(g(n))$. A set $\mathcal{F} \subseteq [\mathbb{N}]^{\mathbb{N}}$ is called *Ramsey* if

$$\exists f \, (\forall g \, (f \cdot g \in \mathcal{F}) \vee \forall g \, (f \cdot g \notin \mathcal{F}));$$

here f and g range over points of $[\mathbb{N}]^{\mathbb{N}}$.

REMARK. In the literature of Ramsey theory, it is usual to identify a strictly increasing function $f \in [\mathbb{N}]^{\mathbb{N}}$ with its range $\mathrm{rng}(f) \subseteq \mathbb{N}$. Thus the Ramsey space $[\mathbb{N}]^{\mathbb{N}}$ is identified with the space $[\mathbb{N}]^{\infty}$ of infinite subsets of $\mathbb{N}$. For $X \in [\mathbb{N}]^{\infty}$ one may write

$$[X]^{\infty} = \{ Y \in [\mathbb{N}]^{\infty} : Y \subseteq X \},$$

and then for $f \in [\mathbb{N}]^{\mathbb{N}}$ we have

$$[\mathrm{rng}(f)]^{\infty} = \{ \mathrm{rng}(f \cdot g) : g \in [\mathbb{N}]^{\mathbb{N}} \}.$$

With this notation, $\mathcal{F} \subseteq [\mathbb{N}]^{\infty}$ is said to be *Ramsey* if and only if there exists $X \in [\mathbb{N}]^{\infty}$ such that either $[X]^{\infty} \subseteq \mathcal{F}$ or $[X]^{\infty} \cap \mathcal{F} = \emptyset$. In our treatment of Ramsey theory here, we shall not make these identifications. However, the reader may find it useful to keep this viewpoint in mind.

The axiom of choice implies that there exist non-Ramsey subsets of $[\mathbb{N}]^{\mathbb{N}}$. However, it is also known that many subsets of $[\mathbb{N}]^{\mathbb{N}}$ are Ramsey. For example, the Galvin/Prikry theorem asserts that all Borel subsets of $[\mathbb{N}]^{\mathbb{N}}$ are Ramsey. See also Mathias [181] and Carlson/Simpson [33]. The purpose of the next definition is to formalize certain special cases of this principle as schemes in the language of second order arithmetic.

DEFINITION V.9.1 (Σ_1^0-RT and Δ_1^0-RT). The Σ_1^0 *Ramsey theorem*, denoted Σ_1^0-RT, is the scheme

$$\exists f \, (\forall g \, \varphi(f \cdot g) \vee \forall g \, \neg \varphi(f \cdot g))$$

where φ is any Σ_1^0 formula. The Δ_1^0 *Ramsey theorem*, denoted Δ_1^0-RT, is the scheme

$$\forall h \, (\varphi(h) \leftrightarrow \psi(h)) \to \exists f \, (\forall g \, \varphi(f \cdot g) \vee \forall g \, \neg \varphi(f \cdot g))$$

where φ and ψ are Σ_1^0 and Π_1^0 respectively. Here f, g, and h range over $[\mathbb{N}]^{\mathbb{N}}$. The Σ_k^i and Δ_k^i *Ramsey theorems*, $i < 2$, $1 \leq k \leq \infty$, denoted Σ_k^i-RT and Δ_k^i-RT respectively, are defined similarly.

Note that Σ^0_∞-RT and Σ^1_∞-RT are not expressible as single sentences of L_2. Rather, they are schemes. (We may call them *Ramsey schemes*.) However, for $k < \infty$, each of Σ^i_k-RT and Δ^i_k-RT for $i \in \{0, 1\}$ may be expressed by means of codes as a single sentence of L_2. The appropriate codes for the Σ^0_1 case are given by the following definition.

DEFINITION V.9.2 (open sets in $[\mathbb{N}]^\mathbb{N}$). Let $[\mathbb{N}]^{<\mathbb{N}} = \bigcup_{m\in\mathbb{N}}[\mathbb{N}]^m =$ the set of all (codes for) strictly increasing finite sequences of natural numbers. (A sequence $\sigma \in \mathbb{N}^{<\mathbb{N}} = $ Seq is said to be *strictly increasing* if $\sigma(i) < \sigma(j)$ for all $i < j < \mathrm{lh}(\sigma)$.) In RCA$_0$, a *code for an open subset of* $[\mathbb{N}]^\mathbb{N}$ is defined to be a subset P of $[\mathbb{N}]^{<\mathbb{N}}$; we then write $f \in P$ to mean that $f \in [\mathbb{N}]^\mathbb{N}$ and $\exists m \, (f[m] \in P)$. In this sense we may write $P \subseteq [\mathbb{N}]^\mathbb{N}$.

Note that, by our formalized Kleene normal form theorem II.2.7, $P \subseteq [\mathbb{N}]^\mathbb{N}$ is open if and only if it is Σ^0_1 definable. This is provable in RCA$_0$.

DEFINITION V.9.3 (open Ramsey theorem). The *open Ramsey theorem* is defined in RCA$_0$ to be the statement that for all (codes for) open sets $P \subseteq [\mathbb{N}]^\mathbb{N}$ there exists f such that $\forall g(f \cdot g \in P) \vee \forall g(f \cdot g \notin P)$. The *clopen Ramsey theorem* is defined in RCA$_0$ to be the statement that for all (codes for) open sets $P, Q \subseteq [\mathbb{N}]^\mathbb{N}$, if $\forall h \, (h \in P \leftrightarrow h \notin Q)$ then

$$\exists f \, (\forall g \, (f \cdot g \in P) \vee \forall g \, (f \cdot g \notin P)).$$

Here f, g and h range over $[\mathbb{N}]^\mathbb{N}$. Clearly (Σ^0_1-RT $\leftrightarrow$ open Ramsey theorem) and (Δ^0_1-RT $\leftrightarrow$ clopen Ramsey theorem) are provable in RCA$_0$.

LEMMA V.9.4. *The open Ramsey theorem is provable in* ATR$_0$.

PROOF. We reason in ATR$_0$. Let P be a code for an open set in $[\mathbb{N}]^\mathbb{N}$ such that for all $f \in [\mathbb{N}]^\mathbb{N}$ there exists $g \in [\mathbb{N}]^\mathbb{N}$ such that $f \cdot g \in P$. We shall prove that there exists $f \in [\mathbb{N}]^\mathbb{N}$ such that $f \cdot g \in P$ for all $g \in [\mathbb{N}]^\mathbb{N}$.

Form the tree $T = \{\sigma \in [\mathbb{N}]^{<\mathbb{N}}$: no subsequence of σ is in $P\}$. Our assumption $\forall f \, \exists g \, f \cdot g \in P$ implies that T is well founded, i.e., T has no infinite path. Hence the Kleene/Brouwer ordering KB(T) is a well ordering.

For infinite sets $U, V \subseteq \mathbb{N}$, let us write $U \subseteq^\infty V$ to mean that U is almost included in V, i.e., $U \setminus V$ is finite.

We shall classify each $\sigma \in [\mathbb{N}]^{<\mathbb{N}}$ as either *good* or *bad*. For $\sigma \notin T$, let σ' be the smallest initial segment of σ such that $\sigma' \notin T$, and classify σ as good if $\sigma' \in P$, bad if $\sigma' \notin P$. For $\sigma \in T$, we shall use arithmetical transfinite recursion along KB(T) to classify σ as good or bad and simultaneously to define an infinite set $U_\sigma \subseteq \mathbb{N}$ with the following properties:

1. if $\tau <_{\mathrm{KB}(T)} \sigma$ then $U_\sigma \subseteq^\infty U_\tau$;
2. if σ is good then $\sigma^\frown\langle n \rangle$ is good for all $n \in U_\sigma$;
3. if σ is bad then $\sigma^\frown\langle n \rangle$ is bad for all $n \in U_\sigma$.

Given $\sigma \in T$, assume inductively that U_τ has been defined and that τ has been classified as good or bad, for each $\tau <_{\mathrm{KB}(T)} \sigma$. By a straightforward diagonal construction, define an infinite set V such that $V \subseteq^\infty U_\tau$ for all $\tau <_{\mathrm{KB}(T)} \sigma$. If there are infinitely many $n \in V$ such that $\sigma^\frown \langle n \rangle$ is good, classify σ as good and define $U_\sigma = \{n \in V : \sigma^\frown \langle n \rangle$ is good$\}$. Otherwise classify σ as bad and define $U_\sigma = \{n \in V : \sigma^\frown \langle n \rangle$ is bad$\}$. This transfinite recursion continues until the empty sequence $\langle \rangle$ has been classified as good or bad and $U_{\langle \rangle}$ has been defined.

We claim that $\langle \rangle$ is good. Suppose not, i.e., $\langle \rangle$ is bad. Define an increasing sequence of integers $k_0 < k_1 < \cdots < k_n < \cdots, n \in \mathbb{N}$. Begin by defining k_0 to be the least element of $U_{\langle \rangle}$. If $k_0 < \cdots < k_n$ have been defined, put $W_n = \bigcap \{U_\sigma : \sigma$ is a subsequence of $\langle k_0, \ldots, k_n \rangle \}$ and let k_{n+1} be the least $m \in W_n$ such that $m > k_n$. Since $\langle \rangle$ is bad, it is clear by induction on n that every subsequence of $\langle k_0, \ldots, k_n \rangle$ is bad, for all $n \in \mathbb{N}$. Now define $f \in [\mathbb{N}]^\mathbb{N}$ by putting $f(n) = k_n$, for all n. Then $f \cdot g \notin P$ for all $g \in [\mathbb{N}]^\mathbb{N}$, a contradiction. This proves our claim.

Since $\langle \rangle$ is good, the same construction can be used to obtain an increasing sequence $k_0 < k_1 < \cdots < k_n < \cdots$ every finite subsequence of which is good. Again define $f \in [\mathbb{N}]^\mathbb{N}$ by putting $f(n) = k_n$ for all n. Since T is well founded, for every $g \in [\mathbb{N}]^\mathbb{N}$ there is a least m such that $(f \cdot g)[m] \notin T$. Since $(f \cdot g)[m]$ is good, we have $f \cdot g \in P$. This completes the proof of lemma V.9.4. □

The rest of this section is devoted to the reversal of lemma V.9.4. We shall in fact show that the clopen Ramsey theorem implies arithmetical transfinite recursion. We begin with:

LEMMA V.9.5. *It is provable in* RCA$_0$ *that the clopen Ramsey theorem implies arithmetical comprehension.*

PROOF. From §III.7 we know that arithmetical comprehension is equivalent over RCA$_0$ to RT(3), i.e., Ramsey's theorem for exponent 3. We shall now prove within RCA$_0$ that the clopen Ramsey theorem implies RT(3).

Given a coloring of 3-tuples $F : [\mathbb{N}]^3 \to \{0, 1\}$, define for $i = 0, 1$

$$P_i = \{\sigma \in [\mathbb{N}]^{<\mathbb{N}} : \mathrm{lh}(\sigma) = 3 \wedge F(\sigma) = i\}.$$

Thus P_0 and P_1 are subsets of $[\mathbb{N}]^{<\mathbb{N}}$, and for all $h \in [\mathbb{N}]^\mathbb{N}$ we have $h \in P_i \leftrightarrow F(h[3]) = i$, hence $h \in P_0 \leftrightarrow h \notin P_1$. Thus, by the clopen Ramsey theorem, there exist $f \in [\mathbb{N}]^\mathbb{N}$ and $i \in \{0, 1\}$ such that $f \cdot g \in P_i$ for all $g \in [\mathbb{N}]^\mathbb{N}$. By Δ^0_1 comprehension there exists $X = \mathrm{rng}(f) = \{n : \exists m \leq n(f(m) = n)\}$, since f is strictly increasing. Clearly $X \subseteq \mathbb{N}$ is infinite and $[X]^3 \subseteq P_i$, i.e., $F(\sigma) = i$ for all $\sigma \in [X]^3$. This proves RT(3). Hence by lemma III.7.5 we have arithmetical comprehension. □

LEMMA V.9.6 (clopen Ramsey reversal). *It is provable in* RCA$_0$ *that the clopen Ramsey theorem implies arithmetical transfinite recursion.*

PROOF. Assume the clopen Ramsey theorem. By lemma V.9.5 we have arithmetical comprehension. We wish to prove arithmetical transfinite recursion. Instead of proving this directly, we shall prove the equivalent Σ_1^1 separation principle V.5.1.2. Reasoning in ACA$_0$, assume that $\neg \exists k \, (\varphi_1(k) \wedge \varphi_0(k))$ where $\varphi_1(k)$ and $\varphi_0(k)$ are Σ_1^1. We seek a set $Z \subseteq \mathbb{N}$ such that $\forall k \, ((\varphi_1(k) \rightarrow k \in Z) \wedge (\varphi_0(k) \rightarrow k \notin Z))$. By theorem V.1.7$'$ there exist sequences of trees $\langle T_k^i : k \in \mathbb{N} \rangle$, $i \in \{0, 1\}$, such that $\forall k \, (\varphi_i(k) \leftrightarrow T_k^i$ has a path). By assumption we have $\forall k \, (T_k^1$ and T_k^0 do not both have a path).

Given any tree $T \subseteq$ Seq, let us say that $f \in [\mathbb{N}]^{\mathbb{N}}$ *majorizes* T if there exists $g \colon \mathbb{N} \rightarrow \mathbb{N}$ such that g is a path through T and $\forall m \, (g(m) \leq f(m))$. Let us say that $\sigma \in [\mathbb{N}]^{<\mathbb{N}}$ *majorizes* T if there exists $\tau \in T$ such that $\text{lh}(\tau) = \text{lh}(\sigma)$ and $\forall j < \text{lh}(\sigma) \, (\tau(j) \leq \sigma(j))$. By König's lemma (theorem III.7.2) it follows that $f \in [\mathbb{N}]^{\mathbb{N}}$ majorizes T if and only if $\forall m \, (f[m]$ majorizes $T)$. Applying this to the T_k^i's from the previous paragraph, we have that $\forall k \, \exists m \, (f[m]$ does not majorize both T_k^1 and $T_k^0)$.

Given $h \in [\mathbb{N}]^{\mathbb{N}}$ define $m_h =$ the least $m > 0$ such that for all $k \leq h(0)$, $\langle h(1), h(2), \ldots, h(m) \rangle$ does not majorize both T_k^1 and T_k^0. Then define $n_h =$ the least $n > m_h$ such that for all $k \leq h(0)$, $\langle h(m_h + 1), h(m_h + 2), \ldots, h(n) \rangle$ does not majorize both T_k^1 and T_k^0. Let P and Q be (codes for) open subsets of $[\mathbb{N}]^{\mathbb{N}}$ such that for all $h \in [\mathbb{N}]^{\mathbb{N}}$, $h \in P \leftrightarrow h \notin Q$, and $h \in P$ if and only if $\forall k \leq h(0) \, (\langle h(1), h(2), \ldots, h(m_h) \rangle$ majorizes $T_k^1 \leftrightarrow \langle h(m_h + 1), h(m_h + 2), \ldots, h(n_h) \rangle$ majorizes $T_k^1)$.

We claim that for all $f \in [\mathbb{N}]^{\mathbb{N}}$ there exists $g \in [\mathbb{N}]^{\mathbb{N}}$ such that $f \cdot g \in P$. To see this, let $f \in [\mathbb{N}]^{\mathbb{N}}$ be given. Define $m_0 < m_1 < \cdots < m_i < \cdots$ by $m_0 = 0$, $m_{i+1} =$ least $n > m_i$ such that $\forall k \leq f(0)(\langle f(m_i + 1), f(m_i + 2), \ldots, f(n) \rangle$ does not majorize both T_k^1 and $T_k^0)$. By the pigeonhole principle there exist i and j such that $i < j$ and $\forall k \leq f(0)(\langle f(m_i + 1), \ldots, f(m_{i+1}) \rangle$ majorizes $T_k^1 \leftrightarrow \langle f(m_j + 1), \ldots, f(m_{j+1}) \rangle$ majorizes $T_k^1)$. Let $g \in [\mathbb{N}]^{\mathbb{N}}$ be such that $g(0) = 0$, $g(1) = m_i + 1, \ldots, g(m_{i+1} - m_i) = m_{i+1}$, $g(m_{i+1} - m_i + 1) = m_j + 1, \ldots, g(m_{i+1} - m_i + m_{j+1} - m_j) = m_{j+1}$. Putting $h = f \cdot g$ we see that $\langle h(1), \ldots, h(m_h) \rangle = \langle f(m_i + 1), \ldots, f(m_{i+1}) \rangle$ and $\langle h(m_h + 1), \ldots, h(n_h) \rangle = \langle f(m_j + 1), \ldots, f(m_{j+1}) \rangle$. It follows that $f \cdot g = h \in P$. This proves our claim.

From the above claim plus the clopen Ramsey theorem, it follows that there exists $f \in [\mathbb{N}]^{\mathbb{N}}$ such that $f \cdot g \in P$ for all $g \in [\mathbb{N}]^{\mathbb{N}}$. For each $k \in \mathbb{N}$ define $f_k \in [\mathbb{N}]^{\mathbb{N}}$ by $f_k(m) = f(k + m)$. Using arithmetical (actually Δ_1^0) comprehension, let Z be the set of all k such that $\langle f_k(1), f_k(2), \ldots, f_k(m_{f_k}) \rangle$ majorizes T_k^1. We claim that $\forall k((\varphi_1(k) \rightarrow k \in Z) \wedge (\varphi_0(k) \rightarrow k \notin Z))$. Suppose first that $\varphi_1(k)$ holds, i.e., T_k^1 has a path, but $k \notin Z$. Let $g \in [\mathbb{N}]^N$ be such that $g(0) = k$, $g(1) = k + 1, \ldots, g(m_{f_k}) = k + m_{f_k}$, and $\langle g(m_{f_k} + 1), g(m_{f_k} + 2), \ldots \rangle$ majorizes T_k^1. Putting $h = f \cdot g$ we see that $k \leq f(k) = h(0)$ and $\langle h(1), \ldots, h(m_h) \rangle = \langle f_k(1), \ldots, f_k(m_{f_k}) \rangle$ does not majorize T_k^1, while $\langle h(m_h + 1), \ldots, h(n_h) \rangle = \langle f(g(m_{f_k} + 1)), f(g(m_{f_k} + 2)), \ldots, f(g(n_h)) \rangle$

does majorize T_k^1. Thus $f \cdot g = h \notin P$, a contradiction. This shows that $\forall k(\varphi_1(k) \to k \in Z)$. A similar argument shows that $\forall k(\varphi_0(k) \to k \notin Z)$. This completes the proof of lemma V.9.6.   □

Summarizing, we have:

THEOREM V.9.7. *The following are pairwise equivalent over* RCA$_0$:

1. ATR$_0$;
2. *the open Ramsey theorem,* Σ_1^0-RT;
3. *the clopen Ramsey theorem,* Δ_1^0-RT.

PROOF. This is immediate from lemmas V.9.4 and V.9.6.   □

Notes for §V.9. Questions concerning effectivity of the open and clopen Ramsey theorems have been considered by Simpson [232], Mansfield [170], Clote [37], and Solovay [252]. The results of this section were first proved in Friedman/McAloon/Simpson [76] using formalized hyperarithmetical theory, pseudohierarchies, and inner models. The greatly simplified proofs of lemmas V.9.4 and V.9.6 presented here are due to Avigad [9] and Jockusch (personal communication), respectively. Some refinements of theorem V.9.7 are in Friedman/McAloon/Simpson [76, appendix] and in Simpson [235]. Other results related to theorem V.9.7 are in §§III.7, VI.6, VI.7.

V.10. Conclusions

In this chapter we have seen that many ordinary mathematical theorems are logically equivalent to ATR$_0$. Among these are: Lusin's separation theorem V.3.9, the Borel domain theorem V.3.11, the perfect set theorem V.4.3, comparability of countable well orderings (§V.6), the existence of Ulm resolutions (§V.7), open and clopen determinacy (§V.8), and the open and clopen Ramsey theorems (§V.9). In order to prove these equivalences, several interesting techniques have been developed. Prominent among the proof techniques are the method of pseudohierarchies (§V.4) and a technique of doing reversals via Σ_1^1 separation (theorem V.5.1) and unique paths (theorem V.5.2).

In §V.8 we obtained the following interesting result: ATR$_0$ proves Σ_1^1 choice. This may be compared to the related results obtained in §§VIII.4–VIII.5.

REMARK V.10.1 (the method of inner models). One important ATR$_0$ technique which has not appeared in this chapter is the method of inner models, where countable coded ω-models of Σ_1^1-AC$_0$ (see §VIII.4) are used to prove mathematical theorems in ATR$_0$. A rather difficult application of this technique will appear in our treatment of Silver's theorem (lemma VI.3.1), below. See also Marcone [172]. Other applications are in the original proof of the open Ramsey theorem in ATR_0 (Friedman/McAloon/Simpson [76]), and in the proof of the countable König duality theorem in ATR_0 (Simpson [247]). A related inner model technique is also useful for proving mathematical theorems in $\Pi_1^1\text{-CA}_0$; see §§VI.5–VI.6.

Chapter VI

Π_1^1 COMPREHENSION

In §I.5 we introduced the formal system Π_1^1-CA$_0$ of Π_1^1 comprehension. In §I.6 we explained how Π_1^1-CA$_0$ is much stronger than ACA$_0$ from the viewpoint of mathematical practice. In §I.9 we saw that Π_1^1-CA$_0$ is one of five basic subsystems of Z$_2$ which are important for Reverse Mathematics.

The purpose of this chapter is to present some details of results which were merely outlined in §§I.6 and I.9. Specifically, we discuss mathematics and Reverse Mathematics for Π_1^1-CA$_0$ and stronger systems. Models of these systems will be considered later, in §§VII.1, VII.5–VII.7, VIII.6, and IX.4.

VI.1. Perfect Kernels

In this section we complete the discussion of perfect trees which was begun in §§V.4 and V.5. We also prove that a certain well known theorem about the structure of closed sets, the Cantor/Bendixson theorem, is equivalent to Π_1^1 comprehension.

The following lemma is useful in showing that various mathematical statements are equivalent to Π_1^1 comprehension. (Compare theorem V.5.2.)

LEMMA VI.1.1 (Π_1^1-CA$_0$ and paths through trees). *The following are equivalent over RCA$_0$.*

1. Π_1^1 *comprehension.*
2. *For any sequence of trees $\langle T_k : k \in \mathbb{N} \rangle$, $T_k \subseteq \mathbb{N}^{<\mathbb{N}}$, there exists a set X such that $\forall k \, (k \in X \leftrightarrow T_k$ has a path).*

PROOF. Obviously Π_1^1-CA$_0$ proves statement 2.

Suppose now that statement 2 holds. We want to prove Π_1^1 comprehension. We first prove arithmetical comprehension. For this, it suffices to show that every function $g : \mathbb{N} \to \mathbb{N}$ has a range (see theorem III.1.3). Given g, use Δ_1^0 comprehension to get the sequence of trees $\langle T_k : k \in \mathbb{N} \rangle$ where $\tau \in T_k$ if and only if $(\forall m < \mathrm{lh}(\tau)) \, (g(m) \neq k)$. Clearly T_k has a

217

path if and only if $\neg\exists m\,(g(m)=k)$. Hence VI.1.1.2 implies the existence of rng(g). Thus we have arithmetical comprehension.

Now let $\varphi(k)$ be a Σ_1^1 formula. We want to prove that $\{k : \varphi(k)\}$ exists. Using arithmetical comprehension, our formal version of the Kleene normal form theorem (lemma V.1.4) gives us an arithmetical formula $\theta(k,\tau)$ such that

$$\forall k\,(\varphi(k) \leftrightarrow \exists f\,\forall m\,\theta(k,f[m])).$$

By arithmetical comprehension, let $\langle T_k : k \in \mathbb{N}\rangle$ be the sequence of trees defined by putting $\tau \in T_k$ if and only if $(\forall m \le \mathrm{lh}(\tau))\,\theta(k,\tau[m])$. Then clearly $\varphi(k)$ holds if and only if T_k has a path. Thus VI.1.1.2 implies the existence of a set X such that $\forall k\,(k \in X \leftrightarrow \varphi(k))$. This proves Σ_1^1 comprehension, which is clearly equivalent to Π_1^1 comprehension.

The proof of lemma VI.1.1 is complete.   □

We now consider what might be called a Cantor/Bendixson theorem for trees.

DEFINITION VI.1.2 (perfect kernel of a tree). Let $T \subseteq \mathbb{N}^{<\mathbb{N}}$ be a tree. The *perfect kernel* of T is defined in RCA$_0$ to be the union of all of the perfect subtrees of T, provided this union exists. Note that the perfect kernel of T, if it exists, is a perfect tree (definition V.4.1), namely the unique largest perfect subtree of T.

THEOREM VI.1.3 (perfect kernels and Π_1^1-CA$_0$). *The following are pairwise equivalent over* RCA$_0$.

1. *Π_1^1 comprehension.*
2. *For any tree $T \subseteq \mathbb{N}^{<\mathbb{N}}$, the perfect kernel of T exists.*
3. *Same as 2 for trees $T \subseteq 2^{<\mathbb{N}}$.*
4. *For any tree $T \subseteq \mathbb{N}^{<\mathbb{N}}$, there is a perfect subtree $P \subseteq T$ such that the set of paths through T which are not paths through P is countable.*
5. *Same as 4 for trees $T \subseteq 2^{<\mathbb{N}}$.*

PROOF. We begin by proving $1 \to 2$ and $1 \to 4$. Reasoning in Π_1^1-CA$_0$, let T be a subtree of $\mathbb{N}^{<\mathbb{N}}$. For each $\sigma \in \mathbb{N}^{<\mathbb{N}}$, put $T_\sigma = \{\tau \in T : \tau \subseteq \sigma \vee \sigma \subseteq \tau\}$. By Π_1^1 comprehension, let P be the set of all $\sigma \in T$ such that T_σ has a nonempty perfect subtree. Clearly P is the perfect kernel of T. This proves 2. By recursive comprehension, form the tree

$$T' = T/P = \{\langle\sigma\rangle^\frown\tau : \sigma \in T \setminus P \wedge \tau \in T_\sigma\}.$$

Clearly T' has no nonempty perfect subtree. Hence, by theorem V.5.5.4, T' has only countably many paths, i.e., there exists $\langle f_n : n \in \mathbb{N}\rangle$ such that $\forall f$ (if f is a path through T' then $\exists n\,(f = f_n)$). It follows that $\forall f$ (if f is a path through T then either f is a path through P or $\exists n(f = f_n')$), where $f_n'(m) = f_n(m+1)$. We have now proved $1 \to 2$ and $1 \to 4$.

Obviously $2 \to 3$ and $4 \to 5$. Reasoning in RCA$_0$, it remains to prove that either 3 or 5 implies Π_1^1 comprehension. We shall use lemma

VI.1.1. Let $\langle T_k : k \in \mathbb{N} \rangle$ be a sequence of trees, $T_k \subseteq \mathbb{N}^{<\mathbb{N}}$. By recursive comprehension, form the sequence of trees $\langle T_k' : k \in \mathbb{N} \rangle$ where T_k' consists of all sequences of the form $\langle (m_0, n_0), \ldots, (m_{j-1}, n_{j-1}) \rangle$ such that $\langle m_0, \ldots, m_{j-1} \rangle \in T_k$. Thus T_k has a path if and only if T_k' has a nonempty perfect subtree. Now put

$$T = \{\langle\rangle\} \cup \{\langle k \rangle^\frown \tau : k \in \mathbb{N} \wedge \tau \in T_k'\} \subseteq \mathbb{N}^{<\mathbb{N}}$$

and form the associated tree $T^* \subseteq 2^{<\mathbb{N}}$ as in the proof of lemma V.5.6. Let $P^* \subseteq T^*$ be a perfect tree as in 3 or 5. By recursive comprehension, let X be the set of all k such that

$$\underbrace{\langle 0, \ldots, 0}_{k}, 1 \rangle \in P^*.$$

Clearly T_k has a path if and only if $k \in X$. By lemma VI.1.1, this proves Π_1^1 comprehension. The proof of theorem VI.1.3 is complete. □

We now turn to our discussion of closed sets.

DEFINITION VI.1.4 (perfect sets). Within RCA$_0$, we define a closed set C in a complete separable metric space to be *perfect* if it has no isolated points, i.e., for any point $x \in C$ and any $\epsilon > 0$ there exists $y \in C$ such that $0 < d(x, y) < \epsilon$.

The relationship between closed sets in $\mathbb{N}^{\mathbb{N}}$ and trees in $\mathbb{N}^{<\mathbb{N}}$ is given by the following lemma.

LEMMA VI.1.5. *The following is provable in* RCA$_0$. *For any closed set* $C \subseteq \mathbb{N}^{\mathbb{N}}$, *there exists a tree* $T \subseteq \mathbb{N}^{<\mathbb{N}}$ *such that*

$$\forall f \, (f \in C \leftrightarrow f \text{ is a path through } T). \tag{17}$$

Conversely, for any tree $T \subseteq \mathbb{N}^{<\mathbb{N}}$, *there exists a closed set* $C = [T] \subseteq \mathbb{N}^{\mathbb{N}}$ *such that* (17) *holds. If* T *is a perfect tree, then* C *is a perfect set.*

PROOF. The formula $f \in C$ is Π_1^0 (using the code of C as a parameter). Hence by the normal form theorem II.2.7 for Π_1^0 formulas, we can find a Σ_0^0 formula $\theta(\tau)$ such that $\forall f \, (f \in C \leftrightarrow \forall m \, \theta(f[m]))$. Let T be the tree of all $\tau \in \mathbb{N}^{<\mathbb{N}}$ such that $(\forall m \leq \mathrm{lh}(\tau)) \, \theta(\tau[m])$. Then clearly (17) holds. Conversely, given a tree T, note that the formula "f is a path through T" is Π_1^0. Hence by lemma II.5.7 there is a closed set C such that (17) holds. If T is perfect, then for all paths f through T we have $\forall n \, \exists g$ (g is a path through T and $g \neq f$ and $g[n] = f[n]$). Hence the closed set C corresponding to T is perfect. This proves the lemma. □

THEOREM VI.1.6 (Cantor/Bendixson and Π_1^1-CA$_0$). *The following are equivalent over* ACA$_0$.

1. Π_1^1 *comprehension.*
2. *Every closed set in* $\mathbb{N}^{\mathbb{N}}$ *is the union of a perfect closed set and a countable set. (This is the Cantor/Bendixson theorem for* $\mathbb{N}^{\mathbb{N}}$.)

3. *Same as 2 for closed sets in* $2^{\mathbb{N}}$. (This is the Cantor/Bendixson theorem for $2^{\mathbb{N}}$.)

PROOF. The implication $1 \to 2$ is immediate from theorem VI.1.3 and lemma VI.1.5. Moreover $2 \to 3$ is trivial. It remains to prove $3 \to 1$. Assume 3. Instead of proving 1 we shall prove the equivalent assertion VI.1.3.5. Let T be a subtree of $2^{<\mathbb{N}}$. By lemma VI.1.5 let $C = [T]$, i.e., C is the closed set in $2^{\mathbb{N}}$ whose points are the paths through T. By our assumption VI.1.6.3, let C_1 be a perfect closed subset of C such that $C \setminus C_1$ is countable. By lemma VI.1.5, let T_1 be a tree whose paths are just the elements of C_1. By weak König's lemma plus arithmetical comprehension, there exists $\widehat{T_1}$ consisting of all $\sigma \in T_1$ such that $\exists f\,(f[\mathrm{lh}(\sigma)] = \sigma \wedge \forall m\,(f[m] \in T_1))$, i.e., ($\exists$ infinitely many $\tau \in T_1$ such that $\tau \supseteq \sigma$). Then clearly $\widehat{T_1}$ is a perfect subtree of T_1 and $[\widehat{T_1}] = [T_1] = C_1$, i.e., $\forall f\,(f \in C_1 \leftrightarrow f$ is a path through $\widehat{T_1})$. This proves VI.1.3.5. Hence by theorem VI.1.3 we have Π^1_1 comprehension. $\square$

EXERCISE VI.1.7. Generalize theorem VI.1.6 to complete separable metric spaces. In other words, show that Π^1_1 comprehension is equivalent over ACA_0 to the assertion that every closed set C in a complete separable metric space can be written as $C = P \cup S$ where P is a perfect closed set and S is countable. (This assertion is known as the *Cantor/Bendixson theorem*, and P is known as the *perfect kernel* of C.)

EXERCISE VI.1.8. For $\sigma, \tau \in \mathbb{N}^{<\mathbb{N}}$ say that σ *majorizes* τ if $\mathrm{lh}(\sigma) = \mathrm{lh}(\tau)$ and $(\forall i < \mathrm{lh}(\sigma))\,(\sigma(i) \geq \tau(i))$. Given a tree $T \subseteq \mathbb{N}^{<\mathbb{N}}$, define

$$T^+ = \{\sigma \in \mathbb{N}^{<\mathbb{N}} : \exists \tau\,(\tau \in T \wedge \sigma \text{ majorizes } \tau)\}.$$

Prove:

1. T^+ is a tree; $T^+ \supseteq T$; $T^{++} = T^+$.
2. T is well founded if and only if T^+ is well founded. (Hint: Use bounded König's lemma.)
3. If T is well founded, then $\mathrm{o}(T) = \mathrm{o}(T^+)$. Here $\mathrm{o}(T)$ denotes the ordinal height of T.

EXERCISE VI.1.9. A tree $T \subseteq \mathbb{N}^{<\mathbb{N}}$ is said to be *smooth* if $T^+ = T$. Show that lemma VI.1.1 holds with "tree" replaced by "smooth tree".

Notes for §VI.1. The equivalence of Π^1_1 comprehension with VI.1.3.2 was announced by Friedman [69]. Exercise VI.1.7 is related to a result of Kreisel [149], who used hyperarithmetical theory to refute a certain predicative analog of the Cantor/Bendixson theorem. Exercises VI.1.8 and VI.1.9 are from Marcone [173, 174]; see also Humphreys/Simpson [127].

VI.2. Coanalytic Uniformization

In this section we continue our exploration of classical descriptive set theory as formalized within subsystems of second order arithmetic. We show that one of the most famous theorems of classical descriptive set theory, Kondo's theorem, is equivalent to Π^1_1 comprehension. The equivalence is proved in ATR_0.

We begin with the following lemma, which is a formal version of the Π^1_1 *uniformization principle.*

LEMMA VI.2.1 (Π^1_1 uniformization in Π^1_1-CA$_0$). *Let $\psi(X)$ be a Π^1_1 formula with a distinguished free set variable X. Then we can find a Π^1_1 formula $\widehat{\psi}(X)$ such that Π^1_1-CA$_0$ proves*

(1) $\forall X\,(\widehat{\psi}(X) \to \psi(X))$,
(2) $\forall Y\,(\psi(Y) \to \exists X\,\widehat{\psi}(X))$,
(3) $\forall X\,\forall Y\,((\widehat{\psi}(X) \wedge \widehat{\psi}(Y)) \to X = Y)$.

PROOF. The proof will be based on the analysis of Π^1_1 formulas given in §V.1, in terms of analytic codes and the Kleene/Brouwer ordering.

By lemma V.1.8, we have an analytic code A such that

$$\forall X(\psi(X) \leftrightarrow \mathrm{WO}(\mathrm{KB}(\mathrm{T}_A(X)))).$$

Let us write $L(X) = \mathrm{KB}(\mathrm{T}_A(X))$. Then we have $\forall X\,\mathrm{LO}(L(X))$ and

$$\forall X(\psi(X) \leftrightarrow \mathrm{WO}(L(X))).$$

For each $k \in \mathbb{N}$ put

$$L_k(X) = \{(i,j) : i \leq_{L(X)} j <_{L(X)} k\}.$$

Thus $L_k(X)$ is the initial segment of $L(X)$ determined by k. (If $k \notin \mathrm{field}(L(X))$ then $L_k(X)$ is defined to be all of $L(X)$.)

Reasoning in Π^1_1-CA$_0$, assume that there exists X such that $\psi(X)$ holds, i.e., $\mathrm{WO}(L(X))$. We claim that there exists X_0 such that $\mathrm{WO}(L(X_0))$ holds and $|L(X_0)|$ is minimal, i.e.,

$$\forall X\,(\mathrm{WO}(L(X)) \to |L(X)| \geq |L(X_0)|).$$

To see this, start with any Y such that $\psi(Y)$ holds. If

$$\forall X\,(\mathrm{WO}(L(X)) \to |L(X)| \geq |L(Y)|)$$

holds, then we may take $X_0 = Y$ and our claim is proved. If not, put

$$K = \{k : k \in \mathrm{field}(L(Y)) \wedge \exists X\,|L(X)| = |L_k(Y)|\}.$$

Here K exists by Σ^1_1 comprehension, with Y as a parameter. By assumption, $K \neq \emptyset$. Let k_0 be the $\leq_{L(Y)}$-least element of K. Let X_0 be such that $|L(X_0)| = |L_{k_0}(Y)|$. Clearly this proves the claim.

Fix X_0 as in the above claim, and let G be the set of all $\sigma \in \mathrm{Seq}$ of the form $\langle(\xi_0, m_0), \ldots, (\xi_{k-1}, m_{k-1})\rangle$ such that

(4) $\exists X\,(|L(X)| = |L(X_0)| \wedge (\forall i < k)\,(X(i) = \xi_i \wedge |L_i(X)| = |L_{m_i}(X_0)|))$.

Here G exists by Σ_1^1 comprehension, with X_0 as a parameter. (G is in fact an analytic code.) Clearly the empty sequence $\langle\rangle$ belongs to G.

We claim that for all $\sigma \in G$ there exist $\xi \in \{0, 1\}$ and $m \in \mathbb{N}$ such that $\sigma^\frown\langle(\xi, m)\rangle$ belongs to G. To see this, let $\sigma = \langle(\xi_0, m_0), \ldots, (\xi_{k-1}, m_{k-1})\rangle \in G$ and pick any X as in (4). Put $\xi = X(k)$ and, by comparability of well orderings, $m =$ the unique m such that $|L_k(X)| = |L_m(X_0)|$. (If $k \notin$ field$(L(X))$ we may take any $m \notin$ field$(L(X_0))$.) Clearly $\sigma^\frown\langle(\xi, m)\rangle \in G$. This proves the claim.

Now define $\sigma_k \in G$, $k \in \mathbb{N}$, by putting $\sigma_0 = \langle\rangle$ and $\sigma_{k+1} = \sigma_k^\frown\langle(\xi_k, m_k)\rangle$, where ξ_k is the least ξ such that $\exists m(\sigma_k^\frown\langle(\xi, m)\rangle \in G)$, and m_k is the $<_{L(X_0)}$-least m such that $\sigma_k^\frown\langle(\xi_k, m)\rangle \in G$. (If there is no such $m \in$ field$(L(X_0))$, we take $m_k =$ least $m \notin$ field$(L(X_0))$.) The sequence $\langle\sigma_k : k \in \mathbb{N}\rangle$ exists by primitive recursion and arithmetical comprehension, with G and $L(X_0)$ as parameters. Define $\widehat{X}: \mathbb{N} \to \{0, 1\}$ and $f: \mathbb{N} \to \mathbb{N}$ by $\widehat{X}(k) = \xi_k$, $f(k) = m_k$ for all $k \in \mathbb{N}$.

We claim that f is an isomorphism of $L(\widehat{X})$ onto a subordering of $L(X_0)$. In other words, we are claiming that $i <_{L(\widehat{X})} j$ implies $m_i <_{L(X_0)} m_j$. Given $i <_{L(\widehat{X})} j$, recall that the field of $L(\widehat{X})$ is the tree $T_A(\widehat{X})$. In view of the way $T_A(X)$ was defined (lemma V.1.8), we see that $i <_{L(X)} j$ holds for any X with $X[l] = \widehat{X}[l]$, $l \geq \max\{\mathrm{lh}(i), \mathrm{lh}(j)\}$. In particular, putting $l = \max\{\mathrm{lh}(i), \mathrm{lh}(j), i+1, j+1\}$, let X be such that $|L(X)| = |L(X_0)|$ and, for all $k < l$, $X(k) = \xi_k$ and $|L_k(X)| = |L_{m_k}(X_0)|$. Then we have $i <_{L(X)} j$, hence $|L_{m_i}(X_0)| = |L_i(X)| < |L_j(X)| = |L_{m_j}(X_0)|$. This proves our claim.

From the above claim, we see immediately that $|L(\widehat{X})| \leq |L(X_0)|$ and, for all $k \in \mathbb{N}$, $|L_k(\widehat{X})| \leq |L_{m_k}(X_0)|$. But then, from the minimality of $|L(X_0)|$ and $|L_{m_k}(X_0)|$, it follows that $|L(\widehat{X})| = |L(X_0)|$ and $|L_k(\widehat{X})| = |L_{m_k}(X_0)|$.

We shall now show how to reformulate the above definition of $\widehat{X}$ so that it does not depend on our choice of the parameter X_0. Define

$$\widehat{\psi}(X) \equiv |L(X)| = |L(X_0)| \wedge \forall k\, (X(k) = \xi_k \wedge |L_k(X)| = |L_{m_k}(X_0)|)$$
$$\equiv \mathrm{WO}(L(X)) \wedge \neg\exists k\, \exists Y\, ((5) \vee (6) \vee (7))$$

where

(5) $|L(Y)| < |L(X)|$,
(6) $|L(Y)| = |L(X)| \wedge Y(k) < X(k) \wedge$
    $(\forall i < k)\,(Y(i) = X(i) \wedge |L_i(Y)| = |L_i(X)|)$,
(7) $|L(Y)| = |L(X)| \wedge Y(k) = X(k) \wedge |L_k(Y)| < |L_k(X)| \wedge$
    $(\forall i < k)\,(Y(i) = X(i) \wedge |L_i(Y)| = |L_i(X)|)$.

Then $\widehat{\psi}(X)$ is equivalent to the assertion that $X = \widehat{X}$ as above. The claims (1), (2) and (3) follow easily. Moreover $\widehat{\psi}(X)$ is explicitly a Π_1^1 formula.

This completes the proof of lemma VI.2.1.   □

The following lemma is a formal version of a theorem of Suzuki. We shall make use of hyperarithmetical theory in ATR_0 as presented in §§VII.1 and VIII.3 below.

LEMMA VI.2.2 (Suzuki theorem in ATR_0). *Let* $\psi(X, Y)$ *be a* Π_1^1 *formula with no free set variables other than* X *and* Y. *The following is provable in* ATR_0. *Suppose that* X *and* Y *are such that*

$$\psi(X, Y) \wedge \forall Z\,(\psi(X, Z) \to Z = Y).$$

Then either $Y \leq_\mathrm{H} X$, *or* $\mathrm{HJ}(X)$ *exists and is* $\leq_\mathrm{H} X \oplus Y$.

Here $Y \leq_\mathrm{H} X$ means that Y is hyperarithmetical in X, and $\mathrm{HJ}(X)$ denotes the hyperjump of X.

PROOF. Reasoning in ATR_0, assume that X and Y are as above. By our formalized version of the Kleene normal form theorem (lemma V.1.4), we have

$$\forall Z\,(\psi(X, Z) \leftrightarrow \forall f\, \exists k\, \theta(X, Z[k], f[k]))$$

where $\theta(X, \sigma, \tau)$ is Σ_0^0 with no free set variables other than X. Let A be the set of all $\langle(\eta_0, m_0), \dots, (\eta_{k-1}, m_{k-1})\rangle$ such that $(\forall j < k)\,\eta_j < 2$ and $(\forall j \leq k)\,\neg\theta(X, \langle\eta_0, \dots, \eta_{j-1}\rangle, \langle m_0, \dots, m_{j-1}\rangle)$. Thus A is X-recursive. Moreover, in the terminology of §V.1, A is an analytic code and we have

$$\forall Z\,(\psi(X, Z) \leftrightarrow \mathrm{WO}(\mathrm{KB}(\mathrm{T}_A(Z)))).$$

In particular $\mathrm{KB}(\mathrm{T}_A(Y))$ is a countable $(X \oplus Y)$-recursive well ordering.

There are now two cases.

Case 1: There exists an X-recursive well ordering R such that $|R| = |\mathrm{KB}(\mathrm{T}_A(Y))|$. In this case, Y can be characterized as the unique Z such that $|\mathrm{KB}(\mathrm{T}_A(Z))| = |R|$. Thus for all $n \in \mathbb{N}$ we have

$$\begin{aligned}n \in Y &\leftrightarrow \exists Z\,(|\mathrm{KB}(\mathrm{T}_A(Z))| = |R| \wedge n \in Z) \\ &\leftrightarrow \forall Z\,(|\mathrm{KB}(\mathrm{T}_A(Z))| = |R| \to n \in Z),\end{aligned}$$

so Y is Δ_1^1 in X. It follows by our formalized Kleene/Souslin theorem VIII.3.19 that Y is hyperarithmetical in X.

Case 2: Case 1 fails. By comparability of well orderings (lemma V.2.9), it follows that $|R| < |\mathrm{KB}(\mathrm{T}_A(Y))|$ for all X-recursive well orderings R. Let $\varphi(k, X)$ be any Σ_1^1 formula with no free set variable other than X. We are going to show that $\{k : \varphi(k, X)\}$ exists and is $\leq_\mathrm{H} X \oplus Y$. By lemma V.1.4 we have

$$\forall k\,(\varphi(k, X) \leftrightarrow \exists f\, \forall m\, \theta(k, X, f[m]))$$

where $\theta(k, X, \tau)$ is Σ_0^0 with no free set variables other than X. Put $R_k = \mathrm{KB}(T_k)$ where T_k is the set of all τ such that $(\forall m \leq \mathrm{lh}(\tau))\,\theta(k, X, \tau[m])$.

Thus $\langle R_k : k \in \mathbb{N} \rangle$ is an X-recursive sequence of X-recursive linear orderings, and for all k we have

$$\varphi(k, X) \leftrightarrow \neg \text{WO}(R_k)$$
$$\leftrightarrow \neg |R_k| < |\text{KB}(\text{T}_A(Y))|.$$

Hence by Δ_1^1 comprehension (lemma VIII.4.1) using $X \oplus Y$ as a parameter, there exists W such that $\forall k \, (k \in W \leftrightarrow \varphi(k, X))$. Since W is Δ_1^1 in $X \oplus Y$, it follows by theorem VIII.3.19 that W is hyperarithmetical in $X \oplus Y$. In particular, taking $\varphi(k, X)$ to be the Σ_1^1 formula which defines the hyperjump (definition VII.1.5), we see that $W = \text{HJ}(X)$ exists and is $\leq_H X \oplus Y$.

This completes the proof of lemma VI.2.2, our formalized Suzuki theorem. $\square$

We now turn to our discussion of coanalytic sets.

DEFINITION VI.2.3 (coanalytic sets). Within RCA$_0$ we define a *coanalytic code* (i.e., a code for a coanalytic set in the Cantor space $2^\mathbb{N}$) to be a set $C \subseteq \text{Seq}$ such that C is the complement of an analytic code (definition V.1.5). In other words, there exists an analytic code A such that $C = \text{Seq} \setminus A$.

If C is a coanalytic code, then for all $X \in 2^\mathbb{N}$ we write $X \in C$ to mean $\forall f \, \exists n \, C(X[n], f[n])$. Here f ranges over $\mathbb{N}^\mathbb{N}$, and $C(X[n], f[n])$ means that $\langle (X(0), f(0)), \ldots, (X(n-1), f(n-1)) \rangle \in C$. Thus $X \in C$ if and only if $X \notin A$.

The relationship between coanalytic sets and Π_1^1 formulas is given by the following lemma.

LEMMA VI.2.4 (coanalytic codes and Π_1^1 formulas). *For a coanalytic code C, the formula $X \in C$ is Π_1^1. Conversely, for any Π_1^1 formula $\psi(X)$,* ACA$_0$ *proves*

$$(\exists \text{ coanalytic code } C) \, \forall X \, (X \in C \leftrightarrow \psi(X)).$$

PROOF. This lemma follows immediately from its dual, theorem V.1.7. $\square$

Recall that $2^\mathbb{N} \times 2^\mathbb{N}$ is homeomorphic to $2^\mathbb{N}$ via the pairing function $(X, Y) \mapsto X \oplus Y$, where $(X \oplus Y)(2n) = X(n)$, $(X \oplus Y)(2n+1) = Y(n)$. Thus any coanalytic set $C \subseteq 2^\mathbb{N}$ may be regarded as a coanalytic relation $C \subseteq 2^\mathbb{N} \times 2^\mathbb{N}$. Formally, we write $C(X, Y)$ to mean that $X \oplus Y \in C$.

We are now ready to state and prove the main result of this section.

DEFINITION VI.2.5 (Kondo's theorem). *Kondo's theorem is the assertion that coanalytic sets in $2^\mathbb{N} \times 2^\mathbb{N}$ have the uniformization property. In other words, for any coanalytic code C, there exists a coanalytic code $\widehat{C}$ such that*

$(1')\ \forall X \, \forall Y \, (\widehat{C}(X, Y) \to C(X, Y)),$

(2') $\forall X \forall Y (C(X, Y) \rightarrow \exists Z \, \widehat{C}(X, Z))$,
(3') $\forall X \forall Y \forall Z ((\widehat{C}(X, Y) \wedge \widehat{C}(X, Z)) \rightarrow Y = Z)$.

THEOREM VI.2.6 (Kondo's theorem and Π^1_1-CA$_0$). *Kondo's theorem is equivalent over* ATR$_0$ *to* Π^1_1 *comprehension*.

PROOF. First, assume Π^1_1 comprehension. Let C be a coanalytic code. Applying lemma VI.2.1 to the Π^1_1 formula $\psi(X, Y) \equiv C(X, Y)$ with the distinguished free set variable Y, we obtain a Π^1_1 formula $\widehat{\psi}(X, Y)$ such that (1'), (2') and (3') hold with ψ, $\widehat{\psi}$ replacing C, $\widehat{C}$. Then lemma VI.2.4 gives us a coanalytic code $\widehat{C}$ with the desired properties. This proves Kondo's theorem.

Now, reasoning in ATR$_0$, assume Kondo's theorem. We want to prove Π^1_1 comprehension. By lemma VII.1.6, it suffices to prove that for all X, $HJ(X)$ exists. Let $\theta(X, Y)$ be the arithmetical formula

$$\forall i \, (X \leq_T (Y)_i \wedge TJ((Y)_{i+1}) \leq_T (Y)_i).$$

By lemmas VIII.3.23 and VIII.3.24 and the proof of lemma VIII.3.25, we have $\forall X \exists Y \, \theta(X, Y)$ and $\forall X \forall Y \, (\theta(X, Y) \rightarrow Y \not\leq_H X)$. By lemma VI.2.4, let C be a coanalytic code such that $\forall X \forall Y \, (C(X, Y) \leftrightarrow \theta(X, Y))$. By Kondo's theorem, we obtain a coanalytic code $\widehat{C}$ such that (1'), (2') and (3') hold. In particular, we have $\forall X \, (\exists \text{ exactly one } Y) \, \widehat{C}(X, Y)$ and $\forall X \forall Y \, (\widehat{C}(X, Y) \rightarrow Y \not\leq_H X)$.

Now given X, put $X' = X \oplus \widehat{C}$ (the recursive join of X with the coanalytic code $\widehat{C}$), and let Y' be the unique Y such that $\widehat{C}(X', Y)$ holds. Since $Y' \not\leq_H X'$, it follows by our formalized Suzuki theorem (lemma VI.2.2) that $HJ(X')$ exists and is $\leq_H X' \oplus Y'$. Hence $HJ(X)$ exists by lemma VII.1.6. This proves Π^1_1 comprehension.

The proof of theorem VI.2.6 is complete. $\square$

Notes for §VI.2. The original source for Kondo's theorem is Kondo [146]. The Π^1_1 uniformization principle is sometimes known as the Kondo/ Addison theorem (see also Moschovakis [191] and Mansfield/Weitkamp [171]). The original source for Suzuki's theorem is Suzuki [258]. Mansfield (unpublished, but see Friedman [64, theorem 6]) has observed that Kondo's theorem and Π^1_1 uniformization are provable in Δ^1_2-CA$_0$ plus full induction. The fact that Π^1_1 uniformization is provable in Π^1_1-CA$_0$ is due to Simpson (unpublished manuscript, January 9–10, 1981), as are the other results of this section.

VI.3. Coanalytic Equivalence Relations

In this section we continue our investigation of classical descriptive set theory in the context of subsystems of Z_2. We show that Π^1_1 comprehension is necessary and sufficient to prove a famous theorem of Silver [226]: For any coanalytic equivalence relation, either the number of equivalence classes is countable or there exists a perfect set of pairwise inequivalent points. Like the perfect set theorem (§V.4), Silver's theorem may be viewed as verifying a special case of the continuum hypothesis.

We begin with the following lemma, which says that a version of Silver's theorem is provable in ATR_0.

LEMMA VI.3.1 (an ATR_0 version of Silver's theorem). *The following is provable in* ATR_0. *Let* E *be a coanalytic equivalence relation. Then either*

(1) $\exists$ *perfect set* P *such that* $\forall X \, \forall Y \, ((X, Y \in P \wedge X \neq Y) \to X \not\mathrel{E} Y)$, *or*
(2) $\exists$ *sequence of Borel codes* $\langle B_n : n \in \mathbb{N} \rangle$ *such that* $\forall X \, \exists n \, (X \in B_n)$ *and* $\forall n \, \forall X \, \forall Y \, ((X, Y \in B_n) \to X E Y)$.

PROOF. We reason in ATR_0. Without loss, we consider only coanalytic equivalence relations on the Cantor space $2^\mathbb{N}$. We prove our lemma only in a lightface form, replacing "coanalytic" by "lightface Π_1^1", and "sequence of Borel sets" by "lightface Δ_1^1 sequence of lightface Δ_1^1 sets", i.e., a lightface Δ_1^1 subset of $\mathbb{N} \times 2^\mathbb{N}$. Here *lightface* means: without parameters. The full lemma is obtained from the lightface version by relativization.

We follow Harrington's [103] unpublished proof of Silver's theorem via Gandy forcing. In order to make Harrington's proof work in ATR_0, we use an inner model technique. Throughout this argument, we use Σ_1^1 to mean lightface Σ_1^1, etc. We use an ATR_0 formalization of hyperarithmetical theory, as presented in §VIII.3, as well as results from §VIII.4 concerning ω-models of $\Sigma_1^1\text{-}AC_0$.

We are given a Π_1^1 equivalence relation, E. Let A be a Σ_1^1 set defined by

$$X \in A \leftrightarrow \forall \Delta_1^1 D \, (X \in D \to \exists Y (Y \in D \wedge X \not\mathrel{E} Y)).$$

To see that this is Σ_1^1, note that by the Kleene/Souslin theorem we can represent Δ_1^1 sets in the form $D = \{X : i \in H_e^X\}$ where $e \in \mathcal{O}$. Thus we have

$$X \in A \leftrightarrow \underbrace{\forall e \, \forall i \, ((e \in \mathcal{O} \wedge i \in H_e^X)}_{\Pi_1^1} \to \underbrace{\exists Y (i \in H_e^Y \wedge X \not\mathrel{E} Y))}_{\Sigma_1^1}$$

which is essentially Σ_1^1 (see definition VIII.6.1), hence Σ_1^1 by the Σ_1^1 axiom of choice (available in ATR_0 by theorem V.8.3).

Case 1: $A = \emptyset$, i.e.,

$$\forall X \, \exists \Delta_1^1 D \, (X \in D \wedge \forall Y (Y \in D \to X E Y)),$$

i.e.,

$$\forall X \, \exists e \, \exists i \, \underbrace{(e \in \mathcal{O} \wedge i \in H_e^X \wedge \forall Y (i \in H_e^Y \to X E Y))}_{\Pi_1^1}.$$

Apply Π_1^1 number uniformization and Σ_1^1 separation to get a Δ_1^1 sequence $\langle (e_n, i_n) : n \in \mathbb{N} \rangle$ such that $\forall n\, (e_n \in \mathcal{O})$ and

$$\forall X\, \exists n\, (i_n \in \mathrm{H}_{e_n}^X \wedge \forall Y\, (i_n \in \mathrm{H}_{e_n}^Y \rightarrow X E Y)).$$

Put $B_n = \{X : i_n \in \mathrm{H}_{e_n}^X\}$. Thus in this case we have conclusion (2).

Case 2: $A \neq \emptyset$. In this case we shall obtain conclusion (1).

Define A in a slightly different but equivalent way:

$$X \in A \leftrightarrow \forall \Delta_1^1 D\, (X \in D \rightarrow \exists X_0, X_1 (X_0, X_1 \in D \wedge X_0 \not\!E X_1))$$

$$\leftrightarrow \forall e\, \forall i\, (\underbrace{(e \in \mathcal{O} \wedge i \in \mathrm{H}_e^X)}_{\Pi_1^1} \rightarrow \underbrace{\exists X_0, X_1 (i \in \mathrm{H}_e^{X_0}, \mathrm{H}_e^{X_1} \wedge X_0 \not\!E X_1)}_{\Sigma_1^1})$$

$$\wedge\, \forall e\, (\underbrace{e \in \mathcal{O}}_{\Pi_1^1} \rightarrow \underbrace{\exists \mathrm{H}_e^X}_{\Sigma_1^1}).$$

By Σ_1^1 choice we can find a countable ω-model

$$M \models \mathrm{ACA}_0 \wedge \underbrace{\exists X (X \in A)}_{\text{essentially } \Sigma_1^1},$$

where we are using the previous definition of A.

Caution: It is probably not the case that $M \models$ "E is an equivalence relation". But this will not matter.

SUBLEMMA VI.3.2. *For any Σ_1^1 set B we have*

$$M \models A \cap B \neq \emptyset \rightarrow \exists X_0, X_1 (X_0, X_1 \in A \cap B \wedge X_0 \not\!E X_1).$$

PROOF. We reason in M. Suppose the conclusion fails, i.e.,

$$\forall X_0, X_1 (X_0, X_1 \in A \cap B \rightarrow X_0 E X_1),$$

i.e.,

$$\forall X_0 (\underbrace{X_0 \in A \cap B}_{\Sigma_1^1} \rightarrow \underbrace{\forall X_1 (X_1 \in A \cap B \rightarrow X_0 E X_1)}_{\Pi_1^1}).$$

By Σ_1^1 separation, there exists a Δ_1^1 interpolant $D_0 = \{X : i_0 \in \mathrm{H}_{e_0}^X\}$, $e_0 \in \mathcal{O}$. Thus we have

$$\forall X_0 (X_0 \in A \cap B \rightarrow X_0 \in D_0)$$

and

$$\forall X_0 (X_0 \in D_0 \rightarrow \forall X_1 (X_1 \in A \cap B \rightarrow X_0 E X_1)),$$

i.e.,

$$\forall X_1 (\underbrace{X_1 \in A \cap B}_{\Sigma_1^1} \rightarrow \underbrace{\forall X_0 (X_0 \in D_0 \rightarrow X_0 E X_1)}_{\Pi_1^1}).$$

By Σ_1^1 separation, there exists a Δ_1^1 interpolant $D_1 = \{X : i_1 \in H_{e_1}^X\}$, $e_1 \in \mathcal{O}$. Thus we have

$$\forall X_1 \, (X_1 \in A \cap B \to X_1 \in D_1)$$

and

$$\forall X_1 \, (X_1 \in D_1 \to \forall X_0 \, (X_0 \in D_0 \to X_0 E X_1)).$$

Put $D = D_0 \cap D_1$. We then have

$$\forall X \, (X \in A \cap B \to X \in D)$$

and

$$\forall X_0, X_1 \, (X_0, X_1 \in D \to X_0 E X_1).$$

Hence by definition of A we have $A \cap D = \emptyset$. Hence $A \cap B = \emptyset$. This proves the sublemma. $\square$

We now define Gandy forcing over M. A *condition* is a Σ_1^1 set B such that $M \models B \neq \emptyset$. The set of all conditions is denoted $\mathcal{C}$. Note that A itself is a condition, i.e., $A \in \mathcal{C}$, by case assumption and our choice of M. Forcing and genericity over M are defined in the usual way. A set $\mathcal{D}$ of conditions is said to be *open* if $B \in \mathcal{D}$, $B' \in \mathcal{C}$, $B' \subseteq B$ imply $B' \in \mathcal{D}$. $\mathcal{D}$ is said to be *dense* if it is open and for all conditions C there exists a condition $B \subseteq C$ such that $B \in \mathcal{D}$. $\mathcal{D}$ is said to be M-*definable* if it is definable over M. $X \in 2^{\mathbb{N}}$ is said to *meet* $\mathcal{D}$ if $X \in B$ for some $B \in \mathcal{D}$. X is said to be *generic* if it meets all dense, M-definable sets of conditions. A condition B is said to *force* $\varphi(X)$, abbreviated $B \Vdash \varphi(X)$, if $\varphi(X)$ holds for all generic $X \in B$. Note that for all conditions B we have $B \Vdash X \in B$. We assume familiarity with basic properties of forcing and genericity.

SUBLEMMA VI.3.3. *If $X_0 \oplus X_1$ is generic, then X_0 and X_1 are generic.*

PROOF. By symmetry we consider only X_0. Given a dense set $\mathcal{D}_0$, we need to show that X_0 meets $\mathcal{D}_0$. Consider

$$\mathcal{D} = \{B \in \mathcal{C} : \{X_0 : \exists X_1 (X_0 \oplus X_1 \in B)\} \in \mathcal{D}_0\}.$$

We claim that $\mathcal{D}$ is dense. Given $C \in \mathcal{C}$, put $C_0 = \{X_0 : \exists X_1 (X_0 \oplus X_1 \in C)\}$. Since $\mathcal{D}_0$ is dense, there exists $B_0 \in \mathcal{D}_0$ such that $B_0 \subseteq C_0$. Put $B = \{X_0 \oplus X_1 : X_0 \oplus X_1 \in C \wedge X_0 \in B_0\}$. Then $B \in \mathcal{D}$ and $B \subseteq C$. This proves the claim. Since $X_0 \oplus X_1$ is generic, there exists $B \in \mathcal{D}$ such that $X_0 \oplus X_1 \in B$. Then $X_0 \in \{X_0 : \exists X_1 (X_0 \oplus X_1 \in B)\} \in \mathcal{D}_0$. This proves the sublemma. $\square$

We consider product Gandy forcing $\mathcal{C} \times \mathcal{C}$ over M. Conditions are now Cartesian products $B \times C$ where $B, C \in \mathcal{C}$. Choose an ω-model N which contains (the code of) M and satisfies the Σ_1^1 axiom of choice. Note that $\mathrm{HYP}(M) \subseteq N$. We consider forcing and genericity over M with respect to dense sets in N. We define $B \times C \Vdash \varphi(X, Y)$ to mean that $\varphi(X, Y)$ holds for all generic $(X, Y) \in B \times C$. Note that $B \times C \Vdash X \in B, Y \in C$.

SUBLEMMA VI.3.4. $A \times A \Vdash X \not\mathrel{E} Y$.

PROOF. Suppose for a contradiction that $(X, Y) \in A \times A$ is generic and XEY, i.e., $T_E(X, Y)$ is well founded. Thus we have $|T_E(X, Y)| \le \alpha$ for some $\alpha < \omega_1^{X \oplus Y} \le \omega_1^M$, which we denote by $XE_\alpha Y$. (Note that E_α need not be an equivalence relation.) By genericity, there exist conditions $B, C \subseteq A$ such that $B \times C \Vdash XE_\alpha Y$. Put $B' = \{X_0 \oplus X_1 : X_0, X_1 \in B, X_0 \not\mathrel{E} X_1\}$. By sublemma VI.3.2 we have $M \models (B' \ne \emptyset)$. Let $(X_0 \oplus X_1, Y) \in B' \times C$ be generic. By a variant of sublemma VI.3.3, (X_0, Y) and (X_1, Y) are generic. Since $(X_0, Y), (X_1, Y) \in B \times C$, we have $X_0 E_\alpha Y$, $X_1 E_\alpha Y$, $X_0 \not\mathrel{E} X_1$, a contradiction. This proves the sublemma. □

Now starting with A we can build a full binary tree of conditions so that each pair of paths is generic. Since each pair of paths belongs to $A \times A$, sublemma VI.3.4 implies that we have a perfect set P such that

$$\forall X \, \forall Y \, ((X \in P \wedge Y \in P \wedge X \ne Y) \to X \not\mathrel{E} Y).$$

This completes the proof of lemma VI.3.1. □

DEFINITION VI.3.5 (Silver's theorem). By *Silver's theorem* we mean the following statement: If E is a coanalytic equivalence relation, then either VI.3.1(1) holds, or

(2') $\exists$ sequence of points $\langle Y_n : n \in \mathbb{N} \rangle$ such that $\forall X \, \exists n \, (XEY_n)$.

THEOREM VI.3.6 (Silver's theorem and Π_1^1-CA$_0$). *The following are pairwise equivalent over* RCA$_0$.

(i) Π_1^1 *comprehension.*
(ii) *Silver's theorem.*
(iii) *Silver's theorem restricted to equivalence relations on* $\mathbb{N}^\mathbb{N}$ *which are* Δ_2^0 *definable (with parameters, of course).*

PROOF. Since Π_1^1-CA$_0$ includes ATR$_0$, lemma VI.3.1 implies that Π_1^1-CA$_0$ proves $(1) \vee (2)$ for any coanalytic equivalence relation E. But if (2) holds, then we can use Π_1^1 comprehension to form $\{n : B_n \ne \emptyset\}$, followed by Σ_1^1 choice to obtain a sequence of points $\langle Y_n : n \in \mathbb{N} \rangle$ such that $\forall n \, (B_n \ne \emptyset \to Y_n \in B_n)$, and this gives (2'). Thus we see that Π_1^1-CA$_0$ proves $(1) \vee (2')$. We have shown (i)→(ii), and (ii)→(iii) is trivial.

For the reversal, we reason in RCA$_0$ and assume (iii). As in the proof of lemma VI.1.1, it is easy to show that (iii) implies arithmetical comprehension. Given a Σ_1^1 formula $\varphi(m)$, use the Kleene normal form theorem (lemma V.1.4) to write $\varphi(m) \equiv \exists f \, \theta(m, f)$, $f \in \mathbb{N}^\mathbb{N}$, where $\theta(m, f)$ is Π_1^0. Define a Δ_2^0 equivalence relation E on $\mathbb{N} \times \mathbb{N}^\mathbb{N} \cong \mathbb{N}^\mathbb{N}$ by putting

$$(m, f)E(n, g) \equiv m = n \wedge (\theta(m, f) \leftrightarrow \theta(n, g)).$$

Clearly (1) does not hold for this equivalence relation, so by $(2')$ let $\langle Y_k : k \in \mathbb{N} \rangle$ be a sequence of points of $\mathbb{N} \times \mathbb{N}^{\mathbb{N}}$ such that $\forall m \, \forall f \, \exists k \, ((m, f) E Y_k)$. Put $Y_k = (m_k, f_k)$. Then $\forall m \, (\exists f \, \theta(m, f) \leftrightarrow \exists k \, (m = m_k \wedge \theta(m, f_k)))$. Hence $\{m : \varphi(m)\} = \{m : \exists f \, \theta(m, f)\}$ exists by arithmetical comprehension, using $\langle (m_k, f_k) : k \in \mathbb{N} \rangle$ as a parameter. This proves Σ_1^1 comprehension, hence (i). □

Notes for §VI.3. Silver's original proof of Silver's theorem [226] used transfinitely many iterations of the power set axiom in ZFC. The fact that Silver's theorem is provable in Z_2 is due to Harrington [103]. Other applications of Harrington's method are in Louveau [164]. The reversal in theorem VI.3.6 is due to Ramez Sami (personal communication, June 1981). Lemma VI.3.1 and theorem VI.3.6 are due to Simpson (unpublished notes, March 17, 1984). For another treatment of lemma VI.3.1 and related results, see Marcone [172]. Other results related to Silver's theorem are in Harrington/Marker/Shelah [105], Louveau [165], and Louveau/Saint-Raymond [166].

VI.4. Countable Abelian Groups

In this section we show that Π_1^1-CA_0 is equivalent over RCA_0 to a well known theorem concerning the structure of countable Abelian groups. Our result is as follows:

THEOREM VI.4.1 (Π_1^1-CA_0 and countable Abelian groups). *The following are equivalent over* RCA_0.

1. Π_1^1 *comprehension.*
2. *Every countable Abelian group is a direct sum of a divisible group and a reduced group.*

PROOF. We shall need the following lemma.

LEMMA VI.4.2. *The following is provable in* ACA_0. *If D is a divisible subgroup of a countable Abelian group G, then $G = D \oplus A$ for some subgroup A.*

PROOF. By injectivity (theorem III.6.5) there is a homomorphism $h : G \to D$ such that $h(d) = d$ for all $d \in D$. Letting A be the kernel of h, i.e., $A = \{a \in G : h(a) = 0\}$, we easily see that $G = D \oplus A$. This proves the lemma. □

Now to prove theorem VI.4.1, assume Π_1^1-CA_0 and let G be an Abelian group. Let us define an element $d \in G$ to be *divisible* if for each prime p there exists $f : \mathbb{N} \to G$ such that $f(0) = d$ and $\forall n \, (pf(n+1) = f(n))$. Being divisible is a Σ_1^1 property, so by Σ_1^1 comprehension, $D = \{d \in G : d$ is divisible$\}$ exists. Clearly D is a subgroup of G and is p-divisible for all primes p. By an easy application of Σ_1^0 induction, it follows that D is divisible. By lemma VI.4.2 we have $G = D \oplus A$, and clearly A is reduced. This proves $1 \rightarrow 2$.

For the converse, we reason in RCA_0. We begin with:

LEMMA VI.4.3. *It is provable in RCA_0 that statement 2 implies arithmetical comprehension.*

PROOF. Reasoning in RCA_0, let $f : \mathbb{N} \rightarrow \mathbb{N}$ be one-to-one. Let G be the Abelian group with generators $x_m, y_{m,i}, m, i \in \mathbb{N}$ and relations $px_m = 0$, $py_{m,i+1} = y_{m,i}$, $py_{m,0} = x_{f(m)}$. The elements of G can be written in normal form as finite sums $\sum k_{m,i} y_{m,i} + \sum l_m x_m$, $0 < k_{m,i} < p$. By our assumption 2 we have $G = D \oplus R$ where D is divisible and R is reduced.

We claim that, for each $n \in \mathbb{N}$, $x_n \in D$ if and only if $n \in \mathrm{rng}(f)$. To see this, suppose $x_n \in D$ and let $d = \sum k_{m,i} y_{m,i} + \sum l_m x_m$ be in D with $pd = x_n$. Note that

$$pd = \sum_{i>0} k_{m,i} y_{m,i-1} + \sum_{i=0} k_{m,i} x_{f(m)} = x_n.$$

By uniqueness of the normal form, we have $k_{m,i} = 0$ for $i > 0$, $k_{m,0} = 0$ for m such that $f(m) \neq n$, and $k_{m,0} = 1$ for m such that $f(m) = n$. Thus $n \in \mathrm{rng}(f)$. Conversely, suppose $n = f(m)$ for some m. Then the sequence $y_{m,0}, y_{m,1}, \dots$ p-divides x_n. If $x_n \notin D$, then using $G = D \oplus R$ we have $x_n = d + r$ and $y_{m,i} = d_i + r_i$ for each i. It follows that $pr_{i+1} = r_i$ for all i, and $pr_0 = r \neq 0$. Let A be the subgroup generated by $r_0, r_1, \dots$. It is easy to see that A exists, A is divisible, and $A \subseteq R$, a contradiction.

The claim implies that $\mathrm{rng}(f)$ exists. By lemma III.1.3 this gives arithmetical comprehension, Q.E.D.   □

Now we use 2 plus arithmetical comprehension to prove Π_1^1 comprehension. Given a tree $T \subseteq \mathbb{N}^{<\mathbb{N}}$, let G be the Abelian group with generators $x_\tau, \tau \in T$, and relations $px_\tau = x_\sigma$, $\tau = \sigma^\wedge\langle i \rangle$, and $x_{\langle\rangle} = 0$. The elements of G can be written in normal form as finite sums $\sum k_\tau x_\tau$ where $0 < k_\tau < p$. By our assumption 2, G can be decomposed as $D \oplus R$ where D is divisible and R is reduced. By lemma VI.4.2, D is the union of all divisible subgroups of G.

We claim that $\tau \in T$ lies on a path of T if and only if $x_\tau \in D$. To see this, note first that if f is a path through T then the subgroup A generated by $x_{f[n]}, n \in \mathbb{N}$ is divisible, hence $A \subseteq D$. Conversely, if $x_\tau \in D$, use primitive recursion to define a sequence $d_n \in D$, $n \in \mathbb{N}$, where $d_0 = x_\tau$ and $pd_{n+1} = d_n$ for all n. If $d_n = \sum k_\sigma x_\sigma$ and $d_{n+1} = \sum l_\rho x_\rho$, then $d_n = \sum k_\sigma x_\sigma = pd_{n+1} = \sum pl_\rho x_\rho$, from which it follows that each σ appearing in d_n is a proper initial segment of some ρ appearing in d_{n+1}. By primitive recursion there exists a sequence $\sigma_n, n \in \mathbb{N}$, such that $\sigma_0 = \tau$ and for all n, σ_n appears in d_n and is a proper initial segment of σ_{n+1}. Thus τ lies on a path through T. This proves our claim.

Since D exists, our claim implies the existence of $\widehat{T} = \{\tau: \tau$ lies on a path of $T\}$. This gives Π_1^1 comprehension, in view of the following easy lemma.

LEMMA VI.4.4. Π_1^1 *comprehension is equivalent over* RCA$_0$ *to the following statement* S: *For any tree* $T \subseteq \mathbb{N}^{<\mathbb{N}}$, *there exists a subtree*

$$\widehat{T} = \{\tau: \tau \text{ lies on a path of } T\}.$$

PROOF. Obviously Π_1^1-CA$_0$ proves statement S. For the converse, assume statement S and let $\langle T_k: k \in \mathbb{N}\rangle$ be an arbitrary sequence of trees. Form a tree

$$T = \{\langle\rangle\} \cup \{\langle k\rangle^\frown\tau: k \in \mathbb{N}, \tau \in T_k\}.$$

By statement S, $\widehat{T}$ exists. We have $\forall k$ ($\langle k\rangle \in \widehat{T} \leftrightarrow T_k$ has a path), hence by Δ_1^0 comprehension $\{k: T_k$ has a path$\}$ exists. Now lemma VI.1.1 gives Π_1^1 comprehension. Lemma VI.4.4 is proved.   □

The proof of theorem VI.4.1 is now complete.   □

REMARK VI.4.5. Combining theorem VI.4.1 with the results of §§III.6 and V.7, we see that Π_1^1-CA$_0$ is necessary and sufficient for the development of the structure theory of countable Abelian groups, although ACA$_0$ and ATR$_0$ suffice for certain parts of the theory. Such conclusions are typical of Reverse Mathematics.

Notes for §VI.4. A nice exposition of the structure theory of countable Abelian groups is in Kaplansky [136]. The construction used in the last part of the proof of theorem VI.4.1 is from Feferman [58]. The theorem itself is due to Friedman/Simpson/Smith [78].

VI.5. $\Sigma_1^0 \wedge \Pi_1^0$ Determinacy

We have seen in §V.8 that arithmetical transfinite recursion is equivalent to Σ_1^0 determinacy. We shall now show that Π_1^1 comprehension is equivalent to a stronger statement, namely $\Sigma_1^0 \wedge \Pi_1^0$ determinacy.

DEFINITION VI.5.1 ($\Sigma_1^0 \wedge \Pi_1^0$ determinacy). A formula θ is $\Sigma_1^0 \wedge \Pi_1^0$ if it is of the form $\varphi \wedge \psi$ where φ is Σ_1^0 and ψ is Π_1^0. $\Sigma_1^0 \wedge \Pi_1^0$ *determinacy* is the scheme

$$\exists S_0 \, \forall S_1 \, \theta(S_0 \otimes S_1) \vee \exists S_1 \, \forall S_0 \, \neg\theta(S_0 \otimes S_1)$$

where $\theta(f)$ is $\Sigma_1^0 \wedge \Pi_1^0$. Here S_0 and S_1 are variables ranging over 0-strategies and 1-strategies respectively, as in §V.8.

LEMMA VI.5.2. Π_1^1-CA$_0$ *proves* $\Sigma_1^0 \wedge \Pi_1^0$ *determinacy*.

PROOF. We reason in Π_1^1-CA$_0$. Let

$$\psi(f) \equiv \varphi_0(f) \wedge \neg\varphi_1(f)$$

be a $\Sigma_1^0 \wedge \Pi_1^0$ formula, where φ_0 and φ_1 are Σ_1^0. We shall prove

$$\exists S_0' \, \forall S_1 \, \psi(S_0' \otimes S_1) \vee \exists S_1' \, \forall S_0 \, \neg\psi(S_0 \otimes S_1').$$

By the Kleene normal form theorem V.1.4, we have $\varphi_i(f) \equiv \exists n \, \theta_i(f[n])$, where $\theta_i(\sigma)$ is arithmetical. Recall from §V.8 that

$$\mathrm{Seq}_0 = \{\sigma \in \mathrm{Seq} : \mathrm{lh}(\sigma) \text{ is even}\}.$$

We may safely assume that $\varphi_i(f) \equiv \exists n \, \theta_i(f[2n])$ and that

$$(\forall\sigma \in \mathrm{Seq}_0) \, \forall n \, ((\theta_i(\sigma) \wedge 2n < \mathrm{lh}(\sigma)) \rightarrow \neg\theta_i(\sigma[2n])).$$

For $\sigma \in \mathrm{Seq} = \mathbb{N}^{<\mathbb{N}}$ and $f \in \mathbb{N}^{\mathbb{N}}$, let $\sigma^\frown f$ be the concatenation, i.e., $\sigma^\frown f \in \mathbb{N}^{\mathbb{N}}$ where

$$(\sigma^\frown f)(n) = \begin{cases} \sigma(n) & \text{if } n < \mathrm{lh}(\sigma), \\ f(n - \mathrm{lh}(\sigma)) & \text{if } n \geq \mathrm{lh}(\sigma). \end{cases}$$

Put $\varphi_i^\sigma(f) \equiv \varphi_i(\sigma^\frown f)$ and $\theta_i^\sigma(\tau) \equiv \theta_i(\sigma^\frown \tau)$. Note that $\varphi_i^\sigma(f) \equiv \exists n \, \theta_i^\sigma(f[2n])$.

Define

$$P = \{\sigma \in \mathrm{Seq}_0 : \theta_0(\sigma) \wedge \exists S_0 \, \forall S_1 \, \neg\varphi_1^\sigma(S_0 \otimes S_1)\}.$$

We claim that P exists by Σ_1^1 comprehension. To see this, it suffices to show that $\forall S_1 \, \neg\varphi_1^\sigma(S_0 \otimes S_1)$ is equivalent to an arithmetical formula. Let us say that $\tau \in \mathrm{Seq}_0$ is *compatible with* S_0 if $\forall n \, (2n < \mathrm{lh}(\tau) \rightarrow \tau(2n) = S_0(\tau[2n]))$. Then $\forall S_1 \, \neg\varphi_1^\sigma(S_0 \otimes S_1)$ is equivalent to $\forall S_1 \, \forall n \, \neg\theta_1^\sigma((S_0 \otimes S_1)[2n])$, i.e.,

$$(\forall\tau \in \mathrm{Seq}_0) \, (\tau \text{ compatible with } S_0 \rightarrow \neg\theta_1^\sigma(\tau)),$$

which is arithmetical. This proves our claim, i.e., P exists.

Now consider the Σ_1^0 formula $\varphi(f) \equiv \exists n \, (f[2n] \in P)$, with parameter P. By theorem V.8.7 we have Σ_1^0 determinacy in Π_1^1-CA$_0$, hence either $\exists S_0 \, \forall S_1 \, \exists n \, ((S_0 \otimes S_1)[2n] \in P)$ or $\exists S_1 \, \forall S_0 \, \forall n \, ((S_0 \otimes S_1)[2n] \notin P)$.

Case 1: $\exists S_0 \, \forall S_1 \, \exists n \, ((S_0 \otimes S_1)[2n] \in P)$. Fix such a 0-strategy S_0. By Σ_1^1 choice, there exists a sequence of 0-strategies $\langle S_0^\sigma : \sigma \in P \rangle$ such that $(\forall\sigma \in P) \, \forall S_1 \, \neg\varphi_1^\sigma(S_0^\sigma \otimes S_1)$. Note that Σ_1^1 choice applies in this situation, because as we have seen above, the formula $\forall S_1 \, \neg\varphi_1^\sigma(S_0 \otimes S_1)$ is equivalent to an arithmetical formula. Now define a 0-strategy S_0' by putting $S_0'(\tau) = S_0(\tau)$ for all $\tau \in \mathrm{Seq}_0$ such that $\forall n \, (2n \leq \mathrm{lh}(\tau) \rightarrow \tau[2n] \notin P)$, and $S_0'(\sigma^\frown\tau) = S_0^\sigma(\tau)$ for all $\sigma \in P$ and all $\tau \in \mathrm{Seq}_0$.

Let S_1 be any 1-strategy. Then there exists a unique n such that $(S_0' \otimes S_1)[2n] \in P$. In particular $\varphi_0(S_0' \otimes S_1)$ holds. Moreover, putting $\sigma = (S_0' \otimes S_1)[2n]$ for this n, we have $S_0' \otimes S_1 = \sigma^\frown f$ where $\forall m \, (f[2m]$ is compatible with S_0^σ). It follows that $\neg\varphi_1^\sigma(f)$ holds, i.e., $\neg\varphi_1(S_0' \otimes S_1)$. Thus in this case we have $\forall S_1 \, \psi(S_0' \otimes S_1)$.

Case 2: $\exists S_1 \, \forall S_0 \, \forall n \, ((S_0 \otimes S_1)[2n] \notin P)$. Fix such a 1-strategy S_1. Define $Q = \{\sigma \in \mathrm{Seq}_0 : \theta_0(\sigma) \wedge \sigma \notin P\}$. By Σ^0_1 determinacy we have that for all $\sigma \in Q$ there exists a 1-strategy $\widetilde{S_1}$ such that $\forall S_0 \, \varphi_1^\sigma(S_0 \otimes \widetilde{S_1})$.

We claim that a choice principle applies, i.e., there exists a sequence of 1-strategies $\langle \widetilde{S_1^\sigma} : \sigma \in Q \rangle$ such that $(\forall \sigma \in Q) \, \forall S_0 \, \varphi_1^\sigma(S_0 \otimes \widetilde{S_1^\sigma})$. To see this, we use an inner model. Let M be a countable coded β-model containing all the parameters of the formula $\varphi_1(f)$ (theorem VII.2.10). Then $M \models$ ATR_0 (theorem VII.2.7). Hence $M \models \Sigma^0_1$ determinacy (theorem V.8.7). For each $\sigma \in Q$ we have $\neg\exists S_0 \, \forall \widetilde{S_1} \, \neg\varphi_1^\sigma(S_0 \otimes \widetilde{S_1})$, and this can be written as a Π^1_1 formula, hence it holds in M, since M is a β-model. It now follows by Σ^0_1 determinacy in M that, for each $\sigma \in Q$, $M \models \exists \widetilde{S_1} \, \forall S_0 \, \varphi_1^\sigma(S_0 \otimes \widetilde{S_1})$. Using the code of M as a parameter, we obtain a sequence of 1-strategies $\langle \widetilde{S_1^\sigma} : \sigma \in Q \rangle$ such that, for each $\sigma \in Q$, $\widetilde{S_1^\sigma} \in M$ and $M \models \forall S_0 \, \varphi_1^\sigma(S_0 \otimes \widetilde{S_1^\sigma})$. Since M is a β-model, it follows that for each $\sigma \in Q$, $\forall S_0 \, \varphi_1^\sigma(S_0 \otimes \widetilde{S_1^\sigma})$ is true. This proves our claim.

Now define a 1-strategy S_1' by putting $S_1'(\tau) = S_1(\tau)$ for all $\tau \in \mathrm{Seq}_1$ such that $\forall n (2n < \mathrm{lh}(\tau) \to \neg\theta_0(\tau[2n]))$, and $S_1'(\sigma^\frown\tau) = \widetilde{S_1^\sigma}(\tau)$ for all $\tau \in \mathrm{Seq}_1$ and all $\sigma \in \mathrm{Seq}_0$ such that $\theta_0(\sigma)$ holds.

Let S_0 be any 0-strategy. We have $\forall n \, (S_0 \otimes S_1')[2n] \notin P$. If $\forall n \, \neg\theta_0(S_0 \otimes S_1')[2n])$ then we have $\neg\varphi_0(S_0 \otimes S_1')$. Otherwise there is a unique n such that $(S_0 \otimes S_1')[2n] \in Q$. Putting $\sigma = (S_0 \otimes S_1')[2n]$ for this n, we have $S_0 \otimes S_1' = \sigma^\frown f$ where $\forall m \, (f[2m]$ is compatible with $\widetilde{S_1^\sigma})$. It follows that $\varphi_1^\sigma(f)$ holds, i.e., $\varphi_1(S_0 \otimes S_1')$. Thus in this case we have $\forall S_0 \, \neg\psi(S_0 \otimes S_1')$.

This completes the proof of lemma VI.5.2. $\qquad\square$

We now turn to the reversal.

LEMMA VI.5.3. *It is provable in* RCA_0 *that* $\Sigma^0_1 \wedge \Pi^0_1$ *determinacy implies* Π^1_1 *comprehension.*

PROOF. Assume $\Sigma^0_1 \wedge \Pi^0_1$ determinacy. By lemma V.8.5 we have arithmetical comprehension. We shall prove Π^1_1 comprehension. Our proof will be analogous to the proof of lemma V.8.5. By lemma VI.1.1 it suffices to show: given any sequence of trees $\langle T_k : k \in \mathbb{N}\rangle$, there exists a set X such that $\forall k \, (k \in X \leftrightarrow T_k$ has a path). Without loss of generality, we may assume $\forall k \, (\langle\rangle \in T_k)$.

Intuitively, consider the following game. Player 0 chooses an integer $k = f(0)$. Then player 1 attempts to build a path $f(1), f(3), \ldots$ through T_k. If player 1 succeeds, he wins the game. Otherwise, player 0 waits until the first n such that $\langle f(1), f(3), \ldots, f(2n+1)\rangle \notin T_k$. Then player 0 attempts to build a path $f(2n+2), f(2n+4), \ldots$ through T_k. If player 0 succeeds, he wins the game. Otherwise, player 1 wins the game.

It is clear that player 0 cannot have a winning strategy. Hence, by $\Sigma_1^0 \wedge \Pi_1^0$ determinacy, player 1 has a winning strategy; call it S_1. Let $g_k(0)$, $g_k(1), \ldots$ be the sequence $f(1), f(3), \ldots$ chosen by player 1 according to S_1 when player 0 begins the game with $f(0) = k$. Then

$$\forall k \ (T_k \text{ has a path} \to g_k \text{ is a path through } T_k).$$

Hence the desired set X exists by arithmetical comprehension with parameters $\langle T_k : k \in \mathbb{N} \rangle$ and $\langle g_k : k \in \mathbb{N} \rangle$. This proves Π_1^1 comprehension.

We now formalize the above intuitive argument.

Formally, let $\theta(f)$ be the following $\Sigma_1^0 \wedge \Pi_1^0$ formula:

$$\exists n \ (\langle f(1), f(3), \ldots, f(2n+1) \rangle \notin T_{f(0)}) \wedge$$
$$\forall m \ \forall n ((\langle f(1), f(3), \ldots, f(2n-1) \rangle \in T_{f(0)} \wedge$$
$$\langle f(1), f(3), \ldots, f(2n+1) \rangle \notin T_{f(0)}) \to$$
$$(\langle f(2n+2), f(2n+4), \ldots, f(2n+2m) \rangle \in T_{f(0)})).$$

We claim that $\forall S_0 \exists S_1 \neg \theta(S_0 \otimes S_1)$. To see this, let S_0 be given and set $k = S_0(\langle \rangle)$. If T_k has a path, let g be such a path and put $S_1(\sigma) = g(n)$ for all σ of length $2n + 1$. Otherwise let $S_1 : \text{Seq}_1 \to \mathbb{N}$ be arbitrary. Putting $f = S_0 \otimes S_1$ we have in the first case

$$\forall n \ (\langle f(1), f(3), \ldots, f(2n+1) \rangle = \langle g(0), g(1), \ldots, g(n) \rangle \in T_k = T_{f(0)}),$$

and in the second case

$$\forall n \ \exists m \ (\langle f(2n+2), f(2n+4), \ldots, f(2n+2m) \rangle \notin T_k = T_{f(0)}).$$

In either case $\neg \theta(f)$. This proves our claim.

Hence by $\Sigma_1^0 \wedge \Pi_1^0$ determinacy there exists S_1 such that $\forall S_0 \ \neg \theta(S_0 \otimes S_1)$. For all k define $g_k : \mathbb{N} \to \mathbb{N}$ recursively by

$$g_k(n) = S_1(\langle k, g_k(0), 0, g_k(1), 0, \ldots, g_k(n-1), 0 \rangle).$$

We claim that $\forall k$ (if T_k has a path, then g_k is a path through T_k). Suppose not. Let k, n, and h be such that $g_k[n] \in T_k$ and $g_k[n+1] \notin T_k$ and $\forall m (h[m] \in T_k)$. Define $S_0(\langle \rangle) = k$, $S_0(\sigma) = 0$ for σ of length $2m + 2 < 2n + 2$, and $S_0(\sigma) = h(m)$ for σ of length $2n + 2m + 2$, for all m. Then clearly $\theta(S_0 \otimes S_1)$ holds, a contradiction. This proves our claim.

By arithmetical comprehension let X be such that $\forall k \ (k \in X \leftrightarrow g_k$ is a path through $T_k)$. The previous claim implies that $\forall k \ (k \in X \leftrightarrow T_k$ has a path). This completes the proof of lemma VI.5.3. $\square$

Summarizing, we have

THEOREM VI.5.4. *The following are equivalent over* RCA$_0$.

1. Π_1^1 *comprehension.*
2. $\Sigma_1^0 \wedge \Pi_1^0$ *determinacy.*

PROOF. This is immediate from lemmas VI.5.2 and VI.5.3. $\square$

Notes for §VI.5. Theorem VI.5.4 is from Tanaka [263]. Earlier Steel [256, page 24] had announced that Π^1_1 comprehension is equivalent to determinacy for Boolean combinations of Σ^0_1 formulas, but the details have not appeared. Other related results are in Tanaka [264]. See also §§V.8 and VI.7.

VI.6. The Δ^0_2 Ramsey Theorem

We have seen in §V.9 that arithmetical transfinite recursion is equivalent to the Σ^0_1 Ramsey theorem. We shall now show that Π^1_1 comprehension is equivalent to the Δ^0_2 Ramsey theorem, and to the arithmetical or Σ^0_∞ Ramsey theorem. See theorem VI.6.4 below.

We begin with the reversal.

LEMMA VI.6.1. *It is provable in RCA$_0$ that Δ^0_2-RT implies Π^1_1 comprehension.*

PROOF. Reasoning in RCA$_0$, assume the Δ^0_2 Ramsey theorem, Δ^0_2-RT (definition V.9.1). Trivially Δ^0_2-RT implies RT(3), hence by lemma III.7.5 we have arithmetical comprehension. We want to prove Π^1_1 comprehension. Let $\langle T_m : m \in \mathbb{N} \rangle$ be a sequence of trees. By lemma VI.1.1 it suffices to prove the existence of the set $\{m : T_m \text{ has a path}\}$.

Recall from §V.9 the notion of a tree T being majorized by a function $f \in [\mathbb{N}]^\mathbb{N}$ or by a finite sequence $\sigma \in [\mathbb{N}]^{<\mathbb{N}}$. Note also that, by bounded König's lemma, f majorizes T if only if $\forall n \, (f[n] \text{ majorizes } T)$. Thus "$f$ majorizes T" can be written as a Π^0_1 formula. Moreover, T has a path if and only if $\exists f \in [\mathbb{N}]^\mathbb{N}$ such that f majorizes T.

For $k \in \mathbb{N}$ and $f \in [\mathbb{N}]^\mathbb{N}$, define $f^{(k)} \in [\mathbb{N}]^\mathbb{N}$ by $f^{(k)}(n) = f(k+n)$. Write

$$\varphi(f) \equiv (\forall m < f(0)) \, (f^{(1)} \text{ majorizes } T_m \leftrightarrow f^{(2)} \text{ majorizes } T_m).$$

Note that $\varphi(f)$ can be written in either Σ^0_2 or Π^0_2 form, i.e., $\varphi(f)$ is Δ^0_2. By Δ^0_2-RT, let $h \in [\mathbb{N}]^\mathbb{N}$ be homogeneous for $\varphi(f)$, i.e., either $(\forall g \in [\mathbb{N}]^\mathbb{N}) \, \varphi(h \cdot g)$ or $(\forall g \in [\mathbb{N}]^\mathbb{N}) \, \neg \varphi(h \cdot g)$.

Claim 1: $\varphi(h \cdot g)$ holds for all $g \in [\mathbb{N}]^\mathbb{N}$.

If not, then $\neg\varphi(h \cdot g)$ holds for all $g \in [\mathbb{N}]^\mathbb{N}$, hence in particular for each $n \in \mathbb{N}$ there exists $m < h(0)$ such that T_m is majorized by $h^{(n+2)}$ but not by $h^{(n+1)}$. Mapping n to the least such m, we would obtain a one-to-one function from $\mathbb{N}$ into $\{0, 1, \ldots, h(0) - 1\}$, contradiction.

Claim 2: For each $m \in \mathbb{N}$, if T_m has a path then T_m is majorized by $h^{(m+2)}$.

Suppose not, i.e., T_m has a path but $h^{(m+2)}$ does not majorize T_m. Let $n \in \mathbb{N}$ be such that $h^{(m+2)}[n]$ does not majorize T_m. Let $g \in [\mathbb{N}]^\mathbb{N}$ be such that g majorizes T_m. Put

$$f = h^{(m+1)}[n+1]^\frown (h^{(m+n+2)} \cdot g).$$

Then T_m is majorized by $f^{(n+1)} = (h^{(m+n+2)} \cdot g)$ but is not majorized by $f^{(1)} = h^{(m+2)}[n]^\frown (h^{(m+n+2)} \cdot g)$. This is a contradiction, since we have $m < h(m+1) = f(0)$ and therefore, by claim 1, for all $k \geq 1$, $f^{(k)}$ majorizes T_m if and only if $f^{(k+1)}$ majorizes T_m. Thus we have proved claim 2.

By claim 2 we have

$$\forall m \ (T_m \text{ has a path} \leftrightarrow T_m \text{ is majorized by } h^{(m+2)}).$$

Hence $\{m \colon T_m \text{ has a path}\}$ exists, by arithmetical comprehension with h as a parameter. This completes the proof of lemma VI.6.1.   □

We shall now show that, for all $k \in \omega$, Σ_k^0-RT is provable in Π_1^1-CA$_0$. We use an inner model technique. Our proof is based on the following lemma, which employs the notion of countable coded β-model from §VII.2. If M_1 and M_2 are countable coded β-models, $M_1 \in M_2$ means that the code of M_1 is an element of M_2.

LEMMA VI.6.2. *The following is provable in* ACA$_0$. *Let* $M_1, \ldots, M_k$ *be a finite sequence of countable coded β-models such that*

$$M_1 \in \cdots \in M_k.$$

Then for any Σ_k^0 formula $\varphi(f)$ with parameters in M_1, there exists $h \in M_k$ such that $\forall g \, \varphi(h \cdot g) \vee \forall g \, \neg \varphi(h \cdot g)$. Here f, g, and h range over $[\mathbb{N}]^\mathbb{N}$.

PROOF. We reason in ACA$_0$ and proceed by induction on $k \geq 1$. For Σ_1^0 formulas, our result follows from Σ_1^0-RT in ATR$_0$ (theorem V.9.7) plus the fact that any countable coded β-model satisfies ATR$_0$ (theorem VII.2.7). We inductively assume our result for Σ_k^0 formulas and prove and prove it for Σ_{k+1}^0 formulas, $k \geq 1$.

Let $M_1 \in \cdots \in M_k \in M_{k+1}$ be countable coded β-models. Let $\varphi(f)$ be a Σ_{k+1}^0 formula with parameters in M_1. Write

$$\varphi(f) \equiv \exists n_1 \forall n_2 \cdots n_k \, \psi(n_1, n_2, \ldots, n_k, f)$$

where $\psi(n_1, \ldots, n_k, f)$ is Σ_1^0 or Π_1^0, depending on whether k is even or odd, with parameters in M_1.

Within M_2, by recursion on $n \in \mathbb{N}$ using the code of M_1 as a parameter, define sequences $\sigma_n \in [\mathbb{N}]^{<\mathbb{N}}$, $f_n \in M_1 \cap [\mathbb{N}]^\mathbb{N}$, $n \in \mathbb{N}$, as follows. We employ the concatenation notation $\sigma^\frown f$ as in §VI.5. For each $n \in \mathbb{N}$ we shall have $\sigma_n^\frown f_n \in [\mathbb{N}]^\mathbb{N}$. Begin with $\sigma_0 = \langle \rangle$ and $f_0 = $ the identity function, i.e., $f_0(m) = m$ for all $m \in \mathbb{N}$. Given $\sigma_n^\frown f_n \in [\mathbb{N}]^\mathbb{N}$, put $\sigma_{n+1} = \sigma_n^\frown \langle f_n(0) \rangle$ and recall that $f_n^{(1)}(m) = f_n(m+1)$ for all $m \in \mathbb{N}$; thus $\sigma_{n+1}^\frown f_n^{(1)} = \sigma_n^\frown f_n \in [\mathbb{N}]^\mathbb{N}$. By finitely many applications of Σ_1^0-RT in M_1, obtain $g_n \in M_1 \cap [\mathbb{N}]^\mathbb{N}$ such that, for all subsequences σ of σ_{n+1}

and all $n_1, \ldots, n_k \le n$,

$$(\forall h \in [\mathbb{N}]^{\mathbb{N}}) \, \psi(n_1, \ldots, n_k, \sigma^\frown(f^{(1)}_n \cdot g_n \cdot h))$$
$$\vee \, (\forall h \in [\mathbb{N}]^{\mathbb{N}}) \, \neg\psi(n_1, \ldots, n_k, \sigma^\frown(f^{(1)}_n \cdot g_n \cdot h)).$$

Put $f_{n+1} = f^{(1)}_n \cdot g_n$. As part of the same recursion, define $p \colon \mathbb{N} \times \mathbb{N}^k \times [\mathbb{N}]^{<\mathbb{N}} \to \{0, 1\}$ such that, for all n and all subsequences σ of σ_{n+1} and $n_1, \ldots, n_k \le n$,

$$p(n, n_1, \ldots, n_k, \sigma) = \begin{cases} 1 & \text{if } (\forall h \in [\mathbb{N}]^{\mathbb{N}}) \, \psi(n_1, \ldots, n_k, \sigma^\frown(f^{(1)}_n \cdot g_n \cdot h)), \\ 0 & \text{if } (\forall h \in [\mathbb{N}]^{\mathbb{N}}) \, \neg\psi(n_1, \ldots, n_k, \sigma^\frown(f^{(1)}_n \cdot g_n \cdot h)). \end{cases}$$

Finally define $\widetilde{f} \in [\mathbb{N}]^{\mathbb{N}}$ by $\widetilde{f}(n) = f_n(0)$ for all $n \in \mathbb{N}$; thus $\widetilde{f}[n] = \sigma_n$ for all $n \in \mathbb{N}$. Note that $\widetilde{f} \in M_2$ and $p \in M_2$.

By construction we have

$$\psi(n_1, \ldots, n_k, \widetilde{f} \cdot g) \leftrightarrow \exists n \, (p(n, n_1, \ldots, n_k, (\widetilde{f} \cdot g)[n]) = 1)$$

and

$$\neg\psi(n_1, \ldots, n_k, \widetilde{f} \cdot g) \leftrightarrow \exists n \, (p(n, n_1, \ldots, n_k, (\widetilde{f} \cdot g)[n]) = 0),$$

for all $n_1, \ldots, n_k \in \mathbb{N}$ and $g \in [\mathbb{N}]^{\mathbb{N}}$. Thus

$$\widetilde{\psi}(n_1, \ldots, n_k, g) \equiv \psi(n_1, \ldots, n_k, \widetilde{f} \cdot g)$$

is Δ^0_1 with parameters in M_2. Hence

$$\widetilde{\varphi}(g) \equiv \varphi(\widetilde{f} \cdot g)$$
$$\equiv \exists n_1 \, \forall n_2 \cdots n_k \, \psi(n_1, \ldots, n_k, \widetilde{f} \cdot g)$$
$$\equiv \exists n_1 \, \forall n_2 \cdots n_k \, \widetilde{\psi}(n_1, \ldots, n_k, g)$$

is Σ^0_k with parameters in M_2. Hence, by inductive hypothesis, there exists $h \in M_{k+1} \cap [\mathbb{N}]^{\mathbb{N}}$ such that $\forall g \, \widetilde{\varphi}(h \cdot g) \vee \forall g \, \neg\widetilde{\varphi}(h \cdot g)$, i.e., $\forall g \, \varphi(\widetilde{f} \cdot h \cdot g) \vee \forall g \, \neg\varphi(\widetilde{f} \cdot h \cdot g)$, where g ranges over $[\mathbb{N}]^{\mathbb{N}}$. This completes the proof. $\square$

LEMMA VI.6.3. Π^1_1-CA$_0$ *proves* Σ^0_∞-RT. *In other words, for each* $k \in \omega$, Π^1_1-CA$_0$ *proves* Σ^0_k-RT.

PROOF. Let $\varphi(f)$ be a Σ^0_k formula, $k \ge 1$. Reasoning in Π^1_1-CA$_0$, let $X \subseteq \mathbb{N}$ be such that all the parameters of $\varphi(f)$ are $\le_T X$. By k applications of theorem VII.2.10, we obtain countable coded β-models $X \in M_1 \in \cdots \in M_k$. Then lemma VI.6.2 gives $\exists h \, (\forall g \, \varphi(h \cdot g) \vee \forall g \, \neg\varphi(h \cdot g))$, i.e., Σ^0_k-RT for $\varphi(f)$. This proves the lemma. $\square$

The main result of this section is:

THEOREM VI.6.4. *The following are pairwise equivalent over* RCA$_0$:

1. Π^1_1 *comprehension*;
2. *the* Δ^0_2 *Ramsey theorem*;
3. *the* Σ^0_∞ *Ramsey theorem*.

PROOF. This is immediate from lemmas VI.6.3 and VI.6.1. □

Notes for §VI.6. Lemma VI.6.1 is due to Simpson (unpublished notes, June 1981). Lemma VI.6.3 is related to results of Solovay [252]. See also Tanaka [262]. Related results are in §§III.7, V.9, VI.7.

VI.7. Stronger Set Existence Axioms

We have seen (in §§V.8 and V.9) that ATR_0 is just strong enough to prove Σ^0_1 determinacy and the Σ^0_1 Ramsey theorem. We have also seen (in §§VI.5 and VI.6) that $\Pi^1_1\text{-}\mathsf{CA}_0$ is just strong enough to prove $\Sigma^0_1 \wedge \Pi^0_1$ determinacy and the Σ^0_∞ Ramsey theorem. The purpose of this section is to point out that stronger forms of determinacy and Ramsey's theorem require stronger set existence axioms.

In analogy with arithmetical transfinite recursion (ATR_0, §V.2), the scheme of Π^1_1 transfinite recursion is defined as follows.

DEFINITION VI.7.1 (Π^1_1 transfinite recursion). We define $\Pi^1_1\text{-}\mathsf{TR}_0$ to be the formal system consisting of ACA_0 plus Π^1_1 *transfinite recursion*, i.e.,

$$\forall X\,(\mathrm{WO}(X) \to \exists Y\,\mathrm{H}_\theta(X,Y))$$

where θ is any Π^1_1 formula.

For $2 \leq k < \infty$, the system $\Pi^1_k\text{-}\mathsf{TR}_0$ is defined similarly, with Π^1_1 replaced by Π^1_k.

REMARK VI.7.2. Some results on models of $\Pi^1_1\text{-}\mathsf{TR}_0$ and related systems are in chapters VII and VIII.

THEOREM VI.7.3. *The following are pairwise equivalent over* RCA_0:

1. Π^1_1 *transfinite recursion*;
2. Δ^0_2 *determinacy*;
3. *the* Δ^1_1 *Ramsey theorem*.

PROOF. We omit the proofs, which can be found in Tanaka [262, 263].
□

REMARK VI.7.4. The previous theorem is due to Tanaka. In addition, Tanaka defined a stronger subsystem of Z_2, $\Sigma^1_1\text{-}\mathsf{MI}_0$ (related to Σ^1_1 monotonic recursion and Σ^1_1 reflecting ordinals), and proved the following theorem.

THEOREM VI.7.5. *The following are pairwise equivalent over* RCA_0:

1. $\Sigma^1_1\text{-}\mathsf{MI}_0$;
2. Σ^0_2 *determinacy*;
3. *the* Σ^1_1 *Ramsey theorem*.

PROOF. See Tanaka [262, 264]. □

REMARK VI.7.6 (stronger forms of Ramsey's theorem). The Borel Ramsey theorem, i.e., the Δ_1^1 Ramsey theorem, is also known as the Galvin/Prikry theorem; see Mathias [181] and Carlson/Simpson [33]. We have seen above that the Galvin/Prikry theorem and indeed the Σ_1^1 Ramsey theorem are provable in Z_2. On the other hand, it is known that the Δ_2^1 Ramsey theorem is not provable in ZFC. This follows from the fact that the canonical well ordering of $P(\omega)$ in $L(X)$ is Σ_2^1 (definition VII.4.20, lemma VII.4.21, sublemma VII.6.8).

REMARK VI.7.7 (stronger forms of determinacy). Friedman [66] has shown that Σ_5^0 determinacy is not provable in Z_2. Martin [177, 178] has shown that Borel determinacy is provable in ZFC. Friedman [66, 71] has shown that the proof of Borel determinacy requires $\aleph_1$ applications of the power set axiom. Friedman [65] has shown that Σ_1^1 determinacy is not provable in ZFC; indeed, it is false in all forcing extensions of $L(X)$. Harrington [104] has improved this by showing that Σ_1^1 determinacy is equivalent to $\forall X (X^\# \text{ exists})$.

Notes for §VI.7. Theorems VI.7.3 and VI.7.5 are due to Tanaka [262, 263, 264]. A result along the lines of $1 \leftrightarrow 2$ of VI.7.3 was announced by Steel [256, page 24], but the proof has not been published. Regarding $1 \leftrightarrow 2$ of VI.7.5, see also Steel [256, pages 24–25] and Moschovakis [191, pages 414–415]. Regarding $1 \leftrightarrow 3$ of VI.7.5, see also Solovay [252]. For more on Σ_1^1 monotonic recursion and Σ_1^1 reflecting ordinals, see Richter/Aczel [206], Aanderaa [1], and Simpson [233].

VI.8. Conclusions

In this chapter we have seen that several mathematical theorems are logically equivalent to $\Pi_1^1\text{-CA}_0$. Among them are: the Cantor/Bendixson theorem for closed sets (§VI.1), Kondo's theorem on coanalytic uniformization (§VI.2), Silver's theorem on Borel equivalence relations (§VI.3), a key structure theorem for countable Abelian groups (§VI.4), the Δ_2^0 Ramsey theorem (§VI.6), and $\Sigma_1^0 \wedge \Pi_1^0$ determinacy (§VI.5). We have also seen (§VI.7) that stronger forms of Ramsey's theorem and determinacy require stronger set existence axioms.

Our proof techniques in this chapter have been based mostly on the Kleene normal form theorem, via lemma VI.1.1 concerning paths through trees. We have also used an inner model technique (see lemmas VI.5.2 and VI.6.2) involving countable coded β-models (§VII.2).

Part B

MODELS OF SUBSYSTEMS OF Z_2

Chapter VII

β-MODELS

A β-*model* is an L_2-model M such that for all Σ_1^1 sentences φ with parameters from M, φ is true if and only if $M \models \varphi$. The purpose of this chapter is to study β-models of various subsystems of second order arithmetic. We concentrate on ATR_0 and Π_1^1-CA_0 and stronger systems. We make extensive use of set-theoretic methods.

Section VII.1 is introductory in nature. In it a recursion-theoretic result, the Kleene basis theorem, is used to obtain a description of the minimum β-model of Π_1^1-CA_0.

In §VII.2 we consider codes for countable β-models as defined *within* subsystems of Z_2. We prove within Π_1^1-CA_0 that for all X there exists a countable coded β-model M such that $X \in M$. We also study certain refinements of this result, involving a transfinite induction scheme.

In §§VII.3 and VII.4 we develop an apparatus whereby set-theoretic methods can be applied to the study of subsystems of Z_2. To any L_2-theory $T_0 \supseteq ATR_0$, we associate in §VII.3 a corresponding set-theoretic theory T_0^{set} in the language L_{set}. We show that T_0^{set} proves the same L_2-sentences as T_0. In other words, T_0^{set} is a conservative extension of T_0. In §VII.4 we introduce constructible sets and show that their basic properties can be proved within ATR_0^{set}. We then go on to show that more advanced properties of constructible sets, e.g., the Shoenfield absoluteness theorem, can be proved within Π_1^1-CA_0^{set}.

The rest of the chapter employs the set-theoretic ideas of §§VII.3 and VII.4 to study β-models of the systems Π_1^1-CA_0, Δ_2^1-CA_0, Π_2^1-CA_0, Δ_3^1-CA_0, Π_3^1-CA_0, ... In §VII.5 we show that these systems have minimum β-models M_1^Π, M_2^Δ, M_2^Π, M_3^Δ, M_3^Π, ..., which can be described in terms of initial segments of the constructible hierarchy. In §VII.6 we show that each of these minimum β-models satisfies an appropriate form of the axiom of choice. In §VII.7 we use reflection to show that these minimum β-models are all distinct.

Throughout this chapter, we formulate our results so as to apply not only to β-models but also to arbitrary models of the systems considered. Nevertheless, it will be clear that the methods are best adapted to the study of minimum β-models. Other methods will be developed in chapters VIII and IX, in order to construct ω-models and non-ω-models, respectively.

VII.1. The Minimum β-Model of Π_1^1-CA$_0$

DEFINITION VII.1.1 (ω-models). An ω-*model* is an L$_2$-model M such that the first order part of M is the standard model $(\omega, +, \cdot, 0, 1, <)$ of Z$_1$. We sometimes identify M with the set $\mathcal{S}_M \subseteq P(\omega)$. Here $P(\omega)$ is the powerset of ω.

DEFINITION VII.1.2 (β-models). A β-*model* is an ω-model M such that for any Σ_1^1 sentence φ with parameters from M, $M \models \varphi$ if and only if φ is true.

The purpose of this chapter is to study β-models of various subsystems of Z$_2$. In the present introductory section, we study β-models of Π_1^1-CA$_0$. We prove that there exists a *minimum* (i.e., unique smallest) β-model of Π_1^1-CA$_0$ (corollary VII.1.10). At the same time we obtain a characterization of β-models of Π_1^1-CA$_0$ by means of the hyperjump (theorem VII.1.8). We also present a more general result which characterizes β-submodels of an arbitrary given model of Π_1^1-CA$_0$ (definition VII.1.11, theorem VII.1.12).

Some of the ideas which are introduced here will be refined and generalized in later sections of this chapter. For instance, in §VII.5 we shall obtain an alternative description of the minimum β-model of Π_1^1-CA$_0$, by means of constructible sets.

Our first goal is to prove a formal version of a well known recursion-theoretic result known as the *Kleene basis theorem*. We begin with definitions of relative recursiveness and the hyperjump. For general background on recursion theory theory and hyperarithmetical theory, see for instance Kleene [142], Rogers [208], Shoenfield [222, chapters 6 and 7], and Sacks [211, part A].

DEFINITION VII.1.3 (universal lightface Π_1^0 formula). Let

$$\pi(e, m_1, \ldots, m_i, X_1, \ldots, X_j)$$

be a Π_1^0 formula with exactly the displayed free variables. (Here $m_1, \ldots, m_i$ are free number variables, $X_1, \ldots, X_j$ are free set variables, and e is a distinguished free number variable.) We say that π is *universal lightface* Π_1^0 if for all Π_1^0 formulas π' with the same free variables as π, RCA$_0$ proves

$$\forall e\, \exists e'\, \forall m_1 \cdots \forall X_1 \cdots (\pi(e', m_1, \ldots, X_1, \ldots) \leftrightarrow \pi'(e, m_1, \ldots, X_1, \ldots)).$$

It is known that for all numbers of variables $i, j < \omega$ there exists a universal lightface Π_1^0 formula. The existence of such formulas is closely related to the *enumeration theorem* in recursion theory.

DEFINITION VII.1.4 (relative recursiveness). The following definition is made in RCA$_0$. Let $\pi(e, m_1, X_1)$ be a fixed universal lightface Π_1^0 formula with exactly the displayed free variables. Given $X, Y \subseteq \mathbb{N}$ we say that Y is *recursive in* X or X-*recursive* (equivalently Y *is Turing reducible to* X, written $Y \leq_T X$), if there exist $e_0, e_1 \in \mathbb{N}$ such that for all m, $m \in Y \leftrightarrow \pi(e_1, m, X)$ and $m \notin Y \leftrightarrow \pi(e_0, m, X)$. In this case we say that $e = (e_0, e_1)$ is an X-*recursive index* of Y.

We say that X is *Turing equivalent to* Y, written $X =_T Y$, if $X \leq_T Y$ and $Y \leq_T X$. This is an equivalence relation on subsets of $\mathbb{N}$. A *Turing degree* is an $=_T$-equivalence class.

DEFINITION VII.1.5 (hyperjump). The following definition is made in RCA$_0$. Let f be a function variable, i.e., f ranges over total functions $f : \mathbb{N} \to \mathbb{N}$. As usual we identify such a function with a set of ordered pairs $f \subseteq \mathbb{N} \times \mathbb{N} \subseteq \mathbb{N}$. Given $X \subseteq \mathbb{N}$, the *hyperjump of* X, denoted HJ(X), is the set of all $(m, e) \in \mathbb{N} \times \mathbb{N} \subseteq \mathbb{N}$ such that $\exists f \, \pi(e, m, f, X)$, if this set exists. Here $\pi(e, m_1, X_1, X_2)$ is a fixed universal lightface Π_1^0 formula with exactly the displayed free variables.

The next lemma is a formal version of the fact that HJ(X) is "complete" among sets which are lightface Σ_1^1 definable from X.

LEMMA VII.1.6. *Let $\varphi(e, m, X)$ be a Σ_1^1 formula with only the displayed free variables. The following is provable in* ACA$_0$. *For all $e \in \mathbb{N}$ and $X \subseteq \mathbb{N}$, if* HJ(X) *exists then $\{m : \varphi(e, m, X)\}$ exists and is recursive in* HJ(X).

PROOF. By the proof of lemma V.1.4 (our formal version of the Kleene normal form theorem), we obtain a Π_1^0 formula

$$\pi'(e, m, f, X)$$

with only the displayed free variables such that ACA$_0$ proves

$$\forall e \, \forall m \, \forall X \, (\varphi(e, m, X) \leftrightarrow \exists f \, \pi'(e, m, f, X)).$$

Now reasoning within ACA$_0$, given e let e' be such that

$$\forall m \, \forall f \, \forall X \, (\pi(e', m, f, X) \leftrightarrow \pi'(e, m, f, X))$$

where π is our fixed universal lightface Π_1^0 formula as in definition VII.1.5. Given X such that HJ(X) exists, let Y be the set of all m such that $(m, e') \in$ HJ(X). Clearly $Y \leq_T$ HJ(X) and $\forall m \, (m \in Y \leftrightarrow \varphi(e, m, X))$. This completes the proof. $\square$

The following lemma is our formal version of the Kleene basis theorem.

LEMMA VII.1.7 (formalized Kleene basis theorem). *Let $\varphi(m, Y, X)$ be a Σ_1^1 formula with only the displayed free variables. The following is provable in* ACA$_0$. *Let $X \subseteq \mathbb{N}$ be given such that* HJ(X) *exists. For all m, if $\exists Y \, \varphi(m, Y, X)$ then $\exists Y \, (Y \leq_T$ HJ$(X) \wedge \varphi(m, Y, X))$.*

PROOF. By the proof of lemma V.1.4, we obtain an arithmetical formula $\theta(m, \sigma, \tau, X)$ with only the displayed free variables, such that ACA_0 proves

$$\forall m\, \forall X\, \forall Y\, (\varphi(m, Y, X) \leftrightarrow \exists f\, \forall n\, \theta(m, Y[n], f[n], X)).$$

Now reasoning within ACA_0, let X be given such that $\mathrm{HJ}(X)$ exists. Let G be the set of all $(m, \sigma, \tau) \in \mathbb{N} \times 2^{<\mathbb{N}} \times \mathbb{N}^{<\mathbb{N}}$ such that

$$\exists Y\, \exists f\, (\forall n\, \theta(m, Y[n], f[n], X) \wedge Y[\mathrm{lh}(\sigma)] = \sigma \wedge f[\mathrm{lh}(\tau)] = \tau).$$

By lemma VII.1.6, G exists and is recursive in $\mathrm{HJ}(X)$. Now let m be given such that $\exists Y\, \varphi(m, Y, X)$. Then clearly $(m, \langle\rangle, \langle\rangle) \in G$. Define $Y(n)$ and $f(n)$ by recursion on n as follows: $Y(n) = 1$ if $(m, Y[n]^\frown\langle 1\rangle, f[n]) \in G$; $Y(n) = 0$ otherwise; $f(n) = $ least j such that $(m, Y[n{+}1], f[n]^\frown\langle j\rangle) \in G$. Clearly Y and f are recursive in G and, by Δ^0_1 induction, $(m, Y[n], f[n]) \in G$ for all $n \in \mathbb{N}$. In particular $\forall n\, \theta(m, Y[n], f[n], X)$ so $\varphi(m, Y, X)$ holds. Also $Y \leq_T \mathrm{HJ}(X)$ by transitivity of $\leq_T$, since $Y \leq_T G$ and $G \leq_T \mathrm{HJ}(X)$. This completes the proof.   □

It can also be shown that lemmas VII.1.6 and VII.1.7 are provable in RCA_0 (rather than ACA_0).

We are now ready to present the following characterization of β-models of $\Pi^1_1\text{-}\mathsf{CA}_0$.

THEOREM VII.1.8 (β-models of $\Pi^1_1\text{-}\mathsf{CA}_0$). *Let M be an ω-model of RCA_0. The following are equivalent.*

1. *M is a β-model of $\Pi^1_1\text{-}\mathsf{CA}_0$.*
2. *M is closed under hyperjump, i.e., $\mathrm{HJ}(X) \in M$ for all $X \in M$.*

PROOF. Suppose first that M is a β-model of $\Pi^1_1\text{-}\mathsf{CA}_0$. Let π be Π^0_1 as in the definition of hyperjump (definition VII.1.5). Given $X \in M$, by Σ^1_1 comprehension within M let $Y \in M$ be the set of all (m, e) such that $M \models \exists f\, \pi(e, m, f, X)$. Since M is a β-model, we have $M \models \exists f\, \pi(e, m, f, X)$ if and only if $\exists f\, \pi(e, m, f, X)$ is true, for all e and m. Hence $Y = \mathrm{HJ}(X)$. Hence $\mathrm{HJ}(X) \in M$. This proves that 1 implies 2.

For the converse, let M be an ω-model of RCA_0 which is closed under hyperjump. We must show that M is a β-model of $\Pi^1_1\text{-}\mathsf{CA}_0$. Let $\varphi(m)$ be Σ^1_1 with no free variables other than m, but with parameters from M. Let $X \in M$ be such that all of the parameters of $\varphi(m)$ are recursive in X. Thus $\varphi(m)$ can be written as $\exists Y\, \theta(m, Y, X)$ where $\theta(m, Y, X)$ is arithmetical with no free variables other than m and Y, and no parameters other than X. By assumption $\mathrm{HJ}(X) \in M$. Hence $Y \in M$ for all $Y \leq_T \mathrm{HJ}(X)$. Hence by the Kleene basis theorem VII.1.7, we see that for each m, $M \models \exists Y\, \theta(m, Y, X)$ if and only if $\exists Y\, \theta(m, Y, X)$ is true. In other words, $M \models \varphi(m)$ if and only if $\varphi(m)$ is true. This shows that M is a β-model. Furthermore, by lemma VII.1.6, the set $Z = \{m : \varphi(m) \text{ is true}\}$ is recursive in $\mathrm{HJ}(X)$. Hence $Z \in M$ and $M \models \forall m\, (m \in Z \leftrightarrow \varphi(m))$. Thus $M \models \Sigma^1_1$ comprehension, or equivalently Π^1_1 comprehension. The proof is complete.   □

We now define iterated hyperjumps HJ(n, X), $n \in \omega$ by recursion on n as follows: HJ$(0, X) = X$ and HJ$(n + 1, X) = $ HJ(HJ(n, X)).

COROLLARY VII.1.9. *Given $X \subseteq \omega$, there exists a minimum (i.e., unique smallest) β-model of Π_1^1-CA$_0$ containing X. This model can be characterized as the ω-model consisting of all sets $Y \subseteq \omega$ such that $Y \leq_T$ HJ(n, X) for some $n \in \omega$.*

COROLLARY VII.1.10. *There exists a minimum β-model of Π_1^1-CA$_0$. It consists of all sets $X \subseteq \omega$ such that X is recursive in HJ$(n, \emptyset)$ for some $n \in \omega$.*

We shall see in chapter VIII that Π_1^1-CA$_0$ does not have a minimum (or even a minimal) ω-model.

We now generalize the previous theorem so as to apply to β-submodels of a given model M' which need not be a β-model.

DEFINITION VII.1.11 (β-submodels). Let M and M' be L$_2$-models. We say that *M is an ω-submodel of M'*, written $M \subseteq_\omega M'$, if M is a submodel of M' and has the same first order part as M'. We say that *M is a β-submodel of M'*, written $M \subseteq_\beta M'$, if $M \subseteq_\omega M'$ and, for all Σ_1^1 sentences φ with parameters from M, $M \models \varphi$ if and only if $M' \models \varphi$.

Thus a β-model is the same thing as a β-submodel of the standard or intended model $(\omega, P(\omega), +, \cdot, 0, 1, <)$ of Z$_2$. But in general, the M and M' in the above definition need not be β-models or even ω-models.

THEOREM VII.1.12. *Let M and M' be given such that $M \subseteq_\omega M'$, $M' \models \Pi_1^1$-CA$_0$, and $M \models$ RCA$_0$. The following are equivalent.*

1. *$M \subseteq_\beta M'$ and $M \models \Pi_1^1$-CA$_0$.*
2. *M is closed under the M'-hyperjump, i.e., for all $X \in M$ there exists $Y \in M$ such that $M' \models (Y$ is the hyperjump of $X)$.*

PROOF. This is a straightforward generalization of theorem VII.1.8.   □

COROLLARY VII.1.13. *Let $X \in M' \models \Pi_1^1$-CA$_0$ be given. Among all β-submodels $M \subseteq_\beta M'$ such that $X \in M \models \Pi_1^1$-CA$_0$, there exists a unique smallest one. It consists of all $Y \in M'$ such that $M' \models Y \leq_T$ HJ(n, X) for some $n \in \omega$.*

COROLLARY VII.1.14. *Let $M' \models \Pi_1^1$-CA$_0$ be given. Among all $M \subseteq_\beta M'$ such that $M \models \Pi_1^1$-CA$_0$, there exists a unique smallest one. It consists of all $X \in M'$ such that $M' \models X \leq_T$ HJ$(n, \emptyset)$ for some $n \in \omega$.*

In the two previous corollaries, note that the restriction $n \in \omega$ applies even if M' is not an ω-model.

EXERCISE VII.1.15. Let M be an ω-model of RCA$_0$. Show that if HJ$(X) \in M$, then $X \in M$ and HJ(X) is satisfied in M to be the hyperjump of X. Generalize this so as to apply to ω-submodels of a given model.

EXERCISE VII.1.16. Show that Π_1^1-CA$_0$ is equivalent over RCA$_0$ to the assertion that for all X, HJ(X) exists.

EXERCISE VII.1.17. Recall from §VI.7 that Π_1^1-TR$_0$ consists of RCA$_0$ plus all axioms $\forall Y\,(\mathrm{WO}(Y) \rightarrow \exists Z\,\mathrm{H}_\theta(Y,Z))$ where $\theta(n,Z)$ is any Π_1^1 formula. Show that Π_1^1-TR$_0$ is equivalent over RCA$_0$ to the assertion that $\forall X\,\forall Y\,(\mathrm{WO}(Y) \rightarrow$ the hyperjump can be iterated along Y starting with X).

EXERCISE VII.1.18. Give a characterization of β-models of Π_1^1-TR$_0$ analogous to theorem VII.1.8. Prove that there exists a minimum β-model of Π_1^1-TR$_0$. Prove that for any model of Π_1^1-TR$_0$ there is a smallest β-submodel of Π_1^1-TR$_0$.

It is natural to ask whether there exists a minimum β-model of ATR$_0$. This question is answered negatively by the following result, which will be proved in chapter VIII; see corollary VIII.6.9.

THEOREM VII.1.19. *Let M' be any countable model of* ATR$_0$. *Then there exists a proper β-submodel $M \subseteq_\beta M'$, $M \neq M'$. For any such M we have also $M \models$* ATR$_0$.

COROLLARY VII.1.20. *There is no minimum β-model of* ATR$_0$.

EXERCISE VII.1.21. Show that any β-model is a model of ATR$_0$. More generally, show that if $M \subseteq_\beta M'$ and $M' \models$ ATR$_0$, then $M \models$ ATR$_0$.

Further results on β-models of ATR$_0$ will be presented in §VII.2 and in chapter VIII. Further results on β-models of Π_1^1-CA$_0$ and stronger theories will be presented in §§VII.5, VII.6 and VII.7.

Notes for §VII.1. The Kleene basis theorem is due to Kleene [143]. Our characterization of β-models of Π_1^1-CA$_0$ in terms of $\leq_T$ and HJ (theorem VII.1.8) is well known. A set-theoretic characterization of such models is given in exercise VII.3.36. The minimum β-model of Π_1^1-CA$_0$ can also be described in terms of constructible sets; see theorem VII.5.17.

VII.2. Countable Coded β-Models

In this section we consider countable β-models which are encoded as single subsets of $\mathbb{N}$. We show that Π_1^1-CA$_0$ is strong enough to prove the existence of such models. We also study a formal theory of transfinite induction which is satisfied by all such models.

We begin by giving a definition within RCA$_0$ of codes for countable ω-models together with the appropriate satisfaction concept. Recall that any set $X \subseteq \mathbb{N}$ can be viewed as a code for a countable sequence of sets $\langle (X)_n : n \in \mathbb{N} \rangle$ where $(X)_n = \{i : (i,n) \in X\}$.

Definition VII.2.1 (countable coded ω-models). The following definition is made within RCA$_0$. A *countable coded ω-model* is a set $W \subseteq \mathbb{N}$, viewed as encoding the L_2-model

$$M = (\mathbb{N}, \mathcal{S}_M, +, \cdot, 0, 1, <)$$

with

$$\mathcal{S}_M = \{(W)_n : n \in \mathbb{N}\}.$$

Let Snt$_M$ be the set of (Gödel numbers of) sentences of L_2 with parameters from $|M| \cup \mathcal{S}_M$, i.e., from $\mathbb{N} \cup \{(W)_n : n \in \mathbb{N}\}$. Given $\varphi \in$ Snt$_M$, let Sub$_M(\varphi)$ be the set of $\psi \in$ Snt$_M$ such that ψ is a substitution instance of a subformula of φ. A *valuation for φ* is a function $f : \text{Sub}_M(\varphi) \to \{0, 1\}$ which obeys the following clauses:

$$f(t_1 = t_2) = \begin{cases} 1 & \text{if } t_1 = t_2, \\ 0 & \text{if } t_1 \neq t_2; \end{cases}$$

$$f(t_1 < t_2) = \begin{cases} 1 & \text{if } t_1 < t_2, \\ 0 & \text{if } t_1 \geq t_2; \end{cases}$$

$$f(\neg\psi) = 1 - f(\psi);$$

$$f(\psi_1 \wedge \psi_2) = \begin{cases} 1 & \text{if } f(\psi_1) = f(\psi_2) = 1, \\ 0 & \text{otherwise}; \end{cases}$$

$$f(\forall m \, \psi(m)) = \begin{cases} 1 & \text{if } f(\psi(m)) = 1 \text{ for all } m \in \mathbb{N}, \\ 0 & \text{otherwise}; \end{cases}$$

$$f(\forall X \, \psi(X)) = \begin{cases} 1 & \text{if } f(\psi((W)_n)) = 1 \text{ for all } n \in \mathbb{N}, \\ 0 & \text{otherwise}. \end{cases}$$

Clearly for any $\varphi \in$ Snt$_M$ there is at most one such valuation. We say that M *satisfies* φ, written $M \models \varphi$, if there exists a valuation f for φ such that $f(\varphi) = 1$. (This concept of satisfaction is similar to the notion of weak model which was introduced in §II.8.)

LEMMA VII.2.2. *Let φ be any sentence of L_2. Then ACA$_0$ proves the following. For all countable coded ω-models M there exists a unique valuation $f : \text{Sub}_M(\varphi) \to \{0, 1\}$.*

PROOF. The proof is straightforward by arithmetical comprehension using the code of M as a parameter. $\square$

Fix a universal lightface Π_1^0 formula $\pi(e, m_1, m_2, X_1, X_2, X_3)$ with exactly the displayed free variables (definition VII.1.3). Let $\varphi_1(e, m, X, Y)$ be the Σ_1^1 formula

$$\exists Z \, \forall n \, \neg\pi(e, m, n, X, Y, Z).$$

Thus $\varphi_1(e, m, X, Y)$ is in some sense a universal lightface Σ_1^1 formula with free variables e, m, X, Y.

DEFINITION VII.2.3 (countable coded β-models). A *countable coded β-model* is defined in RCA$_0$ to be a countable coded ω-model M such that for all $e, m \in \mathbb{N}$ and $X, Y \in S_M$, $\varphi_1(e, m, X, Y)$ if and only if $M \models \varphi_1(e, m, X, Y)$.

The following lemma will be superseded by theorem VII.2.7.

LEMMA VII.2.4. *It is provable in* ACA$_0$ *that, for any countable coded β-model M, we have $M \models$ ACA$_0$.*

PROOF. ACA$_0$ is axiomatized by finitely many Π_1^1 sentences plus the sentence

$$\forall e \, \forall m \, \forall X \, \forall Y \, \exists Z \, \forall n \, (n \in Z \leftrightarrow \pi(e, m, n, X, Y, Y)) \qquad (18)$$

where π is as above. It suffices to show that ACA$_0$ proves that all countable coded β-models satisfy (18). Reasoning in ACA$_0$, let M be a countable coded β-model and let $e, m \in \mathbb{N}$ and $X, Y \in S_M$ be given. Let e' be such that

$$\forall X_1 \, \forall X_2 \, \forall X_3 \, (\forall n \, \neg \pi(e', m, n, X_1, X_2, X_3) \leftrightarrow$$
$$\forall n \, (n \in X_1 \leftrightarrow \pi(e, m, n, X_2, X_3, X_3))).$$

By arithmetical comprehension we have

$$\exists Z \, \forall n \, (n \in Z \leftrightarrow \pi(e, m, n, X, Y, Y)).$$

Hence $\varphi_1(e', m, X, Y)$ holds. Hence

$$M \models \varphi_1(e', m, X, Y).$$

Hence $M \models \exists Z \, \forall n \, (n \in Z \leftrightarrow \pi(e, m, n, X, Y, Y))$. This completes the proof. □

DEFINITION VII.2.5 ($A\Pi_1^1$ formulas). $A\Pi_1^1$ is the smallest class of L$_2$-formulas which contains all Σ_1^1 formulas and is closed under number quantifiers and propositional connectives. (The notation $A\Pi_1^1$ stands for *arithmetical-in-Π_1^1*.)

LEMMA VII.2.6. *Let $\varphi(m_1, \ldots, m_i, X_1, \ldots, X_j)$ be an $A\Pi_1^1$ formula with exactly the displayed free variables. Then* ACA$_0$ *proves the following. For all countable coded β-models M and $m_1, \ldots, m_i \in \mathbb{N}$ and $X_1, \ldots, X_j \in S_M$, $\varphi(m_1, \ldots, m_i, X_1, \ldots, X_j)$ if and only if $M \models \varphi(m_1, \ldots, m_i, X_1, \ldots, X_j)$.*

PROOF. First assume that φ is Σ_1^1. Let $e < \omega$ be such that ACA$_0$ proves

$$\forall m_1 \cdots \forall m_i \, \forall X_1 \cdots \forall X_j \, (\varphi(m_1, \ldots, m_i, X_1, \ldots, X_j)$$
$$\leftrightarrow \varphi_1(e, \langle m_1, \ldots, m_i \rangle, X_1 \oplus \cdots \oplus X_j, \emptyset)).$$

Then the desired conclusion follows easily from definition VII.2.3 and lemma VII.2.4. The result for arbitrary $A\Pi_1^1$ formulas φ follows by a straightforward induction on the complexity of φ. $\square$

We shall now prove (within ACA_0) that every countable β-model is a model of ATR_0.

Theorem VII.2.7. *For any countable coded β-model M, we have $M \models$ ATR_0. This fact is provable in ACA_0.*

Proof. We reason in ACA_0. Let M be a countable coded β-model. By lemma VII.2.4 we have $M \models \mathsf{ACA}_0$. Let $\theta(n, Y)$ be any arithmetical formula with parameters in M. We must show that $M \models \forall X (\mathrm{WO}(X) \to \exists Y\, H_\theta(X, Y))$ (see §V.2.). Let $X \in M$ be such that $M \models \mathrm{WO}(X)$. By lemma VII.2.6 we have $\mathrm{WO}(X)$. Letting W be a code for M, we claim that for each $j \in \mathrm{field}(X)$ there exists m such that $H_\theta(j, X, (W)_m)$ (see definitions VII.2.1 and V.2.2). This claim will now be proved by arithmetical transfinite induction along X (lemma V.2.1). Suppose $j \in \mathrm{field}(X)$ and $\forall i(i <_X j \to \exists m\, H_\theta(i, X, (W)_m))$. By arithmetical comprehension let

$$Z = \{(n, i) : i <_X j \wedge \theta(n, (W)_{f(i)})\}$$

where $f(i) = $ least m such that $H_\theta(i, X, (W)_m)$. Thus we have $H_\theta(j, X, Z)$. Since M is a β-model, it follows by lemma VII.2.6 that $M \models \exists Y$ $H_\theta(j, X, Y)$. In other words, $H_\theta(j, X, (W)_m)$ for some m. This proves the claim. Now by arithmetical comprehension let

$$Z = \{(n, j) : j \in \mathrm{field}(X) \wedge \theta(n, (W)_{f(j)})\}$$

where $f(j) = $ least m such that $H_\theta(j, X, (W)_m)$. Thus we have $H_\theta(X, Z)$. Since M is a β-model, it follows by lemma VII.2.6 that $M \models \exists Y\, H_\theta(X, Y)$. This completes the proof. $\square$

Corollary VII.2.8. ATR_0 *does not prove the existence of a countable coded β-model.*

Proof. Suppose that ATR_0 proves the existence of a countable β-model. By theorem VII.2.7 it follows that ATR_0 proves the consistency of ATR_0. This contradicts Gödel's second incompleteness theorem [94, 115, 55, 222]. $\square$

We shall now show that the existence of countable coded β-models is provable in $\Pi_1^1\text{-}\mathsf{CA}_0$. Recall from definition VII.1.5 that the hyperjump of X is denoted $\mathrm{HJ}(X)$.

Lemma VII.2.9. *The following is provable in ACA_0. For all $X \subseteq \mathbb{N}$, $\mathrm{HJ}(X)$ exists if and only if there exists a countable coded β-model M such that $X \in M$.*

Proof. We reason in ACA_0. Suppose first that $X \in M$ for some countable coded β-model M. Let W be a code for M (definition VII.2.1). Let $\pi(e, m, f, X)$ be Π_1^0 as in the definition of hyperjump. By arithmetical comprehension using W as a parameter, let Y be the set of (e, m) such that $\exists n\, ((W)_n$ is a total function from $\mathbb{N}$ into $\mathbb{N}$ such that $\pi(e, m, (W)_n, X)$ holds). Thus $Y = \{(e, m) : M \models \exists f\, \pi(e, m, f, X)\}$. Since M is a β-model, it follows by lemma VII.2.6 and the definition of hyperjump that $Y = \{(e, m) : \exists f\, \pi(e, m, f, X)\} = \mathrm{HJ}(X)$. This proves the easy direction of the lemma.

We now prove the hard direction. Suppose that $\mathrm{HJ}(X)$ exists. Let

$$\pi(e, m_1, X_1, X_2, X_3)$$

be a universal lightface Π^0_1 formula with exactly the displayed free variables (definition VII.1.3). Write $\pi^*(n, h, g)$ as an abbreviation for $\forall e\, \forall m\, (n = (e, m) \rightarrow \pi(e, m, X, h, g))$. We are going to define a function $f : \mathbb{N} \rightarrow \mathbb{N}$, $f \leq_T \mathrm{HJ}(X)$, such that

$$\forall n\, \forall g\, (\pi^*(n, (f)^n, g) \rightarrow \pi^*(n, (f)^n, (f)_n)) \tag{19}$$

where $(f)_n : \mathbb{N} \rightarrow \mathbb{N}$ and $(f)^n : \mathbb{N} \rightarrow \mathbb{N}$ are given by

$$(f)_n(i) = f((n, i)),$$

$$(f)^n(j) = \begin{cases} f(j) & \text{if } j = (m, i) \text{ for some } m < n \text{ and } i \leq j, \\ 0 & \text{otherwise.} \end{cases}$$

Suppose for a moment that this f has been found. Set $W = \{(i, n) : f((n, i)) = 1\}$. Let M be the countable ω-model which is encoded by W (definition VII.2.1). We claim that $X \in M$ and that M is a β-model. To see that $X \in M$, let n_0 be such that

$$\forall g\, \forall h\, (\pi^*(n_0, h, g) \leftrightarrow \forall i\, (g(i) = 1 \leftrightarrow i \in X)).$$

Then clearly $(W)_{n_0} = X$ so $X \in M$. To see that M is a β-model, let $e, m \in \mathbb{N}$ and $Y_1, Y_2 \in \mathcal{S}_M$ be given such that $\varphi_1(e, m, Y_1, Y_2)$ holds. We must show that $M \models \varphi_1(e, m, Y_1, Y_2)$. Write $\varphi_1(e, m, Y_1, Y_2)$ as

$$\exists Z\, \forall m_1\, \exists m_2\, \theta(e, m, m_1, m_2, Y_1, Y_2, Z)$$

where θ is Σ^0_0 with exactly the displayed free variables. Fix n_1 and n_2 such that $(W)_{n_1} = Y_1$ and $(W)_{n_2} = Y_2$. Let $n_3 > \max(n_1, n_2)$ be such that, for all g and all h, $\pi^*(n_3, h, g)$ if and only if

$$\forall m_1\, \theta(e, m, m_1, (g)_0(m_1), \{i : (h)_{n_1}(i) = 1\},$$
$$\{i : (h)_{n_2}(i) = 1\}, \{i : (g)_1(i) = 1\})$$

holds. Then clearly

$$\forall m_1\, \exists m_2\, \theta(e, m, m_1, m_2, Y_1, Y_2, \{i : ((f)_{n_3})_1(i) = 1\})$$

holds. Let $n_4 > n_3$ be such that

$$\forall h\, \forall g\, (\pi^*(n_4, h, g) \leftrightarrow g = ((h)_{e_1})_1).$$

Then clearly $(f)_{n_4} = ((f)_{e_1})_1$, hence $\forall m_1 \exists m_2 \; \theta(e, m, m_1, m_2, Y_1, Y_2, (W)_{n_4})$ holds. Thus $M \models \varphi_1(e, m, Y_1, Y_2)$. This shows that M is a β-model.

It remains to find $f \leq_T \mathrm{HJ}(X)$ satisfying (19). We shall construct f by finite approximations, as in the proof of the Kleene basis theorem (lemma VII.1.7). Let G be the set of $(\sigma, \tau) \in 2^{<\mathbb{N}} \times \mathbb{N}^{<\mathbb{N}}$ such that

$$\exists h \, (h[\mathrm{lh}(\tau)] = \tau \wedge \forall n \, ((n < \mathrm{lh}(\sigma) \wedge \sigma(n) = 1) \rightarrow \pi^*(n, (h)^n, (h)_n))).$$

By lemma VII.1.6, G exists and is recursive in $\mathrm{HJ}(X)$. Clearly $(\langle\rangle, \langle\rangle) \in G$. Furthermore, if $(\sigma, \tau) \in G$ then $(\sigma^\frown\langle 0\rangle, \tau) \in G$ and also $(\sigma, \tau^\frown\langle j\rangle) \in G$ for at least one $j \in \mathbb{N}$. Recursively in G define $s : \mathbb{N} \rightarrow \{0, 1\}$ and $f : \mathbb{N} \rightarrow \mathbb{N}$ as follows: $s(n) = 1$ if $(s[n]^\frown\langle 1\rangle, f[n]) \in G$; $s(n) = 0$ otherwise; $f(n) = $ least j such that $(s[n+1], f[n]^\frown\langle j\rangle) \in G$. By Δ_1^0 induction, $(s[n], f[n]) \in G$ for all n. Now having defined f, we claim that (19) holds. Let n be given. If $s(n) = 1$, then by construction we have

$$\forall m \, \exists h \, (h[m] = f[m] \wedge \pi^*(n, (h)^n, (h)_n)),$$

hence $\pi^*(n, (f)^n, (f)_n)$. If $s(n) = 0$, then by construction

$$\neg \exists g \, \pi^*(n, (f)^n, g).$$

(We used here the fact that $(n, i) \geq n$ for all i.) In either case we get (19). This completes the proof of lemma VII.2.9.   □

The following theorem says that Π_1^1 comprehension is equivalent to the existence of "sufficiently many" countable coded β-models.

THEOREM VII.2.10 (existence of countable coded β-models). Π_1^1-CA$_0$ is equivalent over ACA$_0$ to the following statement. For all X there exists a countable coded β-model M such that $X \in M$.

PROOF. This follows immediately from lemmas VII.1.6 and VII.2.9.   □

COROLLARY VII.2.11. There exists a β-model of ATR$_0$ which is not a model of Π_1^1-CA$_0$.

PROOF. By corollary VII.1.10 let M' be the minimum β-model of Π_1^1-CA$_0$. By theorem VII.2.10 let $W \in M'$ be such that $M' \models (W$ is a code for a countable β-model). Let M be the countable β-model of which W is a code. Then clearly $M \subseteq_\beta M'$ and $M \neq M'$. Hence M is not a model of Π_1^1-CA$_0$. By theorem VII.2.7, $M \models$ ATR$_0$. This completes the proof.   □

We can sharpen the previous corollary as follows:

COROLLARY VII.2.12. Given $X \subseteq \omega$, there exists a countable β-model M such that $X \in M$ and, for all $Y \in M$, $\mathrm{HJ}(Y) \leq_T \mathrm{HJ}(X)$. In particular $\mathrm{HJ}(X) \notin M$ so M is not closed under hyperjump. Hence M is not a model of Π_1^1-CA$_0$.

PROOF. Given X, let M be the countable β-model which was constructed in the proof of lemma VII.2.9. Thus $X \in M$. Let f and s be as in that construction. Then for all n, we have $s(n) = 1$ if and only if

$\exists g \; \pi^*(n, (f)^n, g)$. Since $s \leq_T \mathrm{HJ}(X)$, it follows that $\mathrm{HJ}(Y) \leq_T \mathrm{HJ}(X)$ for all $Y \in M$. Since $\mathrm{HJ}(\mathrm{HJ}(X)) \not\leq_T \mathrm{HJ}(X)$, it follows that $\mathrm{HJ}(X) \notin M$. By theorem VII.1.8 it follows that M is not a model of $\Pi_1^1\text{-CA}_0$.   □

COROLLARY VII.2.13. *There exists a countable β-model M such that $M \models$ (there is no countable coded β-model).*

PROOF. By the previous corollary, let M be a countable β-model such that $\mathrm{HJ}(\emptyset) \notin M$. By lemma VII.2.9, it follows that $M \models$ (there is no countable coded β-model).   □

We shall now introduce a formal theory of transfinite induction along arbitrary countable well orderings with respect to arbitrary formulas of L_2.

DEFINITION VII.2.14 (transfinite induction). Recall from §V.1 the Π_1^1 formula $\mathrm{WO}(X)$, which says that X is a (code for a) countable well ordering. Given an L_2-formula $\psi(j)$ with a distinguished free number variable j, let $\mathrm{TI}(X, \psi)$ be the formula

$$\forall j \, (\forall i \, (i <_X j \to \psi(i)) \to \psi(j)) \to \forall j \, \psi(j)$$

expressing induction along X with respect to ψ. For $0 \leq k < \omega$ we define $\Pi_k^1\text{-TI}_0$ to be the subsystem of Z_2 whose axioms are those of ACA_0 plus the scheme of Π_k^1 *transfinite induction*:

$$\forall X \, (\mathrm{WO}(X) \to \mathrm{TI}(X, \psi))$$

where $\psi(j)$ is any Π_k^1 formula. We define $\Sigma_k^1\text{-TI}_0$ similarly. We also set

$$\Pi_\infty^1\text{-TI}_0 = \bigcup_{k \in \omega} \Pi_k^1\text{-TI}_0.$$

It is easy to see that any β-model is a model of $\Pi_\infty^1\text{-TI}_0$. The following lemma expresses two formal versions of this observation.

LEMMA VII.2.15 (β-models and $\Pi_\infty^1\text{-TI}_0$).
   1. *For each $k < \omega$, ACA_0 proves that all countable coded β-models satisfy $\Pi_k^1\text{-TI}_0$.*
   2. *ATR_0 proves that all countable coded β-models satisfy $\Pi_\infty^1\text{-TI}_0$.*

PROOF. For part 1 we reason in ACA_0. Let M be a countable coded β-model. By lemma VII.2.4 we have $M \models \mathrm{ACA}_0$. Suppose that $X \in M$ and $M \models \mathrm{WO}(X)$. Given an L_2-formula $\psi(j)$ with parameters from M, we must show that $M \models \mathrm{TI}(X, \psi)$. Suppose that $M \models \forall j \, (\forall i \, (i <_X j \to \psi(i)) \to \psi(j))$. By lemma VII.2.2 let $f \colon \mathrm{Sub}_M(\forall j \, \psi(j)) \to \{0, 1\}$ be a valuation for $\forall j \, \psi(j)$. Put $Y = \{j \colon f(\psi(j)) = 1\}$. Thus we have $\forall j \, (\forall i \, (i <_X j \to i \in Y) \to j \in Y)$. Since $M \models \mathrm{WO}(X)$, it follows by lemma VII.2.6 that $\mathrm{WO}(X)$ is true. Hence $Y = \mathbb{N}$. Hence $f(\forall j \, \psi(j)) = 1$, i.e., $M \models \forall j \, \psi(j)$. Thus $M \models \mathrm{TI}(X, \psi)$. We have now shown that, for each L_2-formula ψ, ACA_0 proves that every countable coded β-model satisfies $\forall X \, (\mathrm{WO}(X) \to \mathrm{TI}(X, \psi))$. Taking ψ to be a universal Π_k^1 formula, we obtain part 1.

For part 2 we reason in ATR_0. Let M be a countable coded β-model. By arithmetical transfinite recursion, there exists a total valuation $f\colon \mathrm{Snt}_M \to \{0, 1\}$. As in the proof of part 1, we can show that $f(\varphi) = 1$ for all $\varphi \in \mathrm{Snt}_M$ of the form $\forall X\, (\mathrm{WO}(X) \to \mathrm{TI}(X, \psi))$. Thus $M \models \Pi^1_\infty\text{-}\mathsf{TI}_0$. (We used arithmetical transfinite recursion only to prove the existence of the valuation f. For this we did need not the full strength of ATR_0. Rather we needed only a single arithmetical recursion along $\mathbb{N}$ using the code of M as a parameter.) □

THEOREM VII.2.16. $\Pi^1_1\text{-}\mathsf{CA}_0$ *proves the existence of a countable coded β-model of $\Pi^1_\infty\text{-}\mathsf{TI}_0$.*

PROOF. By theorem VII.2.10, $\Pi^1_1\text{-}\mathsf{CA}_0$ proves the existence of a countable coded β-model. Since $\Pi^1_1\text{-}\mathsf{CA}_0 \supseteq \mathsf{ATR}_0$, lemma VII.2.15.2 applies to show that any such model satisfies $\Pi^1_\infty\text{-}\mathsf{TI}_0$. □

Our next goal is to obtain a sort of weak converse to the previous theorem.

LEMMA VII.2.17. *Let M be any model of $\Pi^1_\infty\text{-}\mathsf{TI}_0$. Then there exists a model M' such that $M \subseteq_\beta M' \models \mathsf{ACA}_0$ and, for all $Y \in M$, $M' \models \mathrm{HJ}(Y)$ exists.*

PROOF. Let M' be the model with the same first order part as M and $\mathcal{S}_{M'} = \mathrm{Def}(M) = $ the set of all $Z \subseteq |M|$ such that Z is definable over M allowing parameters from M. Clearly $M \subseteq_\omega M'$ and $M' \models \mathsf{ACA}_0$. Since $M \models \Pi^1_\infty\text{-}\mathsf{TI}_0$, we have

$$M \models \mathrm{WO}(X) \quad \text{if and only if} \quad M' \models \mathrm{WO}(X)$$

for all $X \in M$. To show that $M \subseteq_\beta M'$, let φ be any Σ^1_1 sentence with parameters from M. By the Kleene normal form theorem (lemma V.1.4), let $\theta(\tau)$ be arithmetical with the same parameters as φ and such that ACA_0 proves $\varphi \leftrightarrow \exists f\, \forall n\, \theta(f[n])$. Let $T \in M$ be the tree of unsecured sequences, i.e.,

$$M \models \forall \tau\, (\tau \in T \leftrightarrow \forall n\, (n \le \mathrm{lh}(\tau) \to \theta(\tau[n]))).$$

Then by lemma V.1.3 we have

$$
\begin{aligned}
M &\models \varphi && \text{if and only if} \\
M &\models T \text{ has a path,} && \text{if and only if} \\
M &\models \neg\mathrm{WO}(\mathrm{KB}(T)), && \text{if and only if} \\
M' &\models \neg\mathrm{WO}(\mathrm{KB}(T)), && \text{if and only if} \\
M' &\models T \text{ has a path,} && \text{if and only if} \\
M' &\models \varphi.
\end{aligned}
$$

Thus $M \subseteq_\beta M'$. Now given $Y \in M$, set

$$Z = \{(e, m)\colon M \models \exists f\, \pi(e, m, f, Y)\}$$

where π is universal lightface Π_1^0 as in the definition of hyperjump (definition VII.1.5). Thus $Z \in M'$ and, since $M \subseteq_\beta M'$, $M' \models Z = \mathrm{HJ}(Y)$. This completes the proof. $\square$

THEOREM VII.2.18. *Let $\varphi(X)$ be an $A\Pi_1^1$ formula with no free variables other than X. The following assertions are pairwise equivalent.*

1. Π_∞^1-TI_0 *proves* $\forall X \varphi(X)$.
2. ACA_0 *proves* $\forall X$ (*if* $\mathrm{HJ}(X)$ *exists then* $\varphi(X)$ *holds*).
3. ACA_0 *proves that all countable coded β-models satisfy* $\forall X \varphi(X)$.

PROOF. The equivalence of 2 and 3 follows from lemmas VII.2.6 and VII.2.9. Suppose now that 1 holds, i.e., Π_∞^1-TI_0 proves $\forall X \varphi(X)$. Then, for some $k < \omega$, Π_k^1-TI_0 proves $\forall X \varphi(X)$. By lemma VII.2.15.1, ACA_0 proves that all countable coded β-models satisfy Π_k^1-TI_0. Hence ACA_0 proves that all such models satisfy $\forall X \varphi(X)$. This is assertion 3. Thus 1 implies 3.

It remains to show that 2 implies 1. Assume 2. Given $M \models \Pi_\infty^1$-TI_0, let M' be as in lemma VII.2.17. For any $X \in M$ we have $M' \models \mathrm{HJ}(X)$ exists. Hence by assumption $M' \models \varphi(X)$. Since $M \subseteq_\beta M'$, it follows as in lemma VII.2.6 that $M \models \varphi(X)$. Thus $M \models \forall X \varphi(X)$. This shows that any model of Π_∞^1-TI_0 satisfies $\forall X \varphi(X)$. Hence by Gödel's completeness theorem, Π_∞^1-TI_0 proves $\forall X \varphi(X)$. This completes the proof of theorem VII.2.18. $\square$

As an application we note:

COROLLARY VII.2.19. ATR_0 *is provable from* Π_∞^1-TI_0.

PROOF. ATR_0 is axiomatized by ACA_0 plus a certain Π_2^1 sentence $\forall X \varphi(X)$, where $\varphi(X)$ is Σ_1^1. By theorem VII.2.7, ACA_0 proves that $\forall X \varphi(X)$ holds in every countable coded β-model. By theorem VII.2.18 it follows that Π_∞^1-TI_0 proves $\forall X \varphi(X)$. $\square$

REMARK VII.2.20. It can be shown that ATR_0 is provable from Σ_1^1-TI_0. In fact, Σ_1^1-TI_0 is equivalent to ATR_0 plus Σ_1^1-IND (definition VII.6.1.2 below). The systems Π_1^1-TI_0 and Σ_1^1-TI_0 will be discussed in chapter VIII. See also Simpson [235].

We now draw some further corollaries.

COROLLARY VII.2.21. *Let φ be an $A\Pi_1^1$ sentence. The following assertions are pairwise equivalent.*

1. φ *is provable in* Π_∞^1-TI_0.
2. φ *is provable in* ACA_0 *assuming the existence of* $\mathrm{HJ}(\emptyset)$.
3. *It is provable in* ACA_0 *that every countable coded β-model satisfies* φ.

PROOF. This is immediate from theorem VII.2.18. $\square$

COROLLARY VII.2.22. *For each $k < \omega$, Π_∞^1-TI_0 proves the existence of a countable coded ω-model of Π_k^1-TI_0.*

PROOF. Let φ_k be the Σ_1^1 sentence which asserts the existence of a countable coded ω-model of Π_k^1-TI_0. By lemmas VII.2.9 and VII.2.15.1, ACA$_0$ proves that if HJ($\emptyset$) exists then φ_k holds. Hence by corollary VII.2.21, Π_∞^1-TI_0 proves φ_k.   □

COROLLARY VII.2.23. Π_∞^1-TI_0 *is not finitely axiomatizable.*

PROOF. If it were, it would be equivalent to one of the finitely axiomatizable theories Π_k^1-TI_0, $k < \omega$. Hence by corollary VII.2.22 and theorem II.8.8 (our formal version of the soundness theorem), Π_k^1-TI_0 would prove its own consistency. This would contradict Gödel's second incompleteness theorem [94, 115, 55, 222].   □

REMARK VII.2.24. Later ($\S$VIII.5) we shall prove the following result. Let T_0 be any finitely axiomatizable L$_2$-theory. Suppose there exists a countable ω-model of T_0. Then there exists a countable ω-model of T_0 which is not a model of Π_∞^1-TI_0. This will provide an alternative proof that Π_∞^1-TI_0 is not finitely axiomatizable.

We end this section with some further results, stated as exercises.

EXERCISE VII.2.25. Let $A\Pi_1^1$-TI_0 be the L$_2$-theory consisting of ACA$_0$ plus the scheme $\forall X\,(\mathrm{WO}(X) \to \mathrm{TI}(X, \varphi))$ for all $A\Pi_1^1$ formulas φ. Thus $A\Pi_1^1$-TI_0 is a subsystem of Π_∞^1-TI_0. Show that $A\Pi_1^1$-TI_0 proves the same Π_2^1 sentences as Π_∞^1-TI_0. Show that lemma VII.2.17, theorem VII.2.18, and corollaries VII.2.19, VII.2.21, and VII.2.22 continue to hold with Π_∞^1-TI_0 weakened to $A\Pi_1^1$-TI_0. Show that $A\Pi_1^1$-TI_0 is not finitely axiomatizable.

EXERCISE VII.2.26. Show that lemma VII.2.4, theorem VII.2.7, lemma VII.2.9, and theorem VII.2.10 can be proved in RCA$_0$ (rather than ACA$_0$).

EXERCISE VII.2.27. Show that RCA$_0$ proves that all countable coded β-models satisfy Π_2^1-TI_0. (This is a variant of lemma VII.2.15.1.)

For the next few exercises, we need the following definition.

DEFINITION VII.2.28. $R\Sigma_1^1$ is the class of Σ_1^1 formulas of the form $\exists X\,\psi$ where ψ is Π_2^0. ($R\Sigma_1^1$ stands for *restricted-Σ_1^1*.) We say that M is a *restricted-β-submodel of M'*, written $M \subseteq_{R\beta} M'$, if $M \subseteq_\omega M'$ and for all $R\Sigma_1^1$ sentences φ with parameters from M, $M \models \varphi$ if and only if $M' \models \varphi$. Let $AR\Pi_1^1$ be the smallest class of L_2 formulas which includes $R\Sigma_1^1$ and is closed under number quantifiers and propositional connectives. ($AR\Pi_1^1$ stands for *arithmetical-in-restricted-Π_1^1*.)

EXERCISE VII.2.29. Show that lemma VII.2.6 remains true if ACA$_0$, $A\Pi_1^1$ are replaced by RCA$_0$, $AR\Pi_1^1$ respectively.

EXERCISE VII.2.30. Let M be any model of Π_2^1-TI_0. Show that there exists a model M' such that $M \subseteq_{R\beta} M' \models$ RCA$_0$ and, for all $Y \in M$, $M' \models$ HJ(Y) exists. (This is a variant of lemma VII.2.17.)

EXERCISE VII.2.31. Prove the following variant of theorem VII.2.18. Let $\varphi(X)$ be an $A R\Pi_1^1$ formula with no free variables other than X. The following assertions are pairwise equivalent.

1. Π_2^1-TI$_0$ proves $\forall X\, \varphi(X)$.
2. RCA$_0$ proves $\forall X$ (if HJ(X) exists then $\varphi(X)$ holds).
3. RCA$_0$ proves that all countable coded β-models satisfy $\forall X\, \varphi(X)$.

EXERCISE VII.2.32. Show that Π_2^1-TI$_0$ proves $\forall X\, \exists M$ (M is countable coded ω-model of ATR$_0$ and $X \in M$).

Notes for §VII.2. The main results of this section are essentially due to Friedman [63]. The notion of $A\Pi_1^1$ formula and theorem VII.2.18 and the results stated as exercises VII.2.25–VII.2.32 are due to Simpson (unpublished notes, 1985). Theorem VII.2.18 and exercise VII.2.31 have been applied in Blass/Hirst/Simpson [21] to show that certain combinatorial theorems are provable in Π_2^1-TI$_0$.

VII.3. A Set-Theoretic Interpretation of ATR$_0$

There is a certain resemblance between (i) β-models for the language of second order arithmetic, and (ii) transitive models for the language of set theory. The purpose of this section is to explicate this resemblance.

Our main result is that there exists a close, precise relationship of mutual interpretability between (i) ATR$_0$ and (ii) a certain finitely axiomatizable system of set theory known as ATR$_0^{\text{set}}$. This result will be used in §VII.4 to show that certain set-theoretic constructions can be carried out "within ATR$_0$" (actually within ATR$_0^{\text{set}}$). Then in §§VII.5, VII.6, and VII.7 those constructions will be applied to study β-models of certain strong subsystems of Z$_2$, and to prove conservation results for those subsystems.

DEFINITION VII.3.1. The *set-theoretic language*, L$_{\text{set}}$, is the one-sorted, first order language with two binary relation symbols $\in$ and $=$. In addition, L$_{\text{set}}$ contains propositional connectives $\wedge$, $\vee$, $\neg$, $\rightarrow$, $\leftrightarrow$, quantifiers $\forall$, $\exists$, and *set-theoretic variables* v_i, $i \in \omega$.

The set-theoretic variables $v_0, v_1, \ldots, v_i, \ldots$ are intended to range over *sets* in the sense of Zermelo/Fraenkel set theory. Thus $v_i \in v_j$ means that v_i is an element of v_j, while $v_i = v_j$ means that v_i and v_j are equal, i.e., have the same elements.

NOTATION. In writing formulas of L$_{\text{set}}$, we shall employ the following notational conventions.

1. We use $u, v, w, x, y, z, \ldots$ as metavariables standing for set-theoretic variables v_i, $i \in \omega$. In any given context, it is assumed that $u, v, w, x, y, z, \ldots$ stand for distinct set-theoretic variables.
2. $u \notin v, u \neq v$ are abbreviations for $\neg u \in v, \neg u = v$ respectively.

3. $\emptyset =$ the empty set $=$ the unique set u such that $\forall x \, (x \notin u)$. (We shall have axioms which imply the existence and uniqueness of $\emptyset$.)

4. $\{x : \varphi(x)\}$ $=$ the unique set u such that $\forall x \, (x \in u \leftrightarrow \varphi(x))$, if such a set u exists; $\{x : \varphi(x)\} = \emptyset$ otherwise. Here $\varphi(x)$ is any formula of L$_{\text{set}}$, and u is a variable which does not occur freely in φ. Thus $\{x : \varphi(x)\}$ behaves as a *term*. If $\varphi(x)$ has free variables other than x, then the term $\{x : \varphi(x)\}$ also has those free variables. The set (denoted by the term) $\{x : \varphi(x)\}$ is said to *exist properly* if $\exists u \, \forall x \, (x \in u \leftrightarrow \varphi(x))$.

5. $\{t : \varphi\} = \{y : \exists x_1 \cdots \exists x_n \, (y = t \wedge \varphi)\}$. Here t is any term, φ is any formula, y is a variable which does not occur freely in t or φ, and $x_1, \ldots, x_n$ are exactly the variables which do occur freely in t.

DEFINITION VII.3.2 (abbreviated terms). Within L$_{\text{set}}$ we use the following abbreviated terms.

1. $\bigcup u = \{y : \exists x \, (y \in x \wedge x \in u)\}$ (union).
2. $\{u, v\} = \{x : x = u \vee x = v\}$ (unordered pair).
3. $u - v = \{x : x \in u \wedge x \notin v\}$ (complement).
4. $u \cap v = u - (u - v)$ (intersection).
5. $u \cup v = \bigcup\{u, v\}$ (union).
6. $\{x\} = \{x, x\}$ (singleton).
7. $\langle y, x \rangle = \{\{y, x\}, \{x\}\}$ (ordered pair).
8. $v \times u = \{\langle y, x \rangle : y \in v \wedge x \in u\}$ (Cartesian product).
9. $\text{dom}(w) = \{x : \exists y \, (\langle y, x \rangle \in w)\}$ (domain).
10. $\text{rng}(w) = \{y : \exists x \, (\langle y, x \rangle \in w)\}$ (range).
11. $\text{field}(w) = \text{dom}(w) \cup \text{rng}(w)$ (field).
12. $w^{-1} = \{\langle x, y \rangle : \langle y, x \rangle \in w\}$ (inverse).
13. $w'x =$ the unique y such that $\langle y, x \rangle \in w$, if such a y exists; $w'x = \emptyset$ otherwise (value of w at x).
14. $w \restriction u = w \cap (\text{rng}(w) \times u)$ (restriction).
15. $w''u = \text{rng}(w \restriction u)$ (range of the restriction).
16. $\in \restriction u = \{\langle y, x \rangle : y \in x \wedge x \in u\}$.

DEFINITION VII.3.3. B$_0^{\text{set}}$ is a finitely axiomatized theory in the language L$_{\text{set}}$. The four axioms of B$_0^{\text{set}}$ are as follows.

1. Axiom of Equality: $\forall u \, \forall v \, \forall w \, (u = u \wedge (u = v \rightarrow v = u) \wedge$ $((u = v \wedge v = w) \rightarrow u = w) \wedge ((u = v \wedge v \in w) \rightarrow u \in w) \wedge$ $((u \in v \wedge v = w) \rightarrow u \in w))$.

2. Axiom of Extensionality: $\forall u \, \forall v \, (\forall x \, (x \in u \leftrightarrow x \in v) \rightarrow u = v)$.

3. Axiom of Infinity: $\exists u \, (\emptyset \in u \wedge \forall x \, \forall y \, ((x \in u \wedge y \in u) \rightarrow x \cup \{y\} \in u))$.

4. Axiom of Rudimentary Closure: We have an axiom which asserts, for all u, v and w, the proper existence of $\{u, v\}$, $u - v$, $u \times v$, $\bigcup u$,

$\in \restriction u$, $\mathrm{dom}(w)$, w^{-1}, and

$$\{\langle y, \langle x, z \rangle \rangle \colon \langle y, x \rangle \in w \wedge z \in u\},$$
$$\{\langle y, \langle z, x \rangle \rangle \colon \langle y, x \rangle \in w \wedge z \in u\},$$
$$\{v \colon \exists x\, (x \in u \wedge v = w''\{x\})\}.$$

DEFINITION VII.3.4 (Σ_k^{set} formulas). The class of Δ_0^{set} formulas of $\mathrm{L}_{\mathrm{set}}$ is defined inductively as follows. The formulas $u = v$, $u \neq v$, $u \in v$, $u \notin v$ are Δ_0^{set}. If φ and ψ are Δ_0^{set} then so are $\varphi \wedge \psi$ and $\varphi \vee \psi$. If φ is Δ_0^{set} then so are $\forall u\, (u \in v \to \varphi)$ and $\exists u\, (u \in v \wedge \varphi)$. The quantifiers $\forall u\, (u \in v \to \cdots)$ and $\exists u\, (u \in v \wedge \cdots)$ are known as *bounded set-theoretic quantifiers*.

For $k < \omega$, a formula of $\mathrm{L}_{\mathrm{set}}$ is called Σ_k^{set} (respectively Π_k^{set} formula) if it is of the form $\exists u_1 \forall u_2 \cdots u_k\, \varphi$ (respectively $\forall u_1 \exists u_2 \cdots u_k\, \varphi$) where φ is Δ_0^{set}. (This hierarchy of formulas will play an important role in §VII.5.)

LEMMA VII.3.5 (Δ_0^{set} comprehension). *The scheme of Δ_0^{set} comprehension is provable in* $\mathsf{B}_0^{\mathrm{set}}$. *In other words,* $\mathsf{B}_0^{\mathrm{set}}$ *proves*

$$\forall u\, \exists v\, \forall x\, (x \in v \leftrightarrow (x \in u \wedge \varphi(x)))$$

where $\varphi(x)$ is any Δ_0^{set} formula and v is a variable which does not ocur freely in $\varphi(x)$.

PROOF. See Jensen [131, §1]. Alternatively, change the definition of $\mathsf{B}_0^{\mathrm{set}}$ so as to include the Δ_0^{set} comprehension scheme. (It is not then obvious that $\mathsf{B}_0^{\mathrm{set}}$ is finitely axiomatizable. However, this will not matter.) □

DEFINITION VII.3.6 (abbreviated formulas). Within $\mathsf{B}_0^{\mathrm{set}}$ we use the following abbreviated formulas.

1. $u \subseteq v \leftrightarrow u$ is a *subset* of v, i.e., $\forall x\, (x \in u \to x \in v)$.
2. $\mathrm{Rel}(r) \leftrightarrow r$ is a *relation*, i.e., $r \subseteq \mathrm{rng}(r) \times \mathrm{dom}(r)$.
3. $\mathrm{Fcn}(f) \leftrightarrow f$ is a *function*, i.e., $\mathrm{Rel}(f) \wedge \forall x \forall y \forall z\, ((\langle y, x \rangle \in f \wedge \langle z, x \rangle \in f) \to y = z)$.
4. $\mathrm{Inj}(f) \leftrightarrow f$ is an *injection*, i.e., $\mathrm{Fcn}(f) \wedge \forall x \forall y \forall z\, ((\langle z, x \rangle \in f \wedge \langle z, y \rangle \in f) \to x = y)$.
5. $u \approx v \leftrightarrow u$ and v are *equinumerous*, i.e., $\exists f\, (\mathrm{Inj}(f) \wedge \mathrm{dom}(f) = u \wedge \mathrm{rng}(f) = v)$.
6. $\mathrm{Trans}(u) \leftrightarrow u$ is *transitive*, i.e., $\forall x \forall y\, ((x \in y \wedge y \in u) \to x \in u)$.
7. $\mathrm{Ord}(u) \leftrightarrow u$ is an *ordinal*, i.e., $\mathrm{Trans}(u) \wedge \forall x \forall y\, ((x \in u \wedge y \in u) \to (x \in y \vee x = y \vee y \in x)) \wedge \forall v\, ((v \subseteq u \wedge v \neq \emptyset) \to \exists x\, (x \in v \wedge \forall y\, (y \in v \to y \notin x)))$.
8. $\mathrm{Succ}(u) \leftrightarrow u$ is a *successor ordinal*, i.e., $\mathrm{Ord}(u) \wedge \exists v\, (u = v \cup \{v\})$.
9. $\mathrm{Lim}(u) \leftrightarrow u$ is a *limit ordinal*, i.e., $\mathrm{Ord}(u) \wedge u \neq \emptyset \wedge \neg\mathrm{Succ}(u)$.
10. $\mathrm{FinOrd}(u) \leftrightarrow u$ is a *finite ordinal*, i.e., $\mathrm{Ord}(u) \wedge \forall v\, (v \in u \cup \{u\} \to (v = \emptyset \vee \mathrm{Succ}(v)))$.
11. $\mathrm{Fin}(u) \leftrightarrow u$ is *finite*, i.e., $\exists v\, (u \approx v \wedge \mathrm{FinOrd}(v))$.
12. $\mathrm{HFin}(u) \leftrightarrow u$ is *hereditarily finite*, i.e., $\exists v\, (u \subseteq v \wedge \mathrm{Trans}(v) \wedge \mathrm{Fin}(v))$.

13. Ctbl(u) $\leftrightarrow$ u is *countable*, i.e., $\exists f$ (Inj(f) $\wedge$ dom(f) $=$ $u \wedge \forall y$ $(y \in$ rng(f) $\rightarrow$ FinOrd(y))).
14. HCtbl(u) $\leftrightarrow$ u is *hereditarily countable*, i.e., $\exists v$ ($u \subseteq v \wedge$ Trans(v) $\wedge$ Ctbl(v)).

We use $\alpha, \beta, \gamma, \delta, \ldots$ as special variables ranging over ordinals. We use $i, j, k, m, n, \ldots$ as special variables ranging over finite ordinals. We write $0 = \emptyset$, $1 = \{0\}$, $2 = \{0, 1\}$, $\ldots$; $\alpha + 1 = \alpha \cup \{\alpha\}$; $\alpha < \beta \leftrightarrow \alpha \in \beta$; $\alpha \leq \beta \leftrightarrow (\alpha < \beta \vee \alpha = \beta)$; $\omega = \{m : \text{FinOrd}(m)\}$; HF $= \{u : \text{HFin}(u)\}$.

LEMMA VII.3.7. *The following facts are provable in* B$_0^{\text{set}}$.

1. $\neg \alpha < \alpha$; $\alpha = \{\beta : \beta < \alpha\}$.
2. $(\alpha < \beta \wedge \beta < \gamma) \rightarrow \alpha < \gamma$.
3. $\alpha < \beta \vee \alpha = \beta \vee \beta < \alpha$.
4. Ord($\alpha + 1$); $\forall \beta$ ($\beta < \alpha + 1 \leftrightarrow \beta \leq \alpha$).
5. $(\text{Lim}(\beta) \wedge \alpha < \beta) \rightarrow \alpha + 1 < \beta$.
6. Lim(ω); $\forall \alpha$ ($\alpha < \omega \leftrightarrow$ Fin(α)).
7. *Let z be a nonempty set of ordinals. Then:* (i) z *has a least element;* (ii) $\bigcup z$ *is an ordinal;* (iii) $\bigcup z$ *is the least upper bound of z.*
8. $u \approx u$; $u \approx v \rightarrow v \approx u$; $(u \approx v \wedge u \approx w) \rightarrow u \approx w$.
9. $\emptyset \approx 0$; $(u \approx m \wedge x \notin u) \rightarrow u \cup \{x\} \approx m + 1$.
10. $(u \approx m \wedge v \approx n \wedge u \subseteq v) \rightarrow m \leq n$.
11. $m \approx n \leftrightarrow m = n$.
12. Fin(u) $\leftrightarrow \exists m$ ($u \approx m$).
13. Fin($\emptyset$); Fin(x) $\rightarrow$ Fin($x \cup \{y\}$).
14. $(u \subseteq v \wedge \text{Fin}(v)) \rightarrow \text{Fin}(u)$.
15. $(\text{Fin}(u) \wedge \text{Fin}(v)) \rightarrow (\text{Fin}(u \cup v) \wedge \text{Fin}(u \times v))$.
16. $(\text{Fin}(u) \wedge \forall v$ $(v \in u \rightarrow \text{Fin}(v))) \rightarrow \text{Fin}(\bigcup u)$.
17. $(\forall v$ $(v \in w \rightarrow v \subseteq u) \wedge \text{Trans}(u)) \rightarrow \text{Trans}(u \cup w)$.
18. *The set* HF $= \{u : \text{HFin}(u)\}$ *exists properly.*
19. $u \in$ HF $\leftrightarrow (\text{Fin}(u) \wedge u \subseteq \text{HF})$.

PROOF. The proof is straightforward using lemma VII.3.5.   □

We are now ready to define the theory ATR$_0^{\text{set}}$.

DEFINITION VII.3.8. ATR$_0^{\text{set}}$ is a finitely axiomatized theory in the set-theoretic language L$_{\text{set}}$. The axioms of ATR$_0^{\text{set}}$ are those of B$_0^{\text{set}}$ plus the following three:

1. Axiom of Regularity:
$$\forall u \; (u \neq \emptyset \rightarrow \exists x \; (x \in u \wedge \forall y \; (y \in u \rightarrow y \notin x))).$$

2. Axiom of Countability: $\forall u$ (u is hereditarily countable).
3. Axiom Beta: A relation r is said to be *regular* if
$$\forall u \; (u \neq \emptyset \rightarrow \exists x \; (x \in u \wedge \forall y \; (y \in u \rightarrow \langle y, x \rangle \notin r))).$$

The axiom asserts that, for all regular relations r, there exists a function f such that dom(f) $=$ field(r) and, for all $x \in$ field(r),

$f'x = f''\{y\colon \langle y, x \rangle \in r\}$. This f is called the *collapsing function* of r.

It can be shown (in ZF, for instance) that the hereditarily countable sets form a transitive model of $\mathrm{ATR}_0^{\mathrm{set}}$.

We shall see that there is a canonical one-to-one correspondence between transitive models of $\mathrm{ATR}_0^{\mathrm{set}}$ and β-models of ATR_0. This is a special case of a more general canonical one-to-one correspondence between arbitrary models of $\mathrm{ATR}_0^{\mathrm{set}}$ and arbitrary models of ATR_0. We now give one direction of the more general correspondence.

THEOREM VII.3.9. *Each axiom of* ATR_0 *is, in its natural translation, a theorem of* $\mathrm{ATR}_0^{\mathrm{set}}$.

PROOF. The *natural translation* of L_2 into L_{set} is defined as follows. Then number variables of L_2 are interpreted in L_{set} as ranging over finite ordinals, i.e., elements of ω. The set variables of L_2 are interpreted in L_{set} as ranging over subsets of ω. Using lemma VII.3.5 and parts 11, 12 and 15 of lemma VII.3.7, $\mathrm{B}_0^{\mathrm{set}}$ proves the existence of functions $+, \cdot$ with $\mathrm{dom}(+) = \mathrm{dom}(\cdot) = \omega \times \omega$ where $m + n =$ the unique $k \in \omega$ such that $(m \times \{0\}) \cup (n \times \{1\}) \approx k$, $m \cdot n =$ the unique $k \in \omega$ such that $m \times n \approx k$. The symbols $+, \cdot, 0, 1, <, =, \in$ of L_2 are then interpreted by means of their obvious counterparts over ω in L_{set}. By lemmas VII.3.5 and VII.3.7 it is clear that each axiom of ACA_0 becomes, under the above translation, a theorem of $\mathrm{B}_0^{\mathrm{set}}$.

Without comment, we shall from now on identify formulas of L_2 with their translations into L_{set} as given above.

It remains to show that the principal axiom of ATR_0 is a theorem of $\mathrm{ATR}_0^{\mathrm{set}}$. As in §V.6, let CWO be the assertion of *comparability of countable well orderings*, i.e.,

$$\forall X \, \forall Y \, ((\mathrm{WO}(X) \wedge \mathrm{WO}(Y)) \to (|X| \le |Y| \vee |X| \ge |Y|)).$$

By theorem V.6.8 it suffices to show that CWO is a theorem of $\mathrm{ATR}_0^{\mathrm{set}}$. We reason within $\mathrm{ATR}_0^{\mathrm{set}}$. Let $X, Y \subseteq \omega$ be (codes for) countable well orderings, i.e., assume $\mathrm{WO}(X)$ and $\mathrm{WO}(Y)$. Set $r_X = \{\langle n, m \rangle \colon n <_X m\}$. Then r_X is a regular relation, so by Axiom Beta let f_X be the collapsing function of r_X, i.e., $\mathrm{dom}(f_X) = \mathrm{field}(r_X)$ and $f'_X m = f''\{n \colon \langle n, m \rangle \in r_X\}$ for all $m \in \mathrm{field}(r_X)$. Put $\alpha_X = \mathrm{rng}(f_X)$. It is easy to check that α_X is an ordinal, the order type of X, and that f_X is the unique isomorphism of X with α_X. Similarly define r_Y and let f_Y be the unique isomorphism of Y with its order type α_Y. By part 3 of VII.3.7, we have either $\alpha_X \le \alpha_Y$ or $\alpha_Y \le \alpha_X$. Suppose for definiteness that $\alpha_X \le \alpha_Y$. Put

$$g = \{(m, n) \colon m \in \mathrm{field}(X) \wedge n \in \mathrm{field}(Y) \wedge f'_X m = f'_Y n\}.$$

Then clearly $g \colon |X| \le |Y|$. Similarly if $\alpha_Y \le \alpha_X$ we have $g \colon |X| \ge |Y|$. This completes the proof. $\qquad \square$

In the previous theorem we exhibited the natural translation of ATR$_0$ into ATR$_0^{\text{set}}$. Our next goal is to obtain an adjoint translation from ATR$_0^{\text{set}}$ into ATR$_0$.

DEFINITION VII.3.10 (suitable trees). In ATR$_0$ we define a *suitable tree* to be a set $T \subseteq \mathbb{N}^{<\mathbb{N}}$ such that

(i) T is a tree, i.e.,
$$\forall \tau \, \forall m \, ((\tau \in T \wedge m \leq \mathrm{lh}(\tau)) \to \tau[m] \in T);$$

(ii) T is nonempty (equivalently $\langle \rangle \in T$ where $\langle \rangle$ is the empty element of $\mathbb{N}^{<\mathbb{N}}$); and

(iii) T is well founded, i.e., has no path, i.e.,
$$\neg(\exists f : \mathbb{N} \to \mathbb{N}) \, \forall m \, (f[m] \in T).$$

If T is a suitable tree and $\sigma \in T$, we put
$$T^\sigma = \{\tau : \sigma^\frown \tau \in T\}.$$
Note that T^σ is again a suitable tree.

DEFINITION VII.3.11. By theorem VII.3.9 the above definition of suitable tree in ATR$_0$ carries over to ATR$_0^{\text{set}}$. Continuing in ATR$_0^{\text{set}}$, given a suitable tree T we put
$$r_T = \{\langle \sigma^\frown \langle n \rangle, \sigma \rangle : \sigma^\frown \langle n \rangle \in T\}.$$
Then r_T is a regular relation. By Axiom Beta let c_T be the collapsing function of r_T. Define
$$|T| = c_T' \langle \rangle = c_T'' \{\langle n \rangle : \langle n \rangle \in T\}.$$
Note that $|T^\sigma| = c_T' \sigma$ for all $\sigma \in T$.

The idea of our translation of ATR$_0^{\text{set}}$ into ATR$_0$ will be that the suitable tree T is a code for the hereditarily countable set $|T|$.

LEMMA VII.3.12. *Within* ATR$_0^{\text{set}}$ *we can prove that for any set u there exists a suitable tree T such that* $|T| = u$.

PROOF. We reason within ATR$_0^{\text{set}}$. Let u be given. By the Axiom of Countability, there exists an injection g such that $\mathrm{dom}(g) \subseteq \omega$, $\mathrm{rng}(g)$ is transitive, and $u \subseteq \mathrm{rng}(g)$. Let $T \subseteq \mathbb{N}^{<\mathbb{N}}$ consist of $\langle \rangle$ plus all $\langle m_0, \ldots, m_k \rangle$ such that $g' m_0 \in u$ and $\forall i \, (i < k \to g' m_{i+1} \in g' m_i)$. It is easy to check that T is a suitable tree and $|T| = u$. □

DEFINITION VII.3.13 ($=^*$ and $\in^*$ for suitable trees). Within ATR$_0^{\text{set}}$, let T be a suitable tree. We write $\mathrm{Iso}(X, T)$ to mean that $X \subseteq T \times T$ and, for all $(\sigma, \tau) \in T \times T$, $(\sigma, \tau) \in X$ if and only if
$$\forall m \, (\sigma^\frown \langle m \rangle \in T \to \exists n \, (\sigma^\frown \langle m \rangle, \tau^\frown \langle n \rangle) \in X)$$
and
$$\forall n \, (\tau^\frown \langle n \rangle \in T \to \exists m \, (\sigma^\frown \langle m \rangle, \tau^\frown \langle n \rangle) \in X).$$

If S and T are suitable trees, we define $S \oplus T$ to be the suitable tree consisting of $\langle\rangle$ plus all $\langle 0 \rangle^\frown \sigma$ and $\langle 1 \rangle^\frown \tau$ such that $\sigma \in S$ and $\tau \in T$. We define

$$S =^* T \leftrightarrow \exists X \left(\mathrm{Iso}(X, S \oplus T) \wedge (\langle 0 \rangle, \langle 1 \rangle) \in X \right)$$

and

$$S \in^* T \leftrightarrow \exists X \left(\mathrm{Iso}(X, S \oplus T) \wedge \exists n \left(\langle 0 \rangle, \langle 1, n \rangle \right) \in X \right).$$

LEMMA VII.3.14. *Within* $\mathrm{ATR}_0^{\mathrm{set}}$ *we can prove that, for all suitable trees* S *and* T,

$$S =^* T \leftrightarrow |S| = |T|,$$

and

$$S \in^* T \leftrightarrow |S| \in |T|.$$

PROOF. Given a suitable tree T, put

$$X = \{ (\sigma, \tau) \colon \sigma \in T \wedge \tau \in T \wedge c_T' \sigma = c_T' \tau \}.$$

Then clearly X is the unique set such that $\mathrm{Iso}(X, T)$. Applying this to the suitable tree $S \oplus T$ instead of T, we obtain the desired conclusions. $\qquad \square$

DEFINITION VII.3.15. Let V_i, $i \in \omega$ be fixed distinct set variables of L_2. We shall use these variables to denote suitable trees. We shall link V_i to the set-theoretic variable v_i (cf. definition VII.3.1). To each formula φ of $\mathsf{L}_{\mathrm{set}}$, we associate a formula $|\varphi|$ of L_2 as follows.

$$|v_i = v_j| \text{ is } V_i =^* V_j;$$
$$|v_i \in v_j| \text{ is } V_i \in^* V_j;$$
$$|\neg\varphi| \text{ is } \neg|\varphi|; |\varphi \wedge \psi| \text{ is } |\varphi| \wedge |\psi|; \text{ etc.};$$
$$|\forall v_i\, \varphi| \text{ is } \forall V_i \, (V_i \text{ suitable} \rightarrow |\varphi|);$$
$$|\exists v_i\, \varphi| \text{ is } \exists V_i \, (V_i \text{ suitable} \wedge |\varphi|).$$

Note that if $v_{i_1}, \ldots, v_{i_k}$ are the free variables of φ, then $V_{i_1}, \ldots, V_{i_k}$ are the free variables of $|\varphi|$.

LEMMA VII.3.16. *Let* φ *be any formula of* $\mathsf{L}_{\mathrm{set}}$. *Let* $v_{i_1}, \ldots, v_{i_k}$ *be the free variables of* φ. *Then* $\mathrm{ATR}_0^{\mathrm{set}}$ *proves the following. For all sets* $v_{i_1}, \ldots, v_{i_k}$ *and all suitable trees* $V_{i_1}, \ldots, V_{i_k}$ *such that* $|V_{i_1}| = v_{i_1}, \ldots, |V_{i_k}| = v_{i_k}$, *we have* $\varphi \leftrightarrow |\varphi|$. *In particular,* $\mathrm{ATR}_0^{\mathrm{set}}$ *proves* $\varphi \leftrightarrow |\varphi|$ *for all sentences* φ *of* $\mathsf{L}_{\mathrm{set}}$.

PROOF. This follows by a straightforward induction on the number of symbols in φ, using lemmas VII.3.14 and VII.3.12. $\qquad \square$

Our next task is to show that the set-theoretic properties of suitable trees can be proved in ATR_0.

LEMMA VII.3.17. *The following is provable in* ATR$_0$. *Let T be a suitable tree. Then there exists a unique set X such that* Iso(X, T). *Furthermore, for all $\sigma \in T$ and $\tau \in T$, we have*

$$T^\sigma =^* T^\tau \leftrightarrow (\sigma, \tau) \in X$$

and

$$T^\sigma \in^* T^\tau \leftrightarrow \exists n \left((\sigma, \tau \frown \langle n \rangle) \in X \right).$$

In particular, X is an equivalence relation on T.

PROOF. The existence of X is proved by arithmetical transfinite recursion along (the Kleene/Brouwer ordering of) T. The uniqueness is proved by arithmetical transfinite induction. The rest is straightforward using the fact that, for each $\rho \in T$, if we define

$$X^\rho = \{ (\sigma, \tau) : (\rho \frown \sigma, \rho \frown \tau) \in X \}$$

then Iso(X^ρ, T^ρ) holds.   □

DEFINITION VII.3.18. In ATR$_0$, for $X \subseteq \mathbb{N}$ we define X^* to be the suitable tree consisting of $\langle \rangle$ and all $\langle m_0, \ldots, m_k \rangle$ such that $m_0 \in X$ and $\forall i \, (i < k \rightarrow m_{i+1} < m_i)$. For $n \in \mathbb{N}$ we define $n^* = X^*$ where $X = \{0, \ldots, n-1\}$.

The point of the previous definition is that, provably in ATR$_0^{\text{set}}$, $|X^*| = X$ and $|n^*| = n$. Thus $n \in X$ if and only if $n^* \in^* X^*$. This is a special case of:

LEMMA VII.3.19. *Let φ be any formula of* L$_2$. *Let $X_1, \ldots, X_i, n_1, \ldots, n_j$ be the free variables of φ. Then* ATR$_0$ *proves the following. For all suitable trees $V_1, \ldots, V_i, V_{i+1}, \ldots, V_{i+j}$ such that $V_1 =^* X_1^*, \ldots, V_i =^* X_i^*$, $V_{i+1} =^* n_1^*, \ldots, V_{i+j} =^* n_j^*$, we have $\varphi \leftrightarrow |\varphi|$.*

PROOF. The proof is by a straightforward induction on the number of symbols in φ. We omit the details.   □

LEMMA VII.3.20. *Let φ be any one of the axioms of* ATR$_0^{\text{set}}$. *Then $|\varphi|$ is a theorem of* ATR$_0$.

PROOF. We reason within ATR$_0$. The proofs of |Axiom of Equality| and |Axiom of Extensionality| are straightforward, using lemma VII.3.17.

In order to handle the Axiom of Rudimentary Closure, we construct appropriate suitable trees. For example, given suitable trees V_0 and V_1, we can construct a suitable tree

$$V_2 = V_0 \oplus V_1 = \{\langle \rangle\} \cup \{\langle 0 \rangle \frown \tau : \tau \in V_0\} \cup \{\langle 1 \rangle \frown \tau : \tau \in V_1\}.$$

It is then straightforward to prove that for all suitable trees V_3,

$$V_3 \in^* V_2 \leftrightarrow (V_3 =^* V_0 \lor V_3 =^* V_1)$$

i.e., $|\forall v_3 \, (v_3 \in v_2 \leftrightarrow (v_3 = v_0 \lor v_3 = v_1))|$. This shows that

$$|\forall v_0 \, \forall v_1 \, \exists v_2 \forall v_3 \, (v_3 \in v_2 \leftrightarrow (v_3 = v_0 \lor v_3 = v_1))|,$$

i.e., $|\forall v_0 \, \forall v_1 \, \exists \{v_0, v_1\}|$. The other parts of |Rudimentary Closure| are proved similarly.

In order to dispose of the Axiom of Infinity, we construct a suitable tree V_0 in accordance with the usual coding of the hereditarily finite sets. Put $n \, \mathrm{E} \, m$ if and only if n occurs as an exponent in the binary expansion of m, i.e., if $n = m_i$ for some i where

$$m = 2^{m_1} + 2^{m_2} + \cdots + 2^{m_j}, \qquad m_1 > m_2 > \cdots > m_j.$$

We then let V_0 be the suitable tree consisting of $\langle \rangle$ plus all $\langle n_0, \ldots, n_k \rangle$ such that $n_{i+1} \, \mathrm{E} \, n_i$ for all $i < k$. With this V_0 it is straightforward to prove

$$|\emptyset \in v_0 \wedge \forall v_1 \, \forall v_2 \, ((v_1 \in v_0 \wedge v_2 \in v_0) \rightarrow v_1 \cup \{v_2\} \in v_0)|,$$

hence |Axiom of Infinity|. We could also prove $|v_0 = \mathrm{HF}|$, but this is not needed.

We now prove |Axiom of Regularity|. Let V_0 be a suitable tree such that $|v_0 \neq \emptyset|$, i.e., $\exists m \, (\langle m \rangle \in V_0)$. By lemma VII.3.17 let X_0 be such that $\mathrm{Iso}(X_0, V_0)$. Suppose for a contradiction that

$$|\forall v_1 \, (v_1 \in v_0 \rightarrow \exists v_2 \, (v_2 \in v_0 \wedge v_2 \in v_1))|.$$

By lemma VII.3.17 this is equivalent to

$$\forall m \, (\langle m \rangle \in V_0 \rightarrow \exists n \, \exists j \, ((\langle n \rangle, \langle m, j \rangle) \in X_0)).$$

Define $f : \mathbb{N} \rightarrow \mathbb{N}$ by recursion as follows. Put $f(0) =$ the least m such that $\langle m \rangle \in V_0$. Given $f[k+1] = \langle f(0), \ldots, f(k) \rangle$, put $f(k+1) =$ the least j such that $\exists n \, ((\langle n \rangle, f[k]^\frown \langle j \rangle) \in X_0)$. Then $\forall k \, (f[k] \in V_0)$ contradicting the well foundedness of V_0.

The remaining two axioms involve ordered pairs. Note that $|\langle v_i, v_j \rangle = v_k|$ is equivalent to

$$(V_i \oplus V_j) \oplus (V_j \oplus V_j) =^* V_k.$$

To prove |Axiom Beta|, let V_0 be a given suitable tree such that

$$|v_0 \text{ is a regular relation}|.$$

Let X be the set of all $\langle k, i, m \rangle$ such that $\langle k, i, m \rangle \in V_0$. Let V_1 consist of $\langle \rangle$ plus all $\langle \sigma \rangle^\frown \tau$ such that $\sigma \in X$ and $\tau \in V_0^\sigma$. It is easy to check that V_1 is a suitable tree, and that

$$\left| v_1 = \bigcup \bigcup \bigcup v_0 = \mathrm{field}(v_0) \right|.$$

Now by lemma VII.3.17 let X_0 be such that $\mathrm{Iso}(X_0, V_0)$. Let R be the set of all $(\sigma, \tau) \in X \times X$ such that for some i, j, k, m, n, p and p we have $(\tau, \langle k, i, n \rangle) \in X_0, (\sigma, \tau) \notin X_0, (\langle k, i \rangle, \langle k, j \rangle) \notin X_0, (\sigma, \langle k, j, p \rangle) \in X_0$. Let V_2 consist of $\langle \rangle$ plus all $\langle \sigma_0, \ldots, \sigma_k \rangle$ such that $\forall i \, (i < k \rightarrow (\sigma_i, \sigma_{i+1}) \in R)$. Let V_3 consist of $\langle \rangle$ plus all $\langle \sigma \rangle^\frown \tau$ such that $\sigma \in X$ and

$$\tau \in (V_2^{\langle \sigma \rangle} \oplus V_0^\sigma) \oplus (V_0^\sigma \oplus V_0^\sigma).$$

It is straightforward to check that V_2 and V_3 are suitable trees, and that

$$|v_2 = \mathrm{rng}(v_3) \text{ where } v_3 \text{ is the collapsing function of } v_0|.$$

It remains to prove |Axiom of Countability|. Let V_0 be a given suitable tree. We may regard each $\sigma \in V_0 \subseteq \mathbb{N}$ as an element of $\mathbb{N}$ and form the corresponding suitable tree σ^* as in definition VII.3.18. Let V_1 consist of $\langle\rangle$ plus all $\langle\sigma\rangle{}^\frown\tau$ such that $\sigma \in V_0$ and $\tau \in (V_0^\sigma \oplus \sigma^*) \oplus (\sigma^* \oplus \sigma^*)$. It is straightforward to check that $|\mathrm{Fcn}(v_1) \wedge \mathrm{dom}(v_1) \subseteq \omega \wedge \mathrm{Trans}(\mathrm{rng}(v_1)) \wedge v_0 \subseteq \mathrm{rng}(v_1)|$.

This completes the proof of lemma VII.3.20.   □

EXERCISE VII.3.21. Show that ATR$_0^{\mathrm{set}}$ proves, for all sets v, the proper existence of

$$TC(v) = \bigcup \left\{ \bigcup{}^n v : n \in \omega \right\}$$

where $\bigcup^0 v = v$ and $\bigcup^{n+1} v = \bigcup\bigcup^n v$ for all $n \in \omega$. Also, ATR$_0^{\mathrm{set}}$ proves that $TC(v)$ is the smallest transitive set u such that $v \subseteq u$. $TC(v)$ is known as the *transitive closure of* v.

We are now ready to deduce the main results of this section.

THEOREM VII.3.22. *Let φ be a sentence of* L$_{\mathrm{set}}$. *Then* ATR$_0^{\mathrm{set}}$ *proves φ if and only if* ATR$_0$ *proves $|\varphi|$.*

PROOF. Suppose first that ATR$_0^{\mathrm{set}}$ proves φ. By lemma VII.3.20 it follows that ATR$_0$ proves $|\varphi|$. Conversely, suppose that ATR$_0$ proves $|\varphi|$. It follows by theorem VII.3.9 that ATR$_0^{\mathrm{set}}$ proves $|\varphi|$. But then by lemma VII.3.16 it follows that ATR$_0^{\mathrm{set}}$ proves φ.   □

THEOREM VII.3.23 (a conservation theorem). ATR$_0^{\mathrm{set}}$ *is a conservative extension of* ATR$_0$. *In other words, for any sentence φ of* L$_2$, ATR$_0^{\mathrm{set}}$ *proves φ if and only if* ATR$_0$ *proves φ.*

PROOF. By theorem VII.3.22, ATR$_0^{\mathrm{set}}$ proves φ if and only if ATR$_0$ proves $|\varphi|$. But by lemma VII.3.19 ATR$_0$ proves $|\varphi| \leftrightarrow \varphi$, so the desired conclusion follows.   □

For use in §§VII.4 and VII.5, we prove the following theorem which relates the set-theoretic hierarchy Σ_k^{set}, $k < \omega$ (definition VII.3.4) to the projective hierarchy Σ_{k+1}^1, $k < \omega$ (definition I.5.1).

THEOREM VII.3.24. *Assume $0 \leq k < \omega$.*

1. *If φ is a Σ_k^{set} formula of* L$_{\mathrm{set}}$, *then $|\varphi|$ is equivalent (provably in* ATR$_0^{\mathrm{set}}$) *to a Σ_{k+1}^1 formula of* L$_2$.
2. *Conversely, each Σ_{k+2}^1 formula of* L$_2$ *is equivalent (provably in* ATR$_0^{\mathrm{set}}$) *to a $\Sigma_{k+1}^{\mathrm{set}}$ formula of* L$_{\mathrm{set}}$.

PROOF. We first show by induction on Δ_0^{set} formulas φ that $|\varphi|$ is equivalent to a Σ_1^1 formula. By definitions VII.3.15 and VII.3.13, $|v_i = v_j|$ and

$|v_i \in v_j|$ are equivalent to the Σ_1^1 formulas

$$\exists X \, (\text{Iso}(X, V_i \oplus V_j) \wedge (\langle 0 \rangle, \langle 1 \rangle) \in X)$$

and

$$\exists X \, (\text{Iso}(X, V_i \oplus V_j) \wedge \exists n \, ((\langle 0 \rangle, \langle 1, n \rangle) \in X))$$

respectively. Hence, by lemmas VII.3.9 and VII.3.17, $|v_i \neq v_j|$ and $|v_i \notin v_j|$ are equivalent to the Σ_1^1 formulas

$$\exists X \, (\text{Iso}(X, V_i \oplus V_j) \wedge (\langle 0 \rangle, \langle 1 \rangle) \notin X)$$

and

$$\exists X \, (\text{Iso}(X, V_i \oplus V_j) \wedge \forall n \, ((\langle 0 \rangle, \langle 1, n \rangle) \notin X)$$

respectively. It is also clear that if $|\varphi|$ and $|\psi|$ are Σ_1^1, then so are $|\varphi| \wedge |\psi|$ and $|\varphi| \vee |\psi|$, i.e., $|\varphi \wedge \psi|$ and $|\varphi \vee \psi|$. Finally, by lemma VII.3.17, $|\forall v_i \, (v_i \in v_j \rightarrow \varphi)|$ and $|\exists v_i \, (v_i \in v_j \wedge \varphi)|$ are equivalent to

$$\forall n \, (\langle n \rangle \in V_j \rightarrow \exists V_i \, (V_i =^* V_j^{\langle n \rangle} \wedge |\varphi|))$$

and

$$\exists n \, (\langle n \rangle \in V_j \wedge \exists V_i \, (V_i =^* V_j^{\langle n \rangle} \wedge |\varphi|))$$

respectively. These formulas are Σ_1^1 if $|\varphi|$ is. At this point we are using the Σ_1^1 axiom of choice, a consequence of ATR_0 (theorem V.8.3).

So far we have shown that $\varphi \, \Sigma_0^{\text{set}}$ implies $|\varphi| \, \Sigma_1^1$. Suppose now that φ is $\Sigma_{k+1}^{\text{set}}$, say $\exists v_i \, \psi$ where ψ is Π_k^{set}. By induction on k, $|\psi|$ is Π_{k+1}^1. Hence $|\varphi|$, i.e.,

$$\exists V_i \, (V_i \text{ suitable} \wedge |\psi|),$$

is Σ_{k+2}^1. This completes the proof of part 1.

We now prove the converse. Let φ be a Σ_2^1 formula of L_2. By lemma V.1.4 (the Kleene normal form theorem), we can write φ in the form

$$\exists X \, \forall f \, \exists n \, \neg \theta(X, f[n])$$

where θ is arithmetical. We may also assume that $\theta(X, f[n])$ implies $\theta(X, f[m])$ for all $m \leq n$. Thus $\forall f \, \exists n \, \neg \theta(X, f[n])$ is equivalent to regularity of the relation $\{ (\tau^\frown \langle k \rangle, \tau) : \theta(X, \tau^\frown \langle k \rangle) \}$. By Axiom Beta and the Axiom of Regularity, this is equivalent to the existence of an appropriate collapsing function. Thus φ is equivalent to the Σ_1^{set} formula

$$\exists X \, \exists g \, (\text{Fcn}(g) \wedge \forall \tau \, \forall k \, (\theta(X, \tau^\frown \langle k \rangle) \rightarrow g' \tau^\frown \langle k \rangle \in g' \tau)).$$

The previous paragraph shows that every Σ_2^1 formula of L_2 is equivalent to a Σ_1^{set} formula of L_{set}. It follows easily that every Σ_{k+2}^1 formula of L_2 is equivalent to a $\Sigma_{k+1}^{\text{set}}$ formula of L_{set}.

This completes the proof of theorem VII.3.24.   □

Theorems VII.3.22 and VII.3.23 established a close and precise relationship of mutual interpretability between ATR$_0$ (a subsystem of second
order arithmetic) and ATR$_0^{\text{set}}$ (a system of set theory). We shall now
reformulate these results in model-theoretic terms.

DEFINITION VII.3.25.

1. A *model for* L$_{\text{set}}$ or L$_{\text{set}}$-*structure* is an ordered pair

$$A = (|A|, \in_A)$$

   where $|A|$ is a nonempty set and $\in_A \subseteq |A| \times |A|$ is a binary relation
   on $|A|$.
2. Let φ be a sentence of L$_{\text{set}}$ with parameters from $|A|$. We say that A
   satisfies φ or *is a model of* φ, written $A \models \varphi$, if φ is true when the
   variables are interpreted as ranging over $|A|$, $\in$ is interpreted as $\in_A$,
   $=$ is interpreted as

$$=_A = \{\langle a, a \rangle : a \in |A|\},$$

   and the parameters are interpreted as themselves.
3. The model A is said to be *well founded* if there is no infinite sequence
   $\langle a_n : n < \omega \rangle$ such that $a_{n+1} \in_A a_n$ for all $n < \omega$.
4. The model A is said to be *transitive* if $|A|$ is a transitive set and

$$\in_A = \in \restriction |A|.$$

It can be shown (in ZF, for instance) that any transitive L$_{\text{set}}$-structure
satisfies the Axiom of Extensionality and is well founded. Conversely, any
well founded L$_{\text{set}}$-structure which satisfies the Axiom of Extensionality is
uniquely isomorphic to a transitive model.

DEFINITION VII.3.26. To any model $A = (|A|, \in_A)$ of B$_0^{\text{set}}$ we canoni-
cally associate a model

$$A^2 = M = (|M|, \mathcal{S}_M, +_M, \cdot_M, 0_M, 1_M, <_M)$$

of ACA$_0$. Namely $|M| = \{a \in |A| : A \models \text{FinOrd}(a)\}$; $\mathcal{S}_M = \{b_A : A \models b \subseteq \omega\}$ where $b_A = \{a \in |A| : A \models a \in b\}$; and $+_M, \cdot_M, 0_M, 1_M, <_M$
are defined in the natural way (cf. the proof of theorem VII.3.9).

THEOREM VII.3.27. *Let A be a model of* ATR$_0^{\text{set}}$. *Then*:
1. A^2 *is a model of* ATR$_0$;
2. A^2 *is a β-model if and only if A is well founded*.

PROOF. Part 1 is an immediate consequence of theorem VII.3.9. Part 2
follows easily. □

In the opposite direction, we have:

DEFINITION VII.3.28. To any model

$$M = (|M|, \mathcal{S}_M, +_M, \cdot_M, 0_M, 1_M, <_M)$$

for L_2 we associate a model M_{set} for L_{set} as follows. Put

$$\mathcal{T}_M = \{T \in \mathcal{S}_M : M \models T \text{ is a suitable tree}\}$$

For $T \in \mathcal{T}_M$ put

$$[T] = \{T' \in \mathcal{T}_M : M \models T =^* T'\}$$

and define

$$|A| = \{[T] : T \in \mathcal{T}_M\}.$$

For $T, T' \in \mathcal{T}_M$ define $[T] \in_A [T']$ if and only if $M \models T \in^* T'$. Thus $A = (|A|, \in_A)$ is a model for L_{set}, and we define

$$M_{\text{set}} = A = (|A|, \in_A).$$

It can be shown that if M is model of ACA_0, then M_{set} is a model of $\mathsf{B}_0^{\text{set}}$.

THEOREM VII.3.29. *Let M be a model of* ATR_0. *Then M_{set} is a model of* $\mathsf{ATR}_0^{\text{set}}$. *Furthermore* $(M_{\text{set}})^2 = M$ *up to a canonical isomorphism. Conversely, if A is a model of* $\mathsf{ATR}_0^{\text{set}}$, *then* $(A^2)_{\text{set}} = A$ *up to a canonical isomorphism.*

PROOF. This is an immediate consequence of lemmas VII.3.20, VII.3.19, VII.3.12 and VII.3.14. □

DEFINITION VII.3.30. Let $A = (|A|, \in_A)$ and $B = (|B|, \in_B)$ be models for L_{set}. We say that A is a *transitive submodel* of B, written $A \subseteq_{\text{trans}} B$, if $|A| \subseteq |B|$ and, for all $a \in |A|$ and $b \in |B|$, $b \in_B a$ if and only if $b \in |A|$ and $b \in_A a$.

The above notion of transitive submodel ($\subseteq_{\text{trans}}$) is similar to the notion of β-submodel ($\subseteq_\beta$, definition VII.1.11). Thus A is transitive if and only if $A \subseteq_{\text{trans}}$ the universe of ZF set theory. But in general, the models A and B in the above definition need not be transitive or even well founded.

THEOREM VII.3.31. *If M' is a model of* ATR_0 *and $M \subseteq_\beta M'$, then M_{set} is (canonically isomorphic to) a transitive submodel of M'_{set}. Conversely, if A and B are models of* $\mathsf{ATR}_0^{\text{set}}$ *and $A \subseteq_{\text{trans}} B$, then A^2 is a β-submodel of B^2.*

PROOF. The formula "T is a suitable tree" is Π_1^1. Hence $\mathcal{T}_M = \mathcal{S}_M \cap \mathcal{T}_{M'}$ and the first part of the theorem follows easily. The second part follows using Axiom Beta in A and the Axiom of Regularity in B. □

Combining this with theorem VII.1.19, we obtain:

THEOREM VII.3.32. *Let B be any countable model of* $\mathsf{ATR}_0^{\text{set}}$. *Then there exists a proper transitive submodel $A \subseteq_{\text{trans}} B$, $A \neq B$, such that A is again a model of* $\mathsf{ATR}_0^{\text{set}}$.

PROOF. Theorems VII.3.27, VII.3.29 and VII.3.31 establish a canonical one-to-one correspondence between models of $\mathsf{ATR}_0^{\text{set}}$ and models of ATR_0. Applying this to theorem VII.1.19, we obtain the desired result. □

We shall now end this section by generalizing its main result so as to apply to systems of second order arithmetic which are stronger than ATR$_0$.

DEFINITION VII.3.33. Let T_0 be any theory in the language L_2 such that ATR$_0 \subseteq T_0$, i.e., each axiom of ATR$_0$ is a theorem of T_0. Define

$$T_0^{\text{set}} = \text{ATR}_0^{\text{set}} + T_0,$$

i.e., T_0^{set} is that theory in the language L_{set} whose axioms are those of ATR$_0^{\text{set}}$ plus (the natural translations into L_{set}, as in theorem VII.3.9, of) those of T_0.

THEOREM VII.3.34. *Let T_0 be any L_2-theory which includes ATR$_0$. Then lemmas VII.3.12, VII.3.14, VII.3.16, VII.3.17, VII.3.19, VII.3.20 and theorems VII.3.9, VII.3.22, VII.3.23, VII.3.24, VII.3.27, VII.3.29, VII.3.31 continue to hold when ATR$_0$ and ATR$_0^{\text{set}}$ are replaced by T_0 and T_0^{set} respectively. In particular, T_0^{set} is a conservative extension of T_0. Also, for all sentences φ of L_{set}, T_0^{set} proves φ if and only if T_0 proves $|\varphi|$.*

For example, if M is any L_2-model, we have $M \models \Pi_1^1\text{-CA}_0$ if and only if $M_{\text{set}} \models \Pi_1^1\text{-CA}_0^{\text{set}}$. Note also that M is a β-model if and only if M_{set} is well founded. (Compare this with our earlier characterization of β-models of $\Pi_1^1\text{-CA}_0$, theorem VII.1.8.)

PROOF. The results for $T_0 \supseteq \text{ATR}_0$ are all immediate corollaries of the special case $T_0 = \text{ATR}_0$.   □

REMARK VII.3.35. Theorem VII.3.32 does not in general hold with ATR$_0^{\text{set}}$ replaced by T_0^{set}. For example, let M be the unique minimum β-model of $\Pi_1^1\text{-CA}_0$ (corollary VII.1.8). Then by theorem VII.3.34, M_{set} is the unique smallest (up to canonical isomorphism) well founded model of $\Pi_1^1\text{-CA}_0^{\text{set}}$. In particular, there is no proper transitive submodel of M_{set} which is again a model of $\Pi_1^1\text{-CA}_0^{\text{set}}$.

EXERCISE VII.3.36. Show that $\Pi_1^1\text{-CA}_0^{\text{set}}$ is equivalent to ATR$_0^{\text{set}}$ plus the axiom

$$\forall v \, \exists u \, (v \in u \wedge \text{Trans}(u) \wedge \langle u, \in \lceil u \rangle \models \text{ATR}_0^{\text{set}}).$$

Hint: Use theorem VII.2.10.

EXERCISE VII.3.37. Give a characterization of $\Pi_1^1\text{-TR}_0^{\text{set}}$ analogous to the above characterization of $\Pi_1^1\text{-CA}_0^{\text{set}}$. (See exercises VII.1.17 and VII.1.18.)

EXERCISE VII.3.38. Recall that $\Pi_\infty^1\text{-TI}_0$ is the L_2 theory consisting of ACA$_0$ plus the transfinite induction scheme (see §VII.2). Show that Π_∞^1-TI$_0^{\text{set}}$ is equivalent to ATR$_0^{\text{set}}$ plus the $\in$-*induction scheme*

$$\forall x \, (\forall y \, (y \in x \rightarrow \varphi(y)) \rightarrow \varphi(x)) \rightarrow \forall x \, \varphi(x),$$

where $\varphi(x)$ is an arbitrary formula of L_{set}.

EXERCISE VII.3.39. Show that $\Sigma_2^1\text{-AC}_0^{\text{set}}$ is equivalent to $\text{ATR}_0^{\text{set}}$ plus the scheme of Σ_1^{set} *collection*, i.e.,

$$\forall x\, \exists y\, \varphi(x, y) \rightarrow \forall u\, \exists v\, \forall x\, (x \in u \rightarrow \exists y\, (y \in v \wedge \varphi(x, y)))$$

where $\varphi(x, y)$ is any Σ_1^{set} formula and v is a variable which does not occur freely in $\varphi(x, y)$.

EXERCISE VII.3.40. Characterize T_0^{set} when T_0 is any of the following L_2-theories: $\Pi_{k+1}^1\text{-CA}_0$, $\Delta_{k+2}^1\text{-CA}_0$, $\Pi_k^1\text{-TR}_0$, $\Sigma_{k+2}^1\text{-AC}_0$, $\Sigma_{k+2}^1\text{-DC}_0$, $\Pi_\infty^1\text{-}$ CA_0, $\Sigma_\infty^1\text{-AC}_0$, $\Sigma_\infty^1\text{-DC}_0$. (See also §§VII.5, VII.6 and VII.7.)

Notes for §VII.3. The ideas of this section can be traced to the work of Gödel [97, note 1] and Addison [4] relating the projective hierarchy to constructible sets. The fact that Axiom Beta is provable in ZF is due to Mostowski [192]. (Note: We use ZF to denote Zermelo/Fraenkel set theory including the Axiom of Regularity but not the axiom of choice.) In this context Axiom Beta is known as the Mostowski collapsing lemma. Also due to Mostowski [193] is the canonical one-to-one correspondence between β-models of $\Sigma_\infty^1\text{-AC}_0$ and well founded models of B_0^{set} plus the Axiom of Regularity plus the Axiom of Countability plus $\Sigma_\infty^{\text{set}}$ collection. Barwise/Fisher [14] (see also Barwise [13, §V.8]) used Axiom Beta in their analysis of Shoenfield's absoluteness theorem (see also theorem VII.4.12 below). See also Abramson/Sacks [3]. The system $\text{ATR}_0^{\text{set}}$ and the idea of considering Axiom Beta as an alternative to Σ_1^{set} collection are due to Simpson [234] and independently to McAloon/Ressayre [183]. The one-to-one correspondence between models of ATR_0 and models of $\text{ATR}_0^{\text{set}}$ is due to Simpson [234]. Theorems VII.1.19 and VII.3.32 are due to Simpson [234] in answer to a question of McAloon/Ressayre [183].

VII.4. Constructible Sets and Absoluteness

We begin this section by developing some of the basic properties of Gödel's hierarchy of constructible sets within $\text{ATR}_0^{\text{set}}$. We then show that some of the more advanced properties, such as the Shoenfield absoluteness theorem, can be proved within $\Pi_1^1\text{-CA}_0^{\text{set}}$.

The reader of this section is assumed to be familiar with the definitions and results of §VII.3 above. In addition, some previous knowledge of constructible sets would be helpful although perhaps not absolutely indispensable.

LEMMA VII.4.1. *The following is provable in* $\text{ATR}_0^{\text{set}}$. *Let* u *be a nonempty transitive set. There exists a unique set* $\text{def}(u)$ *consisting of all* $v \subseteq u$ *such that* v *is definable over the model* $\langle u, \in \restriction u \rangle$ *by a formula of* L_{set} *with parameters from* u.

Proof. We reason within ATR_0^{set}. Let u be a given nonempty transitive set. By the Axiom of Countability, let g be an injection such that $dom(g) = u$ and $rng(g) \subseteq \omega$.

We shall employ a language L_{set}^u which consists of L_{set} augmented by constant symbols denoting the elements of u. We shall identify terms and formulas with their Gödel numbers. For each $i < \omega$, we have a variable $(0, i)$ denoted v_i and intended to range over u. For each $a \in u$ we have a constant symbol $(1, g'a)$ denoted $\underline{a}$ and intended to denote a. The terms of L_{set}^u are the variables v_i, $i < \omega$ and the constant symbols $\underline{a}$, $a \in u$. For all terms s and t we have formulas $(2, (s, t))$ and $(3, (s, t))$ denoted $s = t$ and $s \in t$ respectively. For all formulas φ and ψ we have formulas $(4, \varphi)$ and $(5, (\varphi, \psi))$ denoted $\neg\varphi$ and $\varphi \wedge \psi$ respectively. For all formulas φ and variables v_i we have a formula $(6, (v_i, \varphi))$ denoted $\forall v_i \varphi$. A sentence of L_{set}^u is a formula of L_{set}^u with no free variables. Let S^u be the set of sentences of L_{set}^u, and let F^u be the set of formulas of L_{set}^u with at most one free variable. If $\varphi(v_i) \in F^u$ with free variable v_i, then for each $a \in u$, $\varphi(\underline{a})$ is the sentence obtained by substituting the constant symbol $\underline{a}$ for each free occurrence of v_i.

By arithmetical transfinite recursion (theorem VII.3.9), there exists a valuation $f : S^u \to \{0, 1\}$ satisfying the following inductive clauses:

$$f(\underline{a} = \underline{b}) = \begin{cases} 1 & \text{if } a = b, \\ 0 & \text{if } a \neq b; \end{cases}$$

$$f(\underline{a} \in \underline{b}) = \begin{cases} 1 & \text{if } a \in b, \\ 0 & \text{if } a \notin b; \end{cases}$$

$$f(\neg\varphi) = 1 - f(\varphi);$$

$$f(\varphi \wedge \psi) = \begin{cases} 1 & \text{if } f(\varphi) = f(\psi) = 1, \\ 0 & \text{otherwise}; \end{cases}$$

$$f(\forall v_i \, \varphi(v_i)) = \begin{cases} 1 & \text{if } f(\varphi(\underline{a})) = 1 \text{ for all } a \in u, \\ 0 & \text{otherwise}. \end{cases}$$

By arithmetical transfinite induction, f is unique. Let T^u be the suitable tree consisting of $\langle\rangle$ plus all $\langle\varphi(v_i)\rangle$ such that $\varphi(v_i) \in F^u$, plus all $\langle\varphi(v_i), \underline{a_0}, \ldots, \underline{a_k}\rangle$ such that $\varphi(v_i) \in F^u$, $f(\varphi(\underline{a_0})) = 1$, and $f(\underline{a_{i+1}} \in \underline{a_i}) = 1$ for all $i < k$. Then clearly $|T^u| = def(u)$. This proves the existence of $def(u)$. The uniqueness is straightforward. (Note that, although we used g to prove the existence of $def(u)$, the set $def(u)$ does not depend on the choice of g.) $\square$

Lemma VII.4.2 (the constructible hierarchy). *The following is provable in ATR_0^{set}. Let u be a nonempty transitive set. Let γ be an ordinal. There* *exists a unique set* L_γ^u *such that*

$$\exists f\ (\mathrm{Fcn}(f) \wedge \mathrm{dom}(f) = \gamma + 1 \wedge f'\gamma = L_\gamma^u$$
$$\wedge\ f'0 = u \wedge \forall \alpha\ (\alpha < \gamma \rightarrow f'\alpha + 1 = \mathrm{def}(f'\alpha))$$
$$\wedge\ \forall \delta\ ((\delta \leq \gamma \wedge \mathrm{Lim}(\delta)) \rightarrow f'\delta = \bigcup f''\delta)).$$

PROOF. We reason within $\mathrm{ATR}_0^{\mathrm{set}}$. Let u be a nonempty transitive set and let γ be an ordinal. By the Axiom of Countability, let g be an injection such that $\mathrm{dom}(g) = \gamma \cup u$ and $\mathrm{rng}(g) \subseteq \omega$.

We shall employ a ramified language. For each $i < \omega$ and $\alpha < \gamma$, we have a variable $(0, (i, g'\alpha))$ denoted v_i^α and intended to range over L_α^u. For each $a \in u$ we have a constant symbol $(1, g'a)$ denoted $\underline{a}$ and intended to denote a. Each variable v_i^α, $i < \omega$, $\alpha < \gamma$ is a term of rank α. Each constant symbol $\underline{a}$, $a \in u$, is a closed term of rank 0. There are other closed terms, to be described below. For all terms s and t, we have formulas $(2, (s, t))$ and $(3, (s, t))$ denoted $s = t$ and $s \in t$ respectively. For all formulas φ and ψ, we have formulas $(4, \varphi)$ and $(5, (\varphi, \psi))$ denoted $\neg\varphi$ and $\varphi \wedge \psi$ respectively. For all formulas φ and variables v_i^α we have a formula $(6, (v_i^\alpha, \varphi))$ denoted $\forall v_i^\alpha\ \varphi$. The rank of a formula is defined to be the maximum of the ranks of the terms occurring in it. If $\varphi(v_i^\alpha)$ is a formula of rank α with unique free variable v_i^α, we have a closed term $(7, (v_i^\alpha, \varphi))$ denoted $\{v_i^\alpha : \varphi(v_i^\alpha)\}$ and intended to denote

$$\{x : x \in L_\alpha^u \wedge \langle L_\alpha^u, \in \restriction L_\alpha^u \rangle \models \varphi(x)\};$$

this will be a typical element of $\mathrm{def}(L_\alpha^u) = L_{\alpha+1}^u$. The rank of the closed term $\{v_i^\alpha : \varphi(v_i^\alpha)\}$ is $\alpha + 1$.

Let S_γ^u be the set of all sentences of rank $< \gamma$. Let F_γ^u be the set of all formulas $\varphi(v_i^\alpha)$ of rank α with at most one free variable v_i^α, $i < \omega$, $\alpha < \gamma$.

By arithmetical transfinite recursion (theorem VII.3.9), there exists a valuation $f : S_\gamma^u \rightarrow \{0, 1\}$ satisfying:

$$f(\neg\varphi) = 1 - f(\varphi);$$

$$f(\varphi \wedge \psi) = \begin{cases} 1 & \text{if } f(\varphi) = f(\psi) = 1, \\ 0 & \text{otherwise;} \end{cases}$$

$$f(\forall v_i^\alpha\ \varphi(v_i^\alpha)) = \begin{cases} 1 & \text{if } f(\varphi(s)) = 1 \text{ for all closed terms } s \text{ of rank } \leq \alpha, \\ 0 & \text{otherwise.} \end{cases}$$

and, for all closed terms s and t, the following inductive clauses for $f(s = t)$ and $f(s \in t)$.

Case 1: $\text{rank}(s) = \alpha + 1$ and $\text{rank}(t) = \beta + 1$. Say $s = \{v_i^\alpha : \varphi(v_i^\alpha)\}$ and $t = \{v_j^\beta : \psi(v_j^\beta)\}$. Assume for convenience that $i \neq j$. Then

$$f(s = t) = \begin{cases} f(\forall v_i^\alpha\, (\varphi(v_i^\beta) \leftrightarrow v_i^\alpha \in t)) & \text{if } \alpha > \beta, \\ f(\forall v_i^\alpha\, \forall v_j^\beta\, (v_i^\alpha = v_j^\beta \rightarrow (\varphi(v_i^\alpha) \leftrightarrow \psi(v_j^\beta)))) & \text{if } \alpha = \beta, \\ f(\forall v_j^\beta\, (v_j^\beta \in s \leftrightarrow \varphi(v_j^\beta))) & \text{if } \alpha < \beta; \end{cases}$$

$$f(s \in t) = \begin{cases} f(\exists v_j^\beta\, (v_j^\beta = s \wedge \psi(v_j^\beta))) & \text{if } \alpha < \beta, \\ f(\exists v_j^\beta\, (\forall v_i^\alpha\, (v_i^\alpha \in v_j^\beta \leftrightarrow \varphi(v_i^\alpha)) \wedge \psi(v_j^\beta))) & \text{if } \alpha \geq \beta. \end{cases}$$

Case 2: $\text{rank}(s) = \alpha + 1$ and $\text{rank}(t) = 0$. Say $s = \{v_i^\alpha : \varphi(v_i^\alpha)\}$. Put $j = i + 1$. Then

$$f(s = t) = f(\forall v_i^\alpha\, (\varphi(v_i^\alpha) \leftrightarrow v_i^\alpha \in t));$$
$$f(s \in t) = f(\exists v_j^0\, (\forall v_i^\alpha\, (v_i^\alpha \in v_j^0 \leftrightarrow \varphi(v_i^\alpha)) \wedge v_j^0 \in t)).$$

Case 3: $\text{rank}(s) = 0$ and $\text{rank}(t) = \beta + 1$. Say $t = \{v_j^\beta : \psi(v_j^\beta)\}$. Then

$$f(s = t) = f(\forall v_j^\beta\, (v_j^\beta \in s \leftrightarrow \psi(v_j^\beta)));$$
$$f(s \in t) = f(\exists v_j^\beta\, (v_j^\beta = s \wedge \psi(v_j^\beta))).$$

Case 4: $\text{rank}(s) = \text{rank}(t) = 0$. Say $s = \underline{a}$ and $t = \underline{b}$ where $a, b \in u$. Then

$$f(s = t) = \begin{cases} 1 & \text{if } a = b, \\ 0 & \text{if } a \neq b; \end{cases}$$

$$f(s \in t) = \begin{cases} 1 & \text{if } a \in b, \\ 0 & \text{if } a \notin b. \end{cases}$$

This completes the definition of the valuation $f : S_\gamma^u \rightarrow \{0, 1\}$. By arithmetical transfinite induction, f is unique.

Let T_γ^u be the suitable tree consisting of $\langle \rangle$ plus all $\langle \varphi(v_i^u) \rangle$ such that $i < \omega$, $\alpha < \gamma$, $\varphi(v_i^\alpha) \in F_\gamma^u$, plus all $\langle \varphi(v_i^\alpha), t_0, \ldots, t_k \rangle$ such that $\varphi(v_i^\alpha) \in F_\gamma^u$, $f(\varphi(t_0)) = 1$, and $f(t_{i+1} \in t_i) = 1$ for all $i < k$. It is straightforward to prove that $|T_\gamma^u| = L_\gamma^u$. This gives the existence of L_γ^u, and the uniqueness is straightforward by transfinite induction on γ. (As in lemma VII.4.1, although g was used to prove the existence of L_γ^u, the set L_γ^u is independent of the choice of g.) □

THEOREM VII.4.3. *The following is provable in* $\text{ATR}_0^{\text{set}}$. *Let u and v be nonempty transitive sets.*

1. $u \cup \{u\} \subseteq \text{def}(u)$.
2. $\text{def}(u)$ *is transitive.*
3. $L_0^u = u$; $L_{\alpha+1}^u = \text{def}(L_\alpha^u)$.
4. $\text{Lim}(\delta) \rightarrow L_\delta^u = \bigcup_{\alpha < \delta} L_\alpha^u$.

5. $\alpha < \beta \rightarrow L^u_\alpha \in L^u_\beta$.
6. $\alpha \subseteq L^u_\alpha$; L^u_α *is transitive.*
7. $\mathrm{Lim}(\delta) \rightarrow L^u_\delta$ *is rudimentarily closed.*
8. $v = L^u_\alpha \rightarrow L^v_\beta = L^u_{\alpha+\beta}$.
9. $v \in L^u_\alpha \rightarrow L^v_\beta \in L^u_{\alpha+\beta}$.

PROOF. The proof is straightforward.   □

DEFINITION VII.4.4 (the inner model L^u). This definition is made in $\mathrm{ATR}^{\mathrm{set}}_0$. Let u be a nonempty transitive set. A set x is said to be *constructible from u*, written $x \in L^u$, if $\exists \alpha \, (x \in L^u_\alpha)$, i.e., $\exists \alpha \, \exists y \, (x \in y \wedge y = L^u_\alpha)$.

DEFINITION VII.4.5 (relativization to L^u). Let φ be any formula of L_{set}. By induction on the complexity of φ we define a formula φ^{L^u}, the *relativization of φ to L^u*, as follows:

$$(x = y)^{L^u} \text{ is } x = y;$$

$$(x \in y)^{L^u} \text{ is } x \in y;$$

$$(\neg\varphi)^{L^u} \text{ is } \neg(\varphi^{L^u});$$

$$(\varphi \wedge \psi)^{L^u} \text{ is } \varphi^{L^u} \wedge \psi^{L^u};$$

$$(\forall x \, \varphi)^{L^u} \text{ is } \forall x \, (x \in L^u \rightarrow \varphi^{L^u}).$$

Intuitively, φ^{L^u} means that φ is true in the transitive model $(L^u, \in {\restriction} L^u)$. We sometimes express φ^{L^u} by saying that L^u *satisfies* φ.

LEMMA VII.4.6. *In* $\mathrm{ATR}^{\mathrm{set}}_0$ *we have*:
1. *The formulas* $v = L^u_\alpha$, $x \in L^u_\alpha$, *and* $x \in L^u$ *are equivalent to* Σ^{set}_1 *formulas.*
2. *If* φ *is equivalent to a* Σ^{set}_k *formula,* $0 \leq k < \omega$, *then* φ^{L^u} *is equivalent to a* Σ^{set}_k *formula.*

PROOF. Part 1 is straightforward. We now deduce part 2. Using the fact that L^u is transitive, we see that for any Σ^{set}_0 formula φ, φ^{L^u} is equivalent to φ itself. Suppose now that φ is $\Sigma^{\mathrm{set}}_{k+1}$. Write φ as $\exists x \, \psi$ where ψ is Π^{set}_k. Then φ^{L^u} is equivalent to

$$\exists x \, (x \in L^u \wedge \psi^{L^u}).$$

By part 1, $x \in L^u$ is equivalent to a Σ^{set}_1 formula. By induction on k, ψ^{L^u} is equivalent to a Π^{set}_k formula. Hence φ^{L^u} is equivalent to a $\Sigma^{\mathrm{set}}_{k+1}$ formula. This completes the proof.   □

DEFINITION VII.4.7 (absoluteness). Within $\mathrm{ATR}^{\mathrm{set}}_0$, we say that φ *is absolute to* L^u if

$$\forall x_1 \cdots \forall x_m \, ((x_1 \in L^u \wedge \cdots \wedge x_m \in L^u) \rightarrow (\varphi \leftrightarrow \varphi^{L^u}))$$

holds, where $x_1, \ldots, x_m$ are the free variables of φ.

We shall sometimes write $V = L^u$ as an abbreviation for the formula

$$\forall x \, (x \in L^u).$$

THEOREM VII.4.8. *The following is provable in* $\mathsf{ATR}_0^{\mathrm{set}}$. *Let u be a nonempty transitive set. The formulas $x = L_\alpha^u$, $x \in L_\alpha^u$, and $x \in L^u$ are absolute to L^u.*

PROOF. By $x = L_\alpha^u$ we mean of course the Σ_1^{set} formula $\exists f \, (\mathrm{Fcn}(f) \wedge \mathrm{dom}(f) = \alpha + 1 \wedge f'\alpha = x \wedge f'0 = u \wedge \forall \beta \, (\beta < \alpha \to f'\beta + 1 = \mathrm{def}(f'\beta)) \wedge \forall \delta \, ((\delta \leq \alpha \wedge \mathrm{Lim}(\delta)) \to f'\delta = \bigcup f''\delta))$. Since L^u is transitive, every Δ_0^{set} formula is absolute to L^u. Using this, it is straightforward to check that each of the component formulas $\mathrm{Fcn}(w)$, $\mathrm{dom}(w) = \alpha + 1$, etc., including $w = \mathrm{def}(v)$ is absolute. It remains to show that the "constructing function" $f = f_\alpha$ is an element of L^u. But, by VII.4.3.7 and transfinite induction on α, it is straightforward to check that in fact $f_\alpha \in L_{\alpha+3}^u$. Hence $x = L_\alpha^u$ is absolute. The absoluteness of $x \in L_\alpha^u$ and of $x \in L^u$ follow immediately. $\square$

COROLLARY VII.4.9. *The following is provable in* $\mathsf{ATR}_0^{\mathrm{set}}$. *Let u be a nonempty transitive set. Then L^u satisfies $V = L^u$.*

PROOF. $V = L^u$ is an abbreviation for $\forall x \, (x \in L^u)$. Hence $(V = L^u)^{L^u}$ is equivalent to $\forall x \, (x \in L^u \to (x \in L^u)^{L^u})$. But by absoluteness of $x \in L^u$ this is equivalent to the tautology $\forall x \, (x \in L^u \to x \in L^u)$. $\square$

We now turn to the Shoenfield absoluteness theorem. We shall see that the formula "r is a regular relation" is, provably in $\Pi_1^1\text{-}\mathsf{CA}_0^{\mathrm{set}}$, absolute to L^u. This will be seen to imply that all Σ_2^1 formulas are absolute to L^u.

LEMMA VII.4.10. *The following is provable in* $\mathsf{ATR}_0^{\mathrm{set}}$. *Let u be a nonempty transitive set. Let r be a regular relation, and let f be the collapsing function of r. If $r \in L^u$, then $f \in L^u$.*

PROOF. Recall from definition VII.3.8 that the collapsing function of r is the unique function f such that $\mathrm{dom}(f) = \mathrm{field}(r)$ and, for all $x \in \mathrm{field}(r)$,

$$f'x = f''\{y \colon \langle y, x \rangle \in r\}.$$

By the Axiom of Countability, let g be an injection such that $\mathrm{dom}(g) = \mathrm{field}(r)$ and $\mathrm{rng}(g) \subseteq \omega$. Let S be the suitable tree consisting of $\langle \rangle$ plus all $\langle g'x_0, \ldots, g'x_k \rangle$ such that $\forall i \, (i < k \to \langle x_{i+1}, x_i \rangle \in r)$. Let $\mathrm{KB}(S)$ be the Kleene-Brouwer ordering of S. By lemma V.1.3, $\mathrm{KB}(S)$ is a well ordering. Hence by Axiom Beta there is a unique function h such that $\mathrm{dom}(h) = S$ and, for all $\sigma \in S$,

$$h'\sigma = h''\{\tau \colon \tau <_{\mathrm{KB}(S)} \sigma\}.$$

Put $\gamma = \mathrm{rng}(h)$. Clearly γ is an ordinal, the order type of $\mathrm{KB}(S)$, and h is the unique order isomorphism of $\mathrm{KB}(S)$ onto γ.

Let r^* be the transitive closure of r, i.e.,

$$r^* = \{\langle y, x\rangle \colon \exists k\, \exists s\, (k \in \omega \wedge \mathrm{Fcn}(s) \wedge \mathrm{dom}(s) = k + 2$$
$$\wedge\, s'0 = x \wedge \forall i\, (i \leq k \rightarrow \langle s'i + 1, s'i\rangle \in r) \wedge s'k + 1 = y\}.$$

For each $x \in \mathrm{field}(r)$, put

$$r^*_x = \{x\} \cup \{y \colon \langle y, x\rangle \in r^*\}.$$

Let α be such that $r \in L^u_\alpha$. The clearly $r^* \in L^u_{\alpha+\omega}$ and $r^*_x \in L^u_{\alpha+\omega}$ for all $x \in \mathrm{field}(r)$.

We claim that, for all $\beta < \gamma$ and $\sigma = \langle g'x_0, \dots, g'x_k\rangle \in S$, if $h'\sigma \leq \beta$ then

$$f \restriction r^*_{x_k} \in L^u_{\alpha+\omega\cdot(1+\beta)+5}.$$

We prove this by transfinite induction on β. Let $\sigma = \langle g'x_0, \dots, g'x_k\rangle \in S$ be such that $h'\sigma \leq \beta$. Let y be such that $\langle y, x_k\rangle \in r$. Then $\sigma^\frown\langle g'y\rangle \in S$ and $h'\sigma^\frown\langle g'y\rangle < \beta$. Hence by inductive hypothesis $f \restriction r^*_y \in L^u_{\alpha+\omega\cdot(1+\beta)}$. From this and VII.4.3.6 it follows that $f \restriction r^*_x \in L^u_{\alpha+\omega\cdot(1+\beta)+5}$. This proves the claim.

In particular, $f \restriction r^*_x \in L^u_{\alpha+\omega\cdot(1+\gamma)}$ for all $x \in \mathrm{field}(r)$. From this it follows that $f \in L^u_{\alpha+\omega\cdot(1+\gamma)+1}$. This proves lemma VII.4.10.   □

LEMMA VII.4.11. *The following is provable in* Π^1_1-$\mathrm{CA}^{\mathrm{set}}_0$. *Let u be a nonempty transitive set. Let r be a relation which is not regular. If $r \in L^u$, then there exists $v \in L^u$ such that $v \neq \emptyset$ and*

$$\forall x\, (x \in v \rightarrow \exists y\, (\langle y, x\rangle \in r \wedge y \in v)). \tag{20}$$

PROOF. Reasoning in Π^1_1-$\mathrm{CA}^{\mathrm{set}}_0$, let $r \in L^u$ be a relation which is not regular. By the Axiom of Countability, let g be an injection such that $\mathrm{dom}(g) = \mathrm{field}(r)$ and $\mathrm{rng}(g) \subseteq \omega$. Let T be the tree consisting of $\langle\rangle$ plus all $\langle g'x_0, \dots, g'x_k\rangle$ such that $\forall i\, (i < k \rightarrow \langle x_{i+1}, x_i\rangle \in r)$. By Π^1_1 comprehension, let W be the set of all m such that $\langle m\rangle \in T$ and $T^{\langle m\rangle}$ is suitable. For each $x \in \mathrm{field}(r)$, put $r_x = r \restriction r^*_x$ where r^*_x is as in the proof of lemma VII.4.10. Thus $g'x \in W$ if and only if r_x is regular. Put

$$w = \{x \colon x \in \mathrm{field}(r) \wedge g'x \in W\}$$

and $v = \mathrm{field}(r) - w$. Thus $v = \{x \colon r_x$ is not regular$\}$. Since r is not regular, we have $v \neq \emptyset$ and (20). It remains to show that $v \in L^u$.

Let S be the suitable tree consisting of $\langle\rangle$ plus all $\langle m\rangle^\frown\sigma$ such that $m \in W$ and $\sigma \in T^{\langle m\rangle}$. As in the proof of lemma VII.4.10, let γ be the order type of $\mathrm{KB}(S)$. Let α be such that $r \in L^u_\alpha$, and put $\beta = \alpha+\omega\cdot(1+\gamma)$. As in the proof of lemma VII.4.10, we see that for all $x \in \mathrm{field}(r)$, r_x is regular if and only if

$$\exists f\, (f \in L^u_{\beta+1} \wedge f \text{ is the collapsing function of } r_x).$$

Hence $v = \{x \colon r_x \text{ is not regular}\} \in L^u_{\beta+2}$. This completes the proof of lemma VII.4.11.   □

THEOREM VII.4.12. *The following is provable in* $\Pi^1_1\text{-CA}^{set}_0$. *Let u be a nonempty transitive set. Then L^u satisfies Axiom Beta. Moreover, for all relations $r \in L^u$, we have*

$$r \text{ is regular} \leftrightarrow (r \text{ is regular})^{L^u}.$$

PROOF. This follows immediately from theorem VII.4.3 and lemmas VII.4.10 and VII.4.11.   □

LEMMA VII.4.13. *The following is provable in* $\Pi^1_1\text{-CA}^{set}_0$. *Let u be a nonempty transitive set. Let $\varphi(X)$ be a Π^1_1 formula with parameters from L^u and no free variables other than X. Then we have*

$$\exists X \, \varphi(X) \rightarrow \exists X \, (X \in L^u \wedge \varphi(X)).$$

PROOF. We reason within $\Pi^1_1\text{-CA}^{set}_0$. By lemma V.1.4 (the Kleene normal form theorem), we can write $\varphi(X)$ in the form $\varphi(X) \equiv \forall f \, \exists n \, \neg\theta(X[n], f[n])$ where $\theta(\sigma, \tau)$ is arithmetical with parameters from L^u and no free variables other than σ and τ. For each $\sigma \in 2^{<\omega}$, let T_σ be the finite tree consisting of $\langle\rangle$ plus all $\tau \in \omega^{<\omega}$ such that

(i) $\tau < \text{lh}(\sigma)$ (viewing τ as an element of ω); and
(ii) $\forall n \, (n \le \text{lh}(\tau) \rightarrow \theta(\sigma[n], \tau[n]))$.

Given $X \in 2^\omega$, put $T_X = \bigcup_{n \in \omega} T_{X[n]}$. Thus by lemma V.1.3 we have

$$\varphi(X) \leftrightarrow T_X \text{ is suitable}$$
$$\leftrightarrow KB(T_X) \text{ is a well ordering.}$$

Fix $X_0 \in 2^\omega$ such that $\varphi(X_0)$ holds. Then $KB(T_{X_0})$ is a well ordering, so by Axiom Beta let α_0 be the ordinal which is the order type of $KB(T_{X_0})$, and let g_0 be the unique order isomorphism of $KB(T_{X_0})$ onto α_0. Thus g_0 is an injection, $\text{dom}(g_0) = T_{X_0}$, $\text{rng}(g_0) = \alpha_0$, and

$$g'_0\tau_1 < g'_0\tau_2 \leftrightarrow \tau_1 <_{KB(T_{X_0})} \tau_2$$

for all $\tau_1, \tau_2 \in T_{X_0}$.

Let d be the set of all ordered pairs $\langle \sigma, s \rangle$ such that $\sigma \in 2^{<\omega}$, $\text{Fcn}(s)$, $\text{dom}(s) = T_\sigma$, $\text{rng}(s) \subseteq \alpha_0$, and

$$s'\tau_1 < s'\tau_2 \leftrightarrow \tau_1 <_{KB(T_\sigma)} \tau_2$$

for all $\tau_1, \tau_2 \in T_\sigma$. Let r be the set of all $\langle\langle \sigma', s' \rangle, \langle \sigma, s \rangle\rangle \in d \times d$ such that $\sigma \subseteq \sigma'$, $\sigma \ne \sigma'$, and $s \subseteq s'$. Since $\alpha_0 \in L^u$, we have $d \in L^u$ and $r \in L^u$.

Let v_0 be the set of all ordered pairs $\langle X_0[n], g_0 \restriction T_{X_0} \rangle$, $n \in \omega$. Then clearly $v_0 \ne \emptyset$ and

$$\forall x \, (x \in v_0 \rightarrow \exists y \, (y \in v_0 \wedge \langle y, x \rangle \in r)).$$

Hence r is not regular. Hence by lemma VII.4.11 there exists $v \in L^u$ such that $v \neq \emptyset$ and

$$\forall x \, (x \in v \rightarrow \exists y \, (\langle y, x \rangle \in r \wedge y \in v)).$$

Since $d = \text{field}(r)$ is well ordered in L^u, we can find a sequence $\langle \langle \sigma_n, s_n \rangle : n < \omega \rangle \in L^u$ such that $\langle \sigma_n, s_n \rangle \in v$ and $\langle \langle \sigma_{n+1}, s_{n+1} \rangle, \langle \sigma_n, s_n \rangle \rangle \in r$ for all $n < \omega$. Putting $X = \bigcup_{n \in \omega} \sigma_n$ and $g = \bigcup_{n \in \omega} s_n$, we see that $X \in 2^\omega$, $\text{dom}(g) = T_X$, $\text{rng}(g) \subseteq \alpha_0$, and

$$g' \tau_1 < g' \tau_2 \leftrightarrow \tau_1 <_{\text{KB}(T_X)} \tau_2$$

for all $\tau_1, \tau_2 \in T_X$. Hence T_X is suitable, i.e., $\varphi(X)$ holds. Also $X \in L^u$ by construction. This completes the proof of lemma VII.4.13. □

The next theorem is our formalized version of the Shoenfield absoluteness theorem within $\Pi_1^1\text{-CA}_0^{\text{set}}$. It will be applied in the next section to prove conservation results.

THEOREM VII.4.14 (Shoenfield absoluteness in $\Pi_1^1\text{-CA}_0^{\text{set}}$). *The following is provable in* $\Pi_1^1\text{-CA}_0^{\text{set}}$. *Let u be a nonempty transitive set. Let φ be any Σ_2^1 sentence with parameters from* L^u. *Then $\varphi \leftrightarrow \varphi^{L^u}$, i.e., φ is absolute to L^u.*

PROOF. This is an immediate consequence of lemma VII.4.13. □

COROLLARY VII.4.15. *The following is provable in* $\Pi_1^1\text{-CA}_0^{\text{set}}$. *Let u be a nonempty transitive set. The transitive model L^u satisfies $V = L^u$ plus all axioms of* $\Pi_1^1\text{-CA}_0^{\text{set}}$ *except possibly the Axiom of Countability.*

PROOF. By corollary VII.4.9 and theorems VII.4.3 and VII.4.12, L^u satisfies $V = L^u$ plus $\text{ATR}_0^{\text{set}}$ except possibly for the Axiom of Countability. It remains to show that L^u satisfies $\Pi_1^1\text{-CA}_0$. By theorem VII.1.12 it suffices to prove that L^u is closed under hyperjump. Let $X \in L^u$, $X \subseteq \omega$ be given. By $\Pi_1^1\text{-CA}_0$ we have $\exists Y \, (Y = \text{HJ}(X))$. This is Σ_2^1 so by the Shoenfield absoluteness theorem VII.4.14, there exists $Y \in L^u$ such that L^u satisfies $Y = \text{HJ}(X)$. By another application of VII.4.14 it follows that $Y = \text{HJ}(X)$. This completes the proof. □

EXERCISE VII.4.16. Show that the following is provable in $\Pi_1^1\text{-TR}_0^{\text{set}}$. Let u be a nonempty transitive set. Then L^u satisfies $V = L^u$ plus all axioms of $\Pi_1^1\text{-TR}_0^{\text{set}}$ except possibly the Axiom of Countability.

REMARKS VII.4.17. Roughly speaking, the content of corollary VII.4.15 is that $\Pi_1^1\text{-CA}_0$ proves its own relativization to the inner model L^u. Exercise VII.4.16 gives the same result for $\Pi_1^1\text{-TR}_0$. In §VII.5 below, we shall obtain a similar result for stronger systems $\Delta_2^1\text{-CA}_0$, $\Pi_2^1\text{-CA}_0$, $\Pi_2^1\text{-TR}_0$, $\Delta_3^1\text{-CA}_0$, $\Pi_3^1\text{-CA}_0$, etc.

The phrase "except possibly the Axiom of Countability" cannot be dropped from corollary VII.4.15 or exercise VII.4.16. Indeed, Feferman and Lévy have exhibited a transitive model of $\Pi_\infty^1\text{-CA}_0^{\text{set}}$ (see definition VII.4.33 below) in which, for all nonempty transitive sets u, the Axiom of Countability fails in L^u. The transitive model exhibited by Feferman and Lévy is M_{set} where M is as in remark VII.6.3 below.

In order to restore the Axiom of Countability, we shall now pass to a smaller inner model $\text{HCL}(X)$, where $X \subseteq \omega$. See definition VII.4.22 and theorem VII.4.27 below. Some of our results about $\text{HCL}(X)$ will also be of use in §VII.5.

DEFINITION VII.4.18. Within $\text{ATR}_0^{\text{set}}$ we make the following definitions.

1. A *linear ordering* is a relation $\prec$ such that, writing $y \prec x$ for $\langle y, x \rangle \in \prec$, one has
   (i) $\forall x \, \forall y \, ((x \in \text{field}(\prec) \wedge y \in \text{field}(\prec)) \rightarrow (x \prec y \vee x = y \vee y \prec x))$;
   (ii) $\forall x \, \forall y \, \forall z \, ((z \prec y \wedge y \prec x) \rightarrow z \prec x)$; and
   (iii) $\neg \exists x \, (x \prec x)$.
2. If $\prec$ and $\prec^1$ are linear orderings, we say that $\prec$ is an *initial segment* of $\prec^1$ if $\prec = \prec^1 \restriction \text{field}(\prec)$. By definition VII.3.2.14 this implies

$$\forall x \, \forall y \, ((y \prec^1 x \wedge x \in \text{field}(\prec)) \rightarrow y \in \text{field}(\prec)).$$

3. A *linear ordering of u* is a linear ordering $\prec$ such that $\text{field}(\prec) = u$.
4. A *well ordering of u* is a linear ordering of u which is regular.

LEMMA VII.4.19. *Within* $\text{ATR}_0^{\text{set}}$, *let u be a nonempty transitive set.*

1. *Given a well ordering $\prec$ of u, we can canonically define a well ordering $\prec^*$ of $\text{def}(u)$. Moreover $\prec$ is an initial segment of $\prec^*$.*
2. *Given a well ordering $<_0$ of u, we can associate to each ordinal α a canonically defined well ordering $<_\alpha$ of L_α^u. Moreover, for all $\beta < \alpha$, $<_\beta$ is an initial segment of $<_\alpha$.*
3. *The definitions of $\prec^*$ and $<_\alpha$ are absolute to L^u (provided $\prec$ and $<_0$ belong to L^u).*

PROOF. This is essentially Gödel's argument for the axiom of choice.

To prove part 1, let F_u be the set of formulas of L_{set} with parameters from u and exactly one free variable. The given well ordering $\prec$ of u canonically induces a well ordering $\prec^F$ of F_u. For all $v \in \text{def}(u)$, let $h'v$ be the $\prec^F$-least formula $\varphi(x) \in F_u$ such that

$$v = \{x : x \in u \wedge \langle u, \in \restriction u \rangle \models \varphi(x)\}.$$

For $v \in \text{def}(u)$ and $w \in \text{def}(u)$, put $w \prec^* v$ if and only if either

(i) $w \in u \wedge v \in u \wedge w \prec v$, or
(ii) $w \in u \wedge v \in \text{def}(u) - u$, or
(iii) $w \in \text{def}(u) - u \wedge v \in \text{def}(u) - u \wedge h'w \prec^F h'v$.

Clearly $\prec^*$ has the desired properties.

For part 2, $<_\alpha$ is defined uniquely so that

$$\exists f\ (\mathrm{Fcn}(f) \wedge \mathrm{dom}(f) = \alpha + 1 \wedge f'\alpha =<_\alpha$$
$$\wedge f'0 =<_0 \wedge \forall \beta\ (\beta < \alpha \to f'\beta + 1 = (f'\beta)^*)$$
$$\wedge \forall \delta\ ((\delta \le \alpha \wedge \mathrm{Lim}(\delta)) \to f'\delta = \bigcup f''\delta)).$$

Here $*$ is as in part 1. The proof of part 3 is straightforward. □

DEFINITION VII.4.20. The following definition is made within $\mathrm{ATR}_0^{\mathrm{set}}$. Given $X \subseteq \omega$, note that $\omega \cup \{X\}$ is a nonempty transitive set and put $\mathrm{L}_\alpha(X) = \mathrm{L}_\alpha^{\omega \cup \{X\}}$ and $\mathrm{L}(X) = \mathrm{L}^{\omega \cup \{X\}}$. Let $<_0^{\mathrm{L}(X)}$ be the well ordering of $\omega \cup \{X\}$ given by

$$<_0^{\mathrm{L}(X)} = \begin{cases} \in \restriction \omega & \text{if } X \in \omega, \\ (\in \restriction \omega) \cup (\omega \times \{X\}) & \text{otherwise.} \end{cases}$$

Then for all ordinals α, let $<_\alpha^{\mathrm{L}(X)}$ be the canonically associated well ordering of $\mathrm{L}_\alpha(X)$ as in lemma VII.4.19. For $u \in \mathrm{L}(X)$ and $v \in \mathrm{L}(X)$, put $v <^{\mathrm{L}(X)} u$ if and only if $\exists \alpha\ (v <_\alpha^{\mathrm{L}(X)} u)$. We refer to $<^{\mathrm{L}(X)}$ as the *canonical well ordering* of $\mathrm{L}(X)$.

LEMMA VII.4.21. *Provably in* $\mathrm{ATR}_0^{\mathrm{set}}$, *the formulas* $u \in \mathrm{L}(X)$, $v <^{\mathrm{L}(X)} u$ *and* $w = \{v : v <^{\mathrm{L}(X)} u\}$ *are* Σ_1^{set} *and absolute to* $\mathrm{L}(X)$.

PROOF. This is similar to the proof of lemma VII.4.6.1 and theorem VII.4.8. □

DEFINITION VII.4.22 (the inner model $\mathrm{HCL}(X)$). Within $\mathrm{ATR}_0^{\mathrm{set}}$, assume $X \subseteq \omega$. We write $u \in \mathrm{HCL}(X)$ to mean that u is *hereditarily constructibly countable* from X, i.e.,

$$\exists f\ (f \in \mathrm{L}(X) \wedge \mathrm{Fcn}(f) \wedge \mathrm{dom}(f) = \omega \wedge u \subseteq \mathrm{rng}(f) \wedge \mathrm{Trans}(\mathrm{rng}(f))).$$

If φ is any formula of $\mathrm{L}_{\mathrm{set}}$, the *relativization of* φ *to* $\mathrm{HCL}(X)$ is written $\varphi^{\mathrm{HCL}(X)}$ and is defined in the obvious way, exactly as for $\varphi^{\mathrm{L}(X)}$ (definition VII.4.5). We sometimes express $\varphi^{\mathrm{HCL}(X)}$ by saying that $\mathrm{HCL}(X)$ *satisfies* φ.

Obviously $\mathrm{HCL}(X)$ is a transitive submodel of $\mathrm{L}(X)$, i.e., $\forall u\ (u \in \mathrm{HCL}(X) \to u \in \mathrm{L}(X))$ and $\forall u\ \forall v\ ((v \in u \wedge u \in \mathrm{HCL}(X)) \to v \in \mathrm{HCL}(X))$. In addition we have $\forall u\ \forall v\ ((u \in \mathrm{HCL}(X) \wedge v \subseteq u \wedge v \in \mathrm{L}(X)) \to v \in \mathrm{HCL}(X))$.

LEMMA VII.4.23. *Provably in* $\mathrm{ATR}_0^{\mathrm{set}}$ *we have:*

1. *The formula* $u \in \mathrm{HCL}(X)$ *is equivalent to a* Σ_1^{set} *formula.*
2. *If* φ *is any* Σ_k^{set} *formula,* $0 \le k < \omega$, *then* $\varphi^{\mathrm{HCL}(X)}$ *is equivalent to a* Σ_k^{set} *formula.*

PROOF. This is exactly like the proof of lemma VII.4.6. □

DEFINITION VII.4.24. We say that φ *is absolute to* $HCL(X)$ if

$$\forall x_1 \cdots \forall x_m \left((x_1 \in HCL(X) \wedge \cdots \wedge x_m \in HCL(X)) \rightarrow (\varphi \leftrightarrow \varphi^{HCL(X)})\right)$$

holds, where $x_1, \ldots, x_m$ are the free variables of φ.

We sometimes write $V = HCL(X)$ as an abbreviation for $\forall u \, (u \in HCL(X))$.

LEMMA VII.4.25. *Within* $\mathsf{ATR}_0^{\mathrm{set}}$, *assume* $X \subseteq \omega$. *If* φ *is any* Σ_1^{set} *sentence with parameters from* $HCL(X)$, *we have*

$$\varphi^{L(X)} \leftrightarrow \varphi^{HCL(X)}.$$

PROOF. Since $HCL(X) \subseteq_{\mathrm{trans}} L(X)$, it is clear $\varphi^{HCL(X)}$ implies $\varphi^{L(X)}$. We shall prove the converse. This will be essentially Gödel's argument for the continuum hypothesis.

Assume $\varphi^{L(X)}$. The parameters of φ belong to $HCL(X)$, so let $u \in HCL(X)$ be transitive and contain these parameters. Write φ as $\exists x \, \theta(x)$ where $\theta(x)$ is Δ_0^{set} with the same parameters as φ. Fix $z \in L(X)$ such that $\theta(z)$ holds in $L(X)$. Let δ be a limit ordinal such that $u \subseteq L_\delta(X)$ and $z \in L_\delta(X)$. Hence $\theta(z)$ holds in $L_\delta(X)$.

Let v be the smallest subset of $L_\delta(X)$ such that $u \subseteq v$, $z \in v$, and v is closed under definability in the language $=, \in, <$ over the model

$$\langle L_\delta(X), \in \upharpoonright L_\delta(X), <_\delta^{L(X)} \rangle.$$

Since $<_\delta^{L(X)}$ is a well ordering of $L_\delta(X)$, v is an elementary submodel of $\langle L_\delta(X), \in \upharpoonright L_\delta(X) \rangle$. In particular v satisfies $\theta(z)$. Also, since $<_\delta^{L(X)} \in L(X)$ and u is countable in $L(X)$, it follows that v is countable in $L(X)$, i.e., there exists a function $g \in L(X)$ such that $\mathrm{dom}(g) = \omega$ and $\mathrm{rng}(g) = v$.

Let f be the collapsing function of v, i.e., f is the unique function such that $\mathrm{dom}(f) = v$ and

$$f'x = f''\{y \colon y \in v \wedge y \in x\}$$

for all $x \in v$. Put $w = \mathrm{rng}(f)$. Thus w is transitive and f is the unique $\in$-isomorphism of v onto w. By lemma VII.4.10 we have $f \in L(X)$, hence $fg \in L(X)$. Since $\mathrm{dom}(fg) = \omega$ and $\mathrm{rng}(fg) = w$, it follows that $w \in HCL(X)$.

The transitivity of u implies that $y = f'y \in w$ for all $y \in u$. In particular this holds for all of the parameters y of $\theta(x)$. Therefore $\langle w, \in \upharpoonright w \rangle$ satisfies $\theta(f'z)$. Since $w \subseteq_{\mathrm{trans}} HCL(X)$, we see that $HCL(X)$ satisfies $\theta(f'z)$. Thus $HCL(X)$ satisfies $\exists x \, \theta(x)$, i.e., φ.

This completes the proof of lemma VII.4.25.   □

LEMMA VII.4.26. *The following is provable in* $\mathsf{ATR}_0^{\mathrm{set}}$. *For* $X \subseteq \omega$, *the formulas* $u \in HCL(X)$, $u = L_\alpha(X)$, $v <^{L(X)} u$, *and* $w = \{v \colon v <^{L(X)} u\}$ *are* Σ_1^{set} *and absolute to* $HCL(X)$. *In particular,* $HCL(X)$ *satisfies* $V = HCL(X)$.

PROOF. By theorem VII.4.8 and lemmas VII.4.6.1, VII.4.21, and VII.4.23.1, the mentioned formulas are Σ_1^{set} and absolute to $L(X)$. Hence by lemma VII.4.25 they are absolute to $\text{HCL}(X)$.   □

THEOREM VII.4.27. *The following is provable in Π_1^1-CA_0^{set}. Assume $X \subseteq \omega$. Then*:

1. $\text{HCL}(X)$ *satisfies* $V = \text{HCL}(X)$ *plus all axioms of Π_1^1-CA_0^{set}.*
2. *All Σ_2^1 and Σ_1^{set} formulas are absolute to $\text{HCL}(X)$. In other words,* $\varphi \leftrightarrow \varphi^{\text{HCL}(X)}$ *holds for all Σ_2^1 or Σ_1^{set} sentences φ with parameters from $\text{HCL}(X)$.*

PROOF. Part 1 follows from corollary VII.4.15 and lemma VII.4.26. Part 2 follows from part 1, theorem VII.4.14 (absoluteness of Σ_2^1 formulas) and theorem VII.3.24 (equivalence of Σ_2^1 with Σ_1^{set}).   □

EXERCISE VII.4.28. Within $\text{ATR}_0^{\text{set}}$, assuming $X \subseteq \omega$, show that $\text{HCL}(X)$ satisfies $V = \text{HCL}(X)$ plus all axioms of $\text{ATR}_0^{\text{set}}$ except possibly Axiom Beta.

EXERCISE VII.4.29. Exhibit a transitive model of $\text{ATR}_0^{\text{set}}$ in which $\text{HCL}(\emptyset)$ does not satisfy Axiom Beta. Show that $\text{ATR}_0^{\text{set}}$ proves the following: Π_1^1-CA_0 if and only if

$$\forall X \, (X \subseteq \omega \rightarrow (\text{Axiom Beta})^{\text{HCL}(X)}).$$

EXERCISE VII.4.30. Exhibit a transitive model of $\text{ATR}_0^{\text{set}}$ in which not all Σ_1^1 sentences are absolute to $\text{HCL}(\emptyset)$. Show that $\text{ATR}_0^{\text{set}}$ proves the following: Π_1^1-CA_0 if and only if all Σ_1^1 formulas are absolute to $\text{HCL}(X)$ for all $X \subseteq \omega$.

EXERCISE VII.4.31. Within $\text{ATR}_0^{\text{set}}$, assuming $X \subseteq \omega$, show that if the hyperjump $Y = \text{HJ}(X)$ exists then $Y \in \text{HCL}(X)$ and $\text{HCL}(X)$ satisfies $Y = \text{HJ}(X)$.

EXERCISE VII.4.32. Let $\varphi(X, Y)$ be a Π_1^1 formula with no free variables other than X and Y. Prove in $\text{ATR}_0^{\text{set}}$ that for all $X \subseteq \omega$, if there exists Y such that $\varphi(X, Y)$ holds and $\text{HJ}(X \oplus Y)$ exists, then there exists $Y \in \text{HCL}(X)$ such that $\varphi(X, Y)$ holds. (This is a refinement of lemma VII.4.13.)

We end this section with a theorem concerning the special situation when $\text{HCL}(X)$ is not all of $L(X)$. The conclusion in this case is rather strong.

DEFINITION VII.4.33. Π_∞^1-CA_0^{set} is the theory in L_{set} which consists of $\text{ATR}_0^{\text{set}}$ plus the *full comprehension scheme*:

$$\forall u \, \exists v \, \forall x \, (x \in v \leftrightarrow (x \in u \wedge \varphi(x)))$$

where $\varphi(x)$ is any formula of L_{set} in which v does not occur freely. (See also lemma VII.5.3.)

THEOREM VII.4.34. *The following is provable in* $\mathrm{ATR}_0^{\mathrm{set}}$. *Assume* $X \subseteq \omega$. *Suppose that* $\mathrm{HCL}(X) \neq \mathrm{L}(X)$. *Then:*

1. $\mathrm{HCL}(X) = \mathrm{L}_\delta(X)$ *for a certain limit ordinal* δ.
2. $\mathrm{HCL}(X)$ *satisfies* $\mathrm{V} = \mathrm{HCL}(X)$ *plus all axioms of* $\Pi_\infty^1\text{-}\mathrm{CA}_0^{\mathrm{set}}$.
3. $\mathrm{HCL}(X)$ *satisfies Axiom Beta. Moreover, for all relations* $r \in \mathrm{HCL}(X)$, *we have*

$$r \text{ is regular} \leftrightarrow (r \text{ is regular})^{\mathrm{HCL}(X)}.$$

4. $\mathrm{HCL}(X)$ *is closed under hyperjump. All* Σ_1^1 *formulas are absolute to* $\mathrm{HCL}(X)$.

PROOF. By lemmas VII.4.25 and VII.4.26, we have $u \in \mathrm{HCL}(X)$ if and only if $\exists \alpha \, (u \in \mathrm{L}_\alpha(X) \wedge \mathrm{L}_\alpha(X) \in \mathrm{HCL}(X))$. If $\mathrm{HCL}(X) \neq \mathrm{L}(X)$, it follows that there exists an ordinal γ such that $\mathrm{L}_\gamma(X) \notin \mathrm{HCL}(X)$. Hence $\mathrm{HCL}(X) \subseteq \mathrm{L}_\gamma(X)$. Let δ be the set of all $\alpha < \gamma$ such that

$$\exists f \, (f \in \mathrm{L}_\gamma(X) \wedge \mathrm{Fcn}(f) \wedge \mathrm{dom}(f) = \omega \wedge \mathrm{rng}(f) = \mathrm{L}_\alpha(X)).$$

Then clearly δ is a limit ordinal and $\mathrm{L}_\delta(X) = \mathrm{HCL}(X)$. This proves part 1. (It can be shown that δ is the smallest uncountable ordinal of $\mathrm{L}(X)$.)

For part 2, let $\varphi(x)$ be a formula of $\mathrm{L}_{\mathrm{set}}$ with parameters from $\mathrm{HCL}(X)$ and no free variables other than x. Given $u \in \mathrm{HCL}(X) = \mathrm{L}_\delta(X)$, put

$$v = \{x : x \in u \wedge \langle \mathrm{L}_\delta(X), \in {\restriction} \mathrm{L}_\delta(X)\rangle \models \varphi(x)\}.$$

Thus $v \in \mathrm{L}_{\delta+1}(X)$. Since $v \subseteq u$ and $u \in \mathrm{HCL}(X)$, it follows trivially that $v \in \mathrm{HCL}(X)$. Then clearly $\mathrm{HCL}(X) = \mathrm{L}_\delta(X)$ satisfies $\forall x \, (x \in v \leftrightarrow (x \in u \wedge \varphi(x)))$. This shows that $\mathrm{HCL}(X)$ is a model of full comprehension.

By lemma VII.4.26, $\mathrm{HCL}(X)$ satisfies $\mathrm{V} = \mathrm{HCL}(X)$. By theorem VII.4.3.7, $\mathrm{HCL}(X) = \mathrm{L}_\delta(X)$ satisfies $\mathrm{B}_0^{\mathrm{set}}$. It is obvious that $\mathrm{HCL}(X)$ satisfies the Axioms of Regularity and Countability. It remains to prove that $\mathrm{HCL}(X)$ satisfies Axiom Beta.

Let $r \in \mathrm{HCL}(X)$ be a relation. If r is regular, lemma VII.4.10 implies that the collapsing function of r belongs to $\mathrm{HCL}(X)$. Suppose now that r is not regular. We shall use the notation r_x which was introduced in the proof of lemma VII.4.11. Let w be the set of $x \in \mathrm{field}(r)$ such that the collapsing function of r_x is an element of $\mathrm{HCL}(X)$. Since $\mathrm{HCL}(X)$ is a transitive model of full comprehension, $w \in \mathrm{HCL}(X)$. Put $v = \mathrm{field}(r) - w$. Then $v \in \mathrm{HCL}(X)$ and, since r is not regular, $v \neq \emptyset$. If $x \in \mathrm{field}(r)$ and $\forall y \, (\langle y, x\rangle \in r \to y \in w)$, then r_x is regular, hence by lemma VII.4.10 $x \in w$. Hence $\forall x \, (x \in v \to \exists y \, (\langle y, x\rangle \in r \wedge y \in v))$. Thus $(r \text{ is not regular})^{\mathrm{HCL}(X)}$.

Combining these observations, we see that $\mathrm{HCL}(X)$ satisfies Axiom Beta. This completes the proof of part 2 and also proves part 3.

To prove part 4, let φ be a Σ_1^1 sentence with parameters from $\mathrm{HCL}(X)$. By the Kleene normal form theorem V.1.4, we can write φ as $\exists f\, \forall n\, \theta(f[n])$ where $\theta(\tau)$ is arithmetical with parameters from $\mathrm{HCL}(X)$. Let r be the set of all $\langle \tau^{\frown}\langle k\rangle, \tau\rangle$ such that $\forall n\, (n \leq \mathrm{lh}(\tau) \to \theta(\tau[n]))$. Using part 3 we see that

$$\varphi \leftrightarrow r \text{ is not regular}$$
$$\leftrightarrow (r \text{ is not regular})^{\mathrm{HCL}(X)}$$
$$\leftrightarrow \varphi^{\mathrm{HCL}(X)}.$$

This shows that all Σ_1^1 formulas are absolute to $\mathrm{HCL}(X)$. In other words, $\mathrm{HCL}(X)$ is a β-model. Also, from part 2 it follows a fortiori that $\mathrm{HCL}(X)$ satisfies Π_1^1-CA_0. Hence by theorem VII.1.12 $\mathrm{HCL}(X)$ is closed under hyperjump.

This completes the proof of theorem VII.4.34. $\square$

EXERCISE VII.4.35. Exhibit a transitive model of $\mathrm{ATR}_0^{\mathrm{set}}$ in which $\mathrm{HCL}(\emptyset) \neq \mathrm{L}(\emptyset)$, yet not all Σ_2^1 or Σ_1^{set} sentences are absolute to $\mathrm{HCL}(\emptyset)$.

Notes for §VII.4. Gödel's original papers [95, 96] on constructible sets are still worth reading. Gödel's subsequent detailed treatment [97] is much less accessible. For further information on constructible sets, see Jensen [131] or any good textbook of axiomatic set theory, e.g., Jech [130]. The original Shoenfield absoluteness theorem is due to Shoenfield [221] and is closely related to Kondo's theorem (§VI.2). Our observation that the Shoenfield absoluteness theorem is provable in Π_1^1-CA_0 (theorem VII.4.14) was inspired by Jensen/Karp [132] and Barwise/Fisher [14] (see also Barwise [13, §V.8]). Our theorem VII.4.27.2 on Σ_1^{set} absoluteness is similar to a result of Lévy [162, theorem 43].

VII.5. Strong Comprehension Schemes

In this and the next two sections, we study β-models of certain subsystems of second order arithmetic which are stronger than Π_1^1-CA_0. We rely on the set-theoretic results of the previous two sections.

DEFINITION VII.5.1 (comprehension schemes). Assume $0 < k < \omega$.

1. Π_k^1-CA_0 is the subsystem of Z_2 which consists of ACA_0 plus the scheme of Π_k^1 *comprehension*:

$$\exists X\, \forall n\, (n \in X \leftrightarrow \psi(n))$$

where $\psi(n)$ is any Π_k^1 formula in which X does not occur freely.

2. Δ_k^1-CA_0 is the subsystem of Z_2 which consists of ACA_0 plus the scheme of Δ_k^1 *comprehension*:

$$\forall n\, (\varphi(n) \leftrightarrow \psi(n)) \to \exists X\, \forall n\, (n \in X \leftrightarrow \psi(n))$$

where $\varphi(n)$ is any Σ_k^1 formula and $\psi(n)$ is any Π_k^1 formula in which X does not occur freely.

3. $\Pi_\infty^1\text{-CA}_0 = \bigcup_{k<\omega} \Pi_k^1\text{-CA}_0$.

REMARKS VII.5.2. Obviously

$$\Delta_k^1\text{-CA}_0 \subseteq \Pi_k^1\text{-CA}_0 \subseteq \Delta_{k+1}^1\text{-CA}_0$$

for all k, $0 \leq k < \omega$. We shall see later that all of these inclusions are proper except for the triviality $\text{ACA}_0 = \Delta_0^1\text{-CA}_0 = \Pi_0^1\text{-CA}_0$. The development of mathematics within ACA_0 has been discussed in chapter III. By theorem V.5.1 we have

$$\Delta_1^1\text{-CA}_0 \subseteq \text{ATR}_0 \subseteq \Pi_1^1\text{-CA}_0.$$

Further results on models of ACA_0 and $\Delta_1^1\text{-CA}_0$ will be presented in chapters VIII and IX. The development of mathematics within $\Pi_1^1\text{-CA}_0$ has been discussed in chapter VI. Models of $\Pi_1^1\text{-CA}_0$ have been discussed in §§VII.1 and VII.4 in the present chapter. Models of $\Pi_{k+1}^1\text{-CA}_0$ and $\Delta_{k+2}^1\text{-CA}_0$ are the principal topic of this and the next two sections. Further results on models of $\Pi_{k+1}^1\text{-CA}_0$ and $\Delta_{k+2}^1\text{-CA}_0$ will be presented in chapters VIII and IX.

In order to prove theorems about models of $\Pi_{k+1}^1\text{-CA}_0$ and $\Delta_{k+2}^1\text{-CA}_0$, it will be convenient to work with the set-theoretic counterparts of these theories. Recall that to each L_2-theory $T_0 \supseteq \text{ATR}_0$, we have associated a set-theoretic counterpart T_0^{set} consisting of $\text{ATR}_0^{\text{set}}$ plus T_0. Thus T_0^{set} is a theory in the language L_{set} and proves the same L_2-sentences as T_0 (theorem VII.3.34). The purpose of the following lemma is to identify each of $\Pi_{k+2}^1\text{-CA}_0^{\text{set}}$ and $\Delta_{k+2}^1\text{-CA}_0^{\text{set}}$.

LEMMA VII.5.3. *Assume $0 \leq k < \omega$. Over $\text{ATR}_0^{\text{set}}$ we have:*

1. $\Pi_{k+2}^1\text{-CA}_0$ *is equivalent to the scheme of Π_{k+1}^{set} comprehension:*

$$\forall u \, \exists v \, \forall x \, (x \subset v \leftrightarrow (x \subset u \wedge \psi(x)))$$

   where $\psi(x)$ is any Π_{k+1}^{set} formula in which v does not occur freely.

2. $\Delta_{k+2}^1\text{-CA}_0$ *is equivalent to the scheme of $\Delta_{k+1}^{\text{set}}$ comprehension:*

$$\forall x \, (\varphi(x) \leftrightarrow \psi(x)) \rightarrow \forall u \, \exists v \, \forall x \, (x \in v \leftrightarrow (x \in u \wedge \psi(x)))$$

   where $\varphi(x)$ is any $\Sigma_{k+1}^{\text{set}}$ formula and $\psi(x)$ is any Π_{k+1}^{set} formula in which v does not occur freely.

PROOF. This follows easily from the intertranslatability of Σ_{k+2}^1 with $\Sigma_{k+1}^{\text{set}}$ (theorem VII.3.24) together with the equivalence of ATR_0 with $|\text{ATR}_0^{\text{set}}|$ (theorem VII.3.22). □

We shall now show that each of $\Pi_{k+1}^1\text{-CA}_0$ and $\Delta_{k+2}^1\text{-CA}_0$, $0 \leq k < \omega$ implies its own relativization to the inner model $\text{HCL}(X)$ (definition VII.4.22).

THEOREM VII.5.4. *In* $\mathsf{ATR}_0^{\mathrm{set}}$, *assume* $0 \leq k < \omega$ *and let* $X \subseteq \omega$ *be given.*
Then:

1. Π_{k+1}^1-CA_0 *implies* $(\Pi_{k+1}^1$-$\mathsf{CA}_0)^{\mathrm{HCL}(X)}$.
2. Δ_{k+2}^1-CA_0 *implies* $(\Delta_{k+2}^1$-$\mathsf{CA}_0)^{\mathrm{HCL}(X)}$.

PROOF. In the case when $\mathrm{HCL}(X)$ is not all of $\mathrm{L}(X)$, the desired result
follows by theorem VII.4.34 (in fact we get the much stronger conclusion
$(\Pi_\infty^1$-$\mathsf{CA}_0)^{\mathrm{HCL}(X)})$. So for the rest of this proof assume that $\mathrm{HCL}(X) = \mathrm{L}(X)$.

We shall prove part 2 first. By lemma VII.5.3.2 it will suffice to prove
that $\Delta_{k+1}^{\mathrm{set}}$ comprehension implies $(\Delta_{k+1}^{\mathrm{set}}$ comprehension$)^{\mathrm{HCL}(X)}$. Assume
$\Delta_{k+1}^{\mathrm{set}}$ comprehension. We first claim that $\mathrm{HCL}(X)$ satisfies the scheme of
$\Sigma_{k+1}^{\mathrm{set}}$ *choice*:

$$\forall x \, \exists y \, \varphi(x, y) \rightarrow \forall u \, \exists f \, \forall x \, (x \in u \rightarrow \varphi(x, f'x))$$

where $\varphi(x, y)$ is any $\Sigma_{k+1}^{\mathrm{set}}$ formula in which f does not occur freely. We
shall now prove this claim by induction on k.

Suppose first that $k = 0$. Assume that $\mathrm{HCL}(X)$ satisfies $\forall x \, \exists y \, \varphi(x, y)$
where $\varphi(x, y)$ is a Σ_1^{set} formula with parameters from $\mathrm{HCL}(X)$. Let
$u \in \mathrm{HCL}(X)$ be given. Let $g \in \mathrm{HCL}(X)$ be an injection such that
$\mathrm{dom}(g) = u$ and $\mathrm{rng}(g) \subseteq \omega$. Write $\varphi(x, y)$ as $\exists z \, \theta(x, y, z)$ where
$\theta(x, y, z)$ is Σ_0^{set} with parameters in $\mathrm{HCL}(X)$. For each $x \in u$, let h_x be
the $<^{\mathrm{L}(X)}$-least injection such that $\mathrm{dom}(h_x) = \omega$ and $\mathrm{rng}(h_x)$ is an ordinal
$\alpha = \alpha_x$ such that

$$u \in \mathrm{L}_\alpha(X) \wedge \exists y \, \exists z \, (y \in \mathrm{L}_\alpha(X) \wedge z \in \mathrm{L}_\alpha(X) \wedge \theta(x, y, z)).$$

Let $\prec$ be the well ordering of $u \times \omega$ such that $\langle x, m \rangle \prec \langle x_1, m_1 \rangle$ if and only
if either (i) $g'x < g'x_1$ or (ii) $x = x_1$ and $h'_x m < h'_x m_1$. This well ordering
$\prec$ exists by Δ_1^{set} comprehension using lemmas VII.4.26 and VII.4.23.2.
Now by Axiom Beta let β be the order type of $\prec$. Thus β is an ordinal
which is the sum of the ordinals α_x, $x \in u$ in the order given by g. In
particular $u \in \mathrm{L}_\beta(X)$ and

$$\forall x \, (x \in u \rightarrow \exists y \, \exists z \, (y \in \mathrm{L}_\beta(X) \wedge z \in \mathrm{L}_\beta(X) \wedge \theta(x, y, z))).$$

Let $f \in \mathrm{L}_{\beta+1}(X)$ be the set of $\langle y, x \rangle$ such that $x \in u$ and y is $<_\beta^{\mathrm{L}(X)}$-least
such that $\exists z \, (z \in \mathrm{L}_\beta(X) \wedge \theta(x, y, z))$. Thus $f \in \mathrm{L}(X) = \mathrm{HCL}(X)$ and
$\mathrm{HCL}(X)$ satisfies $\forall x \, (x \in u \rightarrow \varphi(x, f'x))$. This proves our claim for
$k = 0$.

Next suppose that $k > 0$. By the inductive hypothesis, $\mathrm{HCL}(X)$ satisfies
Σ_k^{set} choice. This implies that, for any Σ_k^{set} formula θ, $\mathrm{HCL}(X)$ satisfies the
equivalence of $\forall x \, (x \in u \rightarrow \exists y \theta)$ with $\exists v \, \forall x \, (x \in u \rightarrow \exists y \, (y \in v \wedge \theta))$.
By repeated application of this quantifier interchange principle, we see that
for any Σ_k^{set} formula θ, $\mathrm{HCL}(X)$ satisfies the equivalence of $\forall x \, (x \in u \rightarrow \theta)$ with a certain Σ_k^{set} formula. In other words, over $\mathrm{HCL}(X)$, the class of

Σ_k^{set} formulas is closed under bounded quantification. Keep in mind also that by lemma VII.4.23, for any Σ_k^{set} formula θ, the relativization $\theta^{\text{HCL}(X)}$ is also Σ_k^{set}.

Assume now that $\text{HCL}(X)$ satisfies $\forall x \, \exists y \, \varphi(x, y)$ where $\varphi(x, y)$ is $\Sigma_{k+1}^{\text{set}}$ with parameters in $\text{HCL}(X)$. Let $u \in \text{HCL}(X)$ be given. Let g be as before. Write $\varphi(x, y)$ as $\exists z \, \theta(x, y, z)$ where $\theta(x, y, z)$ is Π_k^{set}. For each $x \in u$, let h_x be the $<^{\text{L}(X)}$-least injection such that $\text{dom}(h_x) = \omega$ and $\text{rng}(h_x)$ is an ordinal $\alpha = \alpha_x$ such that $\text{HCL}(X)$ satisfies $\exists y \, \exists z \, (y \in \text{L}_\alpha(X) \wedge z \in \text{L}_\alpha(X) \wedge \theta(x, y, z))$. Define $\prec$ as before. This well ordering $\prec$ exists by $\Delta_{k+1}^{\text{set}}$ comprehension using lemma VII.4.26 and the above observations concerning bounded quantifiers. Define β as before. Let f be the set of all $\langle y, x \rangle$ such that $x \in u$ and y is $<_\beta^{\text{L}(X)}$-least such that $\text{HCL}(X)$ satisfies $\exists z \, (z \in \text{L}_\beta(X) \wedge \theta(x, y, z))$. By our observations on bounded quantifiers, this definition of f is Δ_k^{set}. By induction on k, $\text{HCL}(X)$ satisfies Δ_k^{set} comprehension. Hence $f \in \text{HCL}(X)$, and clearly $\text{HCL}(X)$ satisfies $\forall x \, (x \in u \to \varphi(x, f'x))$. This proves our claim.

To complete the proof of part 2, assume that $\text{HCL}(X)$ satisfies $\forall x \, (\varphi(x) \leftrightarrow \psi(x))$ where $\varphi(x)$ and $\psi(x)$ are $\Sigma_{k+1}^{\text{set}}$, respectively Π_{k+1}^{set} formulas with parameters in $\text{HCL}(X)$. Let $\eta(x, y)$ be the $\Sigma_{k+1}^{\text{set}}$ formula

$$(y = 1 \wedge \varphi(x)) \vee (y = 0 \wedge \neg\psi(x)).$$

Then $\text{HCL}(X)$ satisfies $\forall x \, \exists y \, \eta(x, y)$. Given $u \in \text{HCL}(X)$, apply $\Sigma_{k+1}^{\text{set}}$ choice in $\text{HCL}(X)$ to obtain $f \in \text{HCL}(X)$ such that $\text{HCL}(X)$ satisfies $\forall x \, (x \in u \to \eta(x, f'x))$. Putting $v = \{x : x \in u \wedge f'x = 1\}$, we see that $v \in \text{HCL}(X)$ and $\text{HCL}(X)$ satisfies $\forall x \, (x \in v \leftrightarrow (x \in u \wedge \psi(x)))$. Thus $\text{HCL}(X)$ satisfies $\Delta_{k+1}^{\text{set}}$ comprehension. This completes the proof of part 2.

We shall now prove part 1. By lemma VII.5.3.1 it will suffice to prove that Π_{k+1}^{set} comprehension implies $(\Pi_{k+1}^{\text{set}}$ comprehension$)^{\text{HCL}(X)}$. Assume Π_{k+1}^{set} comprehension. Let $\psi(x)$ be a Π_{k+1}^{set} formula with parameters in $\text{HCL}(X)$ and no free variables other than x. Let $u \in \text{HCL}(X)$ be given. By lemma VII.4.23 and Π_{k+1}^{set} comprehension, let v be the set of $x \in u$ such that $\text{HCL}(X)$ satisfies $\psi(x)$. Write $\psi(x)$ as $\forall y \, \theta(x, y)$ where $\theta(x, y)$ is Σ_k^{set}. As before, let β be a sum of ordinals α_x, $x \in u - v$, where $\alpha = \alpha_x$ is chosen so that $\exists y \, (y \in \text{L}_\alpha(X) \wedge \neg\theta(x, y))$. Thus v may be described as the set of $x \in u$ such that $\text{HCL}(X)$ satisfies $\forall y \, (y \in \text{L}_\beta(X) \to \theta(x, y))$. By our observations concerning bounded quantifiers, the latter formula is equivalent over $\text{HCL}(X)$ to a Σ_k^{set} formula with parameters in $\text{HCL}(X)$. Applying Σ_k^{set} comprehension within $\text{HCL}(X)$, it follows that $v \in \text{HCL}(X)$.

This completes the proof of theorem VII.5.4. □

EXERCISE VII.5.5. In $\text{ATR}_0^{\text{set}}$, show that for all $X \subseteq \omega$, $\text{L}(X)$ and $\text{HCL}(X)$ satisfy Σ_1^{set} choice.

EXERCISE VII.5.6. Show that Δ_1^{set} comprehension is equivalent over $\text{ATR}_0^{\text{set}}$ to Σ_1^{set} choice.

EXERCISE VII.5.7. Show that Δ_1^{set} comprehension fails in the minimum transitive model of Π_1^1-CA_0^{set}. (By lemma VII.5.3 we may restate this as follows: Δ_2^1-CA_0 fails in the minimum β-model of Π_1^1-CA_0.)

We shall now reformulate the previous theorem so as to apply directly to subsystems of Z_2 and models of same. Some of our results will be stated as *conservation theorems* (see definition VII.5.12 below).

DEFINITION VII.5.8. In the language L_2, we write $Y \in L(X)$ and $Z <^{L(X)} Y$ as abbreviations for

$$\exists V_0 \exists V_1 (V_0 = X^* \wedge V_1 = Y^* \wedge |v_1 \in L(v_0)|)$$

and

$$\exists V_0 \exists V_1 \exists V_2 (V_0 = X^* \wedge V_1 = Y^* \wedge V_2 = Z^* \wedge |v_2 <^{L(v_0)} v_1|)$$

respectively. By lemma VII.4.21 and theorem VII.3.24.1, the above formulas are Σ_2^1, provably in ATR_0.

REMARK VII.5.9. The point of the above definition is as follows. Let M be a model of ATR_0. Given $X, Y \in \mathcal{S}_M$, identify X and Y with the corresponding elements $[X^*]$ and $[Y^*]$ of $|M_{\text{set}}|$ (definitions VII.3.18 and VII.3.28). Then $M \models Y \in L(X)$ if and only if $M_{\text{set}} \models Y \in L(X)$. A similar remark applies to the formula $Z <^{L(X)} Y$.

THEOREM VII.5.10. *Let M' be any model of Π_1^1-CA_0. Given $X \in \mathcal{S}_{M'}$, let M be the ω-submodel of M' consisting of all $Y \in \mathcal{S}_{M'}$ such that $M' \models Y \in L(X)$. Then:*

1. *M is a model of Π_1^1-CA_0.*
2. *$X \in \mathcal{S}_M$, and M satisfies $\forall Y (Y \in L(X))$.*
3. *M is a β_2-submodel of M'. This means that for any Σ_2^1 sentence φ with parameters from M, $M \models \varphi$ if and only if $M' \models \varphi$.*
4. *If M' is a β-model, then so is M.*
5. *If M' is an ω-model, then so is M.*

Furthermore, for all $k \geq 0$, we have:

6. *To any Σ_{k+2}^1 formula $\varphi(n_1, \ldots, n_i, X_1, \ldots, X_j)$ with parameters from M, we can associate a Σ_{k+2}^1 formula φ' such that, for all $n_1, \ldots, n_i \in |M|$ and $X_1, \ldots, X_j \in \mathcal{S}_M$,*

$$M \models \varphi(n_1, \ldots, n_i, X_1, \ldots, X_j)$$

if and only if

$$M' \models \varphi'(n_1, \ldots, n_i, X_1, \ldots, X_j).$$

7. *If M' is model of Δ_{k+2}^1-CA_0, then so is M.*

8. *If M' is a model of Π^1_{k+2}-CA$_0$, then so is M.*
9. *If M' is a model of Π^1_∞-CA$_0$, then so is M.*

PROOF. Clearly M_{set} is the transitive submodel of M'_{set} consisting of all $a \in |M'_{\text{set}}|$ such that $M'_{\text{set}} \models a \in \text{HCL}(X)$. It follows by theorem VII.4.27.1 that M_{set} satisfies $V = \text{HCL}(X)$ plus all axioms of Π^1_1-CA$^{\text{set}}_0$. Hence M satisfies $\forall Y\,(Y \in \text{L}(X))$ plus Π^1_1-CA$_0$. This gives parts 1 and 2 of the theorem. Part 3 follows from theorem VII.4.14. Part 4 follows immediately from part 3. Part 5 is trivial. For part 6, let $\varphi(n_1,\ldots,n_i,X_1,\ldots,X_j)$ be a given Σ^1_{k+2} formula. By theorem VII.3.24.2, we may regard φ as a $\Sigma^{\text{set}}_{k+1}$ formula of L_{set}. Hence by lemma VII.4.23.2, $\varphi^{\text{HCL}(X)}$ is $\Sigma^{\text{set}}_{k+1}$. Let $\varphi'(n_1,\ldots,n_i,X_1,\ldots,X_j)$ be the L_2-formula

$$\exists V_1 \cdots \exists V_i\, \exists V_{i+1} \cdots \exists V_{i+j}$$
$$(V_1 = n_1^* \wedge \cdots \wedge V_i = n_i^* \wedge V_{i+1} = X_1^* \wedge \cdots \wedge V_{i+j} = X_j^*$$
$$\wedge\, |\varphi^{\text{HCL}(X)}(v_1,\ldots,v_i,v_{i+1},\ldots,v_{i+j})|).$$

By theorem VII.3.24.1, φ' is Σ^1_{k+2}, and clearly φ' satisfies the conclusion of part 6. Parts 7 and 8 follow from theorems VII.5.4.2 and VII.5.4.1 respectively. Part 9 is an immediate consequence of part 8. □

From the above theorem we can deduce the following key result.

COROLLARY VII.5.11 (conservation theorems). *Let T_0 be any one of the L_2-theories Π^1_∞-CA$_0$, Π^1_{k+1}-CA$_0$, or Δ^1_{k+2}-CA$_0$, $0 \le k < \omega$. Let ψ be any Π^1_4 sentence. Suppose that ψ is provable from T_0 plus $\exists X\,\forall Y\,(Y \in \text{L}(X))$. Then ψ is provable from T_0 alone.*

PROOF. Let ψ be a Π^1_4 sentence which is not provable from T_0. By Gödel's completeness theorem, let M' be a model of T_0 plus $\neg\psi$. Write $\neg\psi$ as $\exists X\,\forall Y\,\varphi(X,Y)$ where $\varphi(X,Y)$ is a Σ^1_2 formula. Let $X \in S_{M'}$ be such that $M' \models \forall Y\,\varphi(X,Y)$. Let $M \subseteq_\omega M'$ consist of all $Y \in S_{M'}$ such that $M' \models Y \in \text{L}(X)$. By theorem VII.5.10, M satisfies T_0 plus $\forall Y\,(Y \in \text{L}(X))$ plus $\forall Y\,\varphi(X,Y)$. Thus M is a model of T_0 plus $\exists X\,\forall Y\,(Y \in \text{L}(X))$ plus $\neg\psi$. Therefore, by the soundness theorem, ψ is not provable from T_0 plus $\exists X\,\forall Y\,(Y \in \text{L}(X))$. This proves the corollary. □

The content of the above corollary is that, when trying to prove a Π^1_4 sentence within T_0, it is harmless to assume $\exists X\,\forall Y\,(Y \in \text{L}(X))$. In other words, using the terminology of the following definition, T_0 plus $\exists X\,\forall Y\,(Y \in \text{L}(X))$ is conservative over T_0 for Π^1_4 sentences. Results of this kind are sometimes known as *conservation theorems*. Other conservation theorems will be presented in the next section and in chapter IX.

DEFINITION VII.5.12 (conservativity). *Let T_0 and T'_0 be theories in the language L_2. We say that T'_0 is conservative over T_0 for Π^1_k sentences if*

$T_0' \supseteq T_0$ and any Π_k^1 sentence which is provable in T_0' is already provable in T_0.

EXERCISE VII.5.13 (more conservation theorems). Assume $0 \leq k \leq m \leq n \leq \infty$. Let T_0 consist of either Π_{k+1}^1-CA$_0$ or Δ_{k+2}^1-CA$_0$, plus Π_{m+1}^1-TI$_0$ plus Σ_{n+1}^1-IND (definitions VII.5.1, VII.2.14, and VII.6.1.2). Show that T_0 plus $\exists X \forall Y (Y \in L(X))$ is conservative over T_0 for Π_4^1 sentences.

EXERCISE VII.5.14 (ω-model conservation theorems). Assume $0 \leq k \leq m \leq \infty$. Let T_0 consist of either Π_{k+1}^1-CA$_0$ or Δ_{k+2}^1-CA$_0$, plus Π_{m+1}^1-TI$_0$. Let ψ be any Π_4^1 sentence. Suppose that ψ holds in all ω-models of T_0 plus $\exists X \forall Y (Y \in L(X))$. Show that ψ holds in all ω-models of T_0.

EXERCISE VII.5.15 (β-model conservation theorems). Let T_0 and ψ be as in corollary VII.5.11. Suppose that ψ holds in all transitive models of T_0 of the form $L_\alpha(X)$, $X \subseteq \omega$, α a countable ordinal. Show that ψ holds in all β-models of T_0.

EXERCISE VII.5.16. Show that the results of corollary VII.5.11 and exercises VII.5.13, VII.5.14 and VII.5.15 do not extend to Σ_4^1 sentences.

Hint: The sentence $\exists X \forall Y (Y \in L(X))$ is Σ_4^1. Consider a transitive model of ZFC in which the continuum hypothesis does not hold.

We shall now apply theorem VII.5.10 to obtain the minimum β-models of Π_{k+1}^1-CA$_0$ and Δ_{k+2}^1-CA$_0$, $0 \leq k < \omega$, and of Π_∞^1-CA$_0$.

THEOREM VII.5.17 (minimum β-models). *Assume $X \subseteq \omega$ and $0 \leq k < \omega$.*

1. *Among all β-models of Π_{k+1}^1-CA$_0$ which contain X, there is a unique smallest one $M = M_{k+1}^\Pi(X)$. Furthermore M_{set} can be characterized (up to canonical isomorphism) as $L_\alpha(X)$ for a certain ordinal $\alpha = \alpha_{k+1}^\Pi(X)$, namely the smallest ordinal α such that $L_\alpha(X)$ satisfies Π_{k+1}^1 comprehension.*
2. *Same as part 1 with Π_{k+1}^1, $M_{k+1}^\Pi(X)$, $\alpha_{k+1}^\Pi(X)$ replaced by Δ_{k+2}^1, $M_{k+2}^\Delta(X)$, $\alpha_{k+2}^\Delta(X)$ respectively.*
3. *Same as part 1 with Π_{k+1}^1, $M_{k+1}^\Pi(X)$, $\alpha_{k+1}^\Pi(X)$ replaced by Π_∞^1, $M_\infty^\Pi(X)$, $\alpha_\infty^\Pi(X)$ respectively.*

PROOF. Let M' be any β-model of Π_{k+1}^1-CA$_0$ which contains X. Let $M \subseteq_\omega M'$ consist of all $Y \in M'$ such that $M' \models Y \in L(X)$. By theorem VII.5.10, M is a β-model of Π_{k+1}^1-CA$_0$ and M satisfies $\forall Y (Y \in L(X))$. Thus M_{set} is isomorphic to a transitive model of Π_{k+1}^1-CA$_0^{\mathrm{set}}$ plus $V = HCL(X)$. By lemma VII.4.21, this model is $L_\beta(X)$ for some ordinal β. Then clearly $\beta \geq \alpha$ where $\alpha = \alpha_{k+1}^\Pi(X)$ is as defined in the statement of part 1. Put $M_{k+1}^\Pi(X) = (L_\alpha(X))^2$. Then $M_{k+1}^\Pi(X)$ satisfies Π_{k+1}^1-CA$_0$ by definition, and is a β-model by theorem VII.3.27.2. Also

$M_{k+1}^{\Pi}(X) \subseteq M \subseteq M'$ so part 1 is proved. The proofs of parts 2 and 3 are similar.   □

REMARK VII.5.18. In §VII.7 we shall see that for all $X \subseteq \omega$ and $0 \le k < \omega$,

$$\alpha_{k+1}^{\Pi}(X) < \alpha_{k+2}^{\Delta}(X) < \alpha_{k+2}^{\Pi}(X).$$

In §§VIII.3 and VIII.4 we shall obtain an analogous ordinal $\alpha_1^{\Delta}(X) < \alpha_1^{\Pi}(X)$. Namely, setting $\alpha = \alpha_1^{\Delta}(X) = \sup\{|a|^X : \mathcal{O}(a, X)\}$ and $M = M_1^{\Delta}(X) = \mathrm{HYP}(X)$, we shall see that $M = L_\alpha(X) \cap P(\omega)$ is the minimum ω-model of Δ_1^1-CA$_0$.

EXERCISE VII.5.19 (minimum β-submodels). Assume $0 \le k < \omega$ and let $X \in M' \models \Pi_{k+1}^1$-CA$_0$ be given. Prove that among all β-submodels $M \subseteq_\beta M'$ such that $X \in M \models \Pi_{k+1}^1$-CA$_0$, there is a unique smallest one. Describe this model. Prove similar results with Π_{k+1}^1-CA$_0$ replaced by Δ_{k+2}^1-CA$_0$ and by Π_∞^1-CA$_0$.

EXERCISE VII.5.20. Recall the system Π_k^1-TR$_0$, definition VI.7.1. Extend VII.5.10–VII.5.19 to encompass Π_{k+1}^1-TR$_0$, $0 \le k < \omega$. In particular we have a result similar to VII.5.17.1 with Π_{k+1}^1-CA$_0$, $M_{k+1}^{\Pi}(X)$, $\alpha_{k+1}^{\Pi}(X)$, Π_{k+1}^1 comprehension replaced by Π_{k+1}^1-TR$_0$, $M_{k+1}^{\Pi^*}(X)$, $\alpha_{k+1}^{\Pi^*}(X)$, Π_{k+1}^1 transfinite recursion respectively. Thus we obtain minimum β-models of Π_{k+1}^1-TR$_0$, $0 \le k < \omega$. (See also exercises VII.7.12.)

Notes for §VII.5. The ideas of this section are probably well known, but we have been unable to find bibliographical references for them. Our sharp formulations VII.5.10–VII.5.16 in terms of conservation results are probably new.

Our results on minimum β-models of Δ_k^1-CA$_0$ and Π_k^1-CA$_0$ are closely related to some well known ideas of Barwise and Jensen. In order to explain this connection, let us use the notation of VII.5.17 and VII.5.18 and write $\alpha_k^{\Delta} = \alpha_k^{\Delta}(\emptyset)$ and $\alpha_k^{\Pi} = \alpha_k^{\Pi}(\emptyset)$, where $\emptyset$ is the empty set. For $k = 1$, the ordinals α_1^{Δ} and α_1^{Π} can be characterized in terms of admissibility theory, as discussed in Barwise [13], Simpson [233], and Sacks [211]. Namely, $\alpha_1^{\Delta} = \omega_1^{CK} =$ the least admissible ordinal $> \omega$, and $\alpha_1^{\Pi} =$ the supremum of the first ω admissible ordinals. (Note however that α_1^{Π} is not itself admissible.) Moreover, for arbitrary $k < \omega$, the ordinals α_{k+2}^{Δ} and α_{k+2}^{Π} can be characterized in terms of Jensen's fine structure theory [131]; see also Simpson [233, §3]. Namely, $\alpha_{k+2}^{\Delta} =$ the least α such that $\eta_{k+1}^\alpha > \omega$, or equivalently the least $\alpha > \omega$ such that $\eta_{k+1}^\alpha = \alpha$; and $\alpha_{k+2}^{\Pi} =$ the least α such that $\rho_{k+1}^\alpha > \omega$, or equivalently the least $\alpha > \omega$ such that $\rho_{k+1}^\alpha = \alpha$. These results are easily deduced from theorems VII.3.24 and VII.5.17.

VII.6. Strong Choice Schemes

The purpose of this section is to study the axiom of choice in the context of second order arithmetic. We shall consider several *choice schemes*, i.e., axiom schemes in the language L_2 which express consequences or special cases of the axiom of choice. We shall obtain some conservation theorems relating strong choice schemes to strong comprehension schemes. We make essential use of the results of the previous two sections.

DEFINITION VII.6.1 (choice schemes). Assume $0 \le k < \omega$.

1. Σ_k^1-AC$_0$ is the L_2-theory whose axioms are those of ACA$_0$ plus the scheme of Σ_k^1 *choice*:

$$\forall n \, \exists Y \, \eta(n, Y) \rightarrow \exists Z \, \forall n \, \eta(n, (Z)_n)$$

where $\eta(n, Y)$ is any Σ_k^1 formula in which Z does not occur. We are using the notation

$$(Z)_n = \{i : (i, n) \in Z\}.$$

2. Σ_k^1-IND is the scheme of Σ_k^1 *induction*:

$$(\varphi(0) \wedge \forall n \, (\varphi(n) \rightarrow \varphi(n + 1))) \rightarrow \forall n \, \varphi(n)$$

where $\varphi(n)$ is any Σ_k^1 formula. We also define Σ_∞^1-IND $= \bigcup_{k<\omega} \Sigma_k^1$-IND. Note that Σ_∞^1-IND automatically holds in all ω-models.

3. Σ_k^1-DC$_0$ is the L_2-theory whose axioms are those of ACA$_0$ plus the scheme of Σ_k^1 *dependent choice*:

$$\forall n \, \forall X \, \exists Y \, \eta(n, X, Y) \rightarrow \exists Z \, \forall n \, \eta(n, (Z)^n, (Z)_n)$$

where $\eta(n, X, Y)$ is any Σ_k^1 formula in which Z does not occur. We are using the notation

$$(Z)^n = \{(i, m) : (i, m) \in Z \wedge m < n\}$$

and $(Z)_n$ as above.

4. *Strong* Σ_k^1-DC$_0$ is the L_2-theory whose axioms are those of ACA$_0$ plus the scheme of *strong* Σ_k^1 *dependent choice*:

$$\exists Z \, \forall n \, \forall Y \, (\eta(n, (Z)^n, Y) \rightarrow \eta(n, (Z)^n, (Z)_n))$$

where $\eta(n, X, Y)$ is as above.

We begin with a number of miscellaneous remarks concerning Σ_k^1-AC$_0$, Σ_k^1-DC$_0$, and strong Σ_k^1-DC$_0$, $0 \le k < \omega$.

REMARKS VII.6.2. Trivially Σ_0^1-AC$_0$ is equivalent to Σ_1^1-AC$_0$ and Σ_0^1-DC$_0$ is equivalent to Σ_1^1-DC$_0$. It is easy to see that

$$\Delta_1^1\text{-CA}_0 \subseteq \Sigma_1^1\text{-AC}_0 \subseteq \Sigma_1^1\text{-DC}_0$$

(lemma VII.6.6) and that Σ_1^1-DC$_0$ holds in all β-models of ATR$_0$. Also, by theorem V.8.3, Σ_1^1-AC$_0$ holds in all models of ATR$_0$. Further results concerning Δ^1_1-CA$_0$, Σ^1_1-AC$_0$ and Σ^1_1-DC$_0$ will be presented in chapters VIII and IX.

REMARKS VII.6.3 (Σ^1_k-AC$_0$). The cases $k = 0$ and $k = 1$ are discussed in remark VII.6.2 and in chapters VIII and IX. It is easy to see that Σ^1_k-AC$_0$ implies Δ^1_k-CA$_0$ (lemma VII.6.6.1). For $k = 2$, we shall see below (theorem VII.6.9.1) that Σ^1_2-AC$_0$ is in fact equivalent to Δ^1_2-CA$_0$. For $k \geq 3$ the situation is more complex. On the one hand, Feferman and Lévy have used forcing to exhibit a β-model M of Π^1_∞-CA$_0$ in which Σ^1_3-AC$_0$ fails. Namely, $M = A \cap P(\omega)$ where A is a transitive model of ZF plus $\aleph_1 = \aleph^L_\omega$. See Lévy [163, theorem 8], Cohen [40, §IV.10], and Jech [130, §21, example IV]. On the other hand, we shall see below (theorem VII.6.16.1) that for all k, Σ^1_{k+3}-AC$_0$ is equivalent to Δ^1_{k+3}-CA$_0$ if we assume $\exists X \, \forall Y \, (Y \in \mathrm{L}(X))$. See also remarks VII.6.12 and VII.6.21 below.

REMARKS VII.6.4 (Σ^1_k-DC$_0$). The cases $k = 0$ and $k = 1$ are discussed in remark VII.6.2 and in chapters VIII and IX. It is easy to see that Σ^1_k-DC$_0$ implies Σ^1_k-AC$_0$ plus Σ^1_k-IND (lemma VII.6.6.2). For $k = 2$, we shall see below (theorem VII.6.9.2) that Σ^1_2-DC$_0$ is in fact equivalent to Σ^1_2-AC$_0$ plus Σ^1_2-IND. Simpson [229] has claimed that there exists a β-model of Σ^1_∞-AC$_0$ ($= \bigcup_{k<\omega} \Sigma^1_k$-AC$_0$) in which Σ^1_3-DC$_0$ fails; the proof of this result has not been published. We shall see below (theorem VII.6.16.2) that for all k, Σ^1_{k+3}-DC$_0$ is equivalent to Σ^1_{k+3}-AC$_0$ plus Σ^1_{k+3}-IND if we assume $\exists X \, \forall Y \, (Y \in \mathrm{L}(X))$.

In some of these results, the role of Σ^1_k-IND is rather delicate. See remarks VII.6.12 and VII.6.21 below. Of course we may ignore this delicate issue if we are interested only in ω-models, since in all such models Σ^1_∞-IND automatically holds.

REMARKS VII.6.5 (strong Σ^1_k-DC$_0$). Trivially strong Σ^1_0-DC$_0$ is equivalent to strong Σ^1_1-DC$_0$. It is easy to see (lemma VII.6.6.3) that strong Σ^1_k-DC$_0$ implies Π^1_k-CA$_0$. We shall see below (theorem VII.6.9) that strong Σ^1_1-DC$_0$ is in fact equivalent to Π^1_1-CA$_0$. Likewise, strong Σ^1_2-DC$_0$ is equivalent to Π^1_2-CA$_0$. Furthermore (theorem VII.6.16.3), for all k, strong Σ^1_{k+3}-DC$_0$ is equivalent to Π^1_{k+3}-CA$_0$ if we assume $\exists X \, \forall Y \, (Y \in \mathrm{L}(X))$. We do not know whether strong Σ^1_{k+3}-DC$_0$ is equivalent to Π^1_{k+3}-CA$_0$ plus Σ^1_{k+3}-DC$_0$.

LEMMA VII.6.6. *Assume* $0 \leq k < \omega$.

1. Σ^1_k-AC$_0$ *implies* Δ^1_k-CA$_0$.
2. Σ^1_k-DC$_0$ *implies* Σ^1_k-AC$_0$ *and* Σ^1_k-IND.
3. *Strong* Σ^1_k-DC$_0$ *implies* Π^1_k-CA$_0$ *and* Σ^1_k-DC$_0$.

PROOF. For part 1, assume Σ^1_k-AC$_0$ and suppose $\forall n \, (\varphi(n) \leftrightarrow \psi(n))$ where $\varphi(n)$ and $\psi(n)$ are Σ^1_k and Π^1_k respectively. Let $\eta(n, Y)$ be the Σ^1_k

formula

$$(\varphi(n) \wedge 1 \in Y) \vee (\neg\psi(n) \wedge 1 \notin Y).$$

By Σ_k^1 choice let Z be such that $\forall n \, \eta(n, (Z)_n)$. Putting $X = \{n : 1 \in (Z)_n\}$ we see that $\forall n \, (n \in X \leftrightarrow \psi(n))$. This proves Δ_k^1 comprehension.

For part 2, assume Σ_k^1-DC_0 and suppose $\forall n \, \exists Y \, \eta(n, Y)$ where $\eta(n, Y)$ is Σ_k^1. Let $\varphi(n, X, Y)$ be the Σ_k^1 formula $\eta(n, Y)$. Applying Σ_k^1 dependent choice, we get Z such that $\forall n \, \varphi(n, (Z)^n, (Z)_n)$, i.e., $\forall n \, \eta(n, (Z)_n)$. This proves Σ_k^1 choice. Now to prove Σ_k^1-IND, let $\varphi(n)$ be Σ_k^1 and assume $\varphi(0)$ and $\forall n \, (\varphi(n) \to \varphi(n+1))$. Without loss we may assume $k \geq 1$ and write $\varphi(n)$ as $\exists X \, \theta(n, X)$ where $\theta(n, X)$ is Π_{k-1}^1. Let $\eta(n, X, Y)$ be the Σ_k^1 formula

$$\forall m \, (m < n \to \theta(m, (X)_m)) \to \theta(n, Y).$$

Our assumptions imply that $\forall n \, \forall X \, \exists Y \, \eta(n, X, Y)$. Applying Σ_k^1 dependent choice, we get Z such that $\forall n \, \eta(n, (Z)^n, (Z)_n)$. By part 1 we have Π_{k-1}^1 comprehension. Using this, let W be such that $\forall n \, (n \in W \leftrightarrow \forall m \, (m < n \to \theta(m, (Z)_m)))$. We can then use quantifier-free induction to show that $W = \mathbb{N}$. This gives $\forall n \, \theta(n, (Z)_n)$, hence $\forall n \varphi(n)$. The proof of part 2 is complete.

For part 3, assume strong Σ_k^1-DC_0. Trivially we have Σ_k^1-DC_0. It remains to prove Π_k^1-CA_0. Instead of Π_k^1 comprehension, we shall prove Σ_k^1 comprehension, which is equivalent. Let $\varphi(n)$ be Σ_k^1. Without loss assume $k \geq 1$ and write $\varphi(n)$ as $\exists Y \, \theta(n, Y)$ where $\theta(n, Y)$ is Π_{k-1}^1. Let $\eta(n, X, Y)$ be $\theta(n, Y)$. Applying strong Σ_k^1 dependent choice to $\eta(n, X, Y)$, we get Z such that $\forall n \, \forall Y \, (\theta(n, Y) \to \theta(n, (Z)_n))$. By Π_{k-1}^1 comprehension, let X be such that $\forall n \, (n \in X \leftrightarrow \theta(n, (Z)_n))$. Then $\forall n \, (n \in X \leftrightarrow \varphi(n))$. This proves Σ_k^1 comprehension.

The proof of lemma VII.6.6 is complete. $\square$

We shall now prove some deeper results concerning the relationship between Σ_k^1 choice schemes and Π_k^1 and Δ_k^1 comprehension schemes.

In order to handle the case $k = 2$, we need the following lemma. This is a formal Π_1^1-CA_0 version of the well known Σ_2^1 uniformization principle.

LEMMA VII.6.7 (Σ_2^1 uniformization in Π_1^1-CA_0). *Let $\varphi(Y)$ be a Σ_2^1 formula with a distinguished free set variable Y. Then we can find a Σ_2^1 formula $\widehat{\varphi}(Y)$ such that Π_1^1-CA_0 proves*

(1) $\forall Y \, (\widehat{\varphi}(Y) \to \varphi(Y))$,
(2) $\forall Z \, (\varphi(Z) \to \exists Y \, \widehat{\varphi}(Y))$,
(3) $\forall Y \, \forall Z \, ((\widehat{\varphi}(Y) \wedge \widehat{\varphi}(Z)) \to Y = Z)$.

PROOF. We give two proofs, one based on Kondo's theorem and the other based on the Shoenfield absoluteness theorem.

For the first proof, write $\varphi(Y)$ as $\exists W \, \psi(Y \oplus W)$ where $\psi(Y)$ is Π_1^1. We are using the notation

$$Y \oplus W = \{2n : n \in Y\} \bigcup \{2n + 1 : n \in W\}.$$

Applying lemma VI.2.1 (a formal version of Kondo's Π_1^1 uniformization theorem), we obtain a Π_1^1 formula $\widehat{\psi}(Y)$ such that Π_1^1-CA$_0$ proves (1), (2) and (3) with φ replaced by ψ. Setting $\widehat{\varphi}(Y) \equiv \exists W \, \widehat{\psi}(Y \oplus W)$ we obtain (1), (2) and (3).

In order to present the second proof we need the following sublemma, which will be used again in the proof of lemma VII.6.15. Recall that the formulas $Y \in L(X)$ and $Z <^{L(X)} Y$ are Σ_2^1 (definition VII.5.8).

SUBLEMMA VII.6.8. *In* ATR$_0$ *the formula*

$$\forall Z \, (Z \leq^{L(X)} Y \leftrightarrow \exists n \, (Z = (W)_n)) \qquad (21)$$

is equivalent to a Σ_2^1 formula.

PROOF. By lemma VII.4.21, the set-theoretic formula

$$v_3 = \{v_2 : v_2 \subseteq \omega \wedge v_2 \leq^{L(v_0)} v_1\}$$

is Σ_1^{set}. Recall that, according to definition VII.3.18, Y^* is a suitable tree which represents Y. Hence

$$W^{**} = \{\langle n \rangle {}^\frown \sigma : n \in \mathbb{N} \wedge \sigma \in (W)_n^*\}$$

is a suitable tree which represents $\{(W)_n : n \in \mathbb{N}\}$. Thus (21) is equivalent to

$$\exists V_0 \, \exists V_1 \, \exists V_3 \, (V_0 = X^* \wedge V_1 = Y^* \wedge V_3 = W^{**}$$
$$\wedge \, |v_3 = \{v_2 : v_2 \subseteq \omega \wedge v_2 \leq^{L(v_0)} v_1\}|).$$

This is Σ_2^1 in view of theorem VII.3.24.1. The sublemma is proved.   □

Now as in the hypothesis of lemma VII.6.7, let $\varphi(Y)$ be a Σ_2^1 formula. Without loss of generality, assume that the only free set variables of $\varphi(Y)$ are Y and X. Write $\varphi(Y)$ as $\exists W \, \psi(X, Y \oplus W)$ where $\psi(X, Y)$ is Π_1^1. Let $\widehat{\psi}(X, Y)$ be the formula

$$\exists W \, (\forall Z \, (Z \leq^{L(X)} Y \leftrightarrow \exists n \, (Z = (W)_n)) \wedge$$
$$\forall n \, (\psi(X, (W)_n) \leftrightarrow (W)_n = Y)).$$

We reason within Π_1^1-CA$_0$. By Σ_2^1 choice, the subformula

$$\forall n \, (\psi(X, (W)_n) \leftrightarrow (W)_n = Y)$$

is equivalent to a Σ_2^1 formula. Hence, by sublemma VII.6.8, $\widehat{\psi}(X, Y)$ is also equivalent to a Σ_2^1 formula. Clearly we have $\forall Y \, (\widehat{\psi}(X, Y) \rightarrow \psi(X, Y))$ and $\forall Y \, \forall Z \, ((\widehat{\psi}(X, Y) \wedge \widehat{\psi}(X, Z)) \rightarrow Y = Z)$. By theorem VII.4.14 (a formal version of Shoenfield absoluteness), we have $\forall Z \, (\psi(X, Z) \rightarrow \exists Y \, (Y \in L(X) \wedge \psi(X, Y)))$, hence $\forall Z \, (\psi(X, Z) \rightarrow \exists Y \, \widehat{\psi}(X, Y))$. Setting

$\widehat{\varphi}(Y) \equiv \exists W \, \widehat{\psi}(X, Y \oplus W)$ we obtain (1), (2) and (3). This completes the second proof of lemma VII.6.7. $\square$

We are now ready to prove the following theorem concerning Σ_2^1 choice schemes.

THEOREM VII.6.9 (Σ_2^1 choice schemes).
1. Σ_2^1-AC$_0$ is equivalent to Δ_2^1-CA$_0$.
2. Σ_2^1-DC$_0$ is equivalent to Δ_2^1-CA$_0$ plus Σ_2^1-IND.
3. Strong Σ_2^1-DC$_0$ is equivalent to Π_2^1-CA$_0$.
4. Strong Σ_1^1-DC$_0$ is equivalent to Π_1^1-CA$_0$.

PROOF. We begin with part 1. By lemma VII.6.6.1, Σ_2^1-AC$_0$ implies Δ_2^1-CA$_0$. For the converse, assume Δ_2^1-CA$_0$. To prove Σ_2^1 choice, suppose $\forall n \, \exists Y \, \eta(n, Y)$ where $\eta(n, Y)$ is Σ_2^1. By lemma VII.6.7, let $\widehat{\eta}(n, Y)$ be Σ_2^1 such that $\forall n \, (\exists \text{ exactly one } Y) \, \widehat{\eta}(n, Y)$ and $\forall n \, \forall Y \, (\widehat{\eta}(n, Y) \to \eta(n, Y))$. Set

$$\varphi(i, n) \equiv \exists Y \, (\widehat{\eta}(n, Y) \wedge i \in Y)$$

and

$$\psi(i, n) \equiv \forall Y \, (\widehat{\eta}(n, Y) \to i \in Y).$$

Thus $\varphi(i, n)$ is Σ_2^1, $\psi(i, n)$ is Π_2^1, and $\forall n \, \forall i \, (\varphi(i, n) \leftrightarrow \psi(i, n))$. By Δ_2^1 comprehension, let $Z = \{(i, n): \psi(i, n)\}$. Then $\forall n \, \eta(n, (Z)_n)$. This proves part 1.

Next we prove part 2. By lemma VII.6.6.2, Σ_2^1-DC$_0$ implies Σ_2^1-AC$_0$ and Σ_2^1-IND. For the converse, assume Σ_2^1-AC$_0$ plus Σ_2^1-IND. To prove Σ_2^1 dependent choice, suppose $\forall n \, \forall X \, \exists Y \, \eta(n, X, Y)$ where $\eta(n, X, Y)$ is Σ_2^1. By lemma VII.6.7 let $\widehat{\eta}(n, X, Y)$ be such that $\forall n \, \forall X \, (\exists \text{ exactly one } Y) \, \widehat{\eta}(n, X, Y)$ and

$$\forall n \, \forall X \, \forall Y \, (\widehat{\eta}(n, X, Y) \to \eta(n, X, Y)).$$

By Σ_2^1 choice, the formula

$$\overline{\eta}(n, W) \equiv \forall m \, (m \leq n \to \widehat{\eta}(m, (W)^m, (W)_m))$$

is equivalent to a Σ_2^1 formula. We can therefore use Σ_2^1 induction to prove $\forall n \, \exists W \, \overline{\eta}(n, W)$. It is also easy to see that

$$\forall n \, \forall W \, \forall Z \, ((\overline{\eta}(n, W) \wedge \overline{\eta}(n, Z)) \to (W)_n = (Z)_n).$$

Setting

$$\varphi(i, n) \equiv \exists W \, (\overline{\eta}(n, W) \wedge i \in (W)_n)$$

and

$$\psi(i, n) \equiv \forall W \, (\overline{\eta}(n, W) \to i \in (W)_n)$$

we see that $\varphi(i, n)$ is Σ_2^1, $\psi(i, n)$ is Π_2^1, and $\forall i \, \forall n \, (\varphi(i, n) \leftrightarrow \psi(i, n))$. By Δ_2^1 comprehension, let $Z = \{(i, n): \psi(i, n)\}$. Then $\forall n \, \overline{\eta}(n, Z)$, hence $\forall n \, \eta(n, (Z)^n, (Z)_n)$. This proves part 2.

We now turn to part 3. By lemma VII.6.6.3, strong Σ_2^1-DC_0 implies Π_2^1-CA_0. For the converse, assume Π_2^1-CA_0 and let $\eta(n, X, Y)$ be Σ_2^1. By lemma VII.6.7 let $\widehat{\eta}(n, X, Y)$ be Σ_2^1 such that

$$\forall n \, \forall X \, \forall Y \, (\widehat{\eta}(n, X, Y) \rightarrow \eta(n, X, Y)),$$

$$\forall n \, \forall X \, \forall Z \, (\eta(n, X, Z) \rightarrow \exists Y \, \widehat{\eta}(n, X, Y)),$$

$$\forall n \, \forall X \, \forall Y \, \forall Z \, ((\widehat{\eta}(n, X, Y) \wedge \widehat{\eta}(n, X, Z)) \rightarrow Y = Z).$$

By Σ_2^1 comprehension, let S be the set of all $\sigma \in 2^{<\mathbb{N}}$ such that $\exists W \, \overline{\eta}(\sigma, W)$ where $\overline{\eta}(\sigma, W)$ is the Σ_2^1 formula

$$\forall m \, (\sigma(m) = 1 \rightarrow \widehat{\eta}(m, (W)^m, (W)_m)) \wedge (\sigma(m) = 0 \rightarrow (W)_m = \emptyset)).$$

Note that $\langle \rangle \in S$ and if $\sigma \in S$ then $\sigma^\frown \langle 0 \rangle \in S$. Define $f : \mathbb{N} \rightarrow \{0, 1\}$ by $f(n) = 1$ if $f[n]^\frown \langle 1 \rangle \in S$, $f(n) = 0$ otherwise. Then $\forall n \, \exists W \, \overline{\eta}(f[n+1], W)$ and it is easy to see that $\forall n \, \forall W \, \forall Z \, ((\overline{\eta}(f[n+1], W) \wedge \overline{\eta}(f[n+1], Z)) \rightarrow (W)_n = (Z)_n)$. As before apply Δ_2^1 comprehension to get $Z = \{(i, n) : \forall W \, (\overline{\eta}(f[n+1], W) \rightarrow i \in (W)_n)\}$. Then $\forall n \, \overline{\eta}(f[n+1], Z)$, hence $\forall n \, \forall Y \, (\eta(n, (Z)^n, Y) \rightarrow \eta(n, (Z)^n, (Z)_n))$. This proves strong Σ_2^1 dependent choice.

It remains to prove part 4. By lemma VII.6.6.3, strong Σ_1^1-DC_0 implies Π_1^1-CA_0. For the converse, assume Π_1^1-CA_0 and let $\eta(n, X, Y)$ be Σ_1^1. By theorem VII.2.10, let M be a countable coded β-model such that M contains the parameters of $\eta(n, X, Y)$. Let Z be a code for M according to definition VII.2.1. By arithmetical comprehension with Z as a parameter, define $f : \mathbb{N} \rightarrow \mathbb{N}$ by $f(n) = $ least j such that

$$M \models \eta(n, \{(i, m) : (i, f(m)) \in Z \wedge m < n\}, (Z)_j)$$

if such j exists, $f(n) = 0$ otherwise. Setting $W = \{(i, n) : (i, f(n)) \in Z\}$ we see that for all n, $(W)^n \in M$ and $(W)_n \in M$ and

$$M \models \forall Y \, (\eta(n, (W)^n, Y) \rightarrow \eta(n, (W)^n, (W)_n)).$$

Since M is a β-model, it follows that $\forall n \, \forall Y \, (\eta(n, (W)^n, Y) \rightarrow \eta(n, (W)^n, (W)_n))$ is true. This proves strong Σ_1^1 dependent choice.

The proof of theorem VII.6.9 is complete. $\qquad\qquad\qquad\square$

COROLLARY VII.6.10. Δ_2^1-CA_0 and Σ_2^1-AC_0 and Σ_2^1-DC_0 are all pairwise equivalent in the presence of Σ_2^1-IND.

PROOF. This follows immediately from parts 1 and 2 of theorem VII.6.9. $\qquad\qquad\qquad\square$

COROLLARY VII.6.11. The ω-models of Δ_2^1-CA_0 and of Σ_2^1-AC_0 are the same as those of Σ_2^1-DC_0.

PROOF. This follows from the previous corollary, since Σ_∞^1-IND is true in all ω-models. $\qquad\qquad\qquad\square$

REMARK VII.6.12. In chapter IX we shall prove:

1. Σ_2^1-AC_0 is conservative over Π_1^1-CA_0 for Π_3^1 sentences:
2. the consistency of Σ_2^1-AC_0 is provable from Π_1^1-CA_0 plus Σ_2^1-IND.

From this it follows that Σ_2^1-IND is not provable in Σ_2^1-AC_0, even if we assume $\forall Y \, (Y \in L(\emptyset))$. Hence the assumption Σ_2^1-IND cannot be dropped from VII.6.9.2 or VII.6.10.

EXERCISE VII.6.13 (Π_2^1 separation). Show that Δ_2^1-CA_0 is equivalent over ACA_0 to the Π_2^1 separation principle:

$$\neg \exists n \, (\psi(n, 1) \wedge \psi(n, 0)) \rightarrow$$
$$\exists X \, \forall n \, ((\psi(n, 1) \rightarrow n \in X) \wedge (\psi(n, 0) \rightarrow n \notin X))$$

where $\psi(n, i)$ is any Π_2^1 formula in which X does not occur freely.

EXERCISE VII.6.14 (Σ_2^1 separation). Show that Π_2^1-CA_0 is equivalent over ACA_0 to the Σ_2^1 separation principle:

$$\neg \exists n \, (\varphi(n, 1) \wedge \varphi(n, 0)) \rightarrow$$
$$\exists X \, \forall n \, ((\varphi(n, 1) \rightarrow n \in X) \wedge (\varphi(n, 0) \rightarrow n \notin X))$$

where $\varphi(n, i)$ is any Σ_2^1 formula in which X does not occur freely.

Our next theorem concerns Σ_{k+3}^1 choice schemes, $0 \le k < \omega$. The theorem will be proved under the assumption $\exists X \, \forall Y \, (Y \in L(X))$. We first need the following lemma which is analogous to lemma VII.6.7.

LEMMA VII.6.15 (Σ_{k+3}^1 uniformization). *Let $\varphi(Y)$ be a Σ_{k+3}^1 formula with a distinguished free set variable Y. Assume that X does not occur freely in $\varphi(Y)$. Then we can find a Σ_{k+3}^1 formula $\widehat{\varphi}(X, Y)$ such that Σ_{k+2}^1-AC_0 plus $\forall Y \, (Y \in L(X))$ proves*

(4) $\forall Y \, (\widehat{\varphi}(X, Y) \rightarrow \varphi(Y))$,
(5) $\forall Z \, (\varphi(Z) \rightarrow \exists Y \, \widehat{\varphi}(X, Y))$,
(6) $\forall Y \, \forall Z \, ((\widehat{\varphi}(X, Y) \wedge \widehat{\varphi}(X, Z)) \rightarrow Y = Z)$.

PROOF. We proceed as in the second proof of lemma VII.6.7. Write $\varphi(Y)$ as $\exists W \, \psi(Y \oplus W)$ where $\psi(Y)$ is Π_{k+2}^1. Let $\widehat{\psi}(X, Y)$ be the formula

$$\exists W \, (\forall Z \, (Z \le^{L(X)} Y \leftrightarrow \exists n \, (Z = (W)_n)) \wedge \forall n \, (\psi((W)_n) \leftrightarrow (W)_n = Y)).$$

We reason in Σ_{k+2}^1-AC_0. By Σ_{k+2}^1 choice, the subformula

$$\forall n \, (\psi((W)_n) \leftrightarrow (W)_n = Y)$$

is equivalent to a Σ_{k+3}^1 formula. Hence by sublemma VII.6.8, $\widehat{\psi}(X, Y)$ is also equivalent to a Σ_{k+3}^1 formula. Clearly we have $\forall Y \, (\widehat{\psi}(X, Y) \rightarrow \psi(Y))$ and $\forall Y \, \forall Z \, ((\widehat{\psi}(X, Y) \wedge \widehat{\psi}(X, Z)) \rightarrow Y = Z)$. Our assumption $\forall Y \, (Y \in L(X))$ implies $\forall Z \, (\psi(Z) \rightarrow \exists Y \, \widehat{\psi}(X, Y))$. Setting $\widehat{\varphi}(X, Y) \equiv \exists W \, \widehat{\psi}(X, Y \oplus W)$ we obtain (4), (5) and (6). This completes the proof of lemma VII.6.15.   □

We are now ready to present our main result concerning Σ^1_{k+3} choice schemes.

THEOREM VII.6.16 (Σ^1_{k+3} choice schemes). *The following is provable in* ATR$_0$. *Assume* $\exists X \, \forall Y \, (Y \in L(X))$. *Then:*

1. Σ^1_{k+3}-AC$_0$ *is equivalent to* Δ^1_{k+3}-CA$_0$.
2. Σ^1_{k+3}-DC$_0$ *is equivalent to* Δ^1_{k+3}-CA$_0$ *plus* Σ^1_{k+3}-IND.
3. *Strong* Σ^1_{k+3}-DC$_0$ *is equivalent to* Π^1_{k+3}-CA$_0$.
4. Σ^1_∞-DC$_0$ $(= \bigcup_{k<\omega} \Sigma^1_k$-DC$_0)$ *is equivalent to* Π^1_∞-CA$_0$.

PROOF. Parts 1, 2 and 3 are proved exactly as for theorem VII.6.9 except that Σ^1_2, Π^1_2, Δ^1_2 are replaced by Σ^1_{k+3}, Π^1_{k+3}, Δ^1_{k+3} and lemma VII.6.7 is replaced by lemma VII.6.15. Part 4 follows immediately from part 3. $\quad \Box$

Applying the above theorem to ω-models and β-models, we obtain the following corollaries.

COROLLARY VII.6.17. *Any* ω-model of Δ^1_{k+3}-CA$_0$ *plus* $\exists X \, \forall Y \, (Y \in L(X))$ *is an* ω-model of Σ^1_{k+3}-DC$_0$. *Any model of* Π^1_{k+3}-CA$_0$ *plus* $\exists X \, \forall Y \, (Y \in L(X))$ *is a model of strong* Σ^1_{k+3}-DC$_0$.

PROOF. This follows immediately from VII.6.16.2 and VII.6.16.3 since Σ^1_{k+3}-IND is true in all ω-models. $\quad \Box$

COROLLARY VII.6.18. *Assume* $X \subseteq \omega$ *and let* α *be an ordinal.*

1. $L_\alpha(X) \models \Delta^1_{k+3}$-CA$_0^{\text{set}}$ *if and only if* $L_\alpha(X) \models \Sigma^1_{k+3}$-DC$_0^{\text{set}}$.
2. $L_\alpha(X) \models \Pi^1_{k+3}$-CA$_0^{\text{set}}$ *if and only if* $L_\alpha(X) \models$ *strong* Σ^1_{k+3}-DC$_0^{\text{set}}$.

PROOF. It follows from lemma VII.4.21 that if $L_\alpha(X) \models$ ATR$_0^{\text{set}}$ then $L_\alpha(X) \models V = HCL(X)$, hence $L_\alpha(X) \models \forall Y \, (Y \in L(X))$. Therefore, parts 1 and 2 follow immediately from VII.6.16.2 and VII.6.16.3. $\quad \Box$

COROLLARY VII.6.19. *The minimum* β-model of Δ^1_{k+3}-CA$_0$ *satisfies* Σ^1_{k+3}-DC$_0$. *The minimum* β-model of Π^1_{k+3}-CA$_0$ *satisfies strong* Σ^1_{k+3}-DC$_0$. *The minimum* β-model of Π^1_∞-CA$_0$ *satisfies* Σ^1_∞-DC$_0$.

PROOF. This follows easily from theorem VII.5.17 and corollary VII.6.18. $\quad \Box$

We now use theorem VII.6.16 to obtain some conservation theorems (definition VII.5.12) for Σ^1_{k+3} choice schemes.

THEOREM VII.6.20 (conservation theorems). *Assume* $0 \le k < \omega$.

1. Σ^1_{k+3}-AC$_0$ *is conservative over* Δ^1_{k+3}-CA$_0$ *for* Π^1_4 *sentences.*
2. Σ^1_{k+3}-DC$_0$ *is conservative over* Δ^1_{k+3}-CA$_0$ *plus* Σ^1_{k+3}-IND *for* Π^1_4 *sentences.*
3. *Strong* Σ^1_{k+3}-DC$_0$ *is conservative over* Π^1_{k+3}-CA$_0$ *for* Π^1_4 *sentences.*

PROOF. Parts 1 and 3 are immediate from theorem VII.6.16 and corollary VII.5.11 (see also definition VII.5.12). For part 2, let ψ be a Π^1_4 sentence which is not provable from Δ^1_{k+3}-CA$_0$ plus Σ^1_{k+3}-IND. Let M' be a model of Δ^1_{k+3}-CA$_0$ plus Σ^1_{k+3}-IND plus $\neg\psi$. Write $\neg\psi$ as $\exists X \, \forall Y \, \varphi(X, Y)$

where $\varphi(X, Y)$ is Σ_2^1. Let $X \in M'$ be such that M' satisfies $\forall Y \, \varphi(X, Y)$. Let $M \subseteq_\omega M'$ consist of all $Y \in M$ such that $M' \models Y \in \mathrm{L}(X)$. By theorem VII.5.10, M satisfies Δ_{k+3}^1-CA$_0$ plus Σ_{k+3}^1-IND plus $\forall Y \, \varphi(X, Y)$ plus $\forall Y \, (Y \in \mathrm{L}(X))$. By theorem VII.6.16.2 it follows that M satisfies Σ_{k+3}^1-DC$_0$. Clearly M satisfies $\neg\psi$ so by the soundness theorem, ψ is not provable from Σ_{k+3}^1-DC$_0$. This completes the proof. $\square$

REMARK VII.6.21. In chapter IX we shall see that Σ_{k+3}^1-AC$_0$ is conservative over Π_{k+2}^1-CA$_0$ for Π_4^1 sentences. This strengthens part 1 of the above theorem VII.6.20. Also in chapter IX, we shall see that the consistency of Σ_{k+3}^1-AC$_0$ is provable from Π_{k+2}^1-CA$_0$ plus Σ_{k+3}^1-IND. Hence Σ_{k+3}^1-IND is not provable from Σ_{k+3}^1-AC$_0$, even if we assume $\forall Y \, (Y \in \mathrm{L}(\emptyset))$. It follows that the assumption Σ_{k+3}^1-IND cannot be dropped from VII.6.16.2 or VII.6.20.2.

EXERCISE VII.6.22 (more conservation theorems). Assume $0 \leq k \leq m \leq n \leq \infty$. Show that T_0' is conservative over T_0 for Π_4^1 sentences, where either:

1. T_0 consists of Π_{k+1}^1-CA$_0$ plus Π_{m+1}^1-TI$_0$ plus Σ_{n+1}^1-IND, and T_0' consists of T_0 plus strong Σ_{k+1}^1-DC$_0$; or
2. T_0 consists of Δ_{k+2}^1-CA$_0$ plus Π_{m+1}^1-TI$_0$ plus Σ_{n+2}^1-IND, and T_0' consists of T_0 plus Σ_{k+2}^1-DC$_0$.

EXERCISE VII.6.23 (ω-model conservation theorems). Assume that $0 \leq k \leq m \leq \infty$. Suppose that either:

1. T_0 consists of Π_{k+1}^1-CA$_0$ plus Π_{m+1}^1-TI$_0$, and T_0' consists of T_0 plus strong Σ_{k+1}^1-DC$_0$; or
2. T_0 consists of Δ_{k+2}^1-CA$_0$ plus Π_{m+1}^1-TI$_0$, and T_0' consists of T_0 plus Σ_{k+2}^1-DC$_0$.

Show that any Π_4^1 sentence which is true in all ω-models of T_0' is true in all ω-models of T_0.

EXERCISE VII.6.24 (β-model conservation theorems). Assume $0 \leq k \leq \infty$. Suppose that either:

1. $T_0 = \Pi_{k+1}^1$-CA$_0$ and $T_0' = $ strong Σ_{k+1}^1-DC$_0$; or
2. $T_0 = \Delta_{k+2}^1$-CA$_0$ and $T_0' = \Sigma_{k+2}^1$-DC$_0$.

Show that any Π_4^1 sentence which is true in all β-models of T_0 is true in all β-models of T_0'.

Notes for §VII.6. The Σ_k^1 choice scheme and the Σ_k^1 dependent choice scheme (parts 1 and 3 of definition VII.6.1) originated with Kreisel [150, 151]. Our strong Σ_k^1 dependent choice scheme (part 4 of definition VII.6.1) appears to be new. Theorem VII.6.9 concerning Σ_2^1 choice schemes is inspired by an unpublished result of Mansfield; see Friedman

[64, theorem 6]. Results such as theorem VII.6.16 on Σ^1_{k+3} choice schemes are probably well known, but we have been unable to find bibliographical references for them. Our sharp formulations VII.6.20–VII.6.24 in terms of conservation results are probably new.

VII.7. β-Model Reflection

In this final section of chapter VII we consider one more topic: β-model reflection principles. An important consequence of the ideas in this section is that for all k,

$$\Sigma^1_{k+1}\text{-DC}_0 < \text{strong } \Sigma^1_{k+1}\text{-DC}_0 < \Sigma^1_{k+2}\text{-DC}_0$$

where $<$ denotes increasing logical strength (theorem VII.7.7).

Our results are most conveniently stated in terms of the notion of β_k-model, $1 \leq k < \omega$.

DEFINITION VII.7.1 (β_k-models). Assume $0 \leq k < \omega$. A β_k-*model* is an ω-model M such that for all Σ^1_k sentences φ with parameters from M, φ is true if and only if $M \models \varphi$.

Thus a β_1-model is the same thing as a β-model. Also, a β_0-model is the same thing as an ω-model.

We shall now formalize the notion of β_k-model within RCA$_0$. The following definition within RCA$_0$ is actually an infinite set of definitions, one for each k. Fix a universal lightface Π^0_1 formula

$$\pi(e, m_1, m_2, X_1, X_2, \ldots, X_k, X_{k+1}, X_{k+2})$$

with exactly the displayed free variables (definition VII.1.3). Let $\varphi_k(e, m, X, Y)$ be the Σ^1_k formula

$$\exists X_1 \forall X_2 \cdots X_k \pm \exists n\, \pi(e, m, n, X_1, X_2, \ldots, X_k, X, Y)$$

where $\pm\exists$ is $\exists$ if k is even, $\neg\exists$ if k is odd. Thus $\varphi_k(e, m, X, Y)$ is in some sense a universal lightface Σ^1_k formula with free variables e, m, X, Y.

DEFINITION VII.7.2 (countable coded β_k-models). Within RCA$_0$, we define a *countable coded β_k-model* to be a countable coded ω-model M (definition VII.2.1) such that for all $e, m \in \mathbb{N}$ and $X, Y \in \mathcal{S}_M$, $\varphi_k(e, m, X, Y)$ if and only if $M \models \varphi_k(e, m, X, Y)$.

LEMMA VII.7.3. *Let* $\psi(m_1, \ldots, m_i, X_1, \ldots, X_j)$ *be any* Π^1_{k+1} *formula with exactly the displayed free variables. The following is provable in* ACA$_0$. *Let* M *be a countable coded* β_k-*model. For all* $m_1, \ldots, m_i \in \mathbb{N}$ *and* $X_1, \ldots, X_j \in M$, *if* $\psi(m_1, \ldots, m_i, X_1, \ldots, X_j)$ *is true then* $M \models \psi(m_1, \ldots, m_i, X_1, \ldots, X_j)$.

PROOF. For $k = 0$ the result is trivial, so assume $k \geq 1$. Then by lemma VII.2.4 we have $M \models \mathsf{ACA}_0$. Let $e < \omega$ be such that ACA_0 proves

$$\psi(m_1, \ldots, m_i, X_1, \ldots, X_j) \leftrightarrow \forall Y \varphi_k(e, \langle m_1, \ldots, m_i \rangle, X_1 \oplus \cdots \oplus X_j, Y).$$

The desired conclusion follows easily.   □

THEOREM VII.7.4. *Assume* $0 \leq k < \omega$. *Over* ACA_0, *strong* Σ^1_{k+1}-DC_0 *is equivalent to the following assertion. For all* $X \subseteq \mathbb{N}$, *there exists a countable coded* β_{k+1}-*model* M *such that* $X \in M$.

PROOF. First, assume strong Σ^1_{k+1}-DC_0. Given $X \subseteq \mathbb{N}$, let $\eta(n, Y, Z)$ be the Σ^1_{k+1} formula

$$\exists e \, \exists m \, (n = (e, m) \wedge \varphi_{k+1}(e, m, X \oplus Y, Z)).$$

By strong Σ^1_{k+1}-dependent choice, let W be such that

$$\forall n \, \forall Z \, (\eta(n, (W)^n, Z) \to \eta(n, (W)^n, (W)_n)).$$

It is straightforward to check that W is a code for a countable β_{k+1}-model M such that $X \in M$. (Compare the proof of lemma VII.2.9.)

For the converse we proceed as in the proof of part 4 of theorem VII.6.9. Reasoning in ACA_0, assume that for all X there exists a countable coded β_{k+1}-model which contains X. Let $\eta(n, X, Y, Z)$ be a Σ^1_{k+1} formula with only the displayed free variables. Given X, let W be a code for a countable β_{k+1}-model M such that $X \in M$. Define $f : \mathbb{N} \to \mathbb{N}$ by $f(n) = \text{least } j$ such that

$$M \models \eta(n, X, \{(i, m) \colon (i, f(m)) \in W \wedge m < n\}, (W)_j)$$

if such j exists, $f(n) = 0$ otherwise. Setting

$$W' = \{(i, n) \colon (i, f(n)) \in W\},$$

we see that for all n,

$$M \models \forall Z \, (\eta(n, X, (W')^n, Z) \to \eta(n, X, (W')^n, (W')_n)).$$

Since M is a β_{k+1}-model, it follows by lemma VII.7.3 that the above formula is true for all n. This proves strong Σ^1_{k+1} dependent choice and completes the proof of theorem VII.7.4.   □

DEFINITION VII.7.5 (β_k-model reflection). Assume $0 \leq k \leq m < \omega$. Within ACA_0, we define β_k-*model reflection for* Σ^1_m *formulas* to be the scheme

$$\forall X \, (\theta(X) \to \exists \text{countable coded } \beta_k\text{-model } M$$

$$\text{such that } X \in M \text{ and } M \models \theta(X))$$

where $\theta(X)$ is any Σ^1_m formula with no free set variables other than X.

The situation as regards ω-*model reflection*, i.e., β_0-model reflection, will be considered separately in §VIII.5. For β_{k+1}-model reflection we have the following result.

THEOREM VII.7.6. *The following is provable in* ACA_0. *Assume* $0 \leq k < \omega$.

1. *Strong* Σ^1_{k+1}-DC_0 *is equivalent to* β_{k+1}-*model reflection for* Σ^1_{k+3} *formulas.*
2. Σ^1_{k+2}-DC_0 *is equivalent to* β_{k+1}-*model reflection for* Σ^1_{k+4} *formulas.*

PROOF. Note first that for any Π^1_{k+2} sentence ψ with parameters in a β_{k+1}-model M, we have by lemma VII.7.3 $\psi \to (M \models \psi)$. This observation combines with theorem VII.7.4 to easily yield part 1.

For part 2, assume Σ^1_{k+2}-DC_0. In particular strong Σ^1_{k+1}-DC_0 holds, so by theorem VII.7.4 we have $\forall Y \exists M$ (M is a countable coded β_{k+1}-model such that $Y \in M$). Now let X_0 be such that $\theta(X_0)$ holds where $\theta(X)$ is Σ^1_{k+4}. Write $\theta(X)$ as $\exists V \forall Y \exists Z \, \psi(V, X, Y, Z)$ where ψ is Π^1_{k+1}. Let V_0 be such that $\forall Y \exists Z \, \psi(V_0, X_0, Y, Z)$ holds. Let $\eta(V, X, Y, Z)$ say that Z is a code for a countable β_{k+1}-model and $\forall m \exists j \, \psi(V, X, (Y)_m, (Z)_j)$. Thus $\eta(V, X, Y, Z)$ is Π^1_{k+1} and $\forall Y \exists Z \, \eta(V_0, X_0, Y, Z)$ holds. By Σ^1_{k+2} dependent choice, let W be such that $((W)_0)_0 = X_0$ and $((W)_0)_1 = V_0$ and $\forall n \, \eta(V_0, X_0, (W)_n, (W)_{n+1})$. Setting $W' = \{(i, (j, n)) : ((i, j), n) \in W\}$, we see that W' is a code for a countable β_{k+1}-model M' such that $X_0 \in M'$ and $V_0 \in M'$ and $M' \models \forall Y \exists Z \, \psi(V_0, X_0, Y, Z)$. This proves β_{k+1}-model reflection for Σ^1_{k+4}-formulas.

For the converse, assume β_{k+1}-model reflection for Σ^1_{k+4} formulas. Suppose $\forall n \forall Y \exists Z \, \eta(n, Y, Z)$ where $\eta(n, Y, Z)$ is Σ^1_{k+2}. Let W be a code for a β_{k+1}-model M such that M contains the parameters of $\eta(n, Y, Z)$ and $M \models \forall n \forall Y \exists Z \, \eta(n, Y, Z)$. Define $f : \mathbb{N} \to \mathbb{N}$ by $f(n) = $ least j such that

$$M \models \eta(n, \{(i, m) : (i, f(m)) \in W \wedge m < n\}, (W)_j).$$

Setting $W' = \{(i, n) : (i, f(n)) \in W\}$, we see that $M \models \eta(n, (W')^n, (W')_n)$ for all n. Since M is a β_{k+1}-model, it follows that $\forall n \, \eta(n, (W')^n, (W')_n)$ is true. This proves Σ^1_{k+2} dependent choice and completes the proof of theorem VII.7.6. $\square$

THEOREM VII.7.7. *Assume* $0 \leq k < \omega$.

1. Σ^1_{k+2}-DC_0 *proves the existence of a countable coded* β_{k+1}-*model of strong* Σ^1_{k+1}-DC_0.
2. *Strong* Σ^1_{k+1}-DC_0 *proves the existence of a countable coded* β_{k+1}-*model of* Σ^1_{k+1}-DC_0.

PROOF. First assume Σ^1_{k+2}-DC_0. In particular strong Σ^1_{k+1}-DC_0 holds, so by theorem VII.7.4 we have $\forall X \exists Y$ (Y is a code for a countable β_{k+1}-model M such that $X \in M$). Applying Σ^1_{k+1}-DC_0 we obtain Z such that $\forall n \, ((Z)_n$ is a code for a countable β_{k+1}-model M_n such that $(Z)^n \in M_n)$. Setting $Z' = \{(i, (j, n)) : ((i, j), n) \in Z\}$ we see that Z' is a code for the countable β_{k+1}-model $M' = \bigcup_{n \in \mathbb{N}} M_n$. By construction $M' \models \forall X \exists Y$

(Y is a code for a countable β_{k+1}-model M such that $X \in M$). Hence by theorem VII.7.4, $M' \models$ strong Σ_{k+1}^1-DC$_0$. This establishes part 1.

For part 2, assume strong Σ_{k+1}^1-DC$_0$. By theorem VII.7.4 there exists a countable coded β_{k+1}-model M. We claim that $M \models \Sigma_{k+1}^1$-DC$_0$. Suppose that $M \models \forall n \, \forall X \, \exists Y \, \eta(n, X, Y)$ where $\eta(n, X, Y)$ is Σ_{k+1}^1. Let W be a code for M and define $f : \mathbb{N} \to \mathbb{N}$ by $f(n) =$ least j such that

$$\eta(n, \{(i, m) : (i, f(m)) \in W \wedge m < n\}, (W)_j\}$$

holds. Setting $W' = \{(i, n) : (i, f(n)) \in W\}$ we get

$$\forall n \, \eta(n, (W')^n, (W')_n),$$

hence $\exists Z \, \forall n \, \eta(n, (Z)^n, (Z)_n)$. Since M is a β_{k+1}-model it follows that $M \models \exists Z \, \forall n \, \eta(n, (Z)^n, (Z)_n)$. This proves the claim and completes the proof of theorem VII.7.7. $\square$

COROLLARY VII.7.8. *Assume* $0 \le k < \omega$.

1. Δ_{k+3}^1-CA$_0$ *plus* Σ_{k+3}^1-IND *proves the existence of a countable coded* β_2-*model of* Π_{k+2}^1-CA$_0$.
2. Π_{k+2}^1-CA$_0$ *proves the existence of a countable coded* β_2-*model of* Δ_{k+2}^1-CA$_0$.

PROOF. The sentence

$$\exists \text{countable coded } \beta_2\text{-model of strong } \Sigma_{k+2}^1\text{-DC}_0$$

is Σ_3^1 and, by VII.7.7.1, provable in Σ_{k+3}^1-DC$_0$. Hence by the conservation theorem VII.6.20.2, this same sentence is provable in Δ_{k+3}^1-CA$_0$ plus Σ_{k+3}^1-IND. Part 1 follows in view of VII.6.6.3. The proof of part 2 is similar using VII.7.7.2, VII.6.20.3, VII.6.9.3, VII.6.6.1 and VII.6.6.2. $\square$

COROLLARY VII.7.9. *Assume* $0 \le k < \omega$.

1. Δ_{k+2}^1-CA$_0$ *plus* Σ_{k+2}^1-IND *proves the existence of a countable coded* β-*model of* Π_{k+1}^1-CA$_0$.
2. Π_{k+1}^1-CA$_0$ *proves the existence of a countable coded* β-*model of* Δ_{k+1}^1-CA$_0$.

PROOF. For $k \ge 1$ this is immediate from corollary VII.7.8. For $k = 0$ the result follows easily from corollary VII.7.8 using lemma VII.6.6 and parts 2 and 4 of theorem VII.6.9. $\square$

COROLLARY VII.7.10 (minimum β-models). *For* $X \subseteq \omega$ *and* $0 \le k < \omega$ *we have*

$$\alpha_{k+1}^{\Pi}(X) < \alpha_{k+2}^{\Delta}(X) < \alpha_{k+2}^{\Pi}(X)$$

where $\alpha_{k+1}^{\Pi}(X)$ *and* $\alpha_{k+2}^{\Delta}(X)$ *are the ordinals of the minimum* β-*models of* Π_{k+1}^1-CA$_0$ *and* Δ_{k+2}^1-CA$_0$ *containing* X, *respectively.*

PROOF. This is immediate from corollary VII.7.9 and theorem VII.5.17.
                                                              $\square$

REMARK VII.7.11. In chapter IX we shall see that Δ^1_{k+3}-CA$_0$ is conservative over Π^1_{k+2}-CA$_0$ for Π^1_4 sentences (corollary IX.4.12). Hence the assumption of Σ^1_{k+3}-IND in corollary VII.7.8 cannot be dropped.

EXERCISES VII.7.12. In this set of exercises we consider β-models of Π^1_{k+1}-TR$_0$, $0 \le k < \omega$. (See also exercise VII.5.20.)
Prove the following:

1. Δ^1_{k+2}-CA$_0$ plus Σ^1_{k+2}-TI$_0$ together imply Π^1_{k+1}-TR$_0$. (The proof is straightforward, using transfinite induction to iterate Π^1_{k+1} comprehension along a given countable well ordering.)
2. Σ^1_{k+2}-DC$_0$ plus Σ^1_{k+2}-TI$_0$ proves the existence of a countable coded β_{k+1}-model of Π^1_{k+1}-TR$_0$. (Hint: Use the previous exercise plus β_{k+1}-model reflection.)
3. Δ^1_{k+3}-CA$_0$ plus Σ^1_{k+3}-TI$_0$ proves the existence of a countable coded β_2-model of Π^1_{k+2}-TR$_0$. (Hint: Use the previous exercise plus results from §§VII.5 and VII.6.)
4. Δ^1_{k+2}-CA$_0$ plus Σ^1_{k+2}-TI$_0$ proves the existence of a countable coded β-model of Π^1_{k+1}-TR$_0$.
5. For any $X \subseteq \omega$ we have

$$\alpha^\Pi_{k+1}(X) < \alpha^{\Pi^*}_{k+1}(X) < \alpha^\Delta_{k+2}(X)$$

where $\alpha^{\Pi^*}_{k+1}$ is the ordinal of the minimum β-model of Π^1_{k+1}-TR$_0$ containing X.

REMARK VII.7.13. In exercises VIII.4.24 we shall see that Π^1_{k+1}-TR$_0$ proves the existence of a countable coded ω-model of Σ^1_{k+2}-DC$_0$. Hence the assumption of Σ^1_{k+2}-TI$_0$ in exercise VII.7.12 cannot be dropped.

Notes for §VII.7. Results such as corollary VII.7.10 on minimum β-models of Δ^1_k-CA$_0$ and Σ^1_k-CA$_0$ are probably very plausible to students of Jensen's fine structure theory [131]; see also our notes at the end of §VII.5. However, they do not seem to be in the previously published literature. The other results of this section are probably new.

VII.8. Conclusions

In this chapter we have studied β-models. We have seen that every β-model is automatically a model of ATR$_0$ and indeed of Π^1_∞-TI$_0$, but not of Π^1_1-CA$_0$ (§VII.2). On the other hand, Π^1_1-CA$_0$ has a minimum β-model obtained by iterating the hyperjump ω times (§VII.1). More generally, for all $k \ge 2$, Π^1_k-CA$_0$ and Δ^1_k-CA$_0$ have minimum β-models (§VII.5) which are described in terms of initial segments of the Gödel's hierarchy of constructible sets. These models are all distinct (§VII.7) and satisfy appropriate forms of the axiom of choice (§VII.6). The proofs of these results yield conservation theorems which are best possible.

An important role is played by β-model reflection. Π_1^1-CA$_0$ is equivalent to the existence of sufficiently many countable coded β-models (§VII.2). More generally, for each $k \geq 1$, strong Σ_k^1 dependent choice is equivalent to the existence of sufficiently many countable coded β_k-models (§VII.7).

Set-theoretic methods have been very useful in this chapter. Our coding of hereditarily countable sets by well founded trees works well in ATR$_0$ (§VII.3) and leads to a good theory of constructible sets, including a Π_1^1-CA$_0$ version of the Shoenfield absoluteness lemma (§VII.4).

Chapter VIII

ω-MODELS

An ω-*model* is an L_2-model M whose first order part is standard. Thus M may be viewed simply as a collection of sets of natural numbers, serving as the range of the set variables in L_2.

The purpose of this chapter is to study countable ω-models of subsystems of second order arithmetic. We concentrate on the subsystems which are by now familiar: RCA_0, WKL_0, ACA_0, $\Sigma_1^1\text{-}AC_0$ and related systems, ATR_0, $\Pi_\infty^1\text{-}TI_0$, and $\Pi_1^1\text{-}CA_0$. We shall also obtain some general results about ω-models of fairly arbitrary L_2-theories, which may be stronger than $\Pi_1^1\text{-}CA_0$.

In §VIII.1 we study countable ω-models of RCA_0 and ACA_0. We point out that an ω-model of RCA_0 is essentially the same thing as an ideal of Turing degrees. An ω-model of ACA_0 is then characterized by the additional property of closure under Turing jump. In particular, each of RCA_0 and ACA_0 has a minimum (i.e., unique smallest) ω-model. The minimum ω-model of RCA_0 is the collection REC of recursive sets, and the minimum ω-model of ACA_0 is the collection ARITH of arithmetical sets. By use of countable coded ω-models, we show that ACA_0 proves the consistency of RCA_0.

In §VIII.2 we consider countable ω-models of WKL_0. We show that any such model has a proper ω-submodel which is again a model of WKL_0. Indeed, we can find such a submodel which is coded in the original model. By further use of countable coded ω-models, we show that the consistency of WKL_0 is provable in ACA_0. We go on to show that WKL_0 has countable coded ω-models which are "close to recursive," in various senses of that phrase. In particular, although REC is not itself an ω-model of WKL_0, it is the intersection of all such models.

In §VIII.3 we develop hyperarithmetical theory in a form needed for later applications. We show that the principal axiom of ATR_0 is equivalent to the assertion that the Turing jump operator can be iterated along any countable well ordering starting with any set. Hyperarithmetical sets are those which can be obtained by iterating the Turing jump operator along a recursive well ordering starting with the empty set. The collection of hyperithmetical sets is denoted HYP. We show that $X \in HYP$ if and only if X is Δ_1^1 definable. By the method of pseudohierarchies (compare §V.4), we show that Σ_1^1 definability over HYP is equivalent to Π_1^1 definability. Thus HYP is an ω-model of Δ_1^1 comprehension but not Π_1^1 comprehension. We show that all of these results are, in a sense, provable in ATR$_0$.

In §§VIII.4, VIII.5 and VIII.6 we study ω-models of Δ_1^1-CA$_0$, ATR$_0$, and stronger systems. The countable ω-model HYP plays a key role. Some of the results can be understood in terms of the following analogy:

$$\frac{\mathsf{RCA}_0}{\Delta_1^1\text{-}\mathsf{CA}_0} = \frac{\mathsf{WKL}_0}{\mathsf{ATR}_0} = \frac{\mathsf{REC}}{\mathsf{HYP}}.$$

For example, just as REC is the minimum ω-model of RCA$_0$, so HYP is the minimum ω-model of Δ_1^1-CA$_0$ and of related systems. Similarly, although HYP is not itelf a model of ATR$_0$, it is the intersection of all countable β-models of ATR$_0$. Furthermore, for any recursively axiomatizable L$_2$-theory $S \supseteq \Delta_1^1$-CA$_0$, HYP is the intersection of all countable ω-models of S. If in addition $S \supseteq$ ATR$_0$, then for any countable ω-model M of S, HYP^M is the intersection of all countable ω-submodels of M which satisfy S.

We also obtain some results which are not covered by the above analogy. In §VIII.4 we use pseudohierarchies and countable coded ω-models to show that ATR$_0$ proves the consistency of Σ_1^1-DC$_0$ plus full induction. In §VIII.5 we show that the transfinite induction scheme (introduced in §VII.2) is equivalent to a reflection principle for countable coded ω-models. We also obtain an ω-model version of Gödel's second incompleteness theorem. These theorems are then used to prove some ω-model independence results. In particular, we obtain a countable ω-model of Π_1^1-CA$_0$ which is not a model of the transfinite induction scheme, Π_∞^1-TI$_0$. We also obtain a countable ω-model of Σ_1^1-AC$_0$ which is not a model of Σ_1^1-DC$_0$.

Throughout this chapter, we formulate our results so as to apply not only to ω-models but also, insofar as possible, to arbitrary models of the systems considered. Nevertheless, it will be clear that our methods are best suited to the study of ω-models which are not β-models. Other results about ω-models have been presented in chapter VII.

VIII.1. ω-Models of RCA$_0$ and ACA$_0$

The formal systems RCA$_0$ and ACA$_0$ were studied extensively in chapters II and III respectively. The purpose of this section is to present some simple results about models, especially ω-models, of RCA$_0$ and ACA$_0$. We discuss both ω-submodels of given models and countable coded ω-models. (Compare §§VII.1 and VII.2.) For additional information on models of RCA$_0$ and ACA$_0$, see §IX.1.

We first consider models of RCA$_0$.

Recall from definition VII.1.4 that $Y \leq_T X$ if and only if Y is *Turing reducible to X*, i.e., *recursive in X*. Recall also that, by definition,

$$X \oplus Y = \{2n \colon n \in X\} \cup \{2n + 1 \colon n \in Y\}.$$

These definitions are made within RCA$_0$.

LEMMA VIII.1.1. *Let M' be a model of* RCA$_0$. *Let M be an ω-submodel of M'. Then M is a model of* RCA$_0$ *if and only if M is nonempty and closed under $\oplus$ and $\leq_T$.*

(M is said to be *closed under $\oplus$* if $X \in M$ and $Y \in M$ imply $X \oplus Y \in M$. M is said to be *closed under $\leq_T$* if $X \in M$ and $Y \leq_T X$ imply $Y \in M$.)

PROOF. The proof is straightforward using the fact that $Y \leq_T X$ if and only if Y is Δ_1^0 in X. $\qquad\square$

LEMMA VIII.1.2. *The following is provable in* RCA$_0$. *The relation $\leq_T$ is transitive, i.e., if $Z \leq_T Y$ and $Y \leq_T X$ then $Z \leq_T X$.*

PROOF. The proof is straightforward using the normal form theorem II.2.7 for Π_1^0 formulas. $\qquad\square$

THEOREM VIII.1.3 (minimum ω-submodels of RCA$_0$). *Let M' be a model of* RCA$_0$. *Let $X \in M'$ be given. Among all ω-submodels M of M' such that $X \in M \models$ RCA$_0$, there exists a unique smallest one, namely that $M \subseteq_\omega M'$ which consists of all Y such that $M' \models Y \leq_T X$.*

PROOF. From lemma VIII.1.2 it is clear that M is a model of RCA$_0$. The rest follows from lemma VIII.1.1. $\qquad\square$

COROLLARY VIII.1.4 (minimum ω-model of RCA$_0$). *There exists a minimum (i.e., unique smallest) ω-model of* RCA$_0$, *namely*

$$\text{REC} = \{X \subseteq \omega \colon X \text{ is recursive}\}.$$

(A set $X \subseteq \omega$ is said to be *recursive* if and only if $X \leq_T \emptyset$.)

PROOF. In the previous theorem, take $M' = P(\omega)$ and $X = \emptyset$. $\qquad\square$

LEMMA VIII.1.5 (finite axiomatizability). *The formal systems* RCA$_0$ *and* ACA$_0$ *are finitely axiomatizable.*

PROOF. Let $\pi(e, m_1, X_1)$ be a universal lightface Π_1^0 formula with exactly the displayed free variables (see definition VII.1.3). The axioms of RCA$_0$ can be taken to consist of the basic axioms I.2.4(i), the pairing axiom

$$\forall X \, \forall Y \, \exists Z \, (Z = X \oplus Y),$$

Δ_1^0 comprehension in the form

$$\forall m \, (\neg \pi(e_0, m, X) \leftrightarrow \pi(e_1, m, X)) \rightarrow \exists Y \, \forall m \, (m \in Y \leftrightarrow \pi(e_1, m, X)),$$

and Σ_1^0 induction in the form

$$(\neg \pi(e, 0, X) \wedge \forall m \, (\neg \pi(e, m, X) \rightarrow \neg \pi(e, m+1, X))) \rightarrow \forall m \, \neg \pi(e, m, X).$$

Then, by lemma III.1.3, the axioms of ACA_0 can be taken to be those of RCA_0 plus Σ_1^0 comprehension in the form

$$\exists Y \, \forall m \, (m \in Y \leftrightarrow \neg \pi(e, m, X)).$$

This proves lemma VIII.1.5.   □

THEOREM VIII.1.6. *The following is provable in* ACA_0. *Given* $X \subseteq \mathbb{N}$, *there exists a unique smallest countable coded ω-model such that* $X \in M$ *and* M *satisfies* RCA_0. *Namely,* M *consists of all* $Y \subseteq \mathbb{N}$ *such that* $Y \leq_T X$.

PROOF. We reason within ACA_0. By arithmetical comprehension, let W be the set of triples $(m, (e_0, e_1))$ such that

$$\forall m \, (\neg \pi(e_0, m, X) \leftrightarrow \pi(e_1, m, X))$$

and $\pi(e_1, m, X)$ holds. Thus W is a code of the countable ω-model M. Clearly $X \in M$ and M is included in all countable coded ω-models M' such that $X \in M'$ and M' satisfies RCA_0. It remains to check that M itself satisfies RCA_0. By lemma VIII.1.5 let φ be the conjunction of the axioms of RCA_0. By lemma VII.2.2 there exists a valuation $f : \mathrm{Sub}_M(\varphi) \to \{0, 1\}$. It remains to show that $f(\varphi) = 1$. Going back to the construction of M, it is straightforward to check that M is closed under $\oplus$. Hence f (pairing axiom) $= 1$. By lemma VII.1.1 M is closed under $\leq_T$. From this it follows easily that $f(\Delta_1^0$ comprehension$) = 1$. It is also easy to check that $f(\Sigma_1^0$ induction$) = 1$. This completes the proof.   □

COROLLARY VIII.1.7 (consistency of RCA_0). ACA_0 *proves the consistency of* RCA_0.

PROOF. We reason within ACA_0. Let M and f be as in the proof of theorem VIII.1.6. Then M and f form a weak model of RCA_0 in the sense of definition II.8.9. Hence by the strong soundness theorem II.8.10 it follows that RCA_0 is consistent. This completes the proof.   □

COROLLARY VIII.1.8. *There exists a* Π_1^0 *sentence* ψ *such that* ψ *is provable in* ACA_0 *but not in* RCA_0.

PROOF. Let ψ be the Π_1^0 sentence which asserts the consistency of RCA_0. By corollary VII.1.7, ψ is provable in ACA_0. The fact that ψ is not provable in RCA_0 is just Gödel's second incompleteness theorem [94, 115, 55, 222] applied to the formal system RCA_0.   □

We now turn to models of ACA_0.

DEFINITION VIII.1.9. The following definition is made in RCA_0. Given $X \subseteq \mathbb{N}$, the *Turing jump of* X, denoted $\mathrm{TJ}(X)$, is the set of all (m, e) such that $\pi(e, m, X)$ holds, if this set exists. (Here $\pi(e, m_1, X_1)$ is a fixed universal lightface Π_1^0 formula as in the proof of lemma VIII.1.5 above.)

We define iterated Turing jumps $\mathrm{TJ}(n, X)$, $n \in \omega$ by recursion on n as follows: $\mathrm{TJ}(0, X) = X$ and $\mathrm{TJ}(n + 1, X) = \mathrm{TJ}(\mathrm{TJ}(n, X))$.

THEOREM VIII.1.10 (minimum ω-submodels of ACA_0). *Let M' be a model of ACA_0. Let $X \in M'$ be given. Among all ω-submodels M of M' such that $X \in M \models \mathsf{ACA}_0$, there exists a unique smallest one, namely the $M \subseteq_\omega M'$ consisting of all Y such that, for some $n \in \omega$, $M' \models Y \leq_T \mathrm{TJ}(n, X)$.*

PROOF. Recall that $|M'|$ is the set of natural numbers of the L_2-model M'. Let M be the set of all $Y \in M'$ such that Y is definable over M' by an arithmetical formula with parameters from $|M'| \cup \{X\}$. Obviously M is the smallest ω-submodel of M' which contains X and satisfies arithmetical comprehension. It is straightforward to check that for each $Y \in M$ there exists $n \in \omega$ such that $M' \models Y \leq_T \mathrm{TJ}(n, X)$. This completes the proof. $\qquad\square$

COROLLARY VIII.1.11 (minimum ω-model of ACA_0). *There exists a minimum (i.e., unique smallest) ω-model of ACA_0, namely*

$$\mathrm{ARITH} = \{X \subseteq \omega \colon \exists n \, (X \leq_T \mathrm{TJ}(n, \emptyset))\}$$
$$= \{X \subseteq \omega \colon X \text{ is arithmetical}\}.$$

(A set $X \subseteq \omega$ is said to be *arithmetical* if it is definable over the standard model $(\omega, +, \cdot, 0, 1, <)$ of Z_1.)

PROOF. In the previous theorem, take $M' = P(\omega)$ and $X = \emptyset$. $\qquad\square$

EXERCISE VIII.1.12. Show that ACA_0 is equivalent over RCA_0 to the assertion that, for all $X \subseteq \mathbb{N}$, $\mathrm{TJ}(X)$ exists.

THEOREM VIII.1.13. *The following is provable in ATR_0. Given $X \subseteq \mathbb{N}$, there exists a unique smallest countable coded ω-model M of ACA_0 such that $X \in M$. Namely, M consists of all $Y \subseteq \mathbb{N}$ such that $\exists n \, (Y \leq_T \mathrm{TJ}(n, X))$.*

PROOF. By arithmetical transfinite recursion (along the standard well ordering $<$ of $\mathbb{N}$), the sequence $\langle \mathrm{TJ}(n, X) \colon n \in \mathbb{N} \rangle$ exists. The rest of the proof is similar to that of theorem VIII.1.6. $\qquad\square$

COROLLARY VIII.1.14 (consistency of ACA_0). *ATR_0 proves the consistency of ACA_0 plus full induction, Σ^1_∞-IND.*

PROOF. We reason within ATR_0. By theorem VIII.1.13 there exists a countable coded ω-model M of ACA_0. By another application of arithmetical transfinite recursion, there exists a full valuation $f \colon \mathrm{Snt}_M \to \{0, 1\}$. (Compare definition VII.2.1.) Thus M and f form a countable model in the sense of definition II.8.3. It is straightforward to check that $f(\mathsf{ACA}_0) = 1$ and that $f(\varphi) = 1$ for all instances φ of induction. Hence by the soundness theorem II.8.8 it follows that ACA_0 plus full induction is consistent. This completes the proof. $\qquad\square$

COROLLARY VIII.1.15. *There exists a Π^0_1 sentence ψ such that ψ is provable in ATR_0, but not in ACA_0 plus full induction.*

PROOF. This follows from corollary VIII.1.14 just as VIII.1.8 followed from VIII.1.7.   □

REMARK VIII.1.16. We shall see in §VIII.2 that ACA_0 proves the consistency of WKL_0. We shall see in §VIII.4 that ATR_0 proves the consistency of $\Sigma_1^1\text{-}DC_0$ plus full induction. In chapter IX, we shall see that RCA_0 and WKL_0 prove the same Π_1^1 sentences, while ACA_0 and $\Sigma_1^1\text{-}AC_0$ prove the same Π_2^1 sentences.

VIII.2. Countable Coded ω-Models of WKL_0

The formal system WKL_0 was studied extensively in chapter IV. The purpose of this section is to present some results which imply the existence of a great many different countable ω-models of WKL_0. In particular we shall show that, although each such ω-model contains nonrecursive sets, the only sets which are common to all such models are the recursive sets.

DEFINITION VIII.2.1 (strict β-submodels). $S\Sigma_1^1$ is the class of Σ_1^1 formulas of the form $\exists X \, \psi$ where ψ is Π_1^0. ($S\Sigma_1^1$ stands for *strict* Σ_1^1.) We say that M is a *strict β-submodel* of M', written $M \subseteq_{S\beta} M'$, if $M \subseteq_\omega M'$ and, for all $S\Sigma_1^1$ sentences φ with parameters from M, $M \models \varphi$ if and only if $M' \models \varphi$. (Compare definitions VII.1.11 and VII.2.28.)

THEOREM VIII.2.2. *Let M' be a model of WKL_0. For any $M \subseteq_\omega M'$, we have $M \subseteq_{S\beta} M'$ if and only if M is a model of WKL_0.*

PROOF. Assume first that $M \subseteq_{S\beta} M'$. In order to show that $M \models WKL_0$, it suffices by lemma IV.4.4 to show that $M \models \Sigma_1^0$ separation. Let $\varphi(i, n)$ be a Σ_1^0 formula with parameters from M and only the free variables shown, such that $\neg\exists n \, (\varphi(0, n) \wedge \varphi(1, n))$ holds. Since M' is a model of WKL_0, it follows that M' satisfies

$$\exists X \, \forall n \, ((\varphi(1, n) \to n \in X) \wedge (\varphi(0, n) \to n \notin X)).$$

This assertion is strict Σ_1^1 and hence by assumption also true in M. This proves Σ_1^0 separation within M.

For the converse, assume that $M \subseteq_\omega M'$ is a model of WKL_0. Let φ be a strict Σ_1^1 sentence with parameters from M. Write φ as $\exists X \, \psi(X)$ where $\psi(X)$ is Π_1^0. By the normal form theorem II.2.7, write $\psi(X)$ as $\forall m \, \theta(X[m])$ where θ is Σ_0^0. By Σ_0^0 comprehension within M, let T be the set of $\tau \in 2^{<\mathbb{N}}$ such that $\forall m \, (m \leq \mathrm{lh}(\tau) \to \theta(\tau[m]))$. Thus $M \models T$ is a tree. Now if $M' \models \varphi$, we have $M' \models \exists X \, \psi(X)$, hence $M' \models \exists X \, \forall m \, \theta(X[m])$, hence $M' \models T$ is infinite. Since $M \subseteq_\omega M'$, it follows that $M \models T$ is infinite. Hence by weak König's lemma within M, we have $M \models \exists X \, (X$ is a path through $T)$, hence $M \models \exists X \, \forall m \, \theta(X[m])$, i.e., $\exists X \, \psi(X)$, i.e., φ. Thus $M \subseteq_{S\beta} M'$. This completes the proof.   □

Our next goal is to show that WKL$_0$ proves the existence of countable coded strict β-models (theorem VIII.2.6 below).

DEFINITION VIII.2.3 (countable coded strict β-models). Within RCA$_0$, let $\pi(e, m_1, X_1, X_2, X_3)$ be a universal lightface Π^0_1 formula with exactly the free variables shown. A *countable coded strict β-model* is a countable coded ω-model M such that, for all $e, m \in \mathbb{N}$ and $X, Y \in M$, $\exists Z\, \pi(e, m, X, Y, Z)$ if and only if $M \models \exists Z\, \pi(e, m, X, Y, Z)$.

(Compare definition VII.2.3.)

LEMMA VIII.2.4.

1. *For any Π^0_1 formula $\psi(k, X)$, WKL$_0$ proves*

$$\forall n\, \exists X\, (\forall k < n)\, \psi(k, X) \to \exists X\, \forall k\, \psi(k, X).$$

2. *For any Π^0_1 formula $\psi(X)$, we can find a Π^0_1 formula $\widehat{\psi}$ such that* WKL$_0$ *proves* $\widehat{\psi} \leftrightarrow \exists X\, \psi(X)$.

Note that part 1 of this lemma amounts to a kind of *compactness principle* for Π^0_1 formulas in WKL$_0$. Moreover, part 2 says any strict Σ^1_1 formula is equivalent over WKL$_0$ to a Π^0_1 formula.

PROOF. We first prove part 1. By the normal form theorem II.2.7, write $\psi(k, X)$ as $\forall m\, \theta(k, X[m])$ where θ is Σ^0_0. Let T be the tree consisting of all $\tau \in 2^{<\mathbb{N}}$ such that $(\forall k \leq \mathrm{lh}(\tau))\, (\forall m \leq \mathrm{lh}(\tau))\, \theta(k, \tau[m])$. The assumption $\forall n\, \exists X\, (\forall k < n)\, \psi(k, X)$ implies that $\forall n\, \exists \tau\, (\mathrm{lh}(\tau) = n \land \tau \in T)$. Hence T is infinite, so by weak König's lemma there exists a path X through T. This implies $\forall k\, \psi(k, X)$. Part 1 is proved.

We now prove part 2. Write $\psi(X)$ as $\forall m\, \theta(X[m])$ where θ is Σ^0_0. Applying weak König's lemma as in the previous paragraph, we see that $\exists X\, \psi(X)$ is equivalent to the Π^0_1 formula $\widehat{\psi} \equiv \forall n\, \exists \tau\, (\mathrm{lh}(\tau) = n \land (\forall m \leq n)\, \theta(\tau[m]))$. This completes the proof of lemma VIII.2.4. □

LEMMA VIII.2.5 (strong Π^0_1 dependent choice). WKL$_0$ *proves the scheme of strong Π^0_1 dependent choice:*

$$\exists Z\, \forall n\, \forall Y\, (\eta(n, (Z)^n, Y) \to \eta(n, (Z)^n, (Z)_n))$$

where $\eta(n, X, Y)$ is any Π^0_1 formula in which Z does not occur.

(Compare definition VII.6.1.4.)

PROOF. By the normal form theorem II.2.7, write

$$\eta(n, X, Y) \equiv \forall k\, \theta(n, X, Y[k])$$

where θ is Σ^0_0. Define

$$\eta^+(n, X, Y) \equiv$$
$$\forall k\, \forall \tau\, ((\mathrm{lh}(\tau) = k \land (\forall i \leq k)\, \theta(n, X, \tau[i])) \to \theta(n, X, Y[k])).$$

By weak König's lemma, we have $\forall n\, \forall X\, \exists Y \eta^+(n, X, Y)$. By lemma VIII.2.4.2, let $\psi(n)$ be a Π^0_1 formula which is equivalent to

$\exists Z \, (\forall m \, < \, n) \, \eta^+(m, (Z)^m, (Z)_m)$. We have $\psi(0)$ and $\forall n \, (\psi(n) \, \to \, \psi(n+1))$, so by Π_1^0 induction (theorem II.3.10) it follows that $\forall n \, \psi(n)$ holds, i.e., $\forall n \, \exists Z \, (\forall m < n) \, \eta^+(m, (Z)^m, (Z)_m)$. Hence by compactness (lemma VIII.2.4.1), there exists Z such that $\forall n \, \eta^+(n, (Z)^n, (Z)_n)$ holds. From this and the definition of η^+ it follows that $\forall n \, \forall Y \, (\eta(n, (Z)^n, Y) \, \to \, \eta(n, (Z)^n, (Z)_n))$. This proves the lemma.   □

THEOREM VIII.2.6. *The following is provable in* WKL$_0$. *For all* $X \subseteq \mathbb{N}$, *there exists a countable coded strict β-model M such that $X \in M$.*

PROOF. Let $\pi(e, m_1, X_1, X_2, X_3)$ be a universal lightface Π_1^0 formula with exactly the free variables shown (definition VII.1.3). We reason within WKL$_0$. Fix $X \subseteq \mathbb{N}$. By strong Π_1^0 dependent choice (lemma VIII.2.5), there exists W such that

$$\forall n \, \forall e \, \forall m \, \forall Z \, ((n = (e, m) \wedge \pi(e, m, X, (W)^n, Z)) \to$$
$$\pi(e, m, X, (W)^n, (W)_n)).$$

It is straightforward to verify that W is a code for a countable strict β-model M, and that $X \in M$. This completes the proof. (Compare the proof of theorem VII.7.4.)   □

COROLLARY VIII.2.7 (ω-submodels of WKL$_0$). *Let M' be any model of* WKL$_0$. *Then for any $X \in M'$ there exists $M \subseteq_\omega M'$ such that $M \neq M'$, $X \in M$, and M is again a model of* WKL$_0$.

PROOF. Let $X \in M' \models$ WKL$_0$ be given. Applying theorem VIII.2.6 within M', we get $W \in M'$ such that $M' \models (W$ is a code for a countable coded strict β-model M such that $X \in M)$. In particular, it follows that $M \subseteq_{S\beta} M'$. Hence by theorem VIII.2.2 $M \models$ WKL$_0$. Also $M \neq M'$ since the set $Y = \{n : n \notin (W)_n\}$ belongs to M' but not to M. This completes the proof.   □

COROLLARY VIII.2.8. *There is no minimal ω-model of* WKL$_0$. *In other words, every ω-model of* WKL$_0$ *has another such model properly contained within it.*

The following consequence of the proof of theorem VIII.2.6 will be useful later in this section.

LEMMA VIII.2.9. *There is a Π_1^0 formula $\psi(X, W)$ such that:*

1. WKL$_0$ *proves* $\forall X \, \exists W \, \psi(X, W)$;
2. RCA$_0$ *proves* $\forall X \, \forall W \, (\psi(X, W) \to W$ *is a code of a countable coded strict β-model M such that $X \in M)$.*

PROOF. As in the proof of theorem VIII.2.6, let $\pi(e, m_1, X_1, X_2, X_3)$ be universal lightface Π_1^0 with exactly the free variables shown. As in the proof of lemma VIII.2.5, write

$$\pi(e, m, X, Y, Z) \equiv \forall k \, \theta(e, m, X, Y, Z[k])$$

where θ is Σ_0^0, and define

$$\pi^+(e, m, X, Y, Z) \equiv$$
$$\forall k \, \forall \tau \, ((\mathrm{lh}(\tau) = k \wedge (\forall i \leq k) \, \theta(e, m, X, Y, \tau[i])) \to \theta(e, m, X, Y, Z[k])).$$

Then define $\psi(X, W)$ to be the Π_1^0 formula

$$\forall n \, \forall e \, \forall m \, (n = (e, m) \to \pi^+(e, m, X, (W)^n, (W)_n)).$$

As in the proof of lemma VIII.2.5, we can argue within WKL$_0$ that

$$\forall X \, \exists W \, \psi(X, W).$$

This proves part 1. For part 2, note that $\psi(X, W)$ implies

$$\forall n \, \forall e \, \forall m \, \forall Z \, ((n = (e, m) \wedge \pi(e, m, X, (W)^n, Z)) \to$$
$$\pi(e, m, X, (W)^n, (W)_n))$$

which, as in the proof of theorem VIII.2.6, implies that W is a code for a countable strict β-model containing X. This completes the proof of lemma VIII.2.9. □

Next, we consider the relationship between WKL$_0$ and ACA$_0$.

LEMMA VIII.2.10 (finite axiomatizability). WKL$_0$ *is finitely axiomatizable.*

PROOF. From §VIII.1 we know that RCA$_0$ is finitely axiomatizable. The axioms of WKL$_0$ can be taken to be those of RCA$_0$ plus the single axiom $\forall T$ (if T is an infinite subtree of $2^{<\mathbb{N}}$ then there exists a path through T). □

THEOREM VIII.2.11. *The following is provable in* ACA$_0$. *For all* $X \subseteq \mathbb{N}$, *there exists a countable ω-model M of* WKL$_0$ *such that* $X \in M$.

PROOF. We reason within ACA$_0$. Fix $X \subseteq \mathbb{N}$. By theorem VIII.2.6 there exists a countable coded strict β-model M such that $X \in M$. By lemma VIII.2.10 let φ be the conjunction of the axioms of WKL$_0$. Lemma VII.2.2 provides a valuation $f : \mathrm{Sub}_M(\varphi) \to \{0, 1\}$. We can then use the method of proof of theorem VIII.2.2 to verify that $f(\varphi) = 1$. This completes the proof. □

As in VIII.1.7 and VIII.1.8 we obtain the following corollaries.

COROLLARY VIII.2.12 (consistency of WKL$_0$). ACA$_0$ *proves the consistency of* WKL$_0$.

COROLLARY VIII.2.13. *There exists a Π_1^0 sentence ψ such that ψ is provable in* ACA$_0$ *but not in* WKL$_0$.

REMARK VIII.2.14. In connection with theorems VIII.2.6 and VIII.2.11, note that we have not claimed that WKL$_0$ proves the existence of a countable coded ω-model of WKL$_0$. Indeed, Gödel's second incompleteness theorem [94, 115, 55, 222] shows that this cannot be the case.

The next lemma is implicit in what we have already done, but we now pause in order to make it explicit. Recall from definition VII.1.4 that Y is X-recursive if and only if $Y \leq_T X$.

LEMMA VIII.2.15. *For any $X \subseteq \mathbb{N}$, there exists an X-recursive infinite tree $T \subseteq 2^{<\mathbb{N}}$ which has no X-recursive path. This is provable in* RCA$_0$.

PROOF. Let M' be any model of RCA$_0$. Given $X \in M'$, let M be the ω-submodel of M' consisting of all $Y \in M'$ such that $M' \models Y \leq_T X$. By theorem VIII.1.3 and corollary VIII.2.7, M is a model of RCA$_0$ but is not a model of WKL$_0$. Hence there exists $T \in M$ such that $M \models (T$ is an infinite subtree of $2^{<\mathbb{N}}$ which has no path). Thus $M' \models (T$ is an infinite X-recursive subtree of $2^{<\mathbb{N}}$ which has no X-recursive path). This shows that our lemma is true in any model of RCA$_0$. Hence, by the soundness theorem, our lemma is provable in RCA$_0$. □

The results which we have presented so far in this section are of fundamental importance. The rest of the section is devoted to results which are of more specialized interest. We consider the so-called *basis problem*: Given an X-recursive infinite tree $T \subseteq 2^{<\mathbb{N}}$, to find a path Y through T such that Y is in some sense "close to being X-recursive." Various solutions of the basis problem will be used to construct ω-models of WKL$_0$ with various properties.

A well known solution of the basis problem is given by the following lemma.

LEMMA VIII.2.16 (low basis theorem). *Let $X \subseteq \mathbb{N}$ be given, and let T be any X-recursive infinite subtree of $2^{<\mathbb{N}}$. Then there exists a path Y through T such that* TJ$(Y \oplus X) \leq_T$ TJ(X). *This result is provable in* ACA$_0$.

PROOF. As in the definition of Turing jump (definition VIII.1.9), let $\pi(e, m_1, X_1)$ be a universal lightface Π_1^0 formula. Fix $X \subseteq \mathbb{N}$ and define $\pi^*(n, X, Z) \equiv \forall e \forall m (n = (e, m) \rightarrow \pi(e, m, Z \oplus X))$. Thus for any Z we have TJ$(Z \oplus X) = \{n : \pi^*(n, X, Z)\}$. Let G be the set of all $\sigma \in \mathbb{N}^{<\mathbb{N}}$ such that $\exists Z (\forall i < \mathrm{lh}(\sigma)) \pi^*(\sigma(i), X, Z)$. By lemma VIII.2.4.2, the formula defining G is equivalent to a Π_1^0 formula. Hence $G \leq_T$ TJ(X). Let n_0 be such that $\forall Z (\pi^*(n_0, X, Z) \leftrightarrow Z$ is a path through $T)$. Define a sequence of finite sequences $\sigma_0 \subseteq \sigma_1 \subseteq \cdots \subseteq \sigma_n \subseteq \cdots$ by $\sigma_0 = \langle n_0 \rangle$, $\sigma_{n+1} = \sigma_n^\frown \langle n \rangle$ if $\sigma_n^\frown \langle n \rangle \in G$, otherwise $\sigma_{n+1} = \sigma_n$. Thus $\sigma_n \in G$ for all $n \in \mathbb{N}$, and the sequence $\langle \sigma_n : n \in \mathbb{N} \rangle$ is recursive in G. By compactness (lemma VIII.2.4.1), let Y be such that $\pi^*(n_0, X, Y)$ and $\pi^*(n, X, Y)$ for all n such that $\sigma_n^\frown \langle n \rangle \in G$. Thus Y is a path through T and, for all $n \in \mathbb{N}, n \in$ TJ$(Y \oplus X)$ if and only if $\sigma_n^\frown \langle n \rangle \in G$. Thus TJ$(Y \oplus X) \leq_T G$. Since $G \leq_T$ TJ(X) it follows that TJ$(Y \oplus X) \leq_T$ TJ(X). This completes the proof. □

THEOREM VIII.2.17. *For any $X \subseteq \mathbb{N}$, there exists a countable coded ω-model M of* WKL$_0$ *such that $X \in M$ and, for all $Y \in M$,* TJ$(Y \oplus X) \leq_T$ TJ(X). *This result is provable in* ACA$_0$.

Proof. Fix $X \subseteq \mathbb{N}$. Let $\psi(X, W)$ be a Π_1^0 formula as in lemma VIII.2.9. By lemma VIII.2.9 and the proof of theorem VIII.2.11, we have $\exists W \, \psi(X, W)$ and $\forall W \, (\psi(X, W) \to W$ is a code for a countable ω-model of WKL$_0$ which contains X). By the normal form theorem II.2.7, write $\psi(X, W)$ as $\forall m \, \theta(X, W[m])$ where θ is Σ_0^0. Let T be the tree of all $\tau \in 2^{<\mathbb{N}}$ such that $(\forall m \leq \mathrm{lh}(\tau)) \, \theta(X, \tau[m])$. Thus T is recursive in X and $\forall W \, (\psi(X, W) \leftrightarrow W$ is a path through T). Hence by lemma VIII.2.16 there exists W such that $\psi(X, W)$ and $\mathrm{TJ}(W \oplus X) \leq_{\mathrm{T}} \mathrm{TJ}(X)$. Let M be the countable ω-model of WKL$_0$ which is encoded by W. Then clearly $\mathrm{TJ}(Y \oplus X) \leq_{\mathrm{T}} \mathrm{TJ}(X)$ for all $Y \in M$. This completes the proof.   □

Corollary VIII.2.18. *There exists an ω-model M of* WKL$_0$ *such that every set $X \in M$ is low.*

(A set $X \subseteq \omega$ is said to be *low* if $\mathrm{TJ}(X) \leq_{\mathrm{T}} \mathrm{TJ}(\emptyset)$.)

A second solution of the basis problem is given by the following definition and lemma. Recall that $g \colon \mathbb{N} \to \mathbb{N}$ is said to be *majorized* by $f \colon \mathbb{N} \to \mathbb{N}$ if $f(m) \geq g(m)$ for all $m \in \mathbb{N}$.

Definition VIII.2.19. For $X, Y \subseteq \mathbb{N}$, we say that Y is *almost X-recursive* if for every $Y \oplus X$-recursive function $g \colon \mathbb{N} \to \mathbb{N}$ there exists an X-recursive function $f \colon \mathbb{N} \to \mathbb{N}$ such that f majorizes g. This definition is made in RCA$_0$.

Lemma VIII.2.20 (almost recursive basis theorem). *Let $X \subseteq \mathbb{N}$ be given. For any infinite X-recursive tree $T \subseteq 2^{<\mathbb{N}}$, there exists an almost X-recursive path Y through T. This result is provable in* ACA$_0$.

Proof. The proof is similar to that of lemma VIII.2.16. Let $\pi(e, m_1, m_2, X_1, X_2)$ be a universal lightface Π_1^0 formula with exactly the displayed free variables (definition VII.1.3). Let G be the set of all finite sequences of pairs

$$\langle (e_0, m_0), (e_1, m_1), \ldots, (e_k, m_k) \rangle$$

such that $\exists Z \, (\forall i \leq k) \forall n \, \pi(e_i, m_i, n, X, Z)$. Define an infinite sequence of finite sequences $\sigma_0 \subseteq \sigma_1 \subseteq \cdots \subseteq \sigma_e \subseteq \cdots$ in G as follows. Begin with $\sigma_0 = \langle (e_0', m_0') \rangle$ where e_0' and m_0' are chosen so that

$$\forall Z \, \forall n \, (\pi(e_0', m_0', n, X, Z) \leftrightarrow Z \text{ is a path through } T).$$

Since T has a path, $\sigma_0 \in G$. Given $\sigma_e \in G$, if there exists m such that $\sigma_e {}^\frown \langle (e, m) \rangle \in G$, let m_e' be the least such m and put $\sigma_{e+1} = \sigma_e {}^\frown \langle (e, m_e') \rangle$. Otherwise put $m_e' = 0$ and $\sigma_{e+1} = \sigma_e$. Finally by compactness (lemma VIII.2.4.1) let Y be such that $\forall n \, \pi(e_0', m_0', n, X, Y)$ and $\forall n \, \pi(e, m_e', n, X, Y)$ hold for all e such that $\sigma_{e+1} = \sigma_e {}^\frown \langle (e, m_e') \rangle$. In particular Y is a path through T.

We claim that Y is almost X-recursive. To see this, let $g \colon \mathbb{N} \to \mathbb{N}$ be $Y \oplus X$-recursive. The graph of g is Σ_1^0 in $Y \oplus X$ so let e be such that

$$\forall m \, \forall n \, (g(m) = n \leftrightarrow \neg\pi(e, m, n, X, Y)).$$

In particular we have $\neg\pi(e, m_e, g(m_e), X, Y)$, hence $m'_e = 0$ and $\sigma_e{}^\frown\langle(e, m'_e)\rangle \notin G$. Write

$$\psi(X, Z) \equiv (\forall i \leq k)\,\forall n\,\pi(e_i, m_i, n, X, Z)$$

where $\sigma_e = \langle(e_0, m_0), (e_1, m_1), \ldots, (e_k, m_k)\rangle$. Thus $\psi(X, Z)$ is Π^0_1 and we have $\psi(X, Y)$ and

$$\forall m\,\forall Z\,(\psi(X, Z) \to \exists n\,\neg\pi(e, m, n, X, Z)).$$

Hence by compactness (lemma VIII.2.4.1), it follows that

$$\forall m\,\exists j\,\forall Z\,(\psi(X, Z) \to (\exists n \leq j)\neg\pi(e, m, n, X, Z)).$$

By lemma VIII.2.4.2 the subformula

$$\forall Z\,(\psi(X, Z) \to (\exists n \leq j)\,\neg\pi(e, m, n, X, Z))$$

is equivalent to a Σ^0_1 formula, say $\exists i\,\theta(i, j, m, X)$ where θ is Σ^0_0. We have $\forall m\,\exists i\,\exists j\,\theta(i, j, m, X)$ so define $f : \mathbb{N} \to \mathbb{N}$ by putting $f(m) = $ least (i, j) such that $\theta(i, j, m, X)$. Then f is X-recursive and we have

$$\forall m\,\forall Z\,(\psi(X, Z) \to (\exists n \leq f(m))\,\neg\pi(e, m, n, X, Z)).$$

In particular $\forall m\,(\exists n \leq f(m))\,\neg\pi(e, m, n, X, Y)$, in other words

$$\forall m\, g(m) \leq f(m).$$

This completes the proof. $\square$

THEOREM VIII.2.21. *For any $X \subseteq \mathbb{N}$, there exists a countable coded ω-model M of* WKL$_0$ *such that $X \in M$ and, for all $Y \in M$, Y is almost X-recursive. This result is provable in* ACA$_0$.

PROOF. This follows from lemma VIII.2.20 just as theorem VIII.2.17 followed from lemma VIII.2.16. $\square$

COROLLARY VIII.2.22. *There exists an ω-model M of* WKL$_0$ *such that, for all $X \in M$, X is almost recursive.*

(A set $X \subseteq \omega$ is said to be *almost recursive* if every X-recursive function $g : \omega \to \omega$ is majorized by some recursive function $f : \omega \to \omega$.)

A third solution of the basis problem is given by the following lemma. Recall that, for $Y \subseteq \mathbb{N}$ and $j \in \mathbb{N}$,

$$(Y)_j = \{m : (m, j) \in Y\}.$$

LEMMA VIII.2.23 (GKT basis theorem). *Let $X \subseteq \mathbb{N}$ and $\langle A_i : i \in \mathbb{N}\rangle$, $A_i \subseteq \mathbb{N}$ be given such that $\forall i\, A_i \not\leq_T X$. For any infinite X-recursive tree $T \subseteq 2^{<\mathbb{N}}$, there exists a path Y through T such that $\forall i\,\forall j\, A_i \neq (Y)_j$. This result is provable in* ACA$_0$.

(Compare lemma VIII.6.4.)

PROOF. The proof is similar to that of lemma VIII.2.20. As usual, we shall identify $Z \subseteq \mathbb{N}$ with $Z \colon \mathbb{N} \to \{0, 1\}$, $Z(n) = 1$ if $n \in Z$, 0 otherwise. We define an infinite sequence of triples $(\varepsilon_k, m_k, j_k)$, $k \in \mathbb{N}$, as follows. Begin by putting $m_0 = j_0 = 0$ and $\varepsilon_0 = (Z)_0(0)$ where Z is some path through T. Now assume inductively that $(\varepsilon_0, m_0, j_0), \ldots, (\varepsilon_k, m_k, j_k)$ have already been defined so that $\exists Z \, \psi_k(X, Z)$ holds, where $\psi_k(X, Z)$ is the Π^0_1 formula

$$(Z \text{ is a path through } T) \wedge (\forall i \le k)\,((Z)_{j_i}(m_i) = \varepsilon_i).$$

We shall now define $(\varepsilon_{k+1}, m_{k+1}, j_{k+1})$. If $k \notin \mathbb{N} \times \mathbb{N}$, put

$$(\varepsilon_{k+1}, m_{k+1}, j_{k+1}) = (\varepsilon_k, m_k, j_k).$$

Otherwise, put $k = (i, j)$. We claim that there exists m such that

$$\exists Z \, (\psi_k(X, Z) \wedge A_i(m) \ne (Z)_j(m)).$$

If this were not so, we would have

$$\forall m \, (m \in A_i \leftrightarrow \exists Z \, (\psi_k(X, Z) \wedge m \in (Z)_j))$$

and

$$\forall m \, (m \in A_i \leftrightarrow \forall Z \, (\psi_k(X, Z) \to m \in (Z)_j))$$

which by lemma VIII.2.4.2 would imply that A_i is Δ^0_1 definable from X, i.e., $A_i \le_T X$, a contradiction. This proves the claim, so let m_{k+1} be the least m such that $\exists Z \, (\psi_k(X, Z) \wedge A_i(m) \ne (Z)_j(m))$, and put $j_{k+1} = j$ and $\varepsilon_{k+1} = 1 - A_i(m_{k+1})$. This completes the definition of $(\varepsilon_k, m_k, j_k)$ for all k. Now by compactness (lemma VIII.2.4.1), let Y be a path through T such that $\forall k \, (Y)_{j_k}(m_k) = \varepsilon_k$. This implies $\forall i \, \forall j \, \exists m \, (Y)_j(m) \ne A_i(m)$ and the proof is complete. $\square$

THEOREM VIII.2.24. *Let $X \subseteq \mathbb{N}$ and $\langle A_i \colon i \in \mathbb{N} \rangle$, $A_i \subseteq \mathbb{N}$ be given such that $\forall i \, A_i \not\le_T X$. Then there exists a countable coded ω-model M of WKL$_0$ such that $X \in M$ and $\forall i \, A_i \notin M$. This result is provable in ACA$_0$.*

PROOF. This follows from lemma VIII.2.23 just as theorem VIII.2.17 followed from lemma VIII.2.16. $\square$

COROLLARY VIII.2.25. *Given countably many nonrecursive sets A_i, $i \in \omega$, there exists a countable ω-model M of WKL$_0$ such that $\forall i \, A_i \notin M$.*

COROLLARY VIII.2.26. *For any countable ω-model M_1 of RCA$_0$, there exists a countable ω-model M_2 of WKL$_0$ such that $M_1 \cap M_2 = \text{REC}$, where REC $= \{X \colon X \text{ is recursive}\}$.*

PROOF. Let A_i, $i \in \omega$, be an enumeration of the nonrecursive sets in M_1 and apply corollary VIII.2.25. $\square$

COROLLARY VIII.2.27. *REC is the intersection of all ω-models of WKL$_0$.*

Notes for §VIII.2. Scott [217] has characterized countable ω-models of WKL$_0$ in the following way: M is a countable ω-model of WKL$_0$ if and only if there exists a complete extension T of PA such that for all $X \subseteq \mathbb{N}$, $X \in M$ if and only if X is representable in T. Such ω-models are sometimes known as *Scott systems*. Corollary VIII.2.7 is essentially due to Scott/Tennenbaum [218]; see also Jockusch/Soare [134]. Our lemma VIII.2.5 on strong Π_1^0 dependent choice appears to be new.

Theorem VIII.2.11 and lemma VIII.2.15 are well known, but their origins seem difficult to trace. See the references in Shoenfield [220], e.g., Kleene [142, §72]. According to Kleene [145, note 1], the use of the term "basis" is due to Kreisel. The GKT basis theorem VIII.2.23 is from Gandy/Kreisel/Tait [89]. The low and almost recursive basis theorems VIII.2.16 and VIII.2.20 are from Jockusch/Soare [134].

VIII.3. Hyperarithmetical Sets

In this section we shall develop a technical tool, *relative hyperarithmeticity*, which will be used later in the chapter to study ω-models of Δ_1^1-CA$_0$, ATR$_0$, and stronger systems.

For $X, Y \subseteq \mathbb{N}$, we shall say that Y *is hyperarithmetical in* X, abbreviated $Y \leq_{\mathrm{H}} X$, if Y can be obtained by starting with X and iterating the Turing jump operator along an X-recursive well ordering. (The details are in definitions VIII.3.5 and VIII.3.16 below.) We shall see that the principal axiom of ATR$_0$ is equivalent to the assertion that these iterations can be carried out (theorem VIII.3.15). The main theorem of this section is that Y is hyperarithmetical in X if and only if Y is Δ_1^1 in X (theorem VIII.3.19). At the end of the section, we shall use the method of pseudohierarchies (previously introduced in §V.4) to prove an important theorem about hyperarithmetical quantifiers.

The next three definitions are made in RCA$_0$.

DEFINITION VIII.3.1. An *X-recursive linear ordering* is a countable linear ordering (definition V.1.1) which is X-recursive (definition VII.1.4). The X-recursive linear ordering with X-recursive index e is denoted $\leq_e^X$.

DEFINITION VIII.3.2. Fix $X \subseteq \mathbb{N}$. We write $\mathcal{O}_+(a, X)$ to mean that $a = (e, i)$ for some e and i such that e is an X-recursive index of an X-recursive linear ordering $\leq_e^X$ and $i \in \mathrm{field}(\leq_e^X)$. The set of all a such that $\mathcal{O}_+(a, X)$ holds is denoted $\mathcal{O}_+^X$ if it exists. (For instance, the existence of $\mathcal{O}_+^X$ for all X is provable in ACA$_0$.) If $\mathcal{O}_+(a, X)$ and $\mathcal{O}_+(b, X)$, we write $b \leq_{\mathcal{O}}^X a$ (respectively $b <_{\mathcal{O}}^X a$) to mean that $a = (e, i)$ and $b = (e, j)$ for some e, i and j such that $j \leq_e^X i$ (respectively $j <_e^X i$). Note that $\leq_{\mathcal{O}}^X$ linearly orders $\{b : b <_{\mathcal{O}}^X a\}$ for each a such that $\mathcal{O}_+(a, X)$ holds.

DEFINITION VIII.3.3 (ordinal notations). Fix $X \subseteq \mathbb{N}$. We write $\mathcal{O}(a, X)$ to mean that $\mathcal{O}_+(a, X)$ and there is no infinite descending sequence $\langle a_n : n \in \mathbb{N}\rangle$, $a = a_0 >^X_{\mathcal{O}} a_1 >^X_{\mathcal{O}} \cdots >^X_{\mathcal{O}} a_n >^X_{\mathcal{O}} \cdots$. The set of all a such that $\mathcal{O}(a, X)$ holds is denoted $\mathcal{O}^X$ if it exists. (For instance, the existence of $\mathcal{O}^X$ for all X is provable in Π^1_1-CA$_0$.) Note that $\leq^X_{\mathcal{O}}$ well orders $\{b : b <^X_{\mathcal{O}} a\}$ for each a such that $\mathcal{O}(a, X)$ holds.

The ideas underlying the above definitions are as follows. Suppose that $a \in \mathcal{O}^X_+$ where $a = (e, i)$. Then we think of e as being an X-recursive *system of ordinal notations*, and we think of i as being one of the notations in the system. If in addition $a \in \mathcal{O}^X$, then this means that the system of notations e is well ordered up to and including the notation i. The ordinal of which i is a notation is then the order type of $\{j : j <^X_e i\}$ under $\leq^X_e$, i.e., the order type of $\{b : b <^X_{\mathcal{O}} a\}$ under $\leq^X_{\mathcal{O}}$. This ordinal might be written as $|i|^X_e$ or more simply $|a|^X$. The least ordinal not expressible as $|a|^X$ for some $a \in \mathcal{O}^X$ is sometimes denoted ω^X_1. In particular $\omega^{CK}_1 = \omega^{\emptyset}_1$.

The following lemma is a refinement of theorem V.1.9. It says in effect that, for any fixed $X \subseteq \mathbb{N}$, $\mathcal{O}^X$ is not Σ^1_1 definable from X.

LEMMA VIII.3.4. *Let $\varphi(n, X)$ be any Σ^1_1 formula with no free set variables other than X. Then* ACA$_0$ *proves*

$$\neg \forall a \, (\varphi(a, X) \leftrightarrow \mathcal{O}(a, X)).$$

PROOF. We reason in ACA$_0$. Fix X and put

$$\psi(m, X) \equiv \forall f \, \neg \pi(m, m, X, f)$$

where $\pi(e, m_1, X_1, X_2)$ is universal lightface Π^0_1 as in the definition of the hyperjump (definition VII.1.5). Note that $\psi(m, X)$ is a Π^1_1 formula obtained by diagonalization.

By the Kleene normal form theorem II.2.7, let $\theta(m, \sigma, \tau)$ be a Σ^0_0 formula such that $\forall m \, (\pi(m, m, X, f) \leftrightarrow \forall k \, \theta(m, X[k], f[k]))$. For each $m \in \mathbb{N}$ we have an X-recursive tree

$$T^X_m = \{\tau \in \mathbb{N}^{<\mathbb{N}} : (\forall k \leq \mathrm{lh}(\tau)) \, \theta(m, X[k], \tau[k])\}.$$

Let $R^X_m = \mathrm{KB}(T^X_m)$ be the Kleene/Brouwer ordering of T^X_m. Then for all $m \in \mathbb{N}$, R^X_m is an X-recursive linear ordering, and we have

$$\psi(m, X) \leftrightarrow \forall f \, \neg \pi(m, m, X, f)$$
$$\leftrightarrow \forall f \, \exists k \, \neg \theta(m, X[k], f[k])$$
$$\leftrightarrow T^X_m \text{ has no path}$$
$$\leftrightarrow R^X_m \text{ is a well ordering.}$$

Suppose now that $\forall a\,(\varphi(a, X) \leftrightarrow \mathcal{O}(a, X))$ where $\varphi(a, X)$ is Σ_1^1 with no free set variables other than X. Then for all $m \in \mathbb{N}$ we have

$\psi(m, X) \leftrightarrow R_m^X$ is a well ordering

$\qquad\qquad \leftrightarrow \exists a\,(\varphi(a, X) \wedge \exists$ isomorphism of R_m^X onto $\{b : b <_{\mathcal{O}}^X a\})$.

This is Σ_1^1 so, as in the proof of lemma VII.1.6, we have

$$\exists e\, \forall m\,(\psi(m, X) \leftrightarrow \exists f\, \pi(e, m, X, f)).$$

For this particular e, $\psi(e, X)$ is equivalent to $\exists f\, \pi(e, e, X, f)$, which is equivalent to $\neg\psi(e, X)$, a contradiction. This proves lemma VIII.3.4. $\square$

We now present the key definition of this section.

DEFINITION VIII.3.5 (H-sets). The following definition is made in ATR_0. Fix $X \subseteq \mathbb{N}$. For each a such that $\mathcal{O}(a, X)$ holds, we define a set $H_a^X \subseteq \mathbb{N}$ by

$$H_a^X = \{(m, 0) : m \in X\} \cup \{(m, b+1) : b <_{\mathcal{O}}^X a \wedge m \in \mathrm{TJ}(H_b^X)\}.$$

Here TJ denotes the Turing jump operator (definition VIII.1.9).

The sets H_a^X where $\mathcal{O}(a, X)$ holds are known as H-*sets*. The idea behind the H-sets is that H_a^X is the result of iterating the Turing jump operator along the X-recursive well ordering $\{b : b <_{\mathcal{O}}^X a\}$ starting with X. (See lemmas VIII.3.9, VIII.3.10 and VIII.3.13 below.) The existence and uniqueness of H_a^X are assured by arithmetical transfinite recursion and arithmetical transfinite induction, respectively. (See §V.2.)

We shall sometimes want to consider H-sets in situations where the full strength of arithmetical transfinite recursion is not available. We therefore generalize the concept of H-set as follows.

DEFINITION VIII.3.6. The following definition is made in ACA_0. Recall the notation $(Y)_k = \{m : (m, k) \in Y\}$. Let $\mathrm{H}(a, X, Y)$ be the arithmetical formula

$\mathcal{O}_+(a, X)$

$\wedge\, Y = \{(m, 0) : m \in X\} \cup \{(m, b+1) : b <_{\mathcal{O}}^X a \wedge m \in (Y)_{b+1}\}$

$\wedge\, \forall b(b <_{\mathcal{O}}^X a \to (Y)_{b+1} = \mathrm{TJ}(\{m, 0) : m \in X\}$

$\qquad\qquad\qquad\qquad \cup \{(m, c+1) : c <_{\mathcal{O}}^X b \wedge m \in (Y)_{c+1}\}))$.

Intuitively, $\mathrm{H}(a, X, Y)$ means that $\mathcal{O}_+(a, X)$ holds and that Y "looks like" H_a^X although $\mathcal{O}(a, X)$ is not assumed.

LEMMA VIII.3.7. *The following is provable in* ACA_0. *If* $\mathcal{O}(a, X)$ *holds, then there is at most one* Y *such that* $\mathrm{H}(a, X, Y)$ *holds.*

(Compare lemma V.2.3.)

PROOF. This is a special case of lemma V.2.3. The proof is immediate by arithmetical transfinite induction (lemma V.2.1). $\square$

DEFINITION VIII.3.8 (existence of H_a^X). This definition is made in ACA_0. Assume that $\mathcal{O}(a, X)$ holds. We write H_a^X *exists* to mean that $\exists Y\, H(a, X, Y)$, in which case we put $H_a^X =$ the unique Y such that $H(a, X, Y)$. (Lemma VIII.3.7 tells us that H_a^X is unique if it exists.)

LEMMA VIII.3.9. *The following is provable in* ACA_0. *Assume that* $\mathcal{O}(a, X)$ *holds.*

1. *If* H_a^X *exists and* $b <_{\mathcal{O}}^X a$, *then* H_b^X *exists and is* $<_T H_a^X$.
2. *Suppose that* $|a|^X = 0$, *i.e., there is no* b *such that* $b <_{\mathcal{O}}^X a$. *Then* H_a^X *exists and* $H_a^X =_T X$.
3. *Suppose that* $b <_{\mathcal{O}}^X a$ *and* $|a|^X = |b|^X + 1$, *i..e., there is no* c *such that* $b <_{\mathcal{O}}^X c <_{\mathcal{O}}^X a$. *Then* H_a^X *exists if and only if* H_b^X *exists, in which case* $H_a^X =_T TJ(H_b^X)$.

PROOF. If $b <_{\mathcal{O}}^X a$, then $TJ(H_b^X) = (H_a^X)_{b+1}$, hence $H_b^X <_T TJ(H_b^X) \leq_T H_a^X$. If $|a|^X = 0$, we have $H_a^X = \{(m, 0) : m \in X\}$ and obviously this is $=_T X$. Suppose now that $b <_{\mathcal{O}}^X a$ and $|a|^X = |b|^X + 1$. In this case, the desired conclusions follow easily from the identities $TJ(H_b^X) = (H_a^X)_{b+1}$ and

$$H_a^X = H_b^X \cup \{(m, b+1) : m \in TJ(H_b^X)\}.$$

This completes the proof of lemma VIII.3.9.   □

LEMMA VIII.3.10. *Assume that* $\mathcal{O}(a, X)$ *holds and that* H_a^X *exists. Suppose that* $|a|^X$ *is a limit ordinal, i.e., VIII.3.9.2 and VIII.3.9.3 do not apply. For any arithmetical formula* $\psi(m, X)$ *with exactly the free variables shown, it is provable in* ACA_0 *that the set*

$$\{(m, b) : b <_{\mathcal{O}}^X a \land \psi(m, H_b^X)\}$$

exists and is $\leq_T H_a^X$.

In order to prove lemma VIII.3.10, we first prove the following sublemmas.

SUBLEMMA VIII.3.11. *Let* $\psi(m, Y)$ *be an arithmetical formula with no free set variables other than* Y. *Then we can find* $k < \omega$ *such that* ACA_0 *proves*

$$\exists e\, \forall m\, \forall Y\, (\psi(m, Y) \leftrightarrow (m, e) \in TJ(k, Y)).$$

PROOF. We shall in fact show that k may be taken to be such that $\psi(m, Y)$ is a Π_k^0 formula. Assume first that $k = 1$, i.e., $\psi(m, Y)$ is Π_1^0. Let $\pi(e, m_1, X_1)$ be the universal lightface Π_1^0 formula with exactly the free variables shown, as in the definition of Turing jump (definition VIII.1.9). Thus ACA_0 (in fact RCA_0) proves $\exists e\, \forall m\, \forall Y\, (\psi(m, Y) \leftrightarrow \pi(e, m, Y))$. For this e we have $\forall m\, (\psi(m, Y) \leftrightarrow (m, e) \in TJ(Y))$ so the sublemma is proved in this case. Assume now that $\psi(m, Y)$ is Π_k^0, $1 < k < \omega$. Write $\psi(m, Y) \equiv \forall n\, \varphi(m, n, Y)$ where $\varphi(m, n, Y)$ is Σ_{k-1}^0. Then $\neg\varphi(m, n, Y)$ is Π_{k-1}^0 so by induction on k we have that ACA_0 proves

$\exists e'\,\forall m\,\forall n\,\forall Y\,(\neg\varphi(m,n,Y)\leftrightarrow((m,n),e')\in \mathrm{TJ}(k-1,Y))$. Note also that ACA_0 (in fact RCA_0) proves

$$\forall e'\,\exists e\,\forall m\,\forall Y\,(\pi(e,m,Y)\leftrightarrow \forall n\,((m,n),e')\notin Y).$$

Thus, reasoning in ACA_0, we have

$$\begin{aligned}
\psi(m,Y) &\leftrightarrow \forall n\,\varphi(m,n,Y)\\
&\leftrightarrow \forall n\,((m,n),e')\notin \mathrm{TJ}(k-1,Y)\\
&\leftrightarrow \pi(e,m,TJ(k-1,Y))\\
&\leftrightarrow (m,e)\in \mathrm{TJ}(k,Y).
\end{aligned}$$

This completes the proof of sublemma VIII.3.11. $\square$

SUBLEMMA VIII.3.12. *The following is provable in* ACA_0. *There is a fixed integer* $i_0\in\mathbb{N}$ *such that, for all* $Y\subseteq\mathbb{N}$ *and all* $k\in\mathbb{N}$,

$$\mathrm{TJ}((Y)_k)=((\mathrm{TJ}(Y))_{i_0})_k.$$

PROOF. Let $\pi(e,m_1,X_1)$ be a universal lightface Π_1^0 formula as in the definition of Turing jump (definition VIII.1.9). Since the formula $m\in \mathrm{TJ}((Y)_k)$ is Π_1^0, we can prove within ACA_0 (in fact within RCA_0) the existence of an integer e such that $\pi(e,(m,k),Y)\leftrightarrow m\in \mathrm{TJ}((Y)_k)$ holds for all $Y\subseteq\mathbb{N}$ and $k,m\in\mathbb{N}$. Letting i_0 be any such e, we have for all Y, k and m

$$\begin{aligned}
m\in \mathrm{TJ}((Y)_k) &\leftrightarrow \pi(i_0,(m,k),Y)\\
&\leftrightarrow ((m,k),i_0)\in \mathrm{TJ}(Y)\\
&\leftrightarrow (m,k)\in (TJ(Y))_{i_0}\\
&\leftrightarrow m\in ((\mathrm{TJ}(Y))_{i_0})_k.
\end{aligned}$$

Hence $\mathrm{TJ}((Y)_k)=((\mathrm{TJ}(Y)_{i_0})_k$ and the sublemma is proved. $\square$

PROOF OF LEMMA VIII.3.10. Let $\psi(m,Y)$ be arithmetical with no free set variables other than Y. By sublemma VIII.3.11, let $k<\omega$ be such that ACA_0 proves $\exists e\,\forall m\,\forall Y\,(\psi(m,Y)\leftrightarrow(m,e)\in \mathrm{TJ}(k,Y))$. Reasoning in ACA_0, assume that $\mathcal{O}(a,X)$ holds, that $|a|^X$ is a limit ordinal, and that H_a^X exists. Let

$$': \{b:b<_{\mathcal{O}}^X a\}\to\{b:b<_{\mathcal{O}}^X a\}$$

be such that $|b'|^X=|b|^X+1$ for all $b<_{\mathcal{O}}^X a$. The function $'$ is clearly $\leq_{\mathrm{T}} \mathrm{TJ}(X)$ and hence $\leq_{\mathrm{T}} \mathrm{H}_a^X$. For each $b<_{\mathcal{O}}^X a$, we have $b<_{\mathcal{O}}^X b'<_{\mathcal{O}}^X a$, hence

$$\text{TJ}(H_b^X) = (H_{b'}^X)_{b+1} = (H_a^X)_{b+1};$$

$$\text{TJ}(2, H_b^X) = \text{TJ}(\text{TJ}(H_b^X)) = \text{TJ}((H_{b'}^X)_{b+1}) = ((\text{TJ}(H_{b'}^X))_{i_0})_{b+1}$$

$$= (((H_{b''}^X)_{b'+1})_{i_0})_{b+1} = (((H_a^X)_{b'+1})_{i_0})_{b+1};$$

$$\text{TJ}(3, H_b^X) = \text{TJ}(\text{TJ}(2, H_b^X)) = \text{TJ}((((H_{b''}^X)_{b'+1})_{i_0})_{b+1})$$

$$= ((((((\text{TJ}(H_{b''}^X))_{i_0})_{b'+1})_{i_0})_{i_0})_{b+1}$$

$$= ((((((H_{b'''}^X)_{b''+1})_{i_0})_{b'+1})_{i_0})_{i_0})_{b+1}$$

$$= ((((((H_a^X)_{b''+1})_{i_0})_{b'+1})_{i_0})_{i_0})_{b+1};$$

$$\text{TJ}(4, H_b^X) = \text{TJ}(\text{TJ}(3, H_b^X)) = \cdots;$$

etc., where i_0 is as in sublemma VIII.3.12. If for instance $k = 2$, then for an appropriate e and all $b <_{\mathcal{O}}^X a$ and all m, we have

$$\psi(m, H_b^X) \leftrightarrow (m, e) \in \text{TJ}(2, H_b^X)$$

$$\leftrightarrow (m, e) \in (((H_a^X)_{b'+1})_{i_0})_{b+1}$$

from which it follows immediately that

$$\{(m, b) : b <_{\mathcal{O}}^X a \wedge \psi(m, H_b^X)\}$$

is recursive in H_a^X. This proves lemma VIII.3.10.   □

The next lemma implies that the Turing degree of H_a^X depends only on the ordinal $|a|^X$. Thus, for every ordinal $\alpha < \omega_1^X$, we may define the αth Turing jump of X by putting $\text{TJ}(\alpha, X) = H_a^X$ for some $a \in \mathcal{O}^X$ with $|a|^X = \alpha$, and this is well defined up to Turing degree.

LEMMA VIII.3.13. *The following is provable in* ACA$_0$. *Suppose that* $\mathcal{O}(a, X)$ *and* $\mathcal{O}(a^*, X)$. *Assume that* $|a|^X = |a^*|^X$, *i.e., there exists an order isomorphism of* $\{b : b <_{\mathcal{O}}^X a\}$ *onto* $\{c : c <_{\mathcal{O}}^X a^*\}$. *If* H_a^X *exists, then* $H_{a^*}^X$ *exists and is* $=_{\text{T}} H_a^X$.

PROOF. Assume that H_a^X exists. We want to show that $H_{a^*}^X$ exists and is $=_{\text{T}} H_a^X$. Let f be an order isomorphism of $\{b : b <_{\mathcal{O}}^X a\}$ onto $\{c : c <_{\mathcal{O}}^X a^*\}$. By arithmetical transfinite induction, we may assume that, for all $b <_{\mathcal{O}}^X a$, $H_{f(b)}^X$ exists and is $=_{\text{T}} H_b^X$. If $|a|^X = 0$, it follows that $|a^*|^X = 0$ and by VIII.3.9.2 we have $H_{a^*}^X =_{\text{T}} X =_{\text{T}} H_a^X$. If $|a|^X = |b|^X + 1$ for some $b <_{\mathcal{O}}^X a$, it follows that $|a^*|^X = |c|^X + 1$ where $c = f(b) <_{\mathcal{O}}^X a^*$. Hence by VIII.3.9.3 we have that $H_{a^*}^X$ exists and is $=_{\text{T}} \text{TJ}(H_c^X) =_{\text{T}} \text{TJ}(H_b^X) =_{\text{T}} H_a^X$. Suppose now that $|a|^X$ is a limit ordinal. By lemma VIII.3.10, the set

$$\{(b, (c, m)) : b <_{\mathcal{O}}^X a \wedge c <_{\mathcal{O}}^X a^* \wedge \exists Y (Y =_{\text{T}} H_b^X \wedge H(c, X, Y) \wedge m \in \text{TJ}(Y))\}$$

$$= \{(b, (f(b), m)) : b <_{\mathcal{O}}^X a \wedge m \in \text{TJ}(H_{f(b)}^X)\}$$

exists and is $\leq_T H_a^X$. From this it follows easily that H_{a*}^X exists and is $\leq_T H_a^X$. By symmetry $H_{a*}^X =_T H_a^X$ and the lemma is proved. $\square$

Recall the discussion of comparability of countable well orderings in §§V.2 and V.6. The following lemma is a refinement of lemma V.2.9.

LEMMA VIII.3.14. *The following is provable in* ACA_0. *Assume that* $\mathcal{O}(a, X)$ *and* $\mathcal{O}(a^*, X)$ *and that* H_a^X *exists. Then the countable well orderings* $\{b : b <_{\mathcal{O}}^X a\}$ *and* $\{c : c <_{\mathcal{O}}^X a^*\}$ *are comparable. Furthermore, the comparison map is* $\leq_T H_a^X$.

PROOF. Let f be the set of pairs (b, c) such that $b <_{\mathcal{O}}^X a$ and $c <_{\mathcal{O}}^X a^*$ and $H_b^X =_T H_c^X$, i.e., $\exists Y (Y =_T H_b^X \wedge H(c, X, Y))$. By lemma VIII.3.10, f exists and is $\leq_T H_a^X$. By lemmas VIII.3.9 and VIII.3.13 and arithmetical transfinite induction (using f as a parameter), it follows that either f is an isomorphism of $\{b : b <_{\mathcal{O}}^X a\}$ onto $\{c : c <_{\mathcal{O}}^X a^*\}$, or f is an isomorphism of $\{b : b <_{\mathcal{O}}^X a\}$ onto some initial section of $\{c : c <_{\mathcal{O}}^X a^*\}$, or f is an isomorphism of some initial section of $\{b : b <_{\mathcal{O}}^X a\}$ onto $\{c : c <_{\mathcal{O}}^X a^*\}$. In any case f is a comparison map from $\{b : b <_{\mathcal{O}}^X a\}$ to $\{c : c <_{\mathcal{O}}^X a^*\}$. This completes the proof. $\square$

THEOREM VIII.3.15. ATR_0 *is equivalent over* ACA_0 *to*

$$\forall X \forall a \, (\mathcal{O}(a, X) \to H_a^X \text{ exists}). \tag{22}$$

PROOF. If $\mathcal{O}(a, X)$ holds, the existence of H_a^X can be proved by a direct application of arithmetical transfinite recursion along the countable well ordering $\{b : b <_{\mathcal{O}}^X a\}$. This shows that ATR_0 implies (22). Conversely, if (22) holds, then by lemma VIII.3.14 any two countable well orderings are comparable, and by theorem V.6.8 this implies arithmetical transfinite recursion. $\square$

DEFINITION VIII.3.16. The following definition is made in ACA_0. Given $X, Y \subseteq \mathbb{N}$, we say that Y *is hyperarithmetical in* X, abbreviated $Y \leq_H X$, if there exists a such that $\mathcal{O}(a, X)$ holds and H_a^X exists and $Y \leq_T H_a^X$. We say that Y *is hyperarithmetical if* $Y \leq_H \emptyset$. (Here $\emptyset$ denotes the empty set.)

The following lemma will be useful in several places.

LEMMA VIII.3.17. *We can find a* Π_1^1 *formula* $\nu(i, X)$ *and a* Σ_1^1 *formula* $\alpha(i, X, Y)$, *with no free variables other than those displayed, such that* ACA_0 *proves*

$$\forall X \forall Y (Y \leq_H X \leftrightarrow \exists i (\nu(i, X) \wedge \alpha(i, X, Y)))$$

and

$$\forall X \forall i \forall Y \forall Y' ((\nu(i, X) \wedge \alpha(i, X, Y) \wedge \alpha(i, X, Y')) \to Y = Y'),$$

while ATR_0 *proves*

$$\forall X \forall i (\nu(i, X) \to \exists Y \alpha(i, X, Y)).$$

PROOF. If $Y \leq_H X$, then $Y \leq_T H_a^X$ where $\mathcal{O}(a, X)$ holds. In particular there exists e such that $Y = (\mathrm{TJ}(H_a^X))_e$. Let $v(i, X)$ say that i is of the form (a, e) where $\mathcal{O}(a, X)$ holds, and let $\alpha(i, X, Y)$ say that $\exists Z\,(\mathrm{H}(a, X, Z) \wedge Y = (\mathrm{TJ}(Z))_e)$ where $i = (a, e)$. The desired properties of the formulas $v(i, X)$ and $\alpha(i, X, Z)$ follow easily from lemma VIII.3.7. $\qquad \square$

DEFINITION VIII.3.18 (Δ_1^1 definability). The following definition is made in ACA$_0$. Given $X, Y \subseteq \mathbb{N}$, we say that Y is Σ_1^1 in X if

$$\exists e \, \forall m \, (m \in Y \leftrightarrow \exists f \, \pi(e, m, X, f)).$$

Here $\pi(e, m_1, X_1, X_2)$ is a fixed universal lightface Π_1^0 formula as in the definition of the hyperjump (definition VII.1.5).

Note that, by the normal form theorem for Σ_1^1 relations (lemma V.1.4), $\exists f \, \pi(e, m, X, f)$ is a universal lightface Σ_1^1 formula. Thus Y is Σ_1^1 in X if and only if $Y = \{m : \varphi(m, X)\}$ for some Σ_1^1 formula $\varphi(m, X)$ with no free set variables (i.e., set parameters) other than X.

We say that Y is Π_1^1 in X if $\mathbb{N} \setminus Y$ is Σ_1^1 in X. We say that Y is Δ_1^1 in X if Y is both Σ_1^1 in X and Π_1^1 in X.

The following theorem is the main result of this section.

THEOREM VIII.3.19 (Kleene/Souslin theorem in ACA$_0$). *The following is provable in* ACA$_0$. *Let* $X \subseteq \mathbb{N}$ *be such that* $\forall a \, (\mathcal{O}(a, X) \rightarrow H_a^X \text{ exists})$. *Then for all* $Z \subseteq \mathbb{N}$, Z *is hyperarithmetical in* X *if and only if* Z *is* Δ_1^1 *in* X.

PROOF. We reason in ACA$_0$. Suppose first that Z is hyperarithmetical in X. Using the notation of lemma VIII.3.17, let i be such that $v(i, X)$ and $\alpha(i, X, Z)$ hold. For all $m \in \mathbb{N}$ we have

$$m \in Z \leftrightarrow \exists Y (\alpha(i, X, Y) \wedge m \in Y)$$
$$\leftrightarrow \forall Y (\alpha(i, X, Y) \rightarrow m \in Y)$$

which shows that Z is Δ_1^1 in X.

For the converse, assume that Z is Δ_1^1 in X, say $Z = \{m : \varphi(m, X)\} = \{m : \psi(m, X)\}$ where $\varphi(m, X)$ and $\psi(m, X)$ are respectively Σ_1^1 and Π_1^1 with no free set variables other than X. As in the proof of lemma VIII.3.4, let $\langle R_m^X : m \in X \rangle$ be an X-recursive sequence of X-recursive linear orderings such that

$$\forall m \, (\psi(m, X) \leftrightarrow R_m^X \text{ is a well ordering}).$$

We claim that the order types of the well ordered R_m^X's are bounded. In other words, we claim that there exists an a such that $\mathcal{O}(a, X)$ holds and $\forall m \, (R_m^X$ well ordered $\rightarrow R_m^X$ is isomorphic to some initial section of $\{b : b <_{\mathcal{O}}^X a\})$. Note first that our assumption $\forall a \, (\mathcal{O}(a, X) \rightarrow H_a^X \text{ exists})$ implies by lemma VIII.3.14 that any two X-recursive well orderings are comparable. Now if our claim were false, then by comparability of X-recursive well orderings we would have $\forall a \, (\mathcal{O}(a, X) \leftrightarrow \varphi'(a, X))$ where $\varphi'(a, X)$ is Σ_1^1, namely $\varphi'(a, X) \equiv (\mathcal{O}_+(a, X) \wedge \exists m \, (\varphi(m, X) \wedge$ there exists an isomorphism of $\{b : b <_{\mathcal{O}}^{X} a\}$ onto some initial section of R_m^X)). This would contradict lemma VIII.3.4.

Let a be as in the previous claim. Then by lemma VIII.3.14 we have $\forall m \, (m \in Z \leftrightarrow (\exists f \leq_{\mathrm{T}} \mathrm{H}_a^X) \, (f \text{ is an isomorphism of } R_m^X \text{ onto some initial section of } \{b : b <_{\mathcal{O}}^{X} a\}))$. Thus Z is arithmetically definable from H_a^X. By sublemma VIII.3.11 it follows that $Z = (\mathrm{TJ}(k, \mathrm{H}_a^X))_e$ for some $k, e \in \mathbb{N}$. Letting a^* be such that $\mathcal{O}(a^*, X)$ and $|a^*|^X = |a|^X + k$, it follows by lemmas VIII.3.13 and VIII.3.9.3 that $Z \leq_{\mathrm{T}} \mathrm{H}_{a^*}^X$. Thus Z is hyperarithmetical in X. This completes the proof of the theorem. □

We shall now end this section by presenting two theorems about hyperarithmetical quantifiers. Given an L_2-formula φ, we shall write $(\forall Y \leq_{\mathrm{H}} X)\varphi$ as an abbreviation for

$$\forall Y (Y \leq_{\mathrm{H}} X \to \varphi).$$

The expression $(\forall Y \leq_{\mathrm{H}} X)$ is known as a *hyperarithmetical quantifier*. The first of our two theorems says that the class of Σ_1^1 formulas is closed under universal hyperarithmetical quantification. The second theorem is a sort of converse to the first. We prove both theorems in ATR_0.

THEOREM VIII.3.20 (hyperarithmetical quantifiers, 1). *For any Σ_1^1 formula $\varphi(X, Y)$, we can find a Σ_1^1 formula $\varphi'(X)$ such that ATR_0 proves*

$$\varphi'(X) \leftrightarrow (\forall Y \leq_{\mathrm{H}} X)\varphi(X, Y).$$

(Note that $\varphi(X, Y)$ may contain free variables other than X and Y. If this is the case, then $\varphi'(X)$ will also contain those free variables.)

PROOF. Using the notation of lemma VIII.3.17, we have

$$(\forall Y \leq_{\mathrm{H}} X)\varphi(X, Y) \leftrightarrow \forall i \, (\nu(i, X) \to \exists Y \, (\alpha(i, X, Y) \wedge \varphi(X, Y)))$$
$$\leftrightarrow \forall i \, \varphi''(i, X)$$

where $\varphi''(i, X)$ is Σ_1^1. Our theorem therefore reduces to the following lemma.

LEMMA VIII.3.21. *For any Σ_1^1 formula $\varphi''(n)$, we can find a Σ_1^1 formula φ' such that $\varphi' \leftrightarrow \forall n \, \varphi''(n)$ is provable in ATR_0 (actually in Σ_1^1-AC_0).*

(Note that $\varphi''(n)$ may contain free variables other than n. In this case φ' will also contain those free variables. See also lemma VIII.6.2.)

PROOF. Let us write $\varphi''(n) \equiv \exists Z \, \theta(n, Z)$ where θ is arithmetical and Z is a set variable. Recall from theorem V.8.3 that the Σ_1^1 axiom of choice is provable in ATR_0. (See also §VII.6.) By Σ_1^1 choice we have

$$\forall n \, \varphi''(n) \leftrightarrow \forall n \, \exists Z \, \theta(n, Z)$$
$$\leftrightarrow \exists W \, \forall n \, \theta(n, (W)_n)$$

and the latter expression is Σ_1^1. This proves lemma VIII.3.21 and theorem VIII.3.20. □ □

The rest of this section is devoted to a proof of a sort of converse to theorem VIII.3.20 (see theorem VIII.3.27 below).

LEMMA VIII.3.22. *The following is provable in* ACA$_0$. *Assume that* $\mathcal{O}(a, X)$ *holds and that* $|a|^X$ *is a limit ordinal. Let A be a set such that* $\forall b \, (b <^X_{\mathcal{O}} a \rightarrow H^X_b$ *exists and is* $\leq_T A)$. *Then* H^X_a *exists and is* $\leq_T TJ(2, A)$.

PROOF. We have $(m, b + 1) \in H^X_a$ if and only if

$$(\exists Y \leq_T A) \, (H(b, X, Y) \wedge m \in TJ(Y)).$$

Thus H^X_a is uniformly arithmetically definable from A. Hence by sublemma VIII.3.11 we can find a fixed $k < \omega$ such that our lemma holds with k in place of 2. A more careful computation shows that H^X_a is Δ^0_3 in A, hence by the proof of sublemma VIII.3.11 our lemma holds with $k = 2$. (In the application of our lemma to be made below, only the finiteness of k is important.) □

LEMMA VIII.3.23. *The following is provable in* ACA$_0$. *Let* $\langle A_n : n \in \mathbb{N} \rangle$ *be a sequence of sets such that* $\forall n \, (TJ(A_{n+1}) \leq_T A_n)$, *and let X be a set such that* $\forall n \, (X \leq_T A_n)$. *Then*

$$\forall a \, (\mathcal{O}(a, X) \rightarrow H^X_a \text{ exists})$$

and

$$\forall Y \, (Y \leq_H X \rightarrow \forall n \, (Y \leq_T A_n)).$$

In particular, none of the sets A_n *is hyperarithmetical in X.*

PROOF. Fix X and let a be such that $\mathcal{O}(a, X)$ holds. We wish to prove that H^X_a exists and is $\leq_T A_n$ for all n. This assertion is arithmetical so we may prove it by arithmetical transfinite induction. If $|a|^X = 0$, we have by VIII.3.9.2 $H^X_a =_T X \leq_T A_n$ for all n. If $|a|^X = |b|^X + 1$ for some $b <^X_{\mathcal{O}} a$, we have inductively $H^X_b \leq_T A_n$ for all n, hence by VIII.3.9.3 $H^X_a =_T TJ(H^X_b) \leq_T TJ(A_{n+1}) \leq_T A_n$ for all n. If $|a|^X$ is a limit ordinal, we have inductively $H^X_b \leq_T A_n$ for all $b <^X_{\mathcal{O}} a$ and all n, hence by the previous lemma $H^X_a \leq_T TJ(2, A_{n+2}) \leq_T A_n$ for all n. This completes the proof. □

LEMMA VIII.3.24. *The following is provable in* ACA$_0$. *Fix* $X \subseteq \mathbb{N}$ *and assume that* $\forall a \, (\mathcal{O}(a, X) \rightarrow H^X_a$ *exists). Then there exists* a^* *such that*

$$\mathcal{O}_+(a^*, X) \wedge \exists Y \, H(a^*, X, Y) \wedge \neg \mathcal{O}(a^*, X).$$

PROOF. (This is really a special case of lemma V.4.12 on the existence of pseudohierarchies.) Let $\mathcal{O}_1(a, X)$ be the formula $\mathcal{O}_+(a, X) \wedge \exists Y \, H(a, X, Y)$. By assumption we have $\forall a \, (\mathcal{O}(a, X) \rightarrow \mathcal{O}_1(a, X))$. Since $\mathcal{O}_1(a, X)$ is Σ^1_1, we have by lemma VIII.3.4 $\neg \forall a \, (\mathcal{O}(a, X) \leftrightarrow \mathcal{O}_1(a, X))$, hence $\exists a \, (\mathcal{O}_1(a, X) \wedge \neg \mathcal{O}(a, X))$. Letting a^* be any such a, we obtain the desired conclusion. □

LEMMA VIII.3.25. *The following is provable in* ACA$_0$. *Suppose that* $\mathcal{O}_+(a, X) \wedge H(a, X, Y) \wedge \neg\mathcal{O}(a, X)$ *holds. Then* Y *is not hyperarithmetical in* X. *In fact, we have* $\forall Z \, (Z \leq_H X \rightarrow Z \leq_T Y)$.

PROOF. For each $b \leq_{\mathcal{O}}^X a$ put

$$Y_b = \{(m, 0) \colon (m, 0) \in Y\} \cup \{(m, c+1) \colon c <_{\mathcal{O}}^X b \wedge (m, c+1) \in Y\}.$$

Then $Y_a = Y$ and, for all $c <_{\mathcal{O}}^X b \leq_{\mathcal{O}}^X a$, $(Y_b)_{c+1} = \mathrm{TJ}(Y_c)$. Since $\neg\mathcal{O}(a, X)$ holds, there exists a descending sequence

$$a = a_0 >_{\mathcal{O}}^X a_1 >_{\mathcal{O}}^X \cdots >_{\mathcal{O}}^X a_n >_{\mathcal{O}}^X \cdots.$$

Setting $A_n = Y_{a_n}$, we have $X \leq_T A_n$ and $\mathrm{TJ}(A_{n+1}) \leq_T A_n$ for all n. By lemma VIII.3.23, the desired conclusions follow. $\square$

LEMMA VIII.3.26. *In* ATR$_0$ *we have*

$$\forall X \, \forall a \, (\mathcal{O}(a, X) \leftrightarrow (\exists Y \leq_H X) \, H(a, X, Y)).$$

PROOF. If $\mathcal{O}(a, X)$ holds, then obviously $Y = H_a^X$ satisfies $Y \leq_H X$ and $H(a, X, Y)$. Conversely, suppose that $H(a, X, Y)$ holds and $Y \leq_H X$. Then $\mathcal{O}(a, X)$ follows by lemma VIII.3.25. This completes the proof. $\square$

THEOREM VIII.3.27 (hyperarithmetical quantifiers, 2). *Let* $\varphi(X)$ *be a* Σ_1^1 *formula with no free set variables other than* X. *Then we can find an arithmetical formula* $\theta(X, Z)$ *such that* ATR$_0$ *proves*

$$\forall X \, (\varphi(X) \leftrightarrow (\forall Z \leq_H X) \, \theta(X, Z)).$$

PROOF. For simplicity, assume that $\varphi(X)$ has only one free number variable, call it m. The formula $\neg\varphi(m, X)$ is Π_1^1, so as in the proof of lemma VIII.3.4 we can find an X-recursive sequence of X-recursive linear orderings $\langle R_m^X \colon m \in \mathbb{N} \rangle$ such that $\forall X \, \forall m \, (\varphi(m, X) \leftrightarrow R_m^X$ is not a well ordering). Now for any particular m and X, R_m^X is isomorphic to $\{b \colon b <_{\mathcal{O}}^X a\}$ for some $a \in \mathcal{O}_+^X$. If moreover R_m^X is a well ordering, then $\mathcal{O}(a, X)$ holds and by lemma VIII.3.14 the isomorphism of R_m^X onto $\{b \colon b <_{\mathcal{O}}^X a\}$ is $\leq_H X$. Thus we have

$\neg\varphi(m, X)$

$\qquad \leftrightarrow R_m^X$ is a well ordering

$\qquad \leftrightarrow \exists a \, (\mathcal{O}(a, X) \wedge (\exists f \leq_H X) \, (f \colon |R_m^X| = |a|^X))$

$\qquad \leftrightarrow \exists a \, ((\exists Y \leq_H X) \, H(a, X, Y) \wedge (\exists f \leq_H X) \, (f \colon |R_m^X| = |a|^X))$

where the last equivalence follows from lemma VIII.3.26. We now have

$$\neg\varphi(m, X) \leftrightarrow (\exists Z \leq_H X) \, \psi(m, X, Z)$$

where $\psi(m, X, Z) \equiv (Z$ encodes a triple (a, Y, f) such that $H(a, X, Y)$ and $f \colon |R_m^X| = |a|^X)$. Note that $\psi(m, X, Z)$ is arithmetical. Setting $\theta(m, X, Z) \equiv \neg\psi(m, X, Z)$, we obtain $\varphi(m, X) \leftrightarrow (\forall Z \leq_H X) \, \theta(m, X, Z)$. This completes the proof. $\square$

Notes for §VIII.3. Our exposition of hyperarithmetical theory here is somewhat idiosyncratic in that it avoids the use of the recursion theorem. An orthodox exposition is in Sacks [211, part A]. Other relevant references are Harrison [106] and Steel [255].

Historically, hyperarithmetical theory is the creation of Davis, Mostowski, and Kleene. (For bibliographical references, see Spector [253].) The fact that $|a|^X = |b|^X$ implies $H_a^X \equiv_{\mathrm{T}} H_b^X$ is due to Spector [253]. The so-called Kleene/Souslin theorem $\Delta_1^1(X) = \mathrm{HYP}(X)$ is due to Kleene [144]. The hyperarithmetical quantifier theorem $\Pi_1^1(X) = (\Sigma_1^1)^{\mathrm{HYP}(X)}$ is due to Spector [254] and Gandy [88].

An important feature of this section is that we have shown how to formalize hyperarithmetical theory within relatively weak subsystems of Z_2. Such formalization was apparently first undertaken by Friedman [62, chapter II]. This eventually led to the discovery of the system ATR_0. See Steel [256], Friedman [68, 69], and Friedman/McAloon/Simpson [76].

VIII.4. ω-Models of Σ_1^1 Choice

Recall from §§VII.5 and VII.6 that Δ_1^1-CA_0, Σ_1^1-AC_0, and Σ_1^1-DC_0 are the subsystems of second order arithmetic with Δ_1^1 comprehension, Σ_1^1 choice, and Σ_1^1 dependent choice. The purpose of this section is to discuss ω-models of these three systems. We show that all three systems have the same minimum (i.e., unique smallest) ω-model, namely

$$\mathrm{HYP} = \{X \subseteq \omega : X \text{ is hyperarithmetical}\}$$

(corollary VIII.4.17). In addition, we show that ATR_0 proves the existence of countable coded ω-models of all three systems (theorem VIII.4.20).

LEMMA VIII.4.1. ATR_0 *proves* Δ_1^1 *comprehension and* Σ_1^1 *choice.*

PROOF. This follows from theorem V.8.3 and lemma VII.6.6.1. □

REMARK VIII.4.2. We shall see later (theorem VIII.5.13) that ATR_0 does not prove Σ_1^1 dependent choice.

DEFINITION VIII.4.3 (relativization to $\mathrm{HYP}(X)$). For any set variable X and any L_2-formula φ in which X does not occur quantified, let $\varphi^{\mathrm{HYP}(X)}$ be the L_2-formula which is obtained from φ by relativizing all of the set quantifiers in φ to sets which are hyperarithmetical in X. Thus each quantifier $\forall Y$ is replaced by $(\forall Y \leq_{\mathrm{H}} X)$, etc. (See the discussion of hyperarithmetical quantifiers, at the end of the previous section.) The formula $\varphi^{\mathrm{HYP}(X)}$ is called *the relativization of φ to* $\mathrm{HYP}(X)$. We sometimes express $\varphi^{\mathrm{HYP}(X)}$ by saying that $\mathrm{HYP}(X)$ *satisfies φ.*

REMARK VIII.4.4. Note that ATR_0 is not strong enough to prove the existence of the countable coded ω-model

$$\text{HYP}(X) = \{\, Y \subseteq \mathbb{N} \colon Y \leq_H X \,\}$$

consisting exactly of those sets Y which are hyperarithmetical in a given set X. (To see this, let M be a countable β-model of ATR_0 such that $X \in M$ and $\text{HJ}(X) \notin M$, as in corollary VII.2.12. If M were to contain a code for the countable ω-model $\text{HYP}(X)$, it would follow by theorem VIII.3.27 that $\text{HJ}(X) \in M$, a contradiction.)

Nevertheless, we can use relativization (definition VIII.4.3) to state the following as a theorem of ATR_0.

THEOREM VIII.4.5 (Δ^1_1 comprehension in $\text{HYP}(X)$). *The following is provable in* ATR_0. *For any* $X \subseteq \mathbb{N}$, $\text{HYP}(X)$ *satisfies* Δ^1_1 *comprehension. In other words, we have*

$$\forall X\, (\Delta^1_1 \text{ comprehension})^{\text{HYP}(X)}.$$

PROOF. Assume

$$\forall n (\varphi(n) \leftrightarrow \psi(n))^{\text{HYP}(X)}$$

where $\varphi(n)$ is Σ^1_1, $\psi(n)$ is Π^1_1, and all of the set parameters in $\varphi(n)$ and $\psi(n)$ are hyperarithmetical in X. Write $\varphi(n) \equiv \exists Y \varphi'(n, Y)$ and $\psi(n) \equiv \forall Y \psi'(n, Y)$, where $\varphi'(n, Y)$ and $\psi'(n, Y)$ are arithmetical. Thus $\varphi(n)^{\text{HYP}(X)} \equiv (\exists Y \leq_H X)\, \varphi'(Y, n)$ and $\psi(n)^{\text{HYP}(X)} \equiv (\forall Y \leq_H X)\, \psi'(Y, n)$. By theorem VIII.3.20, $(\exists Y \leq_H X)\, \varphi'(Y, n)$ is equivalent to a Π^1_1 formula $\varphi''(X, n)$, and similarly $(\forall Y \leq_H X)\, \psi'(Y, n)$ is equivalent to a Σ^1_1 formula $\psi''(X, n)$, where $\varphi''(X, n)$ and $\psi''(X, n)$ contain no free set variables other than X. Our assumption $\forall n\, (\varphi(n) \leftrightarrow \psi(n))^{\text{HYP}(X)}$ now reads $\forall n\, (\varphi''(X, n) \leftrightarrow \psi''(X, n))$. Hence by Δ^1_1 comprehension (lemma VIII.4.1) there exists Z such that $\forall n\, (n \in Z \leftrightarrow \varphi''(X, n))$, and by theorem VIII.3.19 this Z is hyperarithmetical in X. Thus we have

$$(\exists Z \,\forall n\, (n \in Z \leftrightarrow \varphi(n)))^{\text{HYP}(X)}$$

and the theorem is proved.   □

Our next goal is to strengthen the previous theorem by showing that $\text{HYP}(X)$ satisfies the Σ^1_1 axiom of choice and indeed Σ^1_1 dependent choice.

LEMMA VIII.4.6 (Π^1_1 uniformization). *Let* $\psi(i)$ *be a* Π^1_1 *formula with a distinguished free number variable* i. *Then we can effectively find a* Π^1_1 *formula* $\widehat{\psi}(i)$ *such that* ATR_0 *proves*

(1) $\forall i\, (\widehat{\psi}(i) \rightarrow \psi(i))$,
(2) $\forall i\, (\psi(i) \rightarrow \exists j\, \widehat{\psi}(j))$,
(3) $\forall i\, \forall j\, ((\widehat{\psi}(i) \wedge \widehat{\psi}(j)) \rightarrow i = j)$.

PROOF. For simplicity, assume that $\psi(i) \equiv \psi(i, X)$ has only one free set variable, X. As in the proof of lemma VIII.3.4, we can effectively find an

X-recursive sequence of X-recursive linear orderings $\langle R_i^X : i \in \mathbb{N} \rangle$ such that

$$\forall i \, (\psi(i, X) \leftrightarrow R_i^X \text{ is a well ordering}).$$

Let $\widehat{\psi}(j, X)$ be the Π_1^1 formula

$$R_j^X \text{ is a well ordering} \wedge \neg \exists k \, |R_k^X| < |R_j^X| \wedge \neg (\exists k < j) \, |R_k^X| = |R_j^X|.$$

(See definition V.2.7.) Trivially we have (1) and (3). To prove (2) within ATR$_0$, fix an i such that $\psi(i)$ holds. Recall that ATR$_0$ proves comparability of countable well orderings (lemma V.2.9). Hence for all j we have: either R_j^X is not a well ordering, or $|R_j^X| \geq |R_i^X|$, or R_j^X is isomorphic to a unique initial section of R_i^X. Hence by Σ_1^1 choice (lemma VIII.4.1), there exists a set A consisting of all $n \in \text{field}(R_i^X)$ such that $\exists j \, (R_j^X$ is isomorphic to the initial section of R_i^X determined by $n)$. If A is the empty set, let j_0 be the least j such that $|R_i^X| = |R_j^X|$. Otherwise, since R_i^X is a well ordering, let n_0 be the R_i^X-least element of A, and then let j_0 be the least j such that R_j^X is isomorphic to the initial section of R_i^X determined by n_0. In either case we clearly have $\widehat{\psi}(j_0, X)$, so (2) is proved. This completes the proof of lemma VIII.4.6. □

LEMMA VIII.4.7. *Let* $\psi(n, i, X)$ *be a* Π_1^1 *formula with no free set variables other than* X. *Then* ATR$_0$ *proves*

$$\forall n \, \exists i \, \psi(n, i, X) \to (\exists f \leq_{\mathrm{H}} X) \, \forall n \, \psi(n, f(n), X).$$

PROOF. By lemma VIII.4.6, let $\widehat{\psi}(n, i, X)$ be a Π_1^1 formula such that ATR$_0$ proves $\widehat{\psi}(n, i, X) \to \psi(n, i, X)$ and $\psi(n, i, X) \to \exists j \, \widehat{\psi}(n, j, X)$ and $(\widehat{\psi}(n, i, X) \wedge \widehat{\psi}(n, j, X)) \to i = j$. Reasoning within ATR$_0$, assume $\forall n \, \exists i \, \psi(n, i, X)$. It follows that $\forall n \, (\exists \text{ exactly one } i) \, \widehat{\psi}(n, i, X)$. Hence, for any pair (n, i), the Π_1^1 assertion $\widehat{\psi}(n, i, X)$ is equivalent to the Σ_1^1 assertion $\forall j \, (j \neq i \to \neg \widehat{\psi}(n, i, X))$. (The latter assertion is Σ_1^1 by lemma VIII.3.21.) Hence by Δ_1^1 comprehension (lemma VIII.4.1) there exists a function $f : \mathbb{N} \to \mathbb{N}$ such that $\forall m \, (f(m)$ is the unique i such that $\widehat{\psi}(m, i, X))$. Furthermore by theorem VIII.3.19 this f is hyperarithmetical in X. The proof of the lemma is complete. □

THEOREM VIII.4.8 (Σ_1^1 choice in HYP(X)). *The following is provable in* ATR$_0$. *For any* $X \subseteq \mathbb{N}$, HYP(X) *satisfies the* Σ_1^1 *axiom of choice. In other words, we have*

$$\forall X \, (\Sigma_1^1 \text{ choice})^{\mathrm{HYP}(X)}.$$

PROOF. Assume $(\forall n \, \exists Y \, \eta(n, Y))^{\mathrm{HYP}(X)}$ where the formula $\eta(n, Y)$ is Σ_1^1 and all set parameters in it are $\leq_{\mathrm{H}} X$. By theorem VIII.3.20 we have

$$\forall n \, (\forall Y \leq_{\mathrm{H}} X) \, (\eta(n, Y)^{\mathrm{HYP}(X)} \leftrightarrow \eta'(n, X, Y))$$

where $\eta'(n, X, Y)$ is Π_1^1 with no free set variables other than X and Y. Thus our assumption becomes $\forall n \, (\exists Y \leq_H X) \, \eta'(n, X, Y)$. Using the notation of lemma VIII.3.17, we may expand this as

$$\forall n \, \exists i \, (\nu(i, X) \wedge \forall Y (\alpha(i, X, Y) \to \eta'(n, X, Y))),$$

or in other words $\forall n \, \exists i \, \psi(n, i, X)$ where $\psi(n, i, X)$ is Π_1^1 with no free set variables other than X. Applying lemma VIII.4.7, we obtain a function $f \leq_H X$ such that $\forall n \, \psi(n, f(n), X)$ holds. Now for all n and k, the Σ_1^1 condition $\exists Y \, (\alpha(f(n), X, Y) \wedge k \in Y)$ is equivalent to the Π_1^1 condition $\forall Y \, (\alpha(f(n), X, Y) \to k \in Y)$. Hence by Δ_1^1 comprehension (lemma VIII.4.1), there exists a set Z such that

$$\forall k \, \forall n \, ((k, n) \in Z \leftrightarrow \exists Y \, (\alpha(f(n), X, Y) \wedge k \in Y)).$$

which implies $\forall n \, \eta'(n, X, (Z)_n)$. Thus, for all n, $(Z)_n$ is the unique Y such that $\alpha(f(n), X, Y)$ holds. Furthermore, by theorem VIII.3.19, Z is hyperarithmetical in X. Thus we conclude $(\exists Z \, \forall n \, \eta(n, (Z)_n))^{\mathrm{HYP}(X)}$ and our theorem is proved. $\qquad \square$

Recall from §VII.6 that Σ_1^1-IND is the scheme of Σ_1^1 induction, i.e.,

$$(\varphi(0) \wedge \forall n \, (\varphi(n) \to \varphi(n + 1))) \to \forall n \, \varphi(n)$$

where $\varphi(n)$ is any Σ_1^1 formula. We define Π_1^1-IND similarly.

LEMMA VIII.4.9. Σ_1^1-IND *is equivalent over* RCA$_0$ *to* Π_1^1-IND.

PROOF. Assume Σ_1^1 induction. Suppose $\forall n \, (\psi(n) \to \psi(n + 1))$ and $\neg \psi(k)$, where $\psi(n)$ is a Π_1^1 formula. Applying Σ_1^1 induction to the formula $n \leq k \to \neg \psi(k - n)$, we obtain $\forall n \, (n \leq k \to \neg \psi(k - n))$ so in particular $\neg \psi(0)$ holds. This proves Π_1^1 induction. The proof of the converse is similar. $\qquad \square$

LEMMA VIII.4.10. *Let* $\varphi(m, X)$ *and* $\psi(m, n, X)$ *be* Π_1^1 *formulas with no free set variable other than* X. *Then* ATR$_0$ *plus* Σ_1^1-IND *proves*

$$\forall m \, [\varphi(m, X) \to \exists n \, [\varphi(n, X) \wedge \psi(m, n, X)]] \to$$
$$\forall m \, [\varphi(m, X) \to (\exists f \leq_H X) \, [f(0) = m \wedge \forall i \, [\varphi(f(i), X) \wedge$$
$$\psi(f(i), f(i + 1), X)]]].$$

PROOF. We reason in ATR$_0$. Assume

$$\forall m \, [\varphi(m, X) \to \exists n \, [\varphi(n, X) \wedge \psi(m, n, X)]].$$

By ATR$_0$ and lemma VIII.4.6, we may also assume $\forall m \, [\varphi(m, X) \to (\exists \text{ exactly one } n) \, \psi(m, n, X)]$. Fix m such that $\varphi(m, X)$ holds. Let $\theta(k, \sigma)$ say that σ is a finite sequence of length $k + 1$ such that $\sigma(0) = m$ and $(\forall i < k) \, [\varphi(\sigma(i), X) \wedge \psi(\sigma(i), \sigma(i + 1), X)]$. By lemma VIII.3.21, $\exists \sigma \, \theta(k, \sigma)$ is equivalent to a Π_1^1 formula. Thus we can use Π_1^1 induction (a consequence of Σ_1^1 induction by lemma VIII.4.9) to prove

$\forall k\,\exists\sigma\,\theta(k,\sigma)$. Moreover, it is easily proved that σ is unique, i.e., $\forall k\,(\exists$ exactly one $\sigma)\,\theta(k,\sigma)$. Hence by Δ_1^1 comprehension (lemma VIII.4.1), there exists a unique function $f\colon \mathbb{N}\ \to\ \mathbb{N}$ such that $f(0)\ =\ m$ and $\forall i\,[\varphi(f(i),X)\wedge\psi(f(i),f(i+1),X)]$. Since X is the only free set variable in the formulas $\varphi(m,X)$ and $\psi(m,n,X)$, it follows by lemma VIII.3.21 that f is Δ_1^1 in X. Hence by theorem VIII.3.19 f is hyperarithmetical in X. This proves the lemma.   □

THEOREM VIII.4.11 (Σ_1^1 dependent choice in $\mathrm{HYP}(X)$).   *The following is provable in* ATR_0 *plus* Σ_1^1-*IND. For any* $X\subseteq\mathbb{N}$, $\mathrm{HYP}(X)$ *satisfies the scheme of* Σ_1^1 *dependent choice. In other words, we have*

$$\forall X\ (\Sigma_1^1\ \text{dependent choice})^{\mathrm{HYP}(X)}.$$

PROOF. We proceed as in the proof of theorem VIII.4.8. Assume

$$(\forall n\,\forall Y\,\exists Z\,\eta(n,Y,Z))^{\mathrm{HYP}(X)}$$

where $\eta(n,Y,Z)$ is Σ_1^1 and all set parameters in it are $\leq_{\mathrm{H}} X$. By theorem VIII.3.20, we have

$$\forall n\,(\forall Y\leq_{\mathrm{H}}X)\,(\forall Z\leq_{\mathrm{H}}X)\,[\eta(n,Y,Z)^{\mathrm{HYP}(X)}\leftrightarrow\eta'(n,X,Y,Z)]$$

where $\eta'(n,X,Y,Z)$ is Π_1^1 with no free set variables other than X, Y and Z. Thus our assumption becomes

$$\forall n\,(\forall Y\leq_{\mathrm{H}}X)\,(\exists Z\leq_{\mathrm{H}}X)\,\eta'(n,X,Y,Z).$$

Using the notation of lemma VIII.3.17, we may expand this as

$$\forall n\,\forall i\,[v(i,X)\rightarrow\exists j\,[v(j,X)\wedge\forall Y\,\forall Z\,[(\alpha(i,X,Y)\wedge\alpha(j,X,Z))\rightarrow$$
$$\eta'(n,X,Y,Z)]]].$$

Applying lemma VIII.4.10, we obtain a function $f\leq_{\mathrm{H}}X$ such that

$$\forall n\,[v(f(n),X)\wedge\forall Y\,\forall Z\,[(\alpha(f(n),X,Y)\wedge\alpha(f(n+1),X,Z))\rightarrow$$
$$[Y=(Z)^n\wedge Z=(Z)^{n+1}\wedge(\forall m\leq n)\,\eta'(m,X,(Z)^m,(Z)_m)]]]$$

holds. As in the last part of the proof of theorem VIII.4.8, we can now use Δ_1^1 comprehension to obtain a set W such that $\forall n\,[(W)^n$ is the unique Z such that $\alpha(f(n),X,Z)]$. By theorem VIII.3.19, W is hyperarithmetical in X. Thus we conclude

$$(\exists W\,\forall n\,\eta(n,(W)^n,(W)_n))^{\mathrm{HYP}(X)}$$

and theorem VIII.4.11 is proved.   □

Having shown that $\mathrm{HYP}(X)$ is an ω-model of Σ_1^1 choice (indeed Σ_1^1 dependent choice), our next goal is to show that $\mathrm{HYP}(X)$ is the smallest such model which contains X. Actually we shall obtain a sharper result by considering a weaker choice scheme, introduced in the following definition.

DEFINITION VIII.4.12 (weak Σ_1^1 choice). *Weak Σ_1^1-AC$_0$ is the L$_2$-theory whose axioms are those of* ACA$_0$ *plus the scheme of weak Σ_1^1 choice, i.e.,*

$$\forall n \, (\exists \text{ exactly one } Y) \, \theta(n, Y) \to \exists Z \, \forall n \, \theta(n, (Z)_n)$$

where $\theta(n, Y)$ is any arithmetical formula in which Z does not occur.

REMARK VIII.4.13. It is clear that Σ_1^1 choice implies weak Σ_1^1 choice. (Compare definition VII.6.1.1.)

EXERCISE VIII.4.14. Show that Δ_1^1 comprehension implies weak Σ_1^1 choice.

LEMMA VIII.4.15. *The following is provable in* ACA$_0$. *Let M be a countable coded ω-model of weak Σ_1^1-AC$_0$. Then for all $X \in M$ we have*

$$\forall a \, (\mathcal{O}(a, X) \to (\mathrm{H}_a^X \text{ exists} \wedge \mathrm{H}_a^X \in M)).$$

Moreover M is closed under relative hyperarithmeticity, i.e., $X \in M$, $Y \leq_{\mathrm{H}} X$ imply $Y \in M$.

PROOF. Let M be a countable coded ω-model of weak Σ_1^1-AC$_0$. Let $X \in M$ be given. If $\mathcal{O}(a, X)$ holds, the statement

$$b \leq_{\mathcal{O}}^X a \wedge \mathrm{H}_b^X \text{ exists} \wedge \mathrm{H}_b^X \in M$$

is arithmetical in the code for M and may therefore be proved by arithmetical transfinite induction along $\{b : b \leq_{\mathcal{O}}^X a\}$. Assume now that H_c^X exists and $\in M$ for all $c <_{\mathcal{O}}^X b$. Hence, by lemma VIII.3.7, M satisfies

$$\forall c \, (c <_{\mathcal{O}}^X b \to (\exists \text{ exactly one } Y) \, \mathrm{H}(c, X, Y)).$$

By weak Σ_1^1 choice within M, it follows that M satisfies

$$\exists Z \, \forall c \, (c <_{\mathcal{O}}^X b \to \mathrm{H}(c, X, (Z)_c)).$$

Hence, by arithmetical comprehension within M, we see that M satisfies $\exists Y \, \mathrm{H}(b, X, Y)$. This implies that H_b^X exists and $\in M$. We have now shown

$$\forall a \, (\mathcal{O}(a, X) \to (\mathrm{H}_a^X \text{ exists} \wedge \mathrm{H}_a^X \in M)).$$

From this and the fact that M is closed under relative recursiveness, it follows that M is closed under relative hyperarithmeticity. Lemma VIII.4.15 is proved. □

THEOREM VIII.4.16. *The following is provable in* Π_1^1-CA$_0$. *For all $X \subseteq \mathbb{N}$, HYP(X) can be characterized as the smallest countable coded ω-model of weak Σ_1^1-AC$_0$ (or of Δ_1^1-CA$_0$, or of Σ_1^1-AC$_0$, or of Σ_1^1-DC$_0$) which contains X.*

PROOF. Since the formula $\mathcal{O}(a, X)$ is Π_1^1, it is clear by Π_1^1 comprehension that the countable coded ω-model

$$\mathrm{HYP}(X) = \{Y : Y \leq_{\mathrm{H}} X\}$$

exists. By theorems VIII.4.8 and VIII.4.11, it follows that $\mathrm{HYP}(X)$ is an ω-model of Σ_1^1-DC_0, etc. By lemma VIII.4.15 $\mathrm{HYP}(X)$ is the smallest such model which contains X. $\square$

COROLLARY VIII.4.17 (minimum ω-model of Σ_1^1-AC_0, etc.). *The systems* Σ_1^1-DC_0, Σ_1^1-AC_0, Δ_1^1-CA_0, *and weak* Σ_1^1-AC_0 *all have the same minimum (i.e., unique smallest) ω-model, namely*

$$\mathrm{HYP} = \{X \subseteq \omega: X \text{ is hyperarithmetical}\}.$$

Our final task in this section is to show that ATR_0 proves the existence of countable coded ω-models of Σ_1^1-AC_0, and indeed of Σ_1^1-DC_0. Theorems VIII.4.8 and VIII.4.11 do not establish this result, since ATR_0 is not strong enough to prove that $\mathrm{HYP}(X)$ is a countable coded ω-model (remark VIII.4.4). Nevertheless, we shall see that $\mathrm{HYP}(X)$ can be characterized within ATR_0 as the intersection of certain countable coded ω-models (theorem VIII.4.23). In §VIII.6 we shall obtain a similar characterization of $\mathrm{HYP}(X)$ in terms of ω-models of stronger theories.

LEMMA VIII.4.18. *The following is provable in* ACA_0. *Let X be such that* $\forall a\,(\mathcal{O}(a, X) \to \mathrm{H}_a^X \text{ exists})$. *Then there exist a^* and M^* such that*

(i) $\mathcal{O}_+(a^*, X)$ *and* $\neg\mathcal{O}(a^*, X)$,
(ii) M^* *is a countable coded ω-model of* ACA_0,
(iii) $X \in M^*$, *and M^* satisfies* $\mathcal{O}(a^*, X) \wedge \exists Y\, \mathrm{H}(a^*, X, Y)$.

PROOF. This is a variant of the proof of lemma VIII.3.24. Let $\mathcal{O}_1(a, X)$ be a Σ_1^1 formula which says: $\mathcal{O}_+(a, X)$ and there exists a countable coded ω-model M of ACA_0 such that $X \in M$ and M satisfies $\mathcal{O}(a, X) \wedge \exists Y\, \mathrm{H}(a, X, Y)$. We claim $\forall a\,(\mathcal{O}(a, X) \to \mathcal{O}_1(a, X))$. If $\mathcal{O}(a, X)$ holds, then by assumption H_a^X exists, and moreover the proof of theorem VIII.1.13 shows that there exists a countable coded ω-model M of ACA_0 such that $\mathrm{H}_a^X \in M$. This implies $\mathcal{O}_1(a, X)$, thus proving our claim. From the claim plus lemma VIII.3.4, we see that $\exists a\,(\mathcal{O}_1(a, X) \wedge \neg\mathcal{O}(a, X))$. Letting a^* be any such a, we obtain the desired conclusion. $\square$

The next lemma may be viewed as a strong converse to lemma VIII.4.15.

LEMMA VIII.4.19. *The following is provable in* ACA_0. *Let X be such that* $\forall a\,(\mathcal{O}(a, X) \to \mathrm{H}_a^X \text{ exists})$. *Then there exists a countable coded ω-model M such that $X \in M$ and M satisfies* Σ_1^1-DC_0 *(hence also* Σ_1^1-AC_0 *and* Δ_1^1-CA_0*)*.

PROOF. We reason in ACA_0. Let X, a^* and M^* be as in the previous lemma. Let $Y \in M^*$ be such that $\mathrm{H}(a^*, X, Y)$ holds. Thus M^* satisfies $Y = \mathrm{H}_{a^*}^X$. For each $b \leq_{\mathcal{O}}^X a^*$, put

$$Y_b = \{(m, 0): (m, 0) \in Y\} \cup \{(m, c+1): c <_{\mathcal{O}}^X b \wedge (m, c+1) \in Y\}.$$

Thus $Y_{a^*} = Y$ and, for each $b \leq_{\mathcal{O}}^X a^*$, M^* satisfies $Y_b = \mathrm{H}_b^X$. For each $b \leq_{\mathcal{O}}^X a^*$, put $M_b = \{Z: Z \leq_{\mathrm{T}} Y_b\}$.

Since $\neg\mathcal{O}(a^*, X)$ holds, there exists $I \subseteq \{b: b <_{\mathcal{O}}^X a^*\}$ such that

$$\forall b \, \forall c \, ((c <_{\mathcal{O}}^X b \wedge b \in I) \rightarrow c \in I)$$

and there is no $a \leq_{\mathcal{O}}^X a^*$ such that $I = \{b: b <_{\mathcal{O}}^X a\}$. Since M^* satisfies $\mathcal{O}(a^*, X)$, we must have $I \notin M$, hence $I \neq \emptyset$ and

$$(\forall b \in I)\,(\exists c \in I)b <_{\mathcal{O}}^X c.$$

Put

$$M = \bigcup_{b \in I} M_b = \{Z: \exists b\,(b \in I \wedge Z \leq_T Y_b)\}.$$

Clearly $X \in M$ and M is a countable coded ω-model of ACA_0. It remains to show that M satisfies Σ_1^1 dependent choice.

Suppose that M satisfies $\forall n \, \forall U \, \exists V \eta(n, U, V)$ where $\eta(n, U, V)$ is a Σ_1^1 formula with parameters from M. Fix $b_0 \in I$ such that all of these parameters belong to M_{b_0}. Put $Z_0 = \emptyset$. Reasoning within M^*, choose a sequence of ordinal notations $b_n <_{\mathcal{O}}^X a^*$, $n \in \mathbb{N}$, and a sequence of sets $Z_n \in M_{b_n}$, as follows. We have already chosen b_0 and Z_0. Given b_n and Z_n, let b_{n+1} be the $<_{\mathcal{O}}^X$-least $b <_{\mathcal{O}}^X a^*$ such that $b_n <_{\mathcal{O}}^X b$ and M_b satisfies $\exists V \, \eta(n, (Z_n)^n, V)$. Then pick $Z_{n+1} \in M_{b_{n+1}}$ such that $(Z_{n+1})^n = (Z_n)^n$ and $M_{b_{n+1}}^X$ satisfies $\eta(n, (Z_n)^n, (Z_{n+1})_n)$. Applying lemma VIII.3.10 inside M^*, we can choose b_n and Z_n so that the sequence $\langle b_n : n \in \mathbb{N} \rangle$ is recursive in Y.

By induction on n, it is clear that $\forall n \, (b_n \in I)$. Put

$$J = \{c: \exists n \, (c \leq_{\mathcal{O}}^X b_n)\}.$$

Then $J \subseteq I$. Moreover, since J is arithmetical in Y, we have $J \in M^*$, hence $J \neq I$. Since M^* satisfies $\mathcal{O}(a^*, X)$, there must exist $b^* \in I$ such that $J = \{c: c <_{\mathcal{O}}^X b^*\}$. Again applying lemma VIII.3.10 inside M^*, we see that the sequence $\langle Z_n : n \in \mathbb{N} \rangle$ is recursive in Y_{b^*}. Hence we can find $Z \in M_{b^*}$ such that $(Z)^n = (Z_n)^n$ for all n. Thus $Z \in M$, and M satisfies $\forall n \, \eta(n, (Z)^n, (Z)_n)$.

This completes the proof of lemma VIII.4.19. $\qquad\qquad\square$

THEOREM VIII.4.20. ATR_0 *proves the existence of a countable coded ω-model of Σ_1^1-DC_0 (hence also of Σ_1^1-AC_0 and Δ_1^1-CA_0).*

PROOF. This follows immediately from the previous lemma. $\qquad\square$

COROLLARY VIII.4.21 (consistency of Σ_1^1-AC_0, etc.). ATR_0 *proves the consistency of Σ_1^1-DC_0 (hence also of Σ_1^1-AC_0 and Δ_1^1-CA_0) plus full induction, Σ_∞^1-IND.*

PROOF. This is like the proof of corollary VIII.1.14, using theorem VIII.4.20 instead of theorem VIII.1.13. $\qquad\qquad\square$

COROLLARY VIII.4.22. *There exists a Π_1^0 sentence ψ such that ψ is provable in ATR_0 but not in Σ_1^1-DC_0 (hence also not in Σ_1^1-AC_0 or Δ_1^1-CA_0) plus full induction.*

PROOF. Let ψ be a sentence asserting the consistency of Σ_1^1-DC$_0$ plus full induction. The result follows from Gödel's second incompleteness theorem [94, 115, 55, 222]. (Compare the proof of corollary VIII.1.8.) $\square$

THEOREM VIII.4.23. *The following is provable in* ATR$_0$. *Given* $W, X \subseteq \mathbb{N}$, *the following are pairwise equivalent.*

1. $W \in \mathrm{HYP}(X)$, *i.e.,* $W \leq_{\mathrm{H}} X$.
2. $W \in M$ *for all countable coded ω-models M such that $X \in M$ and M satisfies weak Σ_1^1-AC$_0$.*
3. *Same as 2 with weak Σ_1^1-AC$_0$ replaced by Δ_1^1-CA$_0$.*
4. *Same as 2 with weak Σ_1^1-AC$_0$ replaced by Σ_1^1-AC$_0$.*
5. *Same as 2 with weak Σ_1^1-AC$_0$ replaced by Σ_1^1-DC$_0$.*

PROOF. The implication $1 \to 2$ follows from lemma VIII.4.15. The implications $2 \to 3$, $3 \to 4$, $4 \to 5$ are immediate since

$$\Sigma_1^1\text{-DC}_0 \supseteq \Sigma_1^1\text{-AC}_0 \supseteq \Delta_1^1\text{-CA}_0 \supseteq \text{weak } \Sigma_1^1\text{-AC}_0.$$

(See exercise VIII.4.14 and lemma VII.6.6.)

In order to prove $5 \to 1$, let X and W be such that $W \not\leq_{\mathrm{H}} X$. We must find a countable coded ω-model M of Σ_1^1-DC$_0$ such that $X \in M$ and $W \notin M$. Let M^*, a^* and Y be as in the proof of lemma VIII.4.19. If $W \notin M^*$, the proof of lemma VIII.4.19 gives a countable coded ω-model M of Σ_1^1-DC$_0$ such that $X \in M$ and $M \subseteq M^*$, hence $W \notin M$ and we are done.

Suppose now that $W \in M^*$. Put

$$K = \{b : b \leq_{\mathcal{O}}^X a^* \wedge W \not\leq_{\mathrm{T}} Y_b\}.$$

Then $K \in M^*$ (because K is arithmetical in X, Y and W). Since $W \not\leq_{\mathrm{H}} X$, we must have $\forall b \, ((b <_{\mathcal{O}}^X a^* \wedge \mathcal{O}(b, X)) \to b \in K)$. But then, since $\neg \mathcal{O}(a^*, X)$ holds while M^* satisfies $\mathcal{O}(a^*, X)$, there must exist $b^* \in K$ such that $\neg \mathcal{O}(b^*, X)$. We can then find I as in the proof of theorem VIII.4.19 with the additional property that $I \subseteq \{b : b <_{\mathcal{O}}^X b^*\}$, hence $I \subseteq K$. Defining $M = \{Z : \exists b \, (Z \leq_{\mathrm{T}} Y_b \wedge b \in I)\}$ as in the proof of theorem VIII.4.19, we see that M is a countable coded ω-model of Σ_1^1-DC$_0$ and $X \in M$ and $W \notin M$.

This completes the proof of theorem VIII.4.23. $\square$

The following exercises provide an analog of theorem VIII.4.20 with ATR$_0$ replaced by Π_k^1-TR$_0$ for arbitrary k.

EXERCISES VIII.4.24. Fix k such that $0 \leq k < \omega$.

1. Show that Π_{k+1}^1-TR$_0$ proves the existence of a countable coded ω-model of Σ_{k+2}^1-DC$_0$, hence also of Σ_{k+2}^1-AC$_0$ and of Δ_{k+2}^1-CA$_0$. (Hint: Imitate the proof of theorem VIII.4.20.)
2. Show that Δ_{k+2}^1-CA$_0$ plus Σ_{k+2}^1-TI$_0$ implies the existence of a countable coded ω-model of Σ_{k+2}^1-DC$_0$, hence also of Σ_{k+2}^1-AC$_0$ and of Δ_{k+2}^1-CA$_0$. (Hint: Use the previous exercise plus VII.7.12.1.)

The following exercise provides a strong converse for lemma VIII.4.7.

EXERCISE VIII.4.25. Show that ATR_0 is equivalent over RCA_0 to the scheme $\forall m \, \exists n \, \psi(m, n) \rightarrow \exists f \, \forall m \, \psi(m, f(m))$ where $\psi(m, n)$ is Π^1_1. (Hint: Use theorem V.5.1 and lemma VIII.4.7.)

Notes for §VIII.4. The fact that HYP is the minimum ω-model of Δ^1_1-CA_0 is due to Kleene [145]. The analogous results for Σ^1_1-CA_0 and Σ^1_1-DC_0 are due to Kreisel [150] and Feferman; see also Harrison [106]. The fact that ATR_0 proves the existence of a countable ω-model of Σ^1_1-DC_0, etc., is due to Friedman [62, chapter II], [64, 68, 69]. The results stated in exercises VIII.4.24 are probably new, but see Friedman [64]. In theorem VIII.4.11, Simpson [235] has shown that the hypothesis Σ^1_1-IND cannot be omitted.

In §VIII.5 we shall see that there exists a countable ω-model of ATR_0 (hence also of Σ^1_1-AC_0) which does not satisfy Σ^1_1-DC_0. This result is due to Friedman [62, chapter II]. Steel [257] has developed a technique known as tagged tree forcing and used it to show that there exists a countable ω-model of Δ^1_1-CA_0 which does not satisfy Σ^1_1-AC_0. Van Wesep [273, §I.1] has used tagged tree forcing to show that there exists a countable ω-model of weak Σ^1_1-AC_0 which does not satisfy Δ^1_1-CA_0.

VIII.5. ω-Model Reflection and Incompleteness

By ω-*model reflection* we mean the principle that any true L_2-sentence (possibly with set parameters) has a countable ω-model. We formalize this in the following definition.

DEFINITION VIII.5.1 (ω-model reflection). Let $\varphi(X_1, \ldots, X_k)$ be an L_2-formula with no free variables other than $X_1, \ldots, X_k$. Then the formula

$$\forall X_1 \cdots \forall X_k \, [\varphi(X_1, \ldots, X_k) \rightarrow \exists \text{ countable coded } \omega\text{-model } M$$

$$\text{such that } X_1, \ldots, X_k \in M \text{ and}$$

$$M \text{ satisfies } ACA_0 \text{ plus } \varphi(X_1, \ldots, X_k)]$$

is an instance of the ω-*model reflection scheme*. We define Σ^1_∞-RFN_0 to be the subsystem of Z_2 whose axioms are those of ACA_0 plus all instances of the ω-model reflection scheme. Also, Σ^1_k-RFN_0 consists of ACA_0 plus all instances of ω-model reflection in which the formula $\varphi(X_1, \ldots, X_k)$ is Σ^1_k.

Recall from §VII.2 that Π^1_∞-TI_0 consists of ACA_0 plus the transfinite induction scheme. We shall prove the following theorem of Friedman: Σ^1_∞-RFN_0 is equivalent to Π^1_∞-TI_0. In other words, ω-model reflection is equivalent to transfinite induction.

LEMMA VIII.5.2. *All instances of the ω-model reflection scheme are provable in Π^1_∞-TI_0.*

PROOF. Without loss, we may restrict our attention to instances of ω-model reflection in which there is only one set parameter. Reasoning in ACA_0, assume that we have a failure of ω-model reflection, i.e., $\varphi(X_0)$ holds but there is no countable coded ω-model M such that $X_0 \in M$ and M satisfies ACA_0 plus $\varphi(X_0)$. Here X_0 is a fixed set, and $\varphi(X_0)$ is an L_2-sentence which contains X_0 as a parameter. Since ACA_0 is finitely axiomatizable (lemma VIII.1.5), we may assume that $\varphi(X_0)$ logically implies the axioms of ACA_0.

Our proof will be based on a model-theoretic construction in the style of Henkin. The idea will be to construct a tree T such that from any path through T we can read off a countable coded ω-model of $\varphi(X_0)$. The non-existence of such a model will imply that T has no path, i.e., is well founded. On the other hand, the fact that $\varphi(X_0)$ is true will yield a failure of transfinite induction along the Kleene/Brouwer ordering of T.

We work with the language $L_2(\underline{C})$ consisting of L_2 plus countably many set constants $\underline{C}_j$, $j \in \mathbb{N}$. We assume that our language has been set up so that it contains no existential quantifiers. Form the $L_2(\underline{C})$-sentence $\varphi(\underline{C}_0)$, and let $\langle \theta_i : i \in \mathbb{N} \rangle$ enumerate all $L_2(\underline{C})$-sentences which are substitution instances of subformulas of $\varphi(\underline{C}_0)$. Let $\langle \eta_i(Y) : i \in \mathbb{N} \rangle$ and $\langle \psi_i(m) : i \in \mathbb{N} \rangle$ enumerate all $L_2(\underline{C})$-formulas which are substitution instances of subformulas of $\varphi(\underline{C}_0)$ and have exactly one free variable, Y or m respectively. (Here of course Y is a set variable and m is a number variable.) We assume that our enumerations have been chosen so that $j \leq i$ whenever $\underline{C}_j$ occurs in θ_i or in $\eta_i(Y)$ or in $\psi_i(m)$.

For each $\tau \in \mathbb{N}^{<\mathbb{N}}$, let S_τ be the finite set of $L_2(\underline{C})$-sentences consisting of $\varphi(\underline{C}_0)$ plus

$$
\begin{array}{ll}
\theta_i & \text{if } 3i < \mathrm{lh}(\tau) \text{ and } \tau(3i) = 0, \\
\neg\theta_i & \text{if } 3i < \mathrm{lh}(\tau) \text{ and } \tau(3i) \neq 0, \\
\forall Y \eta_i(Y) & \text{if } 3i+1 < \mathrm{lh}(\tau) \text{ and } \tau(3i+1) = 0, \\
\neg\eta_i(\underline{C}_{i+1}) & \text{if } 3i+1 < \mathrm{lh}(\tau) \text{ and } \tau(3i+1) \neq 0, \\
\forall m \psi_i(m) & \text{if } 3i+2 < \mathrm{lh}(\tau) \text{ and } \tau(3i+2) = 0, \\
\neg\psi_i(\underline{n}) & \text{if } 3i+2 < \mathrm{lh}(\tau) \text{ and } \tau(3i+2) = n+1.
\end{array}
$$

(Here $\underline{n}$ is a constant term denoting the number n.) The tree $T \subseteq \mathbb{N}^{<\mathbb{N}}$ is defined by putting τ into T if and only if there is no "obvious inconsistency" in S_τ. The "obvious inconsistencies" are of two kinds: (1) a propositional inconsistency; (2) a quantifier-free sentence which is false when $\underline{C}_0$ is interpreted as X_0.

If f were a path through T, then there would exist a countable coded ω-model M such that $X_0 \in M$ and M satisfies $\varphi(X_0)$, namely

$$M = \{(W)_i : i \in \mathbb{N}\}$$

where $n \in (W)_i$ if and only if the sentence $\underline{n} \in \underline{C}_i$ belongs to $\bigcup_{k \in \mathbb{N}} S_{f[k]}$. (See the formal definition of satisfaction for countable coded ω-models, definition VII.2.1.) Since no such ω-model exists, T is well founded. Hence, by lemma V.1.3, the Kleene/Brouwer ordering $KB(T)$ is a well ordering.

Let us say that $\tau \in T$ is *good* if there exists $W \subseteq \mathbb{N}$ such that the sentences in S_τ are all true when, for all i, $\underline{C}_i$ is interpreted as $(W)_i$. Since the sentences in question are all substitution instances of subformulas of a fixed sentence $\varphi(\underline{C}_0)$, the property of goodness is expressible by a single L_2-formula (with X_0 appearing as a parameter). The empty sequence $\langle \rangle$ is good since $S_{\langle \rangle} = \{\varphi(\underline{C}_0)\}$ is true with $\underline{C}_0$ interpreted as X_0. It is also easy to see that if τ is good, then $\tau ^\frown \langle k \rangle$ is good for some k. Hence there is no $KB(T)$-least good $\tau \in T$. Thus we have a failure of transfinite induction along $KB(T)$.

This completes the proof of lemma VIII.5.2. $\square$

LEMMA VIII.5.3. *All instances of the transfinite induction scheme are provable in* Σ^1_∞-RFN$_0$.

PROOF. Let $\psi(j)$ be any L_2-formula with a distinguished free number variable j. Reasoning in Σ^1_∞-RFN$_0$, we want to prove $\forall X (\text{WO}(X) \rightarrow \text{TI}(X, \psi))$. Assume $\text{WO}(X) \wedge \neg\text{TI}(X, \psi)$. By ω-model reflection, let M be a countable coded ω-model such that $X \in M$ and M satisfies $\neg\text{TI}(X, \psi)$. By arithmetical comprehension using the code of M as a parameter, let Y be the set of all $j \in \mathbb{N}$ such that M satisfies $\psi(j)$. Since X is a well ordering, we have

$$\forall j (\forall i (i <_X j \rightarrow i \in Y) \rightarrow j \in Y) \rightarrow \forall j (j \in Y).$$

Hence M satisfies $\text{TI}(X, \psi)$, a contradiction. Lemma VIII.5.3 is proved.
                    $\square$

THEOREM VIII.5.4. Σ^1_∞-RFN$_0$ *is equivalent to* Π^1_∞-TI$_0$.

PROOF. Immediate from lemmas VIII.5.2 and VIII.5.3. $\square$

COROLLARY VIII.5.5. Π^1_∞-CA$_0$ *proves all instances of* ω-*model reflection.*

PROOF. This is immediate from lemma VIII.5.2, since obviously Π^1_∞-CA$_0$ proves all instances of transfinite induction. (In fact, Π^1_k-CA$_0$ includes both Π^1_k-TI$_0$ and Σ^1_k-TI$_0$.) $\square$

Our next theorem, also due to Friedman, is essentially an ω-model version of Gödel's second incompleteness theorem. It will be seen to imply the existence of many countable ω-models in which reflection fails.

THEOREM VIII.5.6 (ω-model incompleteness). *Let S be a recursive set of L_2-sentences which includes the axioms of* ACA$_0$. *If there exists a countable coded ω-model of S, then there exists a countable coded ω-model of*

(1) $S \cup \{\neg\exists$ countable coded ω-model of $S\}$.

Proof. We are given a recursive set of L_2-sentences $S \supseteq$ ACA$_0$. Let S^* be an L_2-theory consisting of ACA$_0$ plus the assertion that our theorem fails for S. Formally, S^* is the finitely axiomatizable L_2-theory consisting of ACA$_0$ plus

(2) $\exists$ countable coded ω-model of S, plus
(3) $\neg\exists$ countable coded ω-model of (1).

We claim that S^* proves its own consistency. To see this, we reason in S^*. By (2), let M be a countable coded ω-model of S. We shall show that M satisfies S^*. By ACA$_0$ and lemma VII.2.2, there exists a truth valuation for (2). Hence M satisfies either (2) or its negation. In view of (3), M does not satisfy (1), hence M satisfies (2). Moreover, by another application of lemma VII.2.2, all countable coded ω-models satisfy all true Π_1^1 sentences. In particular, since (3) is a true Π_1^1 sentence, M satisfies (3). Thus M is a countable coded ω-model of S^*. Hence, by the soundness theorem, S^* is consistent. This proves our claim.

Since S^* proves its own consistency, it follows by Gödel's second incompleteness theorem [94, 115, 55, 222] that S^* must be inconsistent. This means that ACA$_0$ proves "if S has a countable coded ω-model, then so does (1)." Hence this statement is true. The proof of theorem VIII.5.6 is complete.   □

REMARK VIII.5.7. In the above proof, some care is needed as regards satisfaction and truth valuations. Note for instance that the theorem fails with $S =$ WKL$_0$, since every countable ω-model of WKL$_0$ contains a code for a countable ω-model of WKL$_0$ (see theorems VIII.2.2 and VIII.2.6 and remark VIII.2.14). Gödel's second incompleteness theorem is not violated because WKL$_0$ is not strong enough to prove the existence of valuations (compare lemma VII.2.2).

COROLLARY VIII.5.8. *Let S be a finite set of L_2-sentences. If there exists a countable ω-model of S, then there exists a countable ω-model of S which does not satisfy Π_∞^1-TI$_0$.*

Proof. Put $S_1 = S \cup$ ACA$_0$. If S_1 has no countable ω-model, there is nothing to prove. So assume that S_1 has a countable ω-model. By theorem VIII.5.6, let M be a countable ω-model of

$$S_1 \cup \{\neg\exists \text{ countable coded } \omega\text{-model of } S_1\}.$$

Thus we have a failure of ω-model reflection in M. Hence, by theorem VIII.5.4, there must be a failure of transfinite induction in M. This completes the proof.   □

REMARK VIII.5.9. The previous corollary provides a second proof that Π_∞^1-TI$_0$ is not finitely axiomatizable. Compare corollary VII.2.23 and remark VII.2.24.

COROLLARY VIII.5.10. *For each $k < \omega$, there exists a countable ω-model of*

$$\Pi_k^1\text{-CA}_0 \cup \{\neg\exists \text{ countable coded } \omega\text{-model of } \Pi_k^1\text{-CA}_0\}.$$

Such a model does not satisfy $\Pi_\infty^1\text{-TI}_0$.

PROOF. Immediate by theorems VIII.5.6 and VIII.5.4, since $\Pi_k^1\text{-CA}_0$ is finitely axiomatizable.   □

Theorem VIII.5.4 says that the general scheme of transfinite induction is equivalent to the general scheme of ω-model reflection. It is natural to ask how much transfinite induction (as measured by formula complexity) is equivalent to how much ω-model reflection. The next theorem gives a sharp result in one case. See also exercise VIII.5.15 below.

LEMMA VIII.5.11. *Over ACA_0, Π_1^1 transfinite induction implies Σ_1^1 dependent choice.*

PROOF. We are trying to prove Σ_1^1 dependent choice, i.e.,

$$\forall i \,\forall X \,\exists Y \,\eta(i, X, Y) \rightarrow \exists Z \,\forall i \,\eta(i, (Z)^i, (Z)_i)$$

where $\eta(n, X, Y)$ is Σ_1^1. Using lemma V.1.4 (our normal form theorem for Σ_1^1 formulas), we can reduce this to

$$\forall i \,\forall f \,\exists g \,\forall n \,\theta(i, f[n], g[n]) \rightarrow \exists h \,\forall i \,\forall n \,\theta(i, (h)^i[n], (h)_i[n])$$

where $\theta(i, \tau_1, \tau_2)$ is arithmetical. Here of course f, g and h range over $\mathbb{N}^{\mathbb{N}}$, $(h)_i(m) = h((i, m))$, and $(h)^i(m) = h((j, k))$ if $m = (j, k)$ and $j < i$, $(h)^i(m) = 0$ otherwise.

Assume the hypothesis $\forall i \,\forall f \,\exists g \,\forall n \,\theta(i, f[n], g[n])$. To prove the conclusion, form a tree T by putting $\tau \in T$ if and only if

$$(\forall i < \text{lh}(\tau)) \,(\forall n \le \min\{\text{lh}((\tau)^i), \text{lh}((\tau)_i)\}) \,\theta(i, (\tau)^i[n], (\tau)_i[n]).$$

Clearly h is a path through T if and only if h satisfies the conclusion

$$\forall i \,\forall n \,\theta(i, (h)^i[n], (h)_i[n]).$$

Assume now that the conclusion fails, i.e., T is well founded. Then, by lemma V.1.3, the Kleene/Brouwer ordering $\text{KB}(T)$ is a well ordering. On the other hand, say that $\tau \in T$ is *good* if

$$\exists h \,(h[\text{lh}(\tau)] = \tau \wedge (\forall i < \text{lh}(\tau)) \,\forall n \,\theta(i, (h)^i[n], (h)_i[n])).$$

Clearly the empty sequence $\langle\rangle$ is good. Moreover, the hypothesis

$$\forall i \,\forall f \,\exists g \,\forall n \,\theta(i, f[n], g[n])$$

implies that for each good τ there exists m such that $\tau^\frown\langle m \rangle$ is good. Since the property of goodness is Σ_1^1, we have a failure of Π_1^1 transfinite induction along $\text{KB}(T)$.

This completes the proof of lemma VIII.5.11.   □

THEOREM VIII.5.12. *The following are pairwise equivalent over* ACA_0.
1. Π_1^1 *transfinite induction*.
2. Σ_1^1 *dependent choice*.
3. ω-*model reflection for* Σ_3^1 *formulas*.

In other words, $\Pi_1^1\text{-}\mathsf{TI}_0 \equiv \Sigma_1^1\text{-}\mathsf{DC}_0 \equiv \Sigma_3^1\text{-}\mathsf{RFN}_0$.

PROOF. The implication $1 \to 2$ is given by lemma VIII.5.11.

For $2 \to 3$, assume Σ_1^1 dependent choice, and let $\varphi(U_0)$ be a true Σ_3^1 sentence with a set parameter U_0. Write

$$\varphi(U_0) \equiv \exists V \, \forall X \, \exists Y \, \theta(U_0, V, X, Y)$$

where $\theta(U, V, X, Y)$ is arithmetical. Fix U_1 such that

$$\forall X \, \exists Y \, \theta(U_0, U_1, X, Y)$$

holds. Let $\pi(e, m_1, X_1)$ be a universal lightface Π_1^0 formula with exactly the displayed free variables (as in the proof of lemma VIII.1.5). By Σ_1^1 dependent choice, we can find W such that $(W)_0 = U_0$, and $(W)_1 = U_1$, and

$$\forall i \, \theta(U_0, U_1, (W)_i, (W)_{2i+2}),$$

and

$$\forall m (\pi(e, m, (W)^i) \leftrightarrow m \in (W)_{2j+3})$$

for all e, i and $j = (e, i)$. Letting $M = \{(W)_i : i \in \mathbb{N}\}$ be the countable ω-model coded by W, we see that M satisfies $\varphi(U_0)$ and ACA_0. This proves $2 \to 3$.

It remains to prove $3 \to 1$. We proceed as in the proof of lemma VIII.5.3. Reasoning in $\Sigma_3^1\text{-}\mathsf{RFN}_0$, let $\psi(j)$ be Π_1^1 and assume $\mathsf{WO}(X) \wedge \neg \mathsf{TI}(X, \psi)$. Since $\neg \mathsf{TI}(X, \psi)$ is equivalent to a Σ_3^1 formula, we can apply reflection to obtain a countable coded ω-model M such that $X \in M$ and M satisfies $\neg \mathsf{TI}(X, \psi)$. By arithmetical comprehension, let Y be the set of j such that M satisfies $\psi(j)$. Then Y witnesses the failure of $\mathsf{WO}(X)$, a contradiction. This completes the proof of theorem VIII.5.12. $\square$

As an application of theorems VIII.5.6 and VIII.5.12, we present the following independence results, due to Friedman.

THEOREM VIII.5.13. *There exists a countable ω-model of* ATR_0 *which does not satisfy* Σ_1^1 *dependent choice*.

PROOF. By theorem VIII.5.6, let M be a countable ω-model of

$$\mathsf{ATR}_0 \cup \{\neg\exists \text{ countable coded } \omega\text{-model of } \mathsf{ATR}_0\}.$$

Let φ be the Π_2^1 sentence $\forall X \, \forall a \, (\mathcal{O}(a, X) \to \exists Y \, \mathrm{H}(a, X, Y))$. By theorem VIII.3.15 we see that M satisfies $\varphi \wedge \neg\exists$ countable coded ω-model of ACA_0 plus φ. Thus ω-model reflection for φ fails in M. Hence, by the implication $2 \to 3$ in theorem VIII.5.12, Σ_1^1 dependent choice fails in M. This completes the proof. $\square$

COROLLARY VIII.5.14. *There exists a countable ω-model of Σ_1^1-AC$_0$ which does not satisfy Σ_1^1-DC$_0$.*

PROOF. This is immediate from theorem VIII.5.13 in view of the fact that ATR$_0$ includes Σ_1^1-AC$_0$ (lemma VIII.4.1). $\square$

EXERCISE VIII.5.15. Show that, for each $k < \omega$, Π_{k+1}^1 transfinite induction is equivalent to ω-model reflection for Σ_{k+3}^1 formulas, over ACA$_0$. This generalizes the equivalence $1 \leftrightarrow 3$ of theorem VIII.5.12.

Notes for §VIII.5. Theorem VIII.5.4 has been announced by Friedman [68]. The ω-model incompleteness theorem VIII.5.6 and corollaries VIII.5.8 and VIII.5.10 are due to Friedman [62, chapter II], [68]. Steel [255] has given a purely recursion-theoretic proof of theorem VIII.5.6, not using Gödel's second incompleteness theorem. See also Friedman [70]. Lemma VIII.5.11 and theorem VIII.5.12 are due to Simpson [235]; see also Sy Friedman [81]. Theorem VIII.5.13 and corollary VIII.5.14 are due to Friedman [62, chapter II]. The result of exercise VIII.5.15 is due to Jäger/Strahm [129].

VIII.6. ω-Models of Strong Systems

In this section we shall prove that, for any countable model M of ATR$_0$, HYP^M is the intersection of all β-submodels of M (corollary VIII.6.10). In addition, we shall prove the following results concerning an arbitrary, recursively axiomatizable theory S in the language of L$_2$. If S includes weak Σ_1^1-AC$_0$ and has a countable ω-model, then HYP is the intersection of all countable ω-models of S (theorem VIII.6.6). If in addition S includes ATR$_0$, then for any countable ω-model M of S, HYP^M is the intersection of all ω-submodels of M which satisfy S (theorem VIII.6.12, exercise VIII.6.23).

DEFINITION VIII.6.1 (essentially Σ_1^1 formulas). The class of *essentially Σ_1^1 formulas* is the smallest class of L$_2$-formulas which contains all arithmetical formulas and is closed under conjunction, disjunction, universal number quantification, existential number quantification, and existential set quantification.

LEMMA VIII.6.2. *For any essentially Σ_1^1 formula φ, we can find a Σ_1^1 formula φ' with the same free variables, such that*
  (i) *Σ_1^1-AC$_0$ proves $\varphi \rightarrow \varphi'$,*
  (ii) *ACA$_0$ proves $\varphi' \rightarrow \varphi$.*
(See also lemma VIII.3.21.)

PROOF. By induction on φ. The most interesting case is when φ is of the form $\forall n\, \psi(n)$. By inductive hypothesis, find a Σ_1^1 formula $\psi'(n)$

such that $\Sigma_1^1\text{-AC}_0$ proves $\psi(n) \rightarrow \psi'(n)$ and ACA_0 proves $\psi'(n) \rightarrow \psi(n)$. Write $\psi'(n) \equiv \exists X \, \theta'(n, X)$ where $\theta'(n, X)$ is arithmetical, and put $\varphi' \equiv \exists Y \, \forall n \, \theta'(n, (Y)_n)$, where Y is a new set variable. Clearly $\Sigma_1^1\text{-AC}_0$ proves $\varphi \rightarrow \varphi'$ and ACA_0 proves $\varphi' \rightarrow \varphi$. $\qquad\square$

The following lemma expresses a simple instance of ω-model reflection (compare §VIII.5).

LEMMA VIII.6.3. *Let $\varphi(X_1, \ldots, X_k)$ be an essentially Σ_1^1 formula with no free set variables other than $X_1, \ldots, X_k$. Then ATR_0 proves*

$$\forall X_1 \cdots \forall X_k \, [\varphi(X_1, \ldots, X_k) \rightarrow \exists \text{ countable coded } \omega\text{-model } M$$
$$\text{such that } X_1, \ldots, X_k \in M \text{ and}$$
$$M \text{ satisfies } \text{ACA}_0 \text{ plus } \varphi(X_1, \ldots, X_k)].$$

PROOF. Given $\varphi(X_1, \ldots, X_k)$, let $\varphi'(X_1, \ldots, X_k)$ be a Σ_1^1 formula as in the previous lemma. Write

$$\varphi'(X_1, \ldots, X_k) \equiv \exists Y \, \theta'(X_1, \ldots, X_k, Y)$$

where θ' is arithmetical. Reasoning in ATR_0, let $X_1, \ldots, X_k$ be such that $\varphi(X_1, \ldots, X_k)$ holds. By lemma VIII.4.1 we have $\Sigma_1^1\text{-AC}_0$, hence $\varphi'(X_1, \ldots, X_k)$ holds. Let Y be such that $\theta'(X_1, \ldots, X_k, Y)$ holds. By theorem VIII.1.13, let M be a countable coded ω-model of ACA_0 such that $X_1, \ldots, X_k, Y \in M$. Then M satisfies $\varphi'(X_1, \ldots, X_k)$. Hence M satisfies $\varphi(X_1, \ldots, X_k)$. Our lemma is proved. $\qquad\square$

Recall that, for $Y \subseteq \mathbb{N}$, $(Y)_i = \{m : (m, i) \in Y\}$.

LEMMA VIII.6.4 (GKT theorem in ATR_0). *The following is provable in ATR_0. Let X and Y be such that $\forall i \, ((Y)_i \not\leq_{\mathrm{H}} X)$. Let $\varphi(W, X)$ be a Σ_1^1 formula with no free set variables other than W and X. If $\exists W \, \varphi(W, X)$, then*

$$\exists W \, (\varphi(W, X) \wedge \forall i \, \forall j \, ((Y)_i \neq (W)_j))).$$

(Compare lemma VIII.2.23.)

PROOF. We use f as a function variable ranging over $\mathbb{N}^{\mathbb{N}}$. As usual, we identify $W \subseteq \mathbb{N}$ with its characteristic function, $W(m) = 1$ if $m \in W$, 0 if $m \notin W$. By lemma V.1.4 (our formalized version of the Kleene normal form theorem), we can find an arithmetical formula $\theta(\sigma, \tau, X)$ with no free set or function variables other than X, such that ACA_0 proves

$$\varphi(W, X) \leftrightarrow \exists f \, \forall n \, \theta(W[n], f[n], X).$$

Let $\theta^*(\sigma, \tau, X)$ be the Σ_1^1 formula

$$\exists W \, \exists f \, (W[\text{lh}(\sigma)] = \sigma \wedge f[\text{lh}(\tau)] = \tau \wedge \forall n \, \theta(W[n], f[n], X)).$$

Reasoning in ATR_0, assume the hypotheses of our lemma. Note that the following true statements are expressible by essentially Σ_1^1 formulas: $\forall a \, (\mathcal{O}(a, X) \rightarrow H_a^X \text{ exists})$; $\forall i \, ((Y)_i \not\leq_{\mathrm{H}} X)$; and $\exists W \varphi(W, X)$. Hence, by lemma VIII.6.3, there exists a countable coded ω-model M of ACA$_0$ such that X, $Y \in M$ and these statements are true in M.

Clearly M satisfies $\theta^*(\langle\rangle, \langle\rangle, X)$. We claim that if M satisfies $\theta^*(\sigma, \tau, X)$, then for all i and j there exists $\sigma' \supseteq \sigma$ such that M satisfies $\theta^*(\sigma', \tau, X)$ and $(Y)_i[\text{lh}((\sigma')_j)] \neq (\sigma')_j$. If this were not so, then for all $m \in \mathbb{N}$ and $k \in \{0, 1\}$, we would have

$$(Y)_i(m) = k \leftrightarrow M \text{ satisfies } (\exists \sigma' \supseteq \sigma)\,(\theta^*(\sigma', \tau, X) \wedge (\sigma')_j(m) = k)).$$

Thus M would satisfy that $(Y)_i$ is Δ^1_1 in X. Hence by theorem VIII.3.19, M would satisfy $(Y)_i \leq_{\text{H}} X$. This contradiction proves the claim.

Now standing outside M and applying the claim repeatedly, we can find sequences $\sigma_0 \subseteq \sigma_1 \subseteq \cdots \subseteq \sigma_k \subseteq \cdots$ and $\tau_0 \subseteq \tau_1 \subseteq \cdots \subseteq \tau_k \subseteq \cdots$ such that, for all k, $\text{lh}(\sigma_k) = \text{lh}(\tau_k) \geq k$ and M satisfies $\theta^*(\sigma_k, \tau_k, X)$ and, if $k = (i, j)$, $(Y)_i[\text{lh}((\sigma_k)_j)] \neq (\sigma_k)_j$. (These sequences are recursive in the satisfaction function of M.) Putting $W = \bigcup_k \sigma_k$ and $f = \bigcup_k \tau_k$, we get $\forall n\, \theta(W[n], f[n], X)$ and $\forall i\, \forall j\, (Y)_i \neq (W)_j$. This completes the proof of lemma VIII.6.4.   □

THEOREM VIII.6.5. *The following is provable in* ATR$_0$. *Let S be an X-recursive set of* L$_2$-*sentences. Suppose there exists a countable coded ω-model M such that $X \in M$ and M satisfies S. Given Y such that $\forall i\,((Y)_i \nleq_{\text{H}} X)$, there exists a countable coded ω-model M such that $X \in M$ and $\forall i\,((Y)_i \notin M)$ and M satisfies S.*

PROOF. Let $\varphi(W, X)$ be a Σ^1_1 formula which says that $(W)_0 = X$ and W is a code for a countable ω-model of S. Applying lemma VIII.6.4, we obtain W such that $\varphi(W, X)$ holds and $\forall i\, \forall j\, (Y)_i \neq (W)_j$. We complete the proof by letting $M = \{(W)_j : j \in \mathbb{N}\}$ be the countable ω-model which is coded by W.   □

The following corollary is sometimes described by saying that HYP is the "hard core" of ω-models of S.

COROLLARY VIII.6.6 (intersection of ω-models). *Let S be a recursive set of* L$_2$-*sentences which includes the axioms of weak Σ^1_1-AC$_0$. If S has a countable ω-model, then*

$$\text{HYP} = \bigcap\{M : M \text{ is a countable } \omega\text{-model of } S\}.$$

PROOF. By lemma VIII.4.15, HYP is included in all ω-models of weak Σ^1_1-AC$_0$. Hence HYP is included in the intersection of all ω-models of S. Now let M_1 be some countable ω-model of S and let $\langle Y_i : i \in \mathbb{N}\rangle$ be an enumeration of the sets in M_1 which are not in HYP. By theorem VIII.6.5 there exists a countable ω-model M_2 of S such that $\forall i\, (Y_i \notin M_2)$. Thus HYP $= M_1 \cap M_2$. This gives our corollary.   □

The next theorem is essentially a reformulation of theorem VIII.6.5. It differs from theorem VIII.6.5 in that it does not mention ω-models.

THEOREM VIII.6.7. *The following is provable in* ATR$_0$. *Let X and Y be such that $\forall i\,((Y)_i \not\leq_H X)$. Let $\varphi(X,Z)$ be a Σ^1_1 formula with no free set variables other than X and Z. If $\exists Z\,\varphi(X,Z)$, then $\exists Z\,(\varphi(X,Z) \wedge \forall i\,((Y)_i \not\leq_H X \oplus Z))$.*

PROOF. Assume the hypotheses. By lemma VIII.6.3, there exists a countable coded ω-model M of ACA$_0$ such that $X \in M$ and M satisfies

$$\exists Z\,(\varphi(X,Z) \wedge \forall a\,(\mathcal{O}(a, X \oplus Z) \rightarrow \mathrm{H}_a^{X \oplus Z}\text{ exists})).$$

By theorem VIII.6.5, there exists M as above with the additional property that $\forall i\,((Y)_i \notin M)$. Letting $Z \in M$ be as above, we clearly have $\varphi(X,Z)$ and $\forall W\,(W \leq_H X \oplus Z \rightarrow W \in M)$, hence $\forall i\,((Y)_i \not\leq_H X \oplus Z)$. This completes the proof. $\square$

The next theorem and its corollaries concern β-models rather than ω-models and are therefore somewhat out of place in this chapter. Our reason for presenting them here is that they constitute a significant application of theorem VIII.6.7.

Recall from §VII.1 that a *β-submodel* of M, $M' \subseteq_\beta M$, is defined to be a submodel of M with the same integers, $M' \subseteq_\omega M$, such that for any Σ^1_1 sentence χ with parameters from M', $M' \models \chi$ if and only if $M \models \chi$.

THEOREM VIII.6.8. *Let M be any countable model of* ATR$_0$. *Let $X, Y \in M$ be such that M satisfies $\forall i\,((Y)_i \not\leq_H X)$. Then there exists a model $M' \subseteq_\beta M$ such that $X \in M'$ and $\forall i\,((Y)_i \notin M')$. Such an M' is again a model of* ATR$_0$.

PROOF. Let $\varphi(e, X, Z)$ be a universal Σ^1_1 formula with only the free variables shown. Let $\{e_n : n \in \mathbb{N}\}$ be an enumeration of the integers of the countable model M (which need not be an ω-model).

Fix $X, Y \in M$ such that M satisfies $\forall i\,((Y)_i \not\leq_H X)$. Since M is a model of ATR$_0$, we can apply theorem VIII.6.7 repeatedly within M to obtain a sequence of sets $Z_0, Z_1, \ldots, Z_n, \ldots \in M$ such that

(i) $Z_0 = X$;
(ii) for all n, M satisfies $\forall i\,((Y)_i \not\leq_H Z_0 \oplus \cdots \oplus Z_n)$;
(iii) if M satisfies $\exists Z\,\varphi(e_n, Z_0 \oplus \cdots \oplus Z_n, Z)$, then Z_{n+1} is such a Z.

Let M' be the ω-submodel of M consisting of $\{Z_n : n \in \mathbb{N}\}$. By construction, M' is a β-submodel of M. Since ATR$_0$ consists of ACA$_0$ plus some Π^1_2 axioms, any β-submodel of a model of ATR$_0$ is again a model of ATR$_0$. In particular M' is a model of ATR$_0$. This completes the proof. $\square$

COROLLARY VIII.6.9 (proper β-submodels). *If M is any countable model of* ATR$_0$, *then M has a proper β-submodel $M' \subseteq_\beta M$, $M' \neq M$. Any such submodel is again a model of* ATR$_0$.

PROOF. Let $Y \in M$ be such that $M \models Y$ is not hyperarithmetical. (The existence of such a Y is implicit in the results of §§VIII.3 and VIII.4. See for example theorem VIII.3.15 and lemmas VIII.3.24 and VIII.3.25.) By theorem VIII.6.8, there exists $M' \subseteq_\beta M$ such that $Y \notin M$. $\square$

COROLLARY VIII.6.10 (intersection of β-submodels). *If M is any countable model of* ATR_0, *then*

$$\mathrm{HYP}^M = \{Y\colon M \models Y \text{ is hyperarithmetical}\}$$

can be characterized as the intersection of all β-submodels of M.

COROLLARY VIII.6.11 (intersection of β-models). HYP *is the intersection of all β-models.*

The rest of this section is concerned with the following theorem of Quinsey.

THEOREM VIII.6.12 (proper ω-submodels). *Let S be a recursive set of L_2-sentences which includes the axioms of* ATR_0. *Let M be a countable ω-model of S. Then M has a proper ω-submodel $M' \subseteq_\omega M$, $M' \neq M$, such that M' is again an ω-model of S.*

As interesting special cases of Quinsey's theorem VIII.6.12, we mention the following corollaries.

COROLLARY VIII.6.13. *For $1 \leq k < \omega$, any countable ω-model M of $\Pi_k^1\text{-}\mathsf{CA}_0$ has a proper ω-submodel $M' \subseteq_\omega M, M' \neq M$, such that M' is again an ω-model of $\Pi_k^1\text{-}\mathsf{CA}_0$.*

COROLLARY VIII.6.14. *Any countable ω-model M of $\Pi_\infty^1\text{-}\mathsf{CA}_0$ has a proper ω-submodel $M' \subseteq_\omega M, M' \neq M$, such that M' is again an ω-model of $\Pi_\infty^1\text{-}\mathsf{CA}_0$.*

Before beginning the proof of theorem VIII.6.12, we present a couple of preliminary lemmas.

LEMMA VIII.6.15. *The scheme of Σ_1^1 transfinite induction (see definition VII.2.14) is provable in* ATR_0 *plus $\Sigma_1^1\text{-}\mathsf{IND}$. Hence Σ_1^1 transfinite induction holds in any ω-model of* ATR_0.

PROOF. We reason within ATR_0 plus Σ_1^1 induction. Let X be a countable linear ordering on which Σ_1^1 transfinite induction fails, i.e.,

$$\forall j \, (\forall i \, (i <_X j \rightarrow \varphi(i)) \rightarrow \varphi(j))$$

but $\neg\forall j \, \varphi(j)$. Here $\varphi(j)$ is a Σ_1^1 formula with a distinguished free number variable j. Put $\psi(j) \equiv \neg\varphi(j)$. Then we have $\exists j \, \psi(j)$ and

$$\forall j \, (\psi(j) \rightarrow \exists i \, (i <_X j \land \psi(i))).$$

By lemma VIII.4.10, there exists $f\colon \mathbb{N} \to \mathbb{N}$ such that

$$\forall n \, (\psi(f(n)) \land f(n+1) <_X f(n)).$$

Hence X is not a well ordering. This proves the first sentence of the lemma. The second sentence follows since any ω-model automatically satisfies full induction. $\square$

The next lemma is a variant of the "pseudohierarchy principle" of §V.4. Recall definition V.1.1 according to which $\mathrm{WO}(X)$ means that X is a countable well ordering.

LEMMA VIII.6.16. *Let M be a countable ω-model of* ATR_0 *which is not a β-model. Let $\varphi(X)$ be a Σ_1^1 formula with parameters from M and no free variables other than X. If*

$$\forall X \,((\mathrm{WO}(X) \wedge X \in M) \to M \text{ satisfies } \varphi(X)),$$

then

$$\exists X \,(\neg\mathrm{WO}(X) \wedge X \in M \wedge M \text{ satisfies } (\mathrm{WO}(X) \wedge \varphi(X))).$$

PROOF. We first claim that we can find $Y \in M$ such that $\neg\mathrm{WO}(Y)$ and $M \models \mathrm{WO}(Y)$. To see this, let χ be a Σ_1^1 sentence with parameters in M such that χ is true but $M \models \neg\chi$. By the Kleene normal form theorem (lemma V.1.4), we can find an arithmetical formula $\theta(\tau)$ such that ACA_0 proves $\chi \leftrightarrow \exists f \,\forall m \,\theta(f[m])$. Let T consist of all $\tau \in \mathbb{N}^{<\mathbb{N}}$ such that $(\forall m \leq \mathrm{lh}(\tau)) \,\theta(\tau[m])$ holds. Then T is a tree, T has a path, $T \in M$, and $M \models T$ has no path. Putting $Y = \mathrm{KB}(T)$, we see by lemma V.1.3 that $Y \in M$, $\neg\mathrm{WO}(Y)$, and $M \models \mathrm{WO}(Y)$. This proves our claim.

For each $i \in \mathrm{field}(Y)$, let Y_i be the initial segment of Y determined by i, i.e., $k \leq_{Y_i} j \leftrightarrow k \leq_Y j <_Y i$. The hypothesis of our lemma implies $\forall i \,(\mathrm{WO}(Y_i) \to M \models \varphi(Y_i))$. On the other hand, since $\neg\forall i \,\mathrm{WO}(Y_i)$, it cannot be the case that $\forall i \,(\mathrm{WO}(Y_i) \leftrightarrow M \models \varphi(Y_i))$. Otherwise we would have

$$M \models \forall j \,(\forall i \,(i <_Y j \to \varphi(Y_i)) \to \varphi(Y_j)) \wedge \neg\forall i \,\varphi(Y_i)$$

contradicting the fact that $M \models \Sigma_1^1$ transfinite induction (lemma VIII.6.15). Hence there exists k such that $\neg\mathrm{WO}(Y_k)$ and $M \models \varphi(Y_k)$. Putting $X = Y_k$ for any such k, we obtain the desired conclusions.

This completes the proof of lemma VIII.6.16.   □

The proof of theorem VIII.6.12 will be based on the notion of *fulfillment*, defined below.

DEFINITION VIII.6.17 (prenex formulas). An L_2-formula is *prenex* if it is of the form

$$\forall X_1 \,\exists Y_1 \cdots \forall X_k \,\exists Y_k \,\theta(X_1, \ldots, X_k, Y_1, \ldots, Y_k) \tag{23}$$

where θ is arithmetical.

A *finite ω-model* is a finite, nonempty collection of subsets of $\mathbb{N}$.

DEFINITION VIII.6.18 (fulfillment). Let φ be a prenex L_2-sentence of the form (23). Let $\langle M_0, M_1, \ldots, M_l \rangle$ be a finite sequence of finite ω-models. We say that $\langle M_0, M_1, \ldots, M_l \rangle$ *fulfills* φ if $M_0 \subseteq M_1 \subseteq \cdots \subseteq M_l$ and

$$\forall i_1 \,(\forall X_1 \in M_{i_1}) \,(\exists Y_1 \in M_{i_1+1}) \,(\forall i_2 \geq i_1) \,(\forall X_2 \in M_{i_2}) \,(\exists Y_2 \in M_{i_2+1}) \cdots$$

$$(\forall i_k \geq i_{k-1}) \,(\forall X_k \in M_{i_k}) \,(\exists Y_k \in M_{i_k+1}) \,\theta(X_1, X_2, \ldots, X_k, Y_1, Y_2, \ldots, Y_k)$$

where $i_1, i_2, \ldots, i_k$ range over $\{0, 1, \ldots, l-1\}$.

The motivation for the concept of fulfillment is explained by the following example.

EXAMPLE VIII.6.19. Let M be a countable ω-model of an L_2-sentence φ. We may assume that φ is prenex of the form (23). Let

$$f_1 \colon M \to M, f_2 \colon M^2 \to M, \dots, f_k \colon M^k \to M$$

be a set of *Skolem functions for φ*, i.e., we have

$$\theta(X_1, X_2, \dots, X_k, f_1(X_1), f_2(X_1, X_2), \dots, f_k(X_1, X_2, \dots, X_k))$$

for all $X_1, X_2, \dots, X_k \in M$. Let $\langle M_0, M_1, \dots, M_l \rangle$ be a finite sequence of finite ω-models such that

$$M_i \cup f_1(M_i) \cup f_2(M_i^2) \cup \dots \cup f_k(M_i^k) \subseteq M_{i+1}$$

for all $i < l$. Then $\langle M_0, M_1, \dots, M_l \rangle$ fulfills φ. (This is easily proved.)

In light of the above example, the following lemma should be plausible.

LEMMA VIII.6.20. *Let $\langle M_i \colon i \in \mathbb{N} \rangle$ be an infinite sequence of finite ω-models such that, for all l, $\langle M_0, M_1, \dots, M_l \rangle$ fulfills φ. Then the countable ω-model $M = \bigcup_{i \in \mathbb{N}} M_i$ satisfies φ.*

PROOF. Straightforward.   □

DEFINITION VIII.6.21 (fulfillment tree). Let $S = \{\varphi_i \colon i \in \mathbb{N}\}$ be a set of L_2-sentences. We may assume that each φ_i is prenex. A *fulfillment tree for S* consists of a tree $T \subseteq \mathbb{N}^{<\mathbb{N}}$ together with a T-indexed collection of finite ω-models $\langle M_\tau \colon \tau \in T \rangle$ such that, for all $\tau \in T$ and $i \leq \mathrm{lh}(\tau)$, $\langle M_{\tau[i]}, M_{\tau[i+1]}, \dots, M_\tau \rangle$ fulfills φ_i.

LEMMA VIII.6.22. *Let S be a recursive set of prenex L_2-sentences. Let M be a countable ω-model of S plus Σ_1^1 choice, Σ_1^1-AC$_0$. For all countable well orderings X such that $X \in M$, M contains a fulfillment tree $\langle M_\tau \colon \tau \in T \rangle$ for S such that $|\mathrm{KB}(T)| = |X|$.*

PROOF. This lemma will be proved by transfinite induction on the countable ordinal number $|X|$. In order to keep the induction going, we shall prove a stronger statement.

Let $S = \{\varphi_i \colon i \in \mathbb{N}\}$. We may assume without loss that each φ_i is prenex. For each φ_i choose a set of Skolem functions

$$f_{i1} \colon M \to M, f_{i2} \colon M^2 \to M, \dots, f_{ik_i} \colon M^{k_i} \to M$$

as in example VIII.6.19. A finite sequence of finite ω-models $\langle M_0, M_1, \dots, M_l \rangle$ is said to *strongly fulfill φ_i* if $\langle M_0, M_1, \dots, M_l \rangle \in M$ and

$$M_j \cup f_{i1}(M_j) \cup f_{i2}(M_j^2) \cup \dots \cup f_{ik_i}(M_j^{k_i}) \subseteq M_{j+1}$$

for all $j < l$. We say that $\langle M_0, M_1, \dots, M_l \rangle$ *strongly fulfills S* if it strongly fulfills φ_i for all $i \leq l$. It is clear that for any $l \in \mathbb{N}$ there exists $\langle M_0, M_1, \dots, M_l \rangle$ which strongly fulfills S. Moreover, if $\langle M_0, M_1, \dots, M_l \rangle$

strongly fulfills S, then there exists M_{l+1} such that $\langle M_0, M_1, \ldots, M_l, M_{l+1} \rangle$ strongly fulfills S.

A fulfillment tree $\langle M'_\tau : \tau \in T \rangle$ is said to *begin with*

$$\langle M_0, M_1, \ldots, M_l \rangle$$

if there is exactly one $\tau \in T$ of length l, and for this τ we have $M'_{\tau[i]} = M_i$ for all $i \leq l$.

Now let X be a countable well ordering such that $X \in M$, and suppose that $\langle M_0, M_1, \ldots, M_l \rangle$ strongly fulfills S. We claim that M contains a fulfillment tree $\langle M'_\tau : \tau \in T \rangle$ for S beginning with $\langle M_0, M_1, \ldots, M_l \rangle$ such that

$$|\mathrm{KB}(T)| = |X| + l.$$

This claim will be proved by induction on the countable ordinal $|X|$. For $|X| = 0$ there is nothing to prove. For $|X| = |Y| + 1$, let M_{l+1} be such that $\langle M_0, M_1, \ldots, M_l, M_{l+1} \rangle$ strongly fulfills S. Applying the claim for Y, we see that M contains a fulfillment tree $\langle M'_\tau : \tau \in T \rangle$ for S beginning with $\langle M_0, M_1, \ldots, M_l, M_{l+1} \rangle$ such that

$$|\mathrm{KB}(T)| = |Y| + l + 1 = |X| + l.$$

Suppose now that $|X|$ is a limit ordinal, say $X = \sum_{n \in \mathbb{N}} X_n$ where $0 < |X_n| < |X|$ (see definition V.6.7). We may assume that each $|X_n|$ is a successor ordinal, say $|X_n| = |Y_n| + 1$. Let M_{l+1} be such that $\langle M_0, M_1, \ldots, M_l, M_{l+1} \rangle$ strongly fulfills S. Applying the claim for each Y_n, we see that for each n, M contains a fulfillment tree $\langle M'_\tau : \tau \in T_n \rangle$ for S beginning with $\langle M_0, M_1, \ldots, M_l, M_{l+1} \rangle$ such that $|\mathrm{KB}(T_n)| = |Y_n| + l + 1 = |X_n| + l$. By Σ^1_1 choice within M, we see that M contains a sequence of fulfillment trees

$$\langle \langle M'_\tau : \tau \in T_n \rangle : n \in \mathbb{N} \rangle$$

as above. We can arrange these trees so that, for each n,

$$\langle 0, 1, \ldots, l-1, n \rangle$$

is the unique element of T_n of length $l+1$. Now put $T = \bigcup_{n \in \mathbb{N}} T_n$. Then $\langle M'_\tau : \tau \in T \rangle$ belongs to M and is a fulfillment tree for S beginning with $\langle M_0, M_1, \ldots, M_l \rangle$, and $|\mathrm{KB}(T)| = |X| + l$. This completes the proof of our claim.

Lemma VIII.6.22 follows immediately from the above claim. $\square$

PROOF OF THEOREM VIII.6.12. Let $S \supseteq \mathrm{ATR}_0$ be a recursive set of L_2-sentences, and let M be a countable ω-model of S. If M is a β-model, then clearly

$$M \models \exists \text{ countable coded } \omega\text{-model of } S,$$

so the theorem holds in this case. Assume now that M is not a β-model. Since $M \models \mathrm{ATR}_0$, we have by lemma VIII.4.1 that $M \models \Sigma^1_1$ choice. Then lemma VIII.6.22 tells us that for all countable well orderings X, $X \in M$ implies

$$M \models \exists \text{ fulfillment tree } \langle M_\tau : \tau \in T \rangle \text{ for } S \text{ such that } |\mathrm{KB}(T)| = |X|. \tag{24}$$

Hence, by lemma VIII.6.16, there exists $X \in M$ such that X is not a well ordering, yet (24) holds and $M \models \mathrm{WO}(X)$. Let $\langle M_\tau : \tau \in T \rangle$ be as in (24). Standing outside M, we see that $\mathrm{KB}(T)$ is not well ordered, hence by lemma V.1.3 there exists a path f through T. Put

$$M' = \bigcup_{n \in \mathbb{N}} M_{f[n]}.$$

Thus $M' \subseteq_\omega \bigcup_{\tau \in T} M_\tau \subseteq_\omega M$. Moreover $M' \neq M$ since $\langle M_\tau : \tau \in T \rangle$ is coded as an element of M. Finally, we have $M' \models S$ by lemma VIII.6.20 since $\langle M_\tau : \tau \in T \rangle$ is a fulfillment tree for S. This completes the proof of theorem VIII.6.12.   □

EXERCISE VIII.6.23 (intersection of ω-submodels). Prove the following refinement of theorem VIII.6.12. Let S be an X-recursive set of L_2-sentences mentioning X as a parameter. Assume that S includes the axioms of ATR_0. Let M be a countable ω-model such that $X \in M$ and M satisfies S. Then $\mathrm{HYP}(X)^M$ is the intersection of all ω-submodels $M' \subseteq_\omega M$ such that $X \in M'$ and M' satisfies S. Moreover, for any $Y \in M$ such that $M \models \forall i \, ((Y)_i \not\leq_\mathrm{H} X)$, there exists $M' \subseteq_\omega M$ such that $X \in M'$ and $\forall i \, ((Y)_i \notin M')$ and M' satisfies S.

Notes for §VIII.6. Corollary VIII.6.6 is essentially due to Gandy/Kreisel/Tait [89]; it can also be derived from a result in model theory known as the omitting types theorem; see e.g., Chang/Keisler [35] and Sacks [210]. The use of the term "hard core" to describe results such as corollary VIII.6.6 is apparently due to Kreisel. Theorems VIII.6.5 and VIII.6.7 are due to Simpson [234], as are theorem VIII.6.8 and its corollaries. Theorem VIII.6.12 and corollary VIII.6.13 are due to Quinsey [204, pages 93–96]. Corollary VIII.6.14 was proved earlier by Friedman [67].

In connection with lemma VIII.6.15, Simpson [235] has shown that ATR_0 plus Σ_1^1-IND is equivalent to Σ_1^1-TI$_0$. This corrects and improves an earlier result of Friedman [69, theorem 8] and Steel [256, page 22].

VIII.7. Conclusions

The main focus of this chapter has been the existence and non-existence of minimum ω-models of particular subsystems of Z_2. We have seen that the minimum ω-models of RCA_0, ACA_0, and Σ_1^1-AC_0 are REC, ARITH, and HYP respectively (§§VIII.1, VIII.4). On the other hand, for recursive

$T_0 = \mathsf{WKL}_0$ or $T_0 \supseteq \mathsf{ATR}_0$, T_0 does not have a minimum ω-model. Indeed, every model of T_0 has a proper ω-submodel which is again a model of T_0 (VIII.2.7, VIII.6.12). Moreover, REC is the intersection of all ω-models of WKL_0 but does not itself satisfy WKL_0 (§VIII.2), and HYP is the intersection of all β-models of ATR_0 but does not itself satisfy ATR_0 (§VIII.6).

In §VIII.4 we used formalized hyperarithmetical theory and pseudo-hierarchies to obtain the following inner model result: ATR_0 proves the existence of sufficiently many countable coded ω-models of $\Sigma^1_1\text{-}\mathsf{AC}_0$. This has been applied to prove mathematical theorems in ATR_0; see remark V.10.1. In §VIII.5 we used ω-model incompleteness and reflection to obtain another interesting result: ATR_0 does not prove Σ^1_1 dependent choice.

Chapter IX

NON-ω-MODELS

The purpose of this chapter is to study certain logical properties of various L_2-theories. The main results of the chapter are *conservation results*, i.e., theorems to the effect that an apparently stronger theory T_0' is *conservative* over an apparently weaker theory T_0 with respect to a certain class of sentences. (See definition VII.5.12.) Of particular interest is the case when the minimum ω-models of T_0 and T_0' are not the same. In such a situation, non-ω-models play an essential role.

A *non-ω-model* is any model M for the language of second order arithmetic whose first order part $(|M|, +_M, \cdot_M, 0_M, 1_M, <_M)$ is not isomorphic to the intended model of first order arithmetic, $(\omega, +, \cdot, 0, 1, <)$. This means that there exists $v \in |M|$ such that for all $n \in \omega$, $M \models n < v$. One important difference between ω-models and non-ω-models is that non-ω-models do not automatically satisfy full induction. For all of the results of the present chapter, it is crucial that our subsystems of Z_2 contain only restricted induction.

Many of the results in chapters VII and VIII, on β-models and ω-models, were formulated in a general way so as to apply to non-ω-models as well. However, the model-theoretic constructions in those chapters always had the feature that the first order part of the constructed model was the same as that of a previously given model. Because of this limitation, many of the deeper conservation theorems cannot be proved by the methods of chapters VII and VIII. Only here, in chapter IX, do we focus on more radical methods of model construction, in which the integers of the new model are different from those of the given one.

In §IX.1 we show that ACA_0 is conservative over PA (first order Peano arithmetic) while RCA_0 is conservative over Σ_1^0-PA (the fragment of PA with induction restricted to Σ_1^0 formulas). These results are proved by a straightforward expansion of the given first order model. The integers are not changed, and the only nontrivial point is to show that the expansion preserves sufficient induction.

In §IX.2 we show that WKL_0 has the same first order part as RCA_0. In fact, WKL_0 is conservative over RCA_0 for Π_1^1 sentences. This result is proved by forcing over a given countable model of RCA_0. The forcing conditions are infinite trees of sequences of 0's and 1's. Again, the first order part of the given model is unchanged, and the key point is to verify that Σ_1^0 induction is preserved.

In §IX.3 we introduce the formal system PRA of primitive recursive arithmetic. We prove that WKL$_0$ is conservative over PRA for Π_2^0 sentences. The proof of this deep result involves a genuine application of non-ω-models. Namely, the integers of the constructed model of WKL$_0$ are obtained as a certain proper initial segment of the integers of the given model of PRA. At the end of the section we point out that this result is of great philosophical significance in connection with Hilbert's foundational program of finitistic reductionism.

In §IX.4 we use saturated models to prove various conservation results involving the comprehension and choice schemes which were studied in §§VII.5–VII.7. We show that, for all $k < \omega$, Δ_{k+1}^1 comprehension is conservative over Π_k^1 comprehension for Π_l^1 sentences, $l = \min\{k+2, 4\}$. This may seem rather surprising in view of the results of §VII.7, according to which the minimum β-model of Δ_{k+1}^1 comprehension is much bigger than that of Π_k^1 comprehension. It is essential here that the systems we are dealing with have only restricted induction, thus allowing greater freedom to construct non-ω-models.

Although the main focus of this chapter is on model-theoretic methods, many of the theorems can be given alternative proofs using syntactical, i.e., proof-theoretic, methods. Such methods involve a direct analysis of the structure of proofs. For example, the proof-theoretic approach to showing that T_0' is conservative over T_0 for Π_k^1 sentences would be to exhibit an explicit, primitive recursive method whereby any given T_0'-proof of a Π_k^1 sentence can be transformed into a T_0-proof of the same sentence. In addition, proof theory can be used to obtain many other results about subsystems of Z_2 which are apparently inaccessible to model-theoretic methods. In §IX.5 we briefly discuss Gentzen-style proof theory, emphasizing provable ordinals and combinatorial independence results.

IX.1. The First Order Parts of RCA$_0$ and ACA$_0$

Recall that L$_1$ and L$_2$ are the languages of first and second order arithmetic, respectively. If T_0 is any L$_2$-theory, the first order part of T_0 is the L$_1$-theory whose theorems are exactly the L$_1$-formulas which are theorems of T_0. In this section and the next, we shall determine the first order parts of RCA$_0$, ACA$_0$ and WKL$_0$. Our methods involve non-ω-models of the theories in question.

LEMMA IX.1.1. ACA$_0$ *proves all instances of arithmetical induction*:
$$(\varphi(0) \wedge \forall n\, (\varphi(n) \rightarrow \varphi(n+1))) \rightarrow \forall n\, \varphi(n) \tag{25}$$

where $\varphi(n)$ *is any arithmetical formula.*

PROOF. We reason in ACA$_0$. By arithmetical comprehension, form the set X consisting of all n such that $\varphi(n)$ holds. Then (25) follows from the induction axiom

$$(0 \in X \wedge \forall n \, (n \in X \rightarrow n + 1 \in X)) \rightarrow \forall n \, (n \in X).$$

This completes the proof.   □

DEFINITION IX.1.2. If

$$M = (|M|, \mathcal{S}_M, +_M, \cdot_M, 0_M, 1_M, <_M)$$

is any L_2-structure, *the first order part of M* is the L_1-structure

$$(|M|, +_M, \cdot_M, 0_M, 1_M, <_M).$$

LEMMA IX.1.3. *Let M be any L_2-structure which satisfies the basic axioms* I.2.4(i) *plus arithmetical induction. Then M is an ω-submodel of some model of* ACA$_0$. *In other words, we can find a model M' of* ACA$_0$ *such that* (1) *M is a submodel of M',* (2) *M and M' have the same first order part.*

PROOF. Let M be as in the hypothesis of the lemma. A set $X \subseteq |M|$ is said to be *arithmetically definable over M* if there exists an arithmetical formula $\varphi(n)$ with parameters from $|M| \cup \mathcal{S}_M$ and no free variables other than n, such that

$$X = \{a \in |M| : M \models \varphi(a)\}. \tag{26}$$

Let M' be the L_2-structure with the same first order part as M and

$$\mathcal{S}_{M'} = \text{Arith-Def}(M)$$
$$= \{X \subseteq |M| : X \text{ is arithmetically definable over } M\}.$$

Obviously $M \subseteq_\omega M'$. We claim that M' is a model of ACA$_0$. Trivially the basic axioms I.2.4(i) are satisfied in M'. Let $X \in \mathcal{S}_{M'}$ be given as in (26). Since

$$M \models (\varphi(0) \wedge \forall n \, (\varphi(n) \rightarrow \varphi(n+1))) \rightarrow \forall n \, \varphi(n),$$

it follows that

$$M' \models (0 \in X \wedge \forall n \, (n \in X \rightarrow n + 1 \in X)) \rightarrow \forall n \, (n \in X).$$

This shows that M' satisfies the induction axiom I.2.4(ii).

It remains to prove that M' satisfies arithmetical comprehension (definition III.1.2). Let $\varphi(n)$ be an arithmetical formula with parameters from $|M| \cup \mathcal{S}_{M'}$ and no free variables other than n. Exhibiting the parameters, we have

$$\varphi(n) \equiv \varphi(n, b_1, \dots, b_k, Y_1, \dots, Y_l)$$

where $b_1, \dots, b_k \in |M|$ and $Y_1, \dots, Y_l \in \mathcal{S}_{M'}$. For $j = 1, \dots, l$, we have

$$Y_j = \{b \in |M| : M \models \varphi_j(b)\}$$

where $\varphi_j(m)$ is arithmetical with parameters from $|M| \cup \mathcal{S}_M$ and no free variables other than m. Let $\widetilde{\varphi}(n)$ be the result of replacing all atomic formulas $t \in Y_j$ in $\varphi(n)$ by $\varphi_j(t)$, for $j = 1, \ldots, l$. Then $\widetilde{\varphi}(n)$ is arithmetical with parameters from $|M| \cup \mathcal{S}_M$ and no free variables other than n. Hence by definition of M' we have

$$M' \models \exists X \, \forall n \, (n \in X \leftrightarrow \widetilde{\varphi}(n)),$$

namely $X = \{a \in |M| : M \models \widetilde{\varphi}(a)\} \in \text{Arith-Def}(M)$. Moreover

$$M' \models \forall n \, (\varphi(n) \leftrightarrow \widetilde{\varphi}(n)).$$

Combining the last two formulas, we get

$$M' \models \exists X \, \forall n \, (n \in X \leftrightarrow \varphi(n)).$$

Thus M' satisfies arithmetical comprehension. The proof of lemma IX.1.3 is complete. $\square$

DEFINITION IX.1.4 (Peano arithmetic). *First order Peano arithmetic*, denoted Z_1 or PA, is the L_1-theory whose axioms are the basic axioms I.2.4(i) plus *full first order induction*, i.e.,

$$(\varphi(0) \wedge \forall n \, (\varphi(n) \to \varphi(n+1))) \to \forall n \, \varphi(n)$$

for all L_1-formulas $\varphi(n)$.

THEOREM IX.1.5. *An* L_1-*structure*

$$(|M|, +_M, \cdot_M, 0_M, 1_M, <_M) \tag{27}$$

is the first order part of some model of ACA$_0$ *if and only if it is a model of* PA.

PROOF. If (27) is the first order part of some model of ACA$_0$, then by lemma IX.1.1 it satisfies PA. Conversely, if (27) is a model of PA, then viewed as an L_2-structure it is a model of arithmetical induction, so by lemma IX.1.3 we can find a model M' of ACA$_0$ whose first order part is (27). This completes the proof. $\square$

COROLLARY IX.1.6 (first order part of ACA$_0$). PA *is the first order part of* ACA$_0$.

PROOF. Lemma IX.1.1 implies that PA is included in the first order part of ACA$_0$. For the converse, let φ be an L_1-sentence which is not a theorem of PA. By Gödel's completeness theorem, there exists a model (27) of PA in which φ fails. By theorem IX.1.5, this (27) is then the first order part of some model of ACA$_0$. Thus we have a model of ACA$_0$ in which φ fails. Hence, by the soundness theorem, φ is not a theorem of ACA$_0$. This completes the proof. $\square$

REMARK IX.1.7. In the style of definition VII.5.12, we may restate corollary IX.1.6 by saying that ACA$_0$ *is a conservative extension of* PA. Similarly, corollary IX.1.11 below says that RCA$_0$ is a conservative extension of Σ_1^0-PA.

LEMMA IX.1.8. *Let M be any L_2-structure which satisfies the basic axioms* I.2.4(i) *plus the Σ_1^0 induction scheme* II.1.3. *Then M is an ω-submodel of some model of* RCA$_0$.

PROOF. Note first that the basic axioms plus Σ_1^0 induction imply Σ_1^0 *bounding principle*:

$$(\forall i < m)\, \exists j\, \varphi(i, j) \rightarrow \exists n\, (\forall i < m)\, (\exists j < n)\, \varphi(i, j)$$

where $\varphi(i, j)$ is any Σ_1^0 formula in which n does not occur freely. This is easily proved by Σ_1^0 induction on m.

Let M be as in the hypothesis of our lemma. We say that $X \subseteq |M|$ is Δ_1^0 *definable over M* if there exist a Σ_1^0 formula $\varphi(n)$ and a Π_1^0 formula $\psi(n)$, with parameters from $|M| \cup S_M$ and no free variables other than n, such that

$$X = \{a \in |M| : M \models \varphi(a)\}$$
$$= \{a \in |M| : M \models \psi(a)\}.$$

We define M' to be the L_2-structure with the same first order part as M and

$$S_{M'} = \Delta_1^0\text{-Def}(M)$$
$$= \{X \subseteq |M| : X \text{ is } \Delta_1^0 \text{ definable over } M\}.$$

Clearly M is an ω-submodel of M'.

In order to prove that M' is a model of RCA$_0$, we first prove two claims.

Our first claim is that, for any Σ_0^0 formula θ with parameters from $|M| \cup S_{M'}$ and no free set variables, we can find a Σ_1^0 formula θ_Σ and a Π_1^0 formula θ_Π with parameters from $|M| \cup S_M$ and with the same free variables as θ, such that θ_Σ and θ_Π are equivalent to θ over M'. In proving this claim, we may assume without loss that θ is built from atomic formulas by means of negation, conjunction, and bounded universal quantification. The claim is proved by induction on θ. If θ is atomic of the form $t_1 = t_2$ or $t_1 < t_2$, we define $\theta_\Sigma \equiv \theta_\Pi \equiv \theta$. If θ is atomic of the form $t \in X$, then X is a parameter from $S_{M'}$ and we define $\theta_\Sigma \equiv \varphi(t)$ and $\theta_\Pi \equiv \psi(t)$, where $\varphi(n)$ and $\psi(n)$ are as in the Δ_1^0 definition of X over M. If $\theta \equiv \neg\theta'$, put $\theta_\Sigma \equiv \neg\theta'_\Pi$ and $\theta_\Pi \equiv \neg\theta'_\Sigma$. If $\theta \equiv \theta' \wedge \theta''$, put

$$\theta_\Sigma \equiv \theta'_\Sigma \wedge \theta''_\Sigma \equiv \exists m\, ((\exists j' < m)\, \theta'_0(j') \wedge (\exists j'' < m)\, \theta''_0(j''))$$

where $\theta'_\Sigma \equiv \exists j\, \theta'_0(j)$ and $\theta''_\Sigma \equiv \exists j\, \theta''_0(j)$. Also, put

$$\theta_\Pi \equiv \theta'_\Pi \wedge \theta''_\Pi \equiv \forall j\, (\theta'_1 \wedge \theta''_1)$$

where $\theta'_\Pi \equiv \forall j\, \theta'_1$ and $\theta''_\Pi \equiv \forall j\, \theta''_1$. Finally, if $\theta \equiv (\forall i < t)\, \theta'$, put

$$\theta_\Sigma \equiv \exists n\, (\forall i < t)\, (\exists j < n)\, \theta'_0$$

and

$$\theta_\Pi \equiv \forall j\, (\forall i < t)\, \theta'_1$$

where $\theta'_\Sigma \equiv \exists j \, \theta'_0$ and $\theta'_\Pi \equiv \forall j \, \theta'_1$. The equivalence of θ_Σ with θ over M follows from the Σ^0_1 bounding principle. This completes the proof of the first claim.

Our second claim is that, for any Σ^0_1 formula φ with parameters from $|M| \cup \mathcal{S}_{M'}$ and no free set variables, we can find an equivalent Σ^0_1 formula φ' with the same free variables and with parameters from $|M| \cup \mathcal{S}_M$ only. To see this, write $\varphi \equiv \exists j \, \theta$ where θ is Σ^0_0, and put

$$\varphi' \equiv \exists j \, \theta_\Sigma \equiv \exists j \, \exists m \, \theta_0 \equiv \exists k \, (\exists j < k) \, (\exists m < k) \, \theta_0$$

where $\theta_\Sigma \equiv \exists m \, \theta_0$ is as in our first claim, above.

We are now ready to prove that M' is a model of RCA$_0$. Trivially M' satisfies the basic axioms I.2.4(i). Given a Σ^0_1 formula $\varphi(n)$, letting $\varphi'(n)$ be as in our second claim, we have

$$M \models (\varphi'(0) \wedge \forall n \, (\varphi'(n) \to \varphi'(n+1))) \to \forall n \, \varphi'(n),$$

hence

$$M' \models (\varphi(0) \wedge \forall n \, (\varphi(n) \to \varphi(n+1))) \to \forall n \, \varphi(n)$$

so M' satisfies Σ^0_1 induction. Now assume that

$$M' \models \forall n \, (\varphi(n) \leftrightarrow \psi(n))$$

where $\varphi(n)$ and $\psi(n)$ are respectively Σ^0_1 and Π^0_1 with parameters from $|M| \cup \mathcal{S}_{M'}$. By our second claim, we can find equivalent Σ^0_1 and Π^0_1 formulas $\varphi'(n)$ and $\psi'(n)$ with parameters from $|M| \cup \mathcal{S}_M$ only. Thus

$$M' \models \forall n \, (\varphi(n) \leftrightarrow \varphi'(n))$$

and

$$M \models \forall n \, (\varphi'(n) \leftrightarrow \psi'(n)).$$

Putting $X = \{a \in |M| : M \models \varphi'(a)\} = \{a \in |M| : M \models \psi'(a)\}$, we see that $X \in \Delta^0_1\text{-Def}(M) = \mathcal{S}_{M'}$, hence

$$M' \models \exists X \, \forall n \, (n \in X \leftrightarrow \varphi(n)).$$

Thus M' satisfies Δ^0_1 comprehension.

This completes the proof of lemma IX.1.8. $\square$

DEFINITION IX.1.9. For $0 \le k < \omega$, Σ^0_k-PA is the L$_1$-theory consisting of the basic axioms I.2.4(i) plus the induction scheme (25) for all Σ^0_k L$_1$-formulas.

THEOREM IX.1.10. *An L$_1$-structure (27) is the first order part of some model of* RCA$_0$ *if and only if it is a model of* Σ^0_1-PA.

PROOF. Obviously the first order part of any model of RCA$_0$ satisfies Σ^0_1-PA. Conversely, if (27) is a model of Σ^0_1-PA, then viewed as an L$_2$-structure it is a model of Σ^0_1 induction, hence by lemma IX.1.8 it is the first order part of some model of RCA$_0$. $\square$

COROLLARY IX.1.11 (first order part of RCA₀). *The first order part of* RCA₀ *is* Σ_1^0-PA.

PROOF. Since the axioms of RCA₀ include those of Σ_1^0-PA, it is obvious that the first order part of RCA₀ includes Σ_1^0-PA. For the converse, proceed as in the proof of corollary IX.1.6 using theorem IX.1.10 instead of theorem IX.1.5. □

Notes for §IX.1. The results of this section are essentially due to Friedman [69]. For more information on models of Σ_k^0-PA, $0 \le k < \omega$, see Hájek/Pudlák [100] and Kaye [137].

IX.2. The First Order Part of WKL₀

In this section we shall show that the first order part of WKL₀ is the same as that of RCA₀. This is a previously unpublished result of Harrington. Our proof will employ a kind of forcing argument in which the forcing conditions are trees.

THEOREM IX.2.1. *Let* M *be any countable model of* RCA₀. *Then* M *is an ω-submodel of some countable model of* WKL₀.

The proof will be based on the following notion of genericity.

DEFINITION IX.2.2. Let M be a model of RCA₀.

1. We define $\mathcal{T}_M$ to be the set of all $T \in \mathcal{S}_M$ such that

$$M \models T \text{ is an infinite subtree of } 2^{<\mathbb{N}}.$$

   For $T \in \mathcal{T}_M$ and $X \subseteq |M|$, we say that X *is a path through* T if, for all $b \in |M|$, $X[b] \in T$. Here $X[b] \in T$ means that there exists $\sigma \in |M|$ such that $M \models \sigma \in T$ and $\mathrm{lh}(\sigma) = b$, and for all $a <_M b$, $a \in X$ if and only if $M \models \sigma(a) = 1$.
2. We say that $\mathcal{D} \subset \mathcal{T}_M$ is *dense* if for all $T \in \mathcal{T}_M$ there exists $T' \in \mathcal{D}$ such that $T' \subseteq T$. We say that $\mathcal{D}$ is *M-definable* if there exists a formula $\varphi(X)$ with parameters from $|M| \cup \mathcal{S}_M$ and no free variables other than X, such that for all $T \in \mathcal{T}_M$, $M \models \varphi(T)$ if and only if $T \in \mathcal{D}$.
3. We say that $G \subseteq |M|$ is *$\mathcal{T}_M$-generic* if for every dense, M-definable $\mathcal{D} \subseteq \mathcal{T}_M$ there exists $T \in \mathcal{D}$ such that G is a path through T.

LEMMA IX.2.3. *Let* M *be a countable model of* RCA₀. *Given* $T \in \mathcal{T}_M$, *we can find a $\mathcal{T}_M$-generic $G \subseteq |M|$ such that G is a path through T.*

PROOF. Since M is countable, the set of all dense M-definable sets $\mathcal{D} \subseteq \mathcal{T}_M$ is countable. Let $\langle \mathcal{D}_i : i < \omega \rangle$ be an enumeration of these dense sets. Given $T \in \mathcal{T}_M$, we can find a sequence of trees T_i, $i < \omega$, such that $T_0 = T$, $T_{i+1} \subseteq T_i$, and $T_{i+1} \in \mathcal{D}_i$ for all $i < \omega$. We are going to show that there is a unique G such that, for all $i < \omega$, G is a path through T_i. To see this, define

$\mathcal{E}_b$ for each $b \in |M|$ to be the set of $T \in \mathcal{T}_M$ such that T contains exactly one sequence of length b. Clearly $\mathcal{E}_b$ is dense and M-definable. Let i_b be such that $\mathcal{E}_b = \mathcal{D}_{i_b}$, and let τ_b be the unique sequence of length b such that $\tau_b \in T_{i_b+1}$. Clearly $b <_M c$ implies $\tau_b \subseteq \tau_c$. Let G be the unique subset of $|M|$ whose characteristic function is $\bigcup_{b \in |M|} \tau_b$. Clearly $\tau_b \in T_i$ for all $b \in |M|$ and all $i < \omega$. Hence, for all $i < \omega$, G is a path through T_i. It follows by construction that G is $\mathcal{T}_M$-generic. This proves lemma IX.2.3. $\qquad\square$

LEMMA IX.2.4. *Let M be a model of* RCA$_0$ *and suppose that $G \subseteq |M|$ is $\mathcal{T}_M$-generic. Let M' be the L_2-structure with the same first order part as M and $\mathcal{S}_{M'} = \mathcal{S}_M \cup \{G\}$. Then M' satisfies Σ_1^0 induction.*

PROOF. It suffices to prove the following. For any $b \in |M|$ and any Σ_1^0 formula $\varphi(i, X)$ with parameters from $|M| \cup \mathcal{S}_M$ and no free variables other than i and X, the set $\{a : a <_M b \wedge M' \models \varphi(a, G)\}$ is M-finite. (From this it is clear that M' satisfies Σ_1^0 induction.)

In order to prove this, assume first that $\varphi(i, X)$ is in *normal form*, i.e., $\varphi(i, X) \equiv \exists j\, \theta(i, X[j])$ where $\theta(i, \tau)$ is Σ_0^0 with parameters from $|M| \cup \mathcal{S}_M$. (Later we shall show how to eliminate this assumption.) Let $\mathcal{D}_b$ be the set of $T \in \mathcal{T}_M$ such that, for each $a <_M b$, M satisfies either

(i) $\forall \tau\, (\tau \in T \rightarrow \neg \theta(a, \tau))$

or

(ii) $\exists k\, \forall \tau\, ((\tau \in T \wedge \mathrm{lh}(\tau) = k) \rightarrow (\exists j \leq k)\, \theta(a, \tau[j]))$

where $\tau[j]$ denotes the initial sequence of τ of length j. The motivation here is that if G is a path through T, then (i) gives $\neg\varphi(a, G)$ while (ii) gives $\varphi(a, G)$.

We claim that $\mathcal{D}_b$ is dense in $\mathcal{T}_M$. To see this, let $T \in \mathcal{T}_M$ be given. Working within M, for each $\sigma \in (2^{<\mathbb{N}})^M$ define a tree T_σ as follows: $T_{\langle\rangle} = T$; $T_{\sigma \frown \langle 0 \rangle} = \{\tau \in T_\sigma : (\forall j \leq \mathrm{lh}(\tau))\, \neg\theta(a, \tau[j])\}$ where $a = \mathrm{lh}(\sigma)$; and $T_{\sigma \frown \langle 1 \rangle} = T_\sigma$. Set $S_b = \{\sigma : \mathrm{lh}(\sigma) = b \wedge T_\sigma \text{ is infinite}\}$. Since M satisfies bounded Σ_1^0 comprehension (theorem II.3.9), S_b is M-finite. Moreover S_b is nonempty since, for instance, $\langle 1, 1, \ldots, 1 \rangle$ (with b 1's) is an element of S_b. Now let σ_b be the lexicographically least element of S_b. We are going to show that T_{σ_b} belongs to $\mathcal{D}_b$. To see this, let $a <_M b$ be given. If $\sigma_b(a) = 0$, then

$$T_{\sigma_b} \subseteq T_{\sigma_b[a] \frown \langle 0 \rangle}$$
$$= \{\tau \in T_{\sigma_b[a]} : (\forall j \leq \mathrm{lh}(\tau))\, \neg\theta(a, \tau[j])\}$$
$$\subseteq \{\tau : \neg\theta(a, \tau)\}$$

so in this case (i) holds. If $\sigma_b(a) = 1$, then $T_{\sigma_b[a] \frown \langle 0 \rangle}$ is M-finite, i.e., M satisfies

$$\exists k\, \forall \tau\, ((\tau \in T_{\sigma_b[a]} \wedge \mathrm{lh}(\tau) = k) \rightarrow (\exists j \leq k)\, \theta(a, \tau[j])),$$

so in this case (ii) holds. This completes the proof that $\mathcal{D}_b$ is dense.

In addition, $\mathcal{D}_b$ is M-definable, so let $T' \in \mathcal{D}_b$ be such that G is a path through T'. By bounded Σ_1^0 comprehension within M, there exists an M-finite set Y consisting of all $a <_M b$ such that M satisfies (ii) for T' and a. We then have:

$$Y = \{a : a <_M b \wedge M' \models \exists j \, \theta(a, G[j])\}$$
$$= \{a : a <_M b \wedge M' \models \varphi(a, G)\}.$$

This completes the proof under the assumption that $\varphi(i, X)$ is in normal form.

It remains to eliminate this assumption. For this, it suffices to prove the following claim: for every Σ_1^0 formula $\varphi(X)$ with parameters from $|M| \cup \mathcal{S}_M$ and no free set variables other than X, we can find another such formula $\overline{\varphi}(X)$ which is in normal form and such that M' satisfies $\varphi(G) \leftrightarrow \overline{\varphi}(G)$. We shall first prove our claim when $\varphi(X)$ is Σ_0^0, by induction on $\varphi(X)$. The most interesting case is when $\varphi(X)$ is of the form $(\forall i < t) \, \varphi'(i, X)$. By inductive hypothesis, we may assume that $\varphi'(i, X)$ is in normal form, say $\varphi'(i, X) \equiv \exists j \, \theta(i, X[j])$ where $\theta(i, \tau)$ is Σ_0^0. Put $\theta'(i, \tau) \equiv (\exists j < \mathrm{lh}(\tau)) \, \theta(i, \tau[j])$. If $(\forall i < t) \exists j \, \theta(i, G[j])$ holds, then by Σ_1^0 induction on $m \leq t$ we can prove $\exists n \, (\forall i < m) \, \theta'(i, G[n])$. Thus

$$\varphi(G) \leftrightarrow (\forall i < t) \exists j \, \theta(i, G[j])$$
$$\leftrightarrow \exists n \, (\forall i < t) \, \theta'(i, G[n]).$$

Hence in this case we may take $\overline{\varphi}(X) \equiv \exists n \, (\forall i < t) \, \theta'(i, X[n])$. This proves our claim provided $\varphi(X)$ is Σ_0^0. Now when $\varphi(X)$ is Σ_1^0, we have $\varphi(G) \equiv \exists k \, \theta(k, G) \equiv \exists k \, \exists j \, \theta'(k, G[j]) \equiv \exists n \, \theta''(G[n])$ where θ' and θ'' are appropriate Σ_0^0 formulas. Thus we may take $\overline{\varphi}(X) \equiv \exists n \, \theta''(X[n])$. This gives our claim.

The proof of lemma IX.2.4 is now complete.   □

LEMMA IX.2.5. *Let M be a countable model of* RCA₀. *Given $T \in \mathcal{T}_M$, there exists a countable model M'' of* RCA₀ *such that M is an ω-submodel of M'', and $M'' \models T$ has a path.*

PROOF. By lemma IX.2.3, let G be a $\mathcal{T}_M$-generic path through T. Let M' be the model with the same first order part as M and $\mathcal{S}_{M'} = \mathcal{S}_M \cup \{G\}$. Then M is an ω-submodel of M', and by lemma IX.2.4, M' satisfies the Σ_1^0 induction scheme. By lemma IX.1.8, we can find a countable model M'' of RCA₀ such that M' is an ω-submodel of M''. This completes the proof.   □

PROOF OF THEOREM IX.2.1. Use lemma IX.2.5 repeatedly to form a sequence of countable ω-models

$$M = M_0 \subseteq_\omega M_1 \subseteq_\omega \cdots \subseteq_\omega M_i \subseteq_\omega \cdots \qquad (i < \omega)$$

where each M_i is a model of RCA$_0$ and for all $T \in \mathcal{T}_{M_i}$ there exists $j > i$ such that $M_j \models T$ has a path. Let M^* be the union of this sequence of models. Then clearly $M \subseteq_\omega M^*$. Moreover $M^* \models$ RCA$_0$ and, for all $T \in \mathcal{T}_{M^*}$, $M^* \models T$ has a path. Thus M^* is a countable ω-model of WKL$_0$. This completes the proof. $\square$

The following corollary may be expressed by saying that WKL$_0$ *is conservative over* RCA$_0$ *for* Π_1^1 *sentences*. (This terminology is explained in definition VII.5.12.)

COROLLARY IX.2.6 (conservation theorem). *For any* Π_1^1 *sentence* ψ, *if* ψ *is a theorem of* WKL$_0$ *then* ψ *is already a theorem of* RCA$_0$.

PROOF. Suppose that ψ is a Π_1^1 sentence which is not a theorem of RCA$_0$. Then by Gödel's completeness theorem, there exists a countable model M of RCA$_0$ in which ψ fails. By theorem IX.2.1, let M^* be a model of WKL$_0$ such that M is an ω-submodel of M^*. Writing ψ as $\forall X\, \theta(X)$ where $\theta(X)$ is arithmetical, there exists $X \in S_M$ such that $M \models \neg\theta(X)$. Since M^* extends M and has the same first order part as M, it follows that $M^* \models \neg\theta(X)$. Hence ψ fails in M^*. By the soundness theorem, it follows that ψ is not a theorem of WKL$_0$. This completes the proof. $\square$

COROLLARY IX.2.7 (first order part of WKL$_0$). *The first order part of* WKL$_0$ *is the same as that of* RCA$_0$, *namely* Σ_1^0-PA.

PROOF. Corollary IX.2.6 implies that WKL$_0$ and RCA$_0$ have the same first order part, since every L$_1$-formula is Π_1^1. The rest follows from corollary IX.1.11. $\square$

EXERCISE IX.2.8. Let $\mathcal{N}_M$ be the set of $f \in S_M$ such that $M \models (f$ is a total function from $\mathbb{N}$ into $\mathbb{N})$. For $f, g \in \mathcal{N}_M$, let us say that f *majorizes* g if $f(b) \geq_M g(b)$ for all $b \in |M|$. Prove the following refinement of theorem IX.2.1. Given a countable model M of RCA$_0$, we can find a countable ω-model M^* of WKL$_0$ such that M is an ω-submodel of M^* and every $g \in \mathcal{N}_{M^*}$ is majorized by some $f \in \mathcal{N}_M$. (See also theorem VIII.2.21.)

Notes for §IX.2. Theorem IX.2.1 is due to Harrington (1977, unpublished, communicated by Friedman). Our proof of theorem IX.2.1 is inspired by the proof of Jockusch/Soare [134, theorem 2.4]. This same Jockusch/Soare construction is also related to the proof of theorem VIII.2.21 and to exercise IX.2.8 above. Simpson (1982, unpublished) has used a different construction to show that theorem IX.2.1 also holds for uncountable models.

IX.3. A Conservation Result for Hilbert's Program

In this section we shall introduce the formal system PRA of *primitive recursive arithmetic*. We shall then use a model-theoretic method to show that WKL$_0$ is conservative over PRA for Π_2^0 sentences. At the end of the section, we shall explain how this conservation result represents a partial realization of Hilbert's program for the foundations of mathematics.

DEFINITION IX.3.1 (language of PRA). The *language of* PRA is a first order language with equality. In addition to the 2-place predicate symbol $=$, it contains a constant symbol $\underline{0}$, number variables $x_0, x_1, \ldots, x_n, \ldots (n < \omega)$, 1-place operation symbols $\underline{Z}$ and $\underline{S}$, k-place operation symbols $\underline{P}_i^k$ for each i and k with $1 \le i \le k < \omega$, and additional operation symbols, which are introduced as follows. If g is an m-place operation symbol and $\underline{h}_1, \ldots, \underline{h}_m$ are k-place operation symbols, then $\underline{f} = C(\underline{g}, \underline{h}_1, \ldots, \underline{h}_m)$ is an k-place operation symbol. If g is a k-place operation symbol and $\underline{h}$ is a $(k + 2)$-place operation symbol, then $\underline{f} = R(\underline{g}, \underline{h})$ is a $(k + 1)$-place operation symbol. The operation symbols of the language of PRA are called *primitive recursive function symbols*.

The *intended model of* PRA consists of the nonnegative integers, $\omega = \{0, 1, 2, \ldots\}$, together with the primitive recursive functions. In detail, the number variables range over ω and we interpret $=$ as equality on ω, $\underline{0}$ as 0, $\underline{Z}$ as the constant zero function Z defined by $Z(x) = 0$, $\underline{S}$ as the successor function S defined by $S(x) = x + 1$, $\underline{P}_i^k$ as the projection function P_i^k defined by

$$P_i^k(x_1, \ldots, x_k) = x_i,$$

$C(\underline{g}, \underline{h}_1, \ldots, \underline{h}_m)$ as the function f defined by composition as

$$f(x_1, \ldots, x_k) = g(h_1(x_1, \ldots, x_k), \ldots, h_m(x_1, \ldots, x_k)),$$

and $R(\underline{g}, \underline{h})$ as the function f defined by primitive recursion as

$$f(0, x_1, \ldots, x_k) = g(x_1, \ldots, x_k)$$
$$f(y + 1, x_1, \ldots, x_k) = h(y, f(y, x_1, \ldots, x_k), x_1, \ldots, x_k).$$

DEFINITION IX.3.2 (axioms of PRA). The *axioms of* PRA are as follows. We have the usual axioms for equality. We have the usual axioms for $\underline{0}$ and the successor function:

$$\underline{Z}(x) = \underline{0},$$
$$\underline{S}(x) = \underline{S}(y) \to x = y,$$
$$x \ne \underline{0} \leftrightarrow \exists y\,(\underline{S}(y) = x).$$

We have defining axioms for the projection functions:

$$\underline{P}_i^k(x_1, \ldots, x_k) = x_i.$$

For each function $f = C(\underline{g}, \underline{h}_1, \ldots, \underline{h}_m)$ given by composition, we have a defining axiom

$$\underline{f}(x_1, \ldots, x_k) = \underline{g}(\underline{h}_1(x_1, \ldots, x_k), \ldots, \underline{h}_m(x_1, \ldots, x_k)).$$

For each function $\underline{f} = R(\underline{g}, \underline{h})$ given by primitive recursion, we have defining axioms

$$\underline{f}(\underline{0}, x_1, \ldots, x_k) = \underline{g}(x_1, \ldots, x_k),$$
$$\underline{f}(\underline{S}(y), x_1, \ldots, x_k) = \underline{h}(y, \underline{f}(y, x_1, \ldots, x_k), x_1, \ldots, x_k).$$

Finally we have the scheme of primitive recursive induction:

$$(\theta(\underline{0}) \wedge \forall x \, (\theta(x) \to \theta(\underline{S}(x)))) \to \forall x \, \theta(x)$$

where $\theta(x)$ is any quantifier-free formula in the language of PRA with a distinguished free number variable x. We define PRA, *primitive recursive arithmetic*, to be the formal system with the above axioms.

In general, a *model of* PRA consists of a set $|M|$, a distinguished element $0_M \in |M|$, and a k-place function $f_M : |M|^k \to |M|$ for each k-place primitive recursive function symbol $\underline{f}$, such that the axioms of PRA are true when the number variables range over $|M|$ and we interpret $=$ as equality on $|M|$, $\underline{0}$ as 0_M, and $\underline{f}$ as f_M.

EXERCISE IX.3.3. Prove that PRA can be axiomatized by a set of quantifier-free formulas. (Hint: There are two ways to prove this. The first is to exhibit a set of quantifier-free formulas and prove that they axiomatize PRA. The second way is to prove that every submodel of a model of PRA is again a model of PRA, then apply the theorem of Tarski according to which any first order theory with this property can be axiomatized by quantifier-free formulas. See the notes at the end of the section.)

The results of this section have to do with the relationship between PRA and subsystems of first order Peano arithmetic, PA. Our first task is to show that PRA is essentially included in Σ_1^0-PA (modulo a certain interpretation). For the definition of Σ_1^0-PA, see §IX.1.

DEFINITION IX.3.4. We define the *canonical interpretation* of the language of first order arithmetic, L_1, into the language of PRA. The constants 0 and 1 are interpreted as $\underline{0}$ and $\underline{1} \equiv \underline{S}(\underline{0})$ respectively. Addition and multiplication are interpreted as primitive recursive functions given by

$$x + \underline{0} = x, \qquad\qquad x + \underline{S}(y) = \underline{S}(x + y),$$
$$x \cdot \underline{0} = \underline{0}, \qquad\qquad x \cdot \underline{S}(y) = (x \cdot y) + x.$$

We introduce predecessor and truncated subtraction, $\underline{P}$ and $\dot{-}$, as primitive recursive functions given by

$$\underline{P}(\underline{0}) = \underline{0}, \qquad \underline{P}(\underline{S}(y)) = y,$$
$$x \dot{-} \underline{0} = x, \qquad x \dot{-} \underline{S}(y) = \underline{P}(x \dot{-} y).$$

We then interpret $t_1 < t_2$ as $t_2 \dot{-} t_1 \neq 0$.

LEMMA IX.3.5. *Any model of Σ_1^0-PA can be expanded to a model of PRA in a way which respects the canonical interpretation of L_1 into the language of PRA.*

PROOF. By theorem IX.1.10, the given model of Σ_1^0-PA is the first order part of a model

$$M = (|M|, S_M, +_M, \cdot_M, 0_M, 1_M, <_M)$$

of RCA_0. We can then use the results of §II.3 (particularly theorem II.3.4) to expand M to a model of PRA. Namely, to each k-place primitive recursive function symbol f we associate a k-place function $f_M : |M|^k \to |M|$ such that $f_M \in S_M$ and these functions obey the defining axioms as in definition IX.3.2 above. Under this interpretation, the PRA induction axioms follow from Σ_1^0 induction within M. Moreover, since the defining axioms for $+$, $\cdot$ and $<$ are theorems of RCA_0, it is clear that $+$, $\cdot$, 0, 1 and $<$ are interpreted as $+_M$, $\cdot_M$, 0_M, 1_M and $<_M$ respectively. This completes the proof. $\square$

THEOREM IX.3.6. *Let θ be any L_1-formula. If θ is provable in PRA under the canonical interpretation, then θ is provable in Σ_1^0-PA (hence also in RCA_0).*

PROOF. Suppose that θ is not provable in Σ_1^0-PA. By the completeness theorem, let $(|M|, +_M, \cdot_M, 0_M, 1_M, <_M)$ be a model of Σ_1^0-PA in which θ fails. By the previous lemma, this can be expanded to a model of PRA in a way that respects the canonical interpretation of L_1 into the language of PRA. Thus we have a model of PRA in which θ fails. Hence, by the soundness theorem, θ is not provable in PRA. This completes the proof of the theorem. $\square$

Having shown that PRA is essentially included in Σ_1^0-PA, we now turn to the converse. We shall show that every Π_2^0 sentences which is provable in Σ_1^0-PA (indeed WKL_0) is provable in PRA.

A formula in the language of PRA is said to be *generalized Σ_0^0* if it is built from atomic formulas of the form $t_1 = t_2$ and $t_1 < t_2$, where t_1 and t_2 are terms in the language of PRA, by means of propositional connectives and bounded quantifiers of the form $(\forall x < t)$ and $(\exists x < t)$, where t is a term in the language of PRA not mentioning x. The following lemma tells us that every generalized Σ_0^0 formula is equivalent to an atomic formula.

LEMMA IX.3.7. *For any generalized Σ_0^0 formula $\theta(x_1, \ldots, x_k)$ with only the displayed free variables, we can find a k-place primitive recursive function symbol $\underline{f} = \underline{f}_\theta$ such that* PRA *proves*

(1) $\underline{f}(x_1, \ldots, x_k) = \underline{1} \leftrightarrow \theta(x_1, \ldots, x_k)$

and

(2) $\underline{f}(x_1, \ldots, x_k) = \underline{0} \leftrightarrow \neg\theta(x_1, \ldots, x_k)$.

PROOF. The proof is straightforward by induction on θ. If $\theta \equiv \theta' \wedge \theta''$, we can take

$$\underline{f}_\theta(x_1, \ldots, x_k) = \underline{f}_{\theta'}(x_1, \ldots, x_k) \cdot \underline{f}_{\theta''}(x_1, \ldots, x_k).$$

If $\theta(x_1, \ldots, x_k) \equiv (\forall y < t)\, \theta'(y, x_1, \ldots, x_k)$ where t is a term whose free variables are among $x_1, \ldots, x_k$, then we can take

$$\underline{f}_\theta(x_1, \ldots, x_k) = \prod_{y<t} \underline{f}_{\theta'}(y, x_1, \ldots, x_k).$$

Here $\underline{g}(z, x_1, \ldots, x_k) = \prod_{y<z} \underline{f}_{\theta'}(y, x_1, \ldots, x_k)$ is defined primitive recursively by

$$\underline{g}(\underline{0}, x_1, \ldots, x_k) = 1,$$
$$\underline{g}(\underline{S}(z), x_1, \ldots, x_k) = \underline{g}(z, x_1, \ldots, x_k) \cdot \underline{f}_{\theta'}(z, x_1, \ldots, x_k).$$

If $\theta \equiv \neg\theta'$, we can take

$$\underline{f}_\theta(x_1, \ldots, x_k) = \underline{\text{neg}}(\underline{f}_{\theta'}(x_1, \ldots, x_k))$$

where $\underline{\text{neg}}(\underline{0}) = \underline{1}$, $\underline{\text{neg}}(\underline{S}(y)) = \underline{0}$. If θ is atomic of the form $t_1 = t_2$, we can take $\underline{f}_\theta(x_1, \ldots, x_k) = \underline{\text{neg}}((t_2 \dot- t_1) + (t_1 \dot- t_2))$. If θ is atomic of the form $t_1 < t_2$, we can take $\underline{f}_\theta(x_1, \ldots, x_k) = \underline{\text{neg}}(\underline{\text{neg}}(t_2 \dot- t_1))$. The details of the verification that PRA proves (1) and (2) are left to the reader. $\square$

For each k-place primitive recursive function symbol $\underline{f}$, we introduce a k-place *primitive recursive predicate symbol* $\underline{R} = \underline{R}_f$ defined by

$$\underline{R}(x_1, \ldots, x_k) \leftrightarrow \underline{f}(x_1, \ldots, x_k) = \underline{1}.$$

In any model M of PRA, R_M is defined as the set of k-tuples $\langle a_1, \ldots, a_k \rangle \in |M|^k$ such that $f_M(a_1, \ldots, a_k) = 1_M$. The previous lemma implies that every generalized Σ_0^0 formula is equivalent to a primitive recursive predicate.

DEFINITION IX.3.8 (M-finite sets and M-cardinality). Let M be a model of PRA.

1. An *M-finite set* is a set $X \subseteq |M|$ such that

$$X = \{a \in |M| : a <_M b \wedge R_M(a, c_1, \ldots, c_k)\}$$

for some primitive recursive predicate symbol $\underline{R}$ and some parameters $b, c_1, \ldots, c_k \in |M|$.

2. If X is an M-finite set, the M-cardinality of X is the number of elements in X as counted within M. Formally, the M-cardinality of X is defined as $\mathrm{card}_M(X) = \mathrm{card}_M(X, b)$ where $X \subseteq \{a : a <_M b\}$ and

$$\mathrm{card}_M(X, 0) = 0,$$

$$\mathrm{card}_M(X, a + 1) = \begin{cases} \mathrm{card}_M(X, a) + 1 & \text{if } a \in X, \\ \mathrm{card}_M(X, a) & \text{if } a \notin X. \end{cases}$$

In working with models of PRA, it will be important to know that M-finite sets can be *encoded* as single elements of $|M|$ in a primitive recursive way. There are several possible methods to accomplish this. For example, we could use the coding scheme of theorem II.2.5. Instead, we shall use the following method. We say that $c \in |M|$ *encodes* the M-finite set X if

$$\forall a \, (a \in X \leftrightarrow M \models (\exists u < c)\,(\exists v < \underline{2}^a)\,(c = \underline{2}^{a+1} \cdot u + \underline{2}^a + v)).$$

Here the primitive recursive function $\underline{\exp}(x) = \underline{2}^x$ is defined by $\underline{\exp}(0) = 1$, $\underline{\exp}(\underline{S}(y)) = \underline{2} \cdot \underline{\exp}(y)$, where $\underline{2} = \underline{S}(\underline{1})$.

LEMMA IX.3.9. *Let M be a model of PRA. Then for every M-finite set X, there is a unique $c \in |M|$ which encodes X. Furthermore $X \subseteq \{a : a <_M b\}$ if and only if $c <_M \underline{2}^b$.*

PROOF. The code of X is $c = \sum_{a \in X} \underline{2}^a$. More formally, we define

$$c = \mathrm{code}_M(X) = \mathrm{code}_M(X, b)$$

where $X \subseteq \{a : a <_M b\}$ and

$$\mathrm{code}_M(X, 0) = 0,$$

$$\mathrm{code}_M(X, a + 1) = \begin{cases} \mathrm{code}_M(X, a) + \underline{2}^a & \text{if } a \in X, \\ \mathrm{code}_M(X, a) & \text{if } a \notin X. \end{cases}$$

It is straightforward to show within PRA that these codes have the desired properties. □

The concept of semiregular cut, defined below, is due to Kirby and Paris.

DEFINITION IX.3.10 (semiregular cuts). Let M be a model of PRA.

1. A *cut* in M is a set $I \subseteq |M|$, $1_M \in I \neq |M|$, such that $c <_M b$, $b \in I$ imply $c \in I$.
2. If I is a cut in M, a set $X \subseteq I$ is said to be M-*coded* if there exists an M-finite set X^* such that $X^* \cap I = X$. The set of all M-coded subsets of I is denoted $\mathrm{Coded}_M(I)$. A set X is said to be *bounded in* I if $X \subseteq \{a : a <_M b\}$ for some $b \in I$.
3. A cut I is said to be *semiregular* if, for all M-finite sets X such that $\mathrm{card}_M(X) \in I$, $X \cap I$ is bounded in I.

LEMMA IX.3.11. *Let M be a model of* PRA, *and let I be a semiregular cut in M. Then*

$$(I, \mathrm{Coded}_M(I), +_M \restriction I, \cdot_M \restriction I, 0_M, 1_M, <_M \restriction I) \tag{28}$$

is a model of WKL₀.

Here $+_M \restriction I$ is the restriction of $+_M$ to I, etc.

PROOF. We first show that I is closed under $+_M$ and $\cdot_M$. If $b, c \in I$ and $b +_M c \notin I$, then the M-finite set $X = \{a : b \leq_M a <_M b +_M c\}$ has M-cardinality c, yet $X \cap I = \{a \in I : b \leq_M a\}$ is unbounded in I, a contradiction. Thus I is closed under $+_M$. Similarly, if $b, c \in I$ and $b \cdot_M c \notin I$, then the M-finite set $Y = \{b \cdot_M a : a <_M c\}$ has M-cardinality c, yet $Y \cap I$ is unbounded in I, a contradiction. Thus I is closed under $\cdot_M$.

Since M satisfies the primitive recursive induction scheme, every nonempty M-finite set has a $<_M$-least element. We shall now use this observation to show that the L_2-structure (28) satisfies Σ_1^0 induction. Let $\varphi(x)$ be a Σ_1^0 formula with parameters from $I \cup \mathrm{Coded}_M(I)$ and no free variables other than x. We are trying to prove that (28) satisfies

$$(\varphi(0) \wedge \forall x \, (\varphi(x) \to \varphi(x+1))) \to \forall x \, \varphi(x). \tag{29}$$

If (28) satisfies $\varphi(c)$ for all $c \in I$, there is nothing to prove. So let $c \in I$ be such that (28) satisfies $\neg\varphi(c)$. Form the set

$$Y = \{a : a <_M c \text{ and (28) satisfies } \varphi(a)\}.$$

We claim that Y is M-finite. To see this, let $\varphi^*(x)$ be the formula which results from $\varphi(x)$ when we replace each set parameter $X \in \mathrm{Coded}_M(I)$ by an M-finite set X^* such that $X^* \cap I = X$. Thus we have $\varphi^*(x) \equiv \exists y \, \theta^*(x, y)$ where $\theta^*(x, y)$ is a generalized Σ_0^0 formula with parameters from $|M|$. Fix $d \in |M|$ such that $d \notin I$, and let Z be the set of all pairs (a, b) such that $a <_M c$, $b <_M d$, and b is the $<_M$-least b' such that M satisfies $\theta^*(a, b')$. By lemma IX.3.7, Z is M-finite. Moreover, the M-cardinality of Z is at most c. Hence, by semiregularity, $Z \cap I$ is bounded in I. Hence $Z \cap I$ is M-finite. From this it follows (again by lemma IX.3.7) that $Y = \{a : \exists b \, ((a, b) \in Z \cap I)\}$ is M-finite. This proves our claim.

Since Y is M-finite and $c \notin Y$, let b be the $<_M$-least element of I such that $b \notin Y$. If $b = 0_M$ we see that (28) satisfies $\neg\varphi(0)$. If $b = S_M(b')$, we see that (28) satisfies $\varphi(b')$ and $\neg\varphi(b'+1)$. In either case, (28) satisfies (29). We have now shown that (28) satisfies Σ_1^0 induction.

Next we shall show that (28) satisfies the Σ_1^0 separation principle IV.4.4.2. Let $\varphi_i(x)$, $i \in \{0, 1\}$ be Σ_1^0 formulas with parameters from $I \cup \mathrm{Coded}_M(I)$ and no free variables other than x. For $i \in \{0, 1\}$ put

$$A_i = \{a \in I : \text{(28) satisfies } \varphi_i(a)\}$$

and assume that $A_0 \cap A_1 = \emptyset$. We must show that A_0 and A_1 can be separated by an M-coded subset of I. Let $\varphi_i^*(x)$ be the formula which results from $\varphi_i(x)$ when we replace each set parameter $X \in \mathrm{Coded}_M(I)$ by an M-finite set X^* such that $X^* \cap I = X$. As before, we have $\varphi_i^*(x) \equiv \exists y \, \theta_i^*(x, y)$ where $\theta_i^*(x, y)$ is a generalized Σ_0^0 formula with parameters from $|M|$. Fix $d \in |M|$ such that $d \notin I$, and put

$$Y^* = \{a : a <_M d \wedge (\exists b <_M d)\,(\theta_1^*(a, b) \wedge (\forall b' <_M b)\, \neg \theta_0^*(a, b'))\}.$$

Clearly $A_1 \subseteq Y^*$ and $Y^* \cap A_0 = \emptyset$. Moreover, by lemma IX.3.7, Y^* is M-finite. Putting $Y = Y^* \cap I$, we see that Y is an M-coded subset of I which separates A_0 and A_1. Thus (28) satisfies Σ_1^0 separation.

Obviously Σ_1^0 separation implies Δ_1^0 comprehension. Hence (28) is a model of RCA_0. By lemma IV.4.4, it follows that (28) is a model of WKL_0. This completes the proof of lemma IX.3.11.   □

DEFINITION IX.3.12. Let M be a model of PRA. For $b, c \in |M|$, we write $b \ll_M c$ to mean that $f_M(b) <_M c$ for all 1-place primitive recursive function symbols f.

LEMMA IX.3.13 (existence of semiregular cuts). *Let M be a countable model of PRA. Suppose that $b, c \in |M|$ are such that $b \ll_M c$. Then there exists a semiregular cut I in M such that $b \in I$ and $c \notin I$.*

PROOF. For any finite interval $[b, c) = \{i : b \le i < c\}$ of nonnegative integers, we define a concept of n-*bigness*, by recursion on n. We say that $[b, c)$ is 0-big if $b < c$. We say that $[b, c)$ is $(n + 1)$-big if for every finite set X of cardinality $\le b$, there exist b' and c' such that $b < b' < c' < c$ and $[b', c')$ is n-big and disjoint from X.

The definition of n-bigness can be carried out within PRA. Formally, we have a primitive recursive predicate $\underline{B}(x, y, z)$, meaning that the interval $[y, z)$ is x-big, defined by recursion on x. We have $\underline{B}(0, y, z) \leftrightarrow y < z$, and $\underline{B}(\underline{S}(x), y, z) \leftrightarrow$ for all $w < 2^z$, if the finite set X encoded by w is of cardinality $\le y$, then there exist y' and z' such that $y < y' < z' < z$ and $\underline{B}(x, y', z')$ and $\forall u \, (y' \le u < z' \rightarrow u \notin X)$. (We are using definition IX.3.8 and lemmas IX.3.7 and IX.3.9, above.)

Also within PRA we have, for each standard nonnegative integer $n < \omega$, a 1-place primitive recursive function symbol $\underline{g}_n$ with defining axioms

$$\underline{g}_0(y) = y + 1,$$
$$\underline{g}_{n+1}(y) = \underbrace{\underline{g}_n \underline{g}_n \cdots \underline{g}_n}_{y+1}(y + 1) + 1.$$

For each $n < \omega$, we can then prove within PRA that, for all y and z, $\underline{g}_n(y) \le z$ implies $\underline{B}(\underline{n}, y, z)$. Here $\underline{n}$ is the constant term $\underline{S} \cdots \underline{S}(\underline{0})$, with n occurrences of $\underline{S}$.

Now let M be a countable model of PRA. As in the hypothesis of our lemma, let $b, c \in |M|$ be such that $f_M(b) <_M c$ for all 1-place primitive recursive functions symbols f. In particular, for each standard nonnegative integer $n < \omega$, we have $g_n^M(b) <_M c$, hence $B_M(n, b, c)$ holds. On the other hand, it is clear that $B_M(\alpha, b, c)$ does not hold for all $\alpha \in |M|$, so let v be the $<_M$-largest element of $|M|$ such that $B_M(v, b, c)$ does hold. Then for all standard n we have $n <_M v$.

Since M is countable, the set of M-finite sets is countable, so let $\langle X_n : n < \omega \rangle$ be an enumeration of these sets. We may assume that each M-finite set occurs infinitely many times in this enumeration. We shall inductively define sequences

$$b = b_0 <_M b_1 <_M \cdots <_M b_n <_M \cdots$$

and

$$c = c_0 >_M c_1 >_M \cdots >_M c_n >_M \cdots,$$

as follows. Begin with $b_0 = b$ and $c_0 = c$. If $\text{card}_M(X_0) \geq_M b_0$, set $b_1 = b_0 + 1$ and $c_1 = c_0 - 1$. If $\text{card}_M(X_0) <_M b_0$, then since $M \models [b_0, c_0]$ is v-big, we can find b_1 and c_1 such that $b_0 <_M b_1 <_M c_1 <_M c_0$ and $M \models [b_1, c_1]$ is $(v - 1)$-big and disjoint from X_0. Suppose now that b_n and c_n have been defined. If $\text{card}_M(X_n) \geq_M b_n$, set $b_{n+1} = b_n + 1$ and $c_{n+1} = c_n - 1$. If $\text{card}_M(X_n) <_M b_n$, then since $M \models [b_n, c_n]$ is $(v - n)$-big, we can find b_{n+1} and c_{n+1} such that $b_n <_M b_{n+1} <_M c_{n+1} <_M c_n$ and $M \models [b_{n+1}, c_{n+1}]$ is $(v - n - 1)$-big and disjoint from X_n.

Finally, let I be the set of $a \in |M|$ such that $a <_M b_n$ for some $n < \omega$. We claim that I is a semiregular cut. To see this, let X be an M-finite set such that $\text{card}_M(X) \in I$. Since $X = X_n$ for infinitely many n, we can find n such that $X = X_n$ and $\text{card}_M(X) <_M b_n$. Then by construction $M \models [b_{n+1}, c_{n+1}]$ is disjoint from X. Hence $X \cap I \subseteq \{a : a <_M b_{n+1}\}$, so $X \cap I$ is bounded in I.

This completes the proof of lemma IX.3.13. $\square$

EXERCISE IX.3.14. Let the primitive recursive function symbols g_n, $n \in \omega$, be as in the proof of lemma IX.3.13. Show that, for each 1-place primitive recursive function symbol f, we can find $n < \omega$ such that $\forall x\, (\underline{f}(x) \leq \underline{g}_n(x))$ is provable in PRA.

EXERCISE IX.3.15. Let M be a countable model of PRA, and let $b, c \in |M|$ be given. Show that $b \ll_M c$ if and only if there exists a semiregular cut I in M such that $b \in I$ and $c \notin I$. Show that if $b \ll_M c$ and X is M-finite with $\text{card}_M(X) <_M b$, then there exist b' and c' such that $b <_M b' \ll_M c' <_M c$ and $\{a : b' \leq_M a <_M c'\}$ is disjoint from X.

THEOREM IX.3.16 (conservation theorem). *Let ψ be a Π_2^0 sentence. If ψ is provable in* WKL$_0$, *then ψ is provable in* PRA (*under the canonical interpretation of* L$_1$ *into the language of* PRA).

PROOF. Suppose that ψ is not provable in PRA. By Gödel's completeness theorem, there is a countable model M' of PRA in which ψ is false. Writing $\psi \equiv \forall y \, \exists z \, \theta(y, z)$ where $\theta(y, z)$ is Σ_0^0, let $b' \in |M'|$ be such that $M' \models \neg \exists z \, \theta(b', z)$.

We now introduce two new constant symbols $\underline{b}$ and $\underline{c}$ and consider the theory T whose axioms are those of PRA, plus $\neg \exists z \, \theta(\underline{b}, z)$, plus $\underline{f}(\underline{b}) < \underline{c}$ for all 1-place primitive recursive function symbols $\underline{f}$. For any finite subset T_0 of the axioms of T, we can choose an element $c_0' \in |M'|$ such that $f_{M'}(b) <_{M'} c_0'$ for all of the finitely many 1-place primitive recursive function symbols $\underline{f}$ which are mentioned in T_0. Thus $M' \models T_0$ with $\underline{b}$ and $\underline{c}$ interpreted as b' and c_0'. This shows that each finite subset of the axioms of T has a countable model. Hence, by the compactness theorem, T has a countable model.

This means that we have a countable model M of PRA and elements $b, c \in |M|$ such that $b \ll_M c$ and $M \models \neg \exists z \, \theta(b, z)$. By lemma IX.3.13, there is a semiregular cut I in M such that $b \in I$ and $c \notin I$. By lemma IX.3.11,

$$(I, \text{Coded}_M(I), +_M \upharpoonright I, \cdot_M \upharpoonright I, 0_M, 1_M, <_M \upharpoonright I) \tag{30}$$

is a model of WKL_0. It is also clear that (30) satisfies $\neg \exists z \, \theta(b, z)$, hence (30) satisfies $\neg \psi$. Hence, by the soundness theorem, ψ is not provable in WKL_0. This completes the proof of theorem IX.3.16.   □

REMARK IX.3.17 (equiconsistency of PRA and WKL_0). Using the methods of §§II.8 and IV.3, the previous theorem can be proved in WKL_0 and hence in PRA. Thus WKL_0 and PRA have the same consistency strength.

REMARK IX.3.18 (Hilbert's program). The results of this section shed considerable light on a very important direction of research in the foundations of mathematics, known as *Hilbert's program* or, more descriptively, *finitistic reductionism*. Hilbert was the foremost mathematician of his time, and his ideas about the "problem of infinity" in the foundations of mathematics are of great interest. We here limit ourselves to a very brief discussion.

Hilbert assigned a special role to a certain restricted kind of mathematical reasoning, known as *finitistic*. Roughly speaking, finitism is that part of mathematics which rejects completed infinite totalities and is indispensable for all scientific reasoning. For example, finitism is adequate for elementary reasoning about strings of symbols, but it is not adequate for reasoning about arbitrary sets of integers. Hilbert never spelled out a precise definition of finitism, but it is generally agreed that the formal system PRA (definition IX.3.2 above) captures this notion.

The essence of Hilbert's program was to show that non-finitistic, set-theoretical mathematics can be reduced to finitism. The reduction was to be accomplished by means of finitistic consistency proofs or, somewhat more generally, by means of conservation results for Π_1^0 sentences.

Unfortunately, Gödel's incompleteness theorems [94, 115, 55, 222] imply that a wholesale realization of Hilbert's program is impossible. There is no hope of proving the consistency of set theory within PRA, nor is there any hope of showing that set theory is conservative over PRA for Π_1^0 sentences.

In view of Gödel's limitative results, it is of interest to ask what part of Hilbert's program can be carried out. In other words, which portions of infinitistic mathematics can be reduced to finitism? The study of subsystems of second order arithmetic makes it possible to give a more precise formulation of this question: *Which interesting subsystems of* Z_2 *are conservative over* PRA *for* Π_1^0 *sentences?* In this context, a subsystem of Z_2 is considered interesting if it accommodates the development of a large part of mathematical practice.

Thus, theorem IX.3.16 emerges as a key result toward a partial realization of Hilbert's program. Theorem IX.3.16 shows that WKL$_0$ is conservative over PRA for Π_1^0 sentences (in fact Π_2^0 sentences). This conservation result, together with the results of chapters II and IV concerning the development of mathematics within WKL$_0$, implies that a significant part of mathematical practice is finitistically reducible, in the precise sense envisioned by Hilbert.

For example, all of the following key theorems of infinitistic mathematics are provable in WKL$_0$ and therefore, by theorem IX.3.16, reducible to finitism. (1) The Heine/Borel covering theorem for closed bounded subsets of $\mathbb{R}^n$ or for closed subsets of any compact metric space. (2) Basic properties of continuous real-valued functions of several real variables. (3) The local existence theorem for solutions of ordinary differential equations. (4) The Hahn/Banach theorem in separable Banach spaces. (5) The existence theorem for prime ideals in countable commutative rings. (6) Existence and uniqueness of the algebraic closure of a countable field. (7) Orderability and existence of the real closure of a countable formally real field.

To summarize, WKL$_0$ embodies a significant and far-reaching partial realization of Hilbert's program of finite reductionism.

Notes for §IX.3. For a thorough introduction to primitive recursive functions, see Kleene [142]. In connection with exercise IX.3.14, note that the functions g_n, $n < \omega$, are essentially the "branches" of the Ackermann function; see also Robinson [207].

The formal system PRA is of fundamental importance for the foundations of mathematics, being the formal analog of Hilbert's informal notion of finitistic provability. See Hilbert [114], Feferman [55], and Tait [259].

The model-theoretic result of Tarski, referred to in the hint for exercise IX.3.3, can be found in any model theory textbook, e.g., Chang/Keisler [35] or Sacks [210].

Parsons [201] used a functional interpretation to show that every Π_2^0 sentence provable in Σ_1^0-PA is provable in PRA. Theorem IX.3.16 may be viewed as a consequence of this theorem of Parsons plus Harrington's result in §IX.2 above. Theorem IX.3.16 is due to Friedman (1976, unpublished). The idea of the model-theoretic proof of theorem IX.3.16, which we have given here, is from Kirby/Paris [140]. Another proof of theorem IX.3.16, via Gentzen-style proof theory, has been given by Sieg [225]. For a fuller discussion of the relationship between Hilbert's program and theorem IX.3.16, see Simpson [246].

IX.4. Saturated Models

In §VII.6 we obtained some conservation theorems involving versions of the axiom of choice in the language of Z_2. In this section we shall prove some more results of this kind. We shall prove: (1) Σ_1^1-AC$_0$ is conservative over ACA$_0$ for Π_2^1 sentences; (2) Σ_2^1-AC$_0$ is conservative over Π_1^1-CA$_0$ for Π_3^1 sentences; (3) for all $k < \omega$, Σ_{k+3}^1-AC$_0$ is conservative over Π_{k+2}^1-CA$_0$ for Π_4^1 sentences. These results depend essentially on the use of non-ω-models. Specifically, our proofs employ a model-theoretic concept known as *saturation*.

DEFINITION IX.4.1 (saturated models). The following concepts are defined in RCA$_0$. Let L be a countable first order language, and let M be a countable model for L (in the sense of definition II.8.3).

1. A 1-*type over M* is a sequence of formulas $\langle \varphi_i(x, b_1, \ldots, b_k) : i \in \mathbb{N} \rangle$ with a finite set of parameters $b_1, \ldots, b_k \in |M|$ and no free variables other than x, such that

$$\forall j \, (\exists a \in |M|) \, (\forall i \leq j) \, M \models \varphi_i(a, b_1, \ldots, b_k).$$

This 1-type is said to be *realized in M* if

$$(\exists a \in |M|) \, \forall i \, M \models \varphi_i(a, b_1, \ldots, b_k).$$

2. A 1-type as above is said to be *X-recursive*, where $X \subseteq \mathbb{N}$, if the sequence of L-formulas $\langle \varphi_i(x, y_1, \ldots, y_k) : i \in \mathbb{N} \rangle$ is X-recursive.

3. We say that M is *X-recursively saturated* if every X-recursive 1-type over M is realized in M. We say that M is *recursively saturated* if it is $\emptyset$-recursively saturated.

LEMMA IX.4.2 (existence of saturated models). *Let L be a countable first order language. Let S be a consistent set of L-sentences. Then for any $X \subseteq \mathbb{N}$ there exists a countable X-recursively saturated model of S. This result is provable in* WKL$_0$.

PROOF. We reason in WKL_0. We shall employ a variant of the Henkin construction which was already used in §§II.8 and IV.3.

We consider an expanded language $L(C) = L \cup C$ where C is a countably infinite set of new constant symbols. We call a sequence $\langle \varphi_i(x): i \in \mathbb{N} \rangle$ of $L(C)$-formulas *acceptable* if it has only the free variable x and mentions only finitely many of the constant symbols in C. Fix a one-to-one enumeration $\langle \underline{c}_n : n \in \mathbb{N} \rangle$ of C. Fix an enumeration $\langle \langle \varphi_{ni}(x): i \in \mathbb{N} \rangle : n \in \mathbb{N} \rangle$ of all X-recursive acceptable sequences of $L(C)$-formulas. We may safely assume that $\varphi_{ni}(x)$ does not mention any $\underline{c}_m$, $m \geq n$. Let $S(C)$ be the set of $L(C)$-sentences consisting of S plus Henkin axioms

$$(\exists x \, (\varphi_{n0}(x) \wedge \cdots \wedge \varphi_{nj}(x))) \rightarrow (\varphi_{n0}(\underline{c}_n) \wedge \cdots \wedge \varphi_{nj}(\underline{c}_n))$$

for all n and j. A syntactical argument from the consistency of S shows that $S(C)$ is consistent. By Lindenbaum's lemma in WKL_0 (theorem IV.3.3), let $S(C)^*$ be a completion of $S(C)$. As in the proof of theorem II.8.4, we can read off a countable L-model M from $S(C)^*$. By construction, M satisfies S and is X-recursively saturated. This completes the proof of lemma IX.4.2.   □

The main results of this section will be obtained by applying lemma IX.4.2 to sets of sentences in the countable first order language $L_1(\underline{A})$ consisting of L_1, the language of first order arithmetic, plus countably many 1-place predicate symbols $\underline{A}_n$, $n \in \omega$. We shall usually treat $\underline{A}_n$ as a set constant, writing $t \in \underline{A}_n$ instead of the more orthodox $\underline{A}_n(t)$. Thus the formulas of $L_1(\underline{A})$ are built up by means of propositional connectives and number quantifiers from atomic formulas $t_1 = t_2$, $t_1 < t_2$, and $t_1 \in \underline{A}_n$, where t_1 and t_2 are numerical terms. An $L_1(\underline{A})$-structure M consists of an L_1-structure $(|M|, +_M, \cdot_M, 0_M, 1_M, <_M)$ together with sets $A_n^M \subseteq |M|$, $n \in \omega$. Here of course A_n^M is used to interpret $\underline{A}_n$.

If M is an $L_1(\underline{A})$-structure, the associated L_2-structure is

$$M_2 = (|M|, \mathcal{S}_M, +_M, \cdot_M, 0_M, 1_M, <_M)$$

where $\mathcal{S}_M$ consists of all subsets of $|M|$ of the form

$$(A_n^M)_b = \{a \in |M|: M \models (a, b) \in \underline{A}_n\}$$

with $n \in \omega$, $b \in |M|$.

LEMMA IX.4.3. *Let M be a recursively saturated $L_1(\underline{A})$-model which satisfies the basic axioms I.2.4(i), induction for all Σ_1^0 formulas in the language $L_1(\underline{A})$, and the axioms $\mathrm{TJ}(\underline{A}_n) = \underline{A}_{n+1}$ for all $n \in \mathbb{N}$. Then the associated L_2-structure M_2 satisfies Σ_1^1-AC_0. This result is provable in ACA_0.*

PROOF. Note first that M_2 is a model of ACA_0. (See §VIII.1 for a discussion of the relationship between ACA_0 and the Turing jump operator TJ.)

Assume now that M_2 satisfies $\forall x \, \exists Y \, \eta(x, Y)$ where $\eta(x, Y)$ is a Σ_1^1 formula with parameters from $|M| \cup \mathcal{S}_M$. Let us write $\eta(x, Y) \equiv \exists Z \, \theta(x, Y, Z)$

where θ is arithmetical with parameters from $|M| \cup \mathcal{S}_M$. Thus M_2 satisfies $\forall x \, \exists Y \, \exists Z \, \theta(x, Y, Z)$. Here x is a number variable while Y and Z are set variables.

We claim that, for some n, M satisfies

$$\forall x \, \exists y \, \exists z \, \theta(x, (\underline{A}_n)_y, (\underline{A}_n)_z).$$

If not, then we have a recursive 1-type consisting of the formulas

$$\neg \exists y \, \exists z \, \theta(x, (\underline{A}_n)_y, (\underline{A}_n)_z)$$

for all n. By recursive saturation, there exists $a \in |M|$ such that for all n, M satisfies $\neg \exists y \, \exists z \, \theta(a, (\underline{A}_n)_y, (\underline{A}_n)_z)$. This implies that M_2 satisfies

$$\neg \exists Y \, \exists Z \, \theta(a, Y, Z),$$

a contradiction. Our claim is proved.

From the above claim, we see that

$$M_2 \models \exists W \, \forall x \, \exists y \, \exists z \, \theta(x, (W)_y, (W)_z).$$

Since M_2 is a model of ACA_0, it follows that

$$M_2 \models \exists W \, \forall x \, \eta(x, (W)_x).$$

Thus M_2 is a model of Σ_1^1 choice. The proof of lemma IX.4.3 is complete.

□

The following theorem stands in contrast to the results of chapter VIII, according to which the minimum ω-model of ACA_0 is the class **ARITH** of arithmetical sets, while the minimum ω-model of $\Sigma_1^1\text{-}\mathsf{AC}_0$ (or of $\Delta_1^1\text{-}\mathsf{CA}_0$) is the much larger class **HYP** of hyperarithmetical sets.

THEOREM IX.4.4 (conservation theorem). $\Sigma_1^1\text{-}\mathsf{AC}_0$ (*hence also* $\Delta_1^1\text{-}\mathsf{CA}_0$) *is conservative over* ACA_0 *for* Π_2^1 *sentences. In other words, any* Π_2^1 *sentence which is provable in* $\Sigma_1^1\text{-}\mathsf{AC}_0$ *is already provable in* ACA_0.

PROOF. Let ψ be a Π_2^1 sentence which is not provable in ACA_0. By Gödel's completeness theorem, let M' be a model of ACA_0 in which ψ is false. Writing $\psi \equiv \forall X \, \exists Y \, \theta(X, Y)$ where $\theta(X, Y)$ is arithmetical, choose $A' \in \mathcal{S}_{M'}$ such that $M' \models \neg \exists Y \, \theta(A', Y)$. Define a sequence of elements $A'_n \in \mathcal{S}_{M'}$, $n \in \omega$, where $A'_0 = A'$ and for all n, A'_{n+1} is the unique $B \in \mathcal{S}_{M'}$ such that $M' \models B = \mathsf{TJ}(A'_n)$. The first order part of M' together with the sets A'_n, $n \in \omega$, form an $\mathsf{L}_1(\underline{A})$-structure. Clearly this structure satisfies the axioms mentioned in lemma IX.4.3, plus additional axioms $\neg \exists y \, \theta(\underline{A}_0, (\underline{A}_n)_y)$ for all $n \in \omega$. Hence, by lemma IX.4.2, there exists a recursively saturated model M of these axioms. By lemma IX.4.3, the associated L_2-structure M_2 satisfies $\Sigma_1^1\text{-}\mathsf{AC}_0$. It is also clear that M_2 satisfies $\neg \exists Y \, \theta(\underline{A}_0, Y)$, hence M_2 satisfies $\neg \psi$. Therefore, by the soundness theorem, ψ is not provable in $\Sigma_1^1\text{-}\mathsf{AC}_0$. Also, it follows by lemma VII.6.6 that ψ is not provable in $\Delta_1^1\text{-}\mathsf{CA}_0$. This completes the proof of theorem IX.4.4.

□

As another interesting application of the above ideas, we present the following result. For any recursively axiomatizable L_2-theory T_0, Π^1_{k+1} *correctness of* T_0 is the assertion that every Π^1_{k+1} sentence provable in T_0 is true. (This assertion is formalized by means of a universal Π^1_{k+1} formula.)

THEOREM IX.4.5.

1. ACA$_0$ *plus* Σ^1_1-IND *proves* Π^1_2 *correctness of* Σ^1_1-AC$_0$.
2. Σ^1_1-AC$_0$ *plus* Σ^1_1-IND *proves* Π^1_3 *correctness of* Σ^1_1-AC$_0$.

PROOF. In order to prove 1, we reason in ACA$_0$ plus Σ^1_1-IND. We are going to show that every Π^1_2 sentence provable in Σ^1_1-AC$_0$ is true. Let ψ be a Π^1_2 sentence which is not true. Writing $\psi \equiv \forall X \exists Y \, \theta(X, Y)$ where $\theta(X, Y)$ is arithmetical, fix $A_0 \subseteq \mathbb{N}$ such that $\neg \exists Y \, \theta(A_0, Y)$ holds. Let S be the set of $L_1(\underline{A})$-sentences mentioned in lemma IX.4.3 plus additional axioms $\neg \exists y \, \theta(A_0, (\underline{A}_n)_y)$, for all n. Let S_n consist of S restricted to the language $L_1 \cup \{\underline{A}_0, \ldots, \underline{A}_n\}$. By arithmetical comprehension plus Σ^1_1 induction on n, we have

$$\forall n \, \exists W ((W)_0 = A_0 \wedge (\forall i < n) \, \mathrm{TJ}((W)_i) = (W)_{i+1}).$$

It follows that for all n there exists an ω-model of S_n. Hence, by the strong soundness theorem, we have that for all n, S_n is consistent. (Concerning the strong soundness theorem, see lemma VII.2.2 and theorem II.8.10.) Hence S is consistent. Arguing as in the proof of theorem IX.4.4, we conclude that ψ is not provable in Σ^1_1-AC$_0$. This establishes part 1 of our theorem.

In order to prove part 2, we shall need the following variant of lemma IX.4.3. Let $\pi(e, m_1, X_1)$ be a universal lightface Π^0_1 formula, as in the definition of Turing jump (definition VIII.1.9). Then lemma IX.4.3 holds if we replace

$$\mathrm{TJ}(\underline{A}_n) = \underline{A}_{n+1}$$

by the weaker condition

$$\forall i \, \exists j \, \forall m \, (\pi(i, m, \underline{A}_n) \leftrightarrow m \in (\underline{A}_{n+1})_j).$$

The proof is the same as for lemma IX.4.3.

Now, reasoning in Σ^1_1-AC$_0$ plus Σ^1_1-IND, we are going to show that any Π^1_3 sentence provable in Σ^1_1-AC$_0$ is true. Let ψ be a Π^1_3 sentence which is not true. Writing $\psi \equiv \forall X \exists Y \forall Z \, \theta(X, Y, Z)$ where $\theta(X, Y, Z)$ is arithmetical, fix $A_0 \subseteq \mathbb{N}$ such that

$$\forall Y \exists Z \, \neg \theta(A_0, Y, Z)$$

holds. Let $\varphi(X, Y, Z)$ be the arithmetical formula

$$\forall i \, \exists j \, \forall m \, (\pi(i, m, Y) \leftrightarrow m \in (Z)_j) \wedge \forall i \, \exists j \, \neg \theta(X, (Y)_i, (Z)_j).$$

Let S be the set of $L_1(\underline{A})$-sentences consisting of the basic axioms I.2.4(i), induction for all Σ^0_1 formulas of $L_1(\underline{A})$, and $\varphi(\underline{A}_0, \underline{A}_n, \underline{A}_{n+1})$ for all n. We are going to show that S is consistent. Let S_n consist of S restricted to $L_1 \cup \{\underline{A}_0, \ldots, \underline{A}_n\}$. By arithmetical comprehension and Σ_1^1 choice, we have $\forall Y \exists Z \, \varphi(A_0, Y, Z)$. Hence, by Σ_1^1 induction on n, we have

$$\forall n \, \exists W \, ((W)_0 = A_0 \wedge (\forall i < n) \, \varphi(A_0, (W)_i, (W)_{i+1})).$$

Thus we see that for all n there exists an ω-model of S_n. Hence by the strong soundness theorem, S is consistent. By lemma IX.4.2, there exists a recursively saturated model M of S. Let M_2 be the associated L_2-structure. By the above-mentioned variant of lemma IX.4.3, M_2 satisfies Σ_1^1-AC_0. It is also clear that M_2 satisfies $\forall Y \exists Z \, \neg \theta(\underline{A}_0, Y, Z)$, hence M_2 satisfies $\neg \psi$. Hence ψ is not provable in Σ_1^1-AC_0. This establishes part 2.

The proof of theorem IX.4.5 is complete. $\square$

COROLLARY IX.4.6 (consistency of Σ_1^1-AC_0). ACA_0 *plus* Σ_1^1-IND *proves the consistency of* Σ_1^1-AC_0.

PROOF. This follows by applying part 1 of theorem IX.4.5 to the sentence $1 = 0$. $\square$

COROLLARY IX.4.7. ATR_0 *plus* Σ_1^1-IND *proves the consistency of* ATR_0. *It follows that* Σ_1^1-IND *is not provable in* ATR_0.

(Compare lemma VIII.6.15.)

PROOF. We reason in ATR_0 plus Σ_1^1-IND. Recall that ATR_0 can be axiomatized by Σ_1^1-AC_0 plus a single Π_2^1 sentence ψ (see theorem VIII.3.15). If ATR_0 were inconsistent, then Σ_1^1-AC_0 would prove $\neg \psi$. Hence, by Π_3^1 soundness of Σ_1^1-AC_0 (part 2 of theorem IX.4.5), $\neg \psi$ would be true, a contradiction. This proves the first sentence of our corollary. The second sentence follows by Gödel's second incompleteness theorem [94, 115, 55, 222]. $\square$

For the next lemma, let $k < \omega$ be fixed.

LEMMA IX.4.8. *Let M be a recursively saturated $L_1(\underline{A})$-model which satisfies the basic axioms* I.2.4(i) *plus arithmetical induction (i.e., induction for all $L_1(\underline{A})$-formulas) plus "the countable ω-model encoded by $\underline{A}_{n+1}$ contains $\underline{A}_n$ as an element, satisfies* ACA_0, *and is a β_{k+1}-submodel of the countable ω-model encoded by $\underline{A}_{n+2}$," for all n. Then the associated L_2-structure M_2 satisfies strong Σ_{k+1}^1-DC_0 plus Σ_{k+2}^1-AC_0. This result is provable in* ACA_0.

PROOF. Clearly M_2 satisfies ACA_0 and, for each $n \in \omega$,

$$M_2 \models \text{ the countable } \omega\text{-model encoded by } \underline{A}_{n+1} \text{ is a } \beta_{k+1}\text{-model}. \quad (31)$$

Hence, by theorem VII.7.4, M_2 satisfies strong Σ_{k+1}^1-DC_0.

Assume now that M_2 satisfies $\forall x \, \exists Y \, \eta(x, Y)$ where $\eta(x, Y)$ is a Σ_{k+2}^1 formula with parameters from $|M| \cup \mathcal{S}_M$. Let us write $\eta(x, Y) \equiv \exists Z \, \psi(x, Y, Z)$ where ψ is Π_{k+1}^1. Thus M_2 satisfies $\forall x \, \exists Y \, \exists Z \, \psi(x, Y, Z)$. (Here x is a number variable while Y and Z are set variables.) This implies that

$$(\forall a \in |M|) \, \exists n \, M_2 \models \exists y \, \exists z \, \psi(a, (\underline{A}_{n+1})_y, (\underline{A}_{n+1})_z).$$

Hence by (31) it follows that

$(\forall a \in |M|) \exists n \, M \models$ the countable ω-model coded by $\underline{A}_{n+1}$

satisfies $\exists y \, \exists z \, \psi(a, (\underline{A}_{n+1})_y, (\underline{A}_{n+1})_z)$.

By recursive saturation, there exists $n \in \omega$ such that

$(\forall a \in |M|) \, M \models$ the countable ω-model coded by $\underline{A}_{n+1}$

satisfies $\exists y \, \exists z \, \psi(a, (\underline{A}_{n+1})_y, (\underline{A}_{n+1})_z)$.

By (31) plus arithmetical comprehension within M_2, it follows that that $M_2 \models \exists W \, \forall x \, \psi(x, ((W)_x)_0, ((W)_x)_1)$. Hence $M_2 \models \exists W \, \forall x \, \eta(x, (W)_x)$. Thus M_2 is a model of Σ^1_{k+2} choice. This completes the proof of lemma IX.4.8.   □

The following theorem stands in contrast to results of §VII.7 according to which the minimum β-model of $\Delta^1_2\text{-CA}_0$ is much larger than that of $\Pi^1_1\text{-CA}_0$.

THEOREM IX.4.9 (conservation theorem). $\Sigma^1_2\text{-AC}_0$ (hence also $\Delta^1_2\text{-CA}_0$) is conservative over $\Pi^1_1\text{-CA}_0$ for Π^1_3 sentences.

PROOF. Let ψ be a Π^1_3 sentence which is not provable in $\Pi^1_1\text{-CA}_0$. By Gödel's completeness theorem, let M' be a model of $\Pi^1_1\text{-CA}_0$ in which ψ is false. Writing $\psi \equiv \forall X \, \varphi(X)$ where $\varphi(X)$ is Σ^1_2, choose $A' \in S_{M'}$ such that $M' \models \neg\varphi(A')$. By theorem VII.2.10, we can find a sequence of sets $A'_n \in S_{M'}$, $n \in \omega$, such that $A' = A'_0$ and, for all n, $M' \models A'_{n+1}$ encodes a countable β-model of ACA_0 which contains A'_n. Thus the first order part of M together with the sets A'_n, $n \in \omega$, form an $L_1(\underline{A})$-model of the axioms mentioned in lemma IX.4.8 (with $k = 0$) plus additional axioms saying "the countable ω-model coded by $\underline{A}_{n+1}$ satisfies $\neg\varphi(\underline{A}_0)$." Hence, by lemma IX.4.2, there exists a recursively saturated model M of these axioms. By lemma IX.4.8 (with $k = 0$), the associated L_2-structure M_2 satisfies $\Sigma^1_2\text{-AC}_0$. It is also clear that M_2 satisfies $\neg\varphi(\underline{A}_0)$, hence M_2 satisfies $\neg\psi$. Therefore, by the soundness theorem, ψ is not provable in $\Sigma^1_2\text{-AC}_0$. It follows by theorem VII.6.9.1 that ψ is not provable in $\Delta^1_2\text{-CA}_0$. This completes the proof of theorem IX.4.9.   □

THEOREM IX.4.10 (consistency of $\Sigma^1_2\text{-AC}_0$).

1. $\Pi^1_1\text{-CA}_0$ plus $\Sigma^1_2\text{-IND}$ proves Π^1_3 correctness of $\Sigma^1_2\text{-AC}_0$.
2. $\Sigma^1_2\text{-AC}_0$ plus $\Sigma^1_2\text{-IND}$ proves Π^1_4 correctness of $\Sigma^1_2\text{-AC}_0$.
3. $\Pi^1_1\text{-CA}_0$ plus $\Sigma^1_2\text{-IND}$ proves the consistency of $\Sigma^1_2\text{-AC}_0$.
4. $\Sigma^1_2\text{-AC}_0$ does not prove $\Sigma^1_2\text{-IND}$.

(Compare theorem VII.6.9.)

PROOF. These results are obtained by imitating the proofs of theorem IX.4.5 and corollaries IX.4.6 and IX.4.7 above, using lemma IX.4.8 instead of lemma IX.4.3.   □

THEOREM IX.4.11 (conservation theorem). *For all $k < \omega$, Σ^1_{k+3}-AC$_0$ plus strong Σ^1_{k+2}-DC$_0$ is conservative over strong Σ^1_{k+2}-DC$_0$ for Π^1_{k+4} sentences.*

PROOF. This result is obtained by imitating the proof of theorem IX.4.9, using theorem VII.7.4 (characterizing strong Σ^1_{k+2}-DC$_0$ in terms of countable coded β_{k+2}-models) instead of theorem VII.2.10 (characterizing Π^1_1-CA$_0$ in terms of countable coded β-models). $\square$

The following corollary stands in contrast to results of §VII.7 according to which the minimum β-model of Δ^1_{k+3}-CA$_0$ is much larger than that of Π^1_{k+2}-CA$_0$.

COROLLARY IX.4.12 (conservation theorem). *For all $k < \omega$, Σ^1_{k+3}-AC$_0$ (hence also Δ^1_{k+3}-CA$_0$) is conservative over Π^1_{k+2}-CA$_0$ for Π^1_4 sentences.*

PROOF. This is immediate from theorem IX.4.11 plus the fact that strong Σ^1_{k+2}-DC$_0$ is conservative over Π^1_{k+2}-CA$_0$ for Π^1_4 sentences (theorems VII.6.9 and VII.6.20), recalling also that Σ^1_{k+3}-AC$_0$ includes Δ^1_{k+3}-CA$_0$ (lemma VII.6.6). $\square$

EXERCISE IX.4.13 (conservation theorem). Show that, for all $k < \omega$, Σ^1_{k+2}-AC$_0$ is conservative over Π^1_{k+1}-CA$_0$ plus Σ^1_{k+1}-AC$_0$ for Π^1_{k+3} sentences.

THEOREM IX.4.14 (consistency of Σ^1_{k+4}-AC$_0$). *Let $k < \omega$ be fixed.*

1. *Strong Σ^1_{k+3}-DC$_0$ plus Σ^1_{k+4}-IND proves Π^1_{k+4} correctness of strong Σ^1_{k+3}-DC$_0$ plus Σ^1_{k+4}-AC$_0$.*
2. *Strong Σ^1_{k+3}-DC$_0$ plus Σ^1_{k+4}-AC$_0$ plus Σ^1_{k+4}-IND proves Π^1_{k+5} correctness of strong Σ^1_{k+3}-DC$_0$ plus Σ^1_{k+4}-AC$_0$.*
3. *Π^1_{k+3}-CA$_0$ plus Σ^1_{k+4}-IND proves Π^1_4 correctness of strong Σ^1_{k+3}-DC$_0$ plus Σ^1_{k+4}-AC$_0$.*
4. *Π^1_{k+3}-CA$_0$ plus Σ^1_{k+4}-IND proves consistency of strong Σ^1_{k+3}-DC$_0$ plus Σ^1_{k+4}-AC$_0$.*
5. *Σ^1_{k+4}-IND is not provable from strong Σ^1_{k+3}-DC$_0$ plus Σ^1_{k+4}-AC$_0$.*

(Compare theorem VII.6.20.)

PROOF. Same as for theorem IX.4.10. $\square$

Notes for §IX.4. For general background on saturated models, see Chang/Keisler [35] and Sacks [210]. Theorems IX.4.4 and IX.4.9 and corollary IX.4.12 are closely related to the conservation results of Friedman [64], with the difference that Friedman was considering systems with full induction. The concept of recursive saturation, as well as lemma IX.4.3 and theorem IX.4.4, are due to Barwise/Schlipf [15]. The fact that the existence of recursively saturated models is provable in WKL$_0$ (lemma IX.4.2) appears to be new. Theorem IX.4.5 appears to be new. Corollaries IX.4.6 and IX.4.7 are due to Simpson [235]. Theorem IX.4.9 and

corollary IX.4.12 are due to Feferman and Feferman/Sieg respectively;
see [29, §II.2]. The result of exercise IX.4.13 has been announced by Sieg
and is proved in Schmerl [213, theorem 2.10]. Theorems IX.4.10, IX.4.11
and IX.4.14 appear to be new.

IX.5. Gentzen-Style Proof Theory

In this section we briefly indicate the relationship between the material
in this book and Gentzen-style proof theory. We state a few results and
provide references to the published literature.

DEFINITION IX.5.1 (provable ordinals). Let T_0 be a subsystem of Z_2
which includes $\mathsf{RCA_0}$. A *provable ordinal* of T_0 is a countable ordinal α
such that, for some primitive recursive well ordering $W \subseteq \mathbb{N}$, $|W| = \alpha$
and T_0 proves $\mathrm{WO}(W)$. The supremum of the provable ordinals of T_0 is
denoted $\mathrm{ord}(T_0)$. Note that if T_0 is any reasonable subsystem of Z_2 then
$\mathrm{ord}(T_0)$ is a recursive ordinal, i.e., $\mathrm{ord}(T_0) < \omega_1^{\mathrm{CK}}$.

REMARK IX.5.2. A principal focus of Gentzen-style proof theory is the
computation of $\mathrm{ord}(T_0)$ for various well known subsystems T_0 of second
order arithmetic. One of the tools used in such computations is cut
elimination. Generally speaking, as T_0 gets stronger, $\mathrm{ord}(T_0)$ gets much
larger and much more difficult to describe. It is interesting to note that
these ordinals are closely related to consistency strength. Usually, if
$\mathrm{ord}(T_0) = \mathrm{ord}(T_0')$ then T_0 and T_0' are equiconsistent, and if $\mathrm{ord}(T_0) >$
$\mathrm{ord}(T_0')$ then T_0 proves the consistency of T_0'.

Clearly non-ω-models are relevant here. To see this, note that if $\alpha =$
$\mathrm{ord}(T_0)$ then $T_0 + \neg\mathrm{WO}(\alpha)$ is consistent but any model of it is necessarily
a non-ω-model.

DEFINITION IX.5.3 (ordinal arithmetic). The operations of ordinal
arithmetic are defined as usual by transfinite induction:

$$
\begin{array}{lll}
\text{addition:} & \alpha + \beta = \sup\{\alpha, (\alpha + \gamma) + 1 : \gamma < \beta\} \\
\text{multiplication:} & \alpha \cdot \beta = \sup\{\alpha \cdot \gamma + \alpha : \gamma < \beta\} \\
\text{exponentiation:} & \alpha^{\beta} = \sup\{1, \alpha^{\gamma} \cdot \alpha : \gamma < \beta\}
\end{array}
$$

Recall also that ω is the smallest infinite ordinal.

THEOREM IX.5.4 (provable ordinals of $\mathsf{RCA_0}$ and $\mathsf{WKL_0}$). *We have*

$$\mathrm{ord}(\mathsf{RCA_0}) = \mathrm{ord}(\mathsf{WKL_0}) = \omega^{\omega}.$$

PROOF. It is straightforward to show that, for each $n < \omega$, $\mathsf{RCA_0}$ proves
$\mathrm{WO}(\omega^n)$. On the other hand, if $\mathrm{WO}(\omega^{\omega})$ were provable in $\mathsf{WKL_0}$, then
this would allow us to prove the totality of the Ackermann function,
contradicting theorem IX.3.16 which says that $\mathsf{WKL_0}$ is conservative over
primitive recursive arithmetic for Π_2^0 sentences.   □

DEFINITION IX.5.5. Let F be a function from ordinals to ordinals. F is said to be *monotone* (respectively *strictly monotone*) if $\alpha < \beta$ implies $F(\alpha) \leq F(\beta)$ (respectively $F(\alpha) < F(\beta)$), and *continuous* if $F(\beta) = \sup\{F(\alpha): \alpha < \beta\}$ for all limit ordinals β. A *fixed point* of F is an ordinal α such that $F(\alpha) = \alpha$.

For each $\alpha > 1$ the functions $\beta \mapsto \alpha + \beta$, $\beta \mapsto \alpha \cdot \beta$, $\beta \mapsto \alpha^\beta$ are strictly monotone and continuous. It is well known that if F is strictly monotone and continuous then F has arbitrarily large fixed points. More generally, if $\{F_i : i \in I\}$ is any family of strictly monotone continuous functions, then there exist arbitrarily large *simultaneous fixed points* of $\{F_i : i \in I\}$, i.e., ordinals α such that $F_i(\alpha) = \alpha$ for all $i \in I$.

DEFINITION IX.5.6 (the ordinals ε_0 and Γ_0). For each ordinal α we define a strictly monotone continuous function φ_α from ordinals to ordinals as follows: $\varphi_0(\beta) = \omega^\beta$ and, for $\alpha > 0$, $\varphi_\alpha(\beta) =$ the βth simultaneous fixed point of the functions φ_γ, $\gamma < \alpha$. We put

$$\varepsilon_0 = \varphi_1(0) = \sup\left(\omega, \omega^\omega, \omega^{\omega^\omega}, \ldots\right)$$

and more generally $\varepsilon_\alpha = \varphi_1(\alpha)$. Γ_0 is defined to be the least ordinal $\gamma > 0$ such that $\varphi_\alpha(\beta) < \gamma$ for all $\alpha, \beta < \gamma$. Note that

$$\omega < \omega^\omega < \omega^{\omega^\omega} < \cdots < \varepsilon_0 < \varepsilon_{\varepsilon_0} < \cdots < \varphi_2(0) < \varphi_2(1) < \cdots < \Gamma_0.$$

It can be shown that ε_0 and Γ_0 are recursive ordinals.

THEOREM IX.5.7 (provable ordinals of ACA_0 and ATR_0). *We have*

$$\mathrm{ord}(ACA_0) = \varepsilon_0$$

and

$$\mathrm{ord}(ATR_0) = \Gamma_0.$$

PROOF. We have seen in §VIII.1 that the first order part of ACA_0 is Peano arithmetic, PA, i.e., first-order arithmetic, Z_1. The proof-theoretic analysis of Z_1 in terms of ε_0 goes back to Gentzen; see for instance Takeuti [261] and Schütte [214]. The fact that $\mathrm{ord}(ATR_0) = \Gamma_0$ is due to Friedman/McAloon/Simpson [76]. Earlier Feferman [56, 57] had introduced his system IR of predicative analysis and showed that $\mathrm{ord}(IR) = \Gamma_0$. See also remark I.11.9.   □

DEFINITION IX.5.8 (collapsing functions). We write $\Omega_0 = 0$ and $\Omega_n = \aleph_n =$ the nth infinite initial ordinal, for $1 \leq n < \omega$. Following Buchholz/Schütte [30] we define *collapsing functions* $\Psi_i(\alpha)$, $i < \omega$, by induction on α. First let $C_i(\alpha)$ be the smallest set of ordinals such that

1. $\{\Omega_n : n < \omega\} \cup \{\xi : \xi < \Omega_i\} \subseteq C_i(\alpha)$;
2. if $\xi, \eta \in C_i(\alpha)$, then also $\xi + \eta, \omega^\xi \in C_i(\alpha)$;

3. if $\xi \in C_i(\alpha)$ and $\xi < \alpha$, then also $\Psi_j(\xi) \in C_i(\alpha)$ for all $j \geq i$, $j < \omega$.

Then $\Psi_i(\alpha)$ is defined to be the smallest β such that $\beta \notin C_i(\alpha)$.

REMARK IX.5.9. Each Ψ_i, $i < \omega$, is monotone and continuous. In particular, we have $\Psi_0(\Omega_\omega) = \sup\{\Psi_0(\Omega_n) : n < \omega\}$. This turns out to be a recursive ordinal. It can also be characterized in terms of Takeuti's ordinal diagrams of finite order; see Takeuti [261, chapter 5].

THEOREM IX.5.10 (provable ordinals of Π_1^1-CA$_0$). *We have*

$$\text{ord}(\Pi_1^1\text{-CA}_0) = \Psi_0(\Omega_\omega).$$

PROOF. This is an advanced result of Gentzen-style proof theory. See Takeuti [261], Schütte [214], Buchholz/Schütte [30], and Buchholz/Feferman/Pohlers/Sieg [29].   □

REMARK IX.5.11 (mathematical independence results). Gentzen-style proof theory has been used to obtain various independence results for subsystems of Z_2. Friedman (see Simpson [239, 240]) used theorem IX.5.7 to show that Kruskal's theorem is not provable in ATR$_0$; Friedman/Robertson/Seymour [77] used theorem IX.5.10 to show that the graph minor theorem is not provable in Π_1^1-CA$_0$; see remark X.3.23 below. There are some closely related finite combinatorial independence results; Simpson [244] provides an overview of this area. Simpson [245] used theorem IX.5.4 to show that the Hilbert basis theorem is not provable in RCA$_0$; see remark X.3.21 below.

IX.6. Conclusions

In §§IX.1–IX.4 we have used non-ω-models to obtain some striking conservation results. The simplest such result is that ACA$_0$ is conservative over first order arithmetic, PA. In addition, RCA$_0$ and WKL$_0$ are conservative over Σ_1^0-PA, which is in turn conservative over primitive recursive arithmetic for Π_2^0 sentences. By remark IX.3.18, the latter result is of great significance with respect to Hilbert's program of finite reductionism. In addition, we obtained some surprising results concerning choice schemes: Σ_1^1-AC$_0$ is conservative over ACA$_0$ for Π_2^1 sentences; Σ_2^1-AC$_0$ is conservative over Π_1^1-CA$_0$ for Π_3^1 sentences; for each $k < \omega$, Σ_{k+3}^1-AC$_0$ is conservative over Π_{k+2}^1-CA$_0$ for Π_4^1 sentences. In §IX.5 we ended the chapter with some very brief remarks on Gentzen-style proof theory, specifically provable ordinals.

APPENDIX

Chapter X

ADDITIONAL RESULTS

This appendix is a supplement to chapters I through IX. We outline various results without proof but with references to the published literature.

X.1. Measure Theory

In this section we discuss measure theory in the context of subsystems of second order arithmetic.

Measure theory is a particularly interesting topic from the viewpoint of the Main Question and Reverse Mathematics (chapter I). Recall from §§I.1 and I.12 that the Main Question concerns the role of set existence axioms. Historically, the subject of measure theory developed hand in hand with the nonconstructive, set-theoretic approach to mathematics. Bishop has remarked that the foundations of measure theory present a special challenge to the constructive mathematician. Although Reverse Mathematics is quite different from Bishop-style constructivism (see remarks I.8.9 and IV.2.8), we feel that Bishop's remark implicitly raises an interesting question: *Which nonconstructive set existence axioms are needed for measure theory?*

Measure Theory in RCA_0. We begin by noting that some basic measure-theoretic notions can be defined in RCA_0.

DEFINITION X.1.1 (Borel measures). Within RCA_0, let X be a compact metric space. Recall from exercise IV.2.13 the separable Banach space $\mathrm{C}(X)$. A *Borel measure on X* (more accurately, a nonnegative Borel probability measure on X) is defined to be a nonnegative bounded linear functional $\mu\colon \mathrm{C}(X) \to \mathbb{R}$ such that $\mu(1) = 1$. See also definition IV.2.14.

DEFINITION X.1.2 (measure of an open set). Within RCA_0, let X be a compact metric space, and let μ be a Borel measure on X. If U is an open set in X, we define

$$\mu(U) = \sup\{\mu(\phi) \mid \phi \in \mathrm{C}(X), 0 \le \phi \le 1, \phi = 0 \text{ on } X \setminus U\}.$$

Note that, within RCA_0, the above supremum need not exist as a real number. Indeed, the existence of $\mu(U)$ for all open sets U is equivalent

391

to ACA$_0$ over RCA$_0$. See also Yu [277], where it is shown that a certain version of the Riesz representation theorem is equivalent over RCA$_0$ to arithmetical comprehension.

Therefore, when working in RCA$_0$ in situations when arithmetical comprehension is not available, we interpret statements about $\mu(U)$ in a "virtual" or comparative sense. For example, $\mu(U) \leq \mu(V)$ is taken to mean that for all $\epsilon > 0$ and all $\phi \in C(X)$ with $0 \leq \phi \leq 1$ and $\phi = 0$ on $X \setminus U$, there exists $\psi \in C(X)$ with $0 \leq \psi \leq 1$ and $\psi = 0$ on $X \setminus V$ such that $\mu(\phi) \leq \mu(\psi) + \epsilon$.

EXAMPLES X.1.3. *Lebesgue measure* measure on the closed unit interval $[0, 1]$ is given by the bounded linear functional $\mu \colon C[0, 1] \to \mathbb{R}$ where $\mu(\phi) = \int_0^1 \phi(x)\, dx$, the Riemann integral of ϕ from 0 to 1. It can be shown that the Lebesgue measure of an open interval is the length of the interval. There is also the obvious generalization to Lebesgue measure on the n-cube $[0, 1]^n$. Another example is the familiar *fair coin measure* on the Cantor space $2^{\mathbb{N}}$, given by $\mu(\{x \mid x(n) = i\}) = 1/2$ for all $n \in \mathbb{N}$ and $i \in \{0, 1\}$.

DEFINITION X.1.4 (countable additivity, etc.). Within RCA$_0$, let X be a compact metric space and let μ be a Borel measure on X. We say that μ is *countably additive* if

$$\mu\left(\bigcup_{n=0}^{\infty} U_n\right) = \lim_{k \to \infty} \mu\left(\bigcup_{n=0}^{k} U_n\right)$$

for any sequence of open sets $U_n \subseteq X$, $n \in \mathbb{N}$; *disjointly countably additive* if

$$\mu\left(\bigcup_{n=0}^{\infty} U_n\right) = \sum_{n=0}^{\infty} \mu(U_n)$$

for any sequence of pairwise disjoint open sets $U_n \subseteq X$, $n \in \mathbb{N}$; *finitely additive* if

$$\mu(U) + \mu(V) = \mu(U \cup V) + \mu(U \cap V)$$

for all open $U, V \subseteq X$.

DEFINITION X.1.5 (nice metric spaces). An open set is said to be *connected* if it is not the union of two disjoint nonempty open sets. A separable metric space X is said to be *nice* if for all sufficiently small $\delta > 0$ and all $x \in X$, the open ball

$$B(x, \delta) = \{y \in X \mid d(x, y) < \delta\}$$

is connected. Such a δ is called a *modulus of niceness* for X.

For example, the unit interval $[0, 1]$ and the n-cube $[0, 1]^n$ are nice, but the Cantor space $2^{\mathbb{N}}$ is not nice.

THEOREM X.1.6 (disjoint countable additivity). *The following is provable in RCA$_0$. Let X be a compact metric space, and let μ be a Borel measure on X. If X is nice, then μ is disjointly countably additive.*

PROOF. See Brown/Giusto/Simpson [26].   □

Measure Theory in WWKL$_0$. In order to obtain countable additivity, we need an axiom which goes beyond RCA$_0$ yet is weaker than weak König's lemma.

DEFINITION X.1.7 (weak weak König's lemma). We define *weak weak König's lemma* to be the following axiom: if T is a subtree of $2^{<\mathbb{N}}$ with no infinite path, then

$$\lim_{n\to\infty} \frac{|\{\sigma \in T \mid \mathrm{lh}(\sigma) = n\}|}{2^n} = 0.$$

Note that weak weak König's lemma is a consequence of weak König's lemma, which reads as follows: if T is a subtree of $2^{<\mathbb{N}}$ with no infinite path, then T is finite.

WWKL$_0$ is the subsystem of Z_2 consisting of RCA$_0$ plus weak weak König's lemma.

REMARK X.1.8 (ω-models of WWKL$_0$). It is known that

$$\text{RCA}_0 \subsetneq \text{WWKL}_0 \subsetneq \text{WKL}_0$$

and there are ω-models for the independence. For the first inequality, note that the ω-model REC consisting of the recursive subsets of ω (see remark I.7.5) does not satisfy WWKL$_0$. For the second inequality, one can easily construct an ω-model of WWKL$_0$, namely a random real model, which does not satisfy WKL$_0$. See for example Yu/Simpson [280]. The study of ω-models of WWKL$_0$ is closely related to the theory of 1-random sequences, as initiated by Martin-Löf [179] and continued by Kučera [156, 157, 158].

THEOREM X.1.9 (countable additivity). *The following assertions are pairwise equivalent over* RCA$_0$.

1. *Weak weak König's lemma.*
2. *For any compact metric space X and any Borel measure μ on X, μ is countably additive.*
3. *For any covering of the closed unit interval $[0, 1]$ by a sequence of open intervals (a_n, b_n), $n \in \mathbb{N}$, we have $\sum_{n=0}^{\infty} |a_n - b_n| \geq 1$.*

PROOF. See Yu/Simpson [280], Brown/Giusto/Simpson [26], and Simpson [248].   □

THEOREM X.1.10 (finite additivity). *The following statements are pairwise equivalent over* RCA$_0$.

1. *Weak weak König's lemma.*
2. *Any Borel measure μ on a compact metric space is finitely additive.*

3. *If μ is the fair coin measure on the Cantor space $X = 2^{\mathbb{N}}$, then for any two open sets $U, V \subseteq X$ with $X = U \cup V$ and $U \cap V = \emptyset$ we have $\mu(U) + \mu(V) = 1$.*

PROOF. See Simpson [248].   □

It turns out that WWKL_0 is sufficient to develop a fair amount of measure theory and to prove several key theorems, as we now show.

REMARK X.1.11 (measurable functions). Let X be a compact metric space and let μ be a Borel measure on X. Recall from exercise IV.2.15 that there is a separable Banach space $\mathsf{L}_1(X, \mu)$ with the L_1-norm given by $\|f\|_1 = \mu(|f|)$. For $f \in \mathsf{L}_1(X, \mu)$ we define $\int f\, d\mu = \mu(f)$. All of this makes sense in RCA_0.

An obvious question is whether elements of the separable Banach space $\mathsf{L}_1(X, \mu)$ can be identified with real-valued measurable functions on X in the usual way. The answer is that this can be done in WWKL_0. Namely, given $f \in \mathsf{L}_1(X, \mu)$, we know in RCA_0 that f is given by a sequence of real-valued continuous functions $\phi_n \in C(X)$, $n \in \mathbb{N}$, which converges in the L_1-norm, indeed $\|\phi_m - \phi_{m'}\|_1 \leq 1/2^n$ for all $m, m', n \in \mathbb{N}$ with $m, m' \geq n$. We can prove in WWKL_0 that this sequence converges pointwise almost everywhere, in the following sense: There is a sequence of closed sets

$$C_0^f \subseteq C_1^f \subseteq \cdots \subseteq C_n^f \subseteq \cdots, \quad n \in \mathbb{N}$$

such that $\mu(X \setminus C_n^f) \leq 1/2^n$ for all n, and $|\phi_m(x) - \phi_{m'}(x)| \leq 1/2^k$ for all $x \in C_n^f$ and all m, m', k such that $m, m' \geq n + 2k + 2$. We then define $f(x) = \lim_{n \to \infty} \phi_n(x)$ for all $x \in \bigcup_{n=0}^{\infty} C_n^f$. Thus we see that $f(x)$ is defined for almost all $x \in X$. Moreover, $f = g$ in $\mathsf{L}_1(X, \mu)$, i.e., if $\|f - g\|_1 = 0$, if and only if $f(x) = g(x)$ for almost all $x \in X$. These facts are provable in WWKL_0.

The above remarks on pointwise values of measurable functions are due to Yu [275, 278].

Our approach to measurable sets within WWKL_0 is to identify them with their characteristic functions in $\mathsf{L}_1(X, \mu)$, according to the following definition.

DEFINITION X.1.12 (measurable sets). We say that $f \in \mathsf{L}_1(X, \mu)$ is a *measurable characteristic function* if $f(x) \in \{0, 1\}$ for almost all $x \in X$, i.e., there exists a sequence of closed sets

$$C_0 \subseteq C_1 \subseteq \cdots \subseteq C_n \subseteq \ldots, \quad n \in \mathbb{N},$$

such that $\mu(X \setminus C_n) \leq 1/2^n$ for all n, and $f(x) \in \{0, 1\}$ for all $x \in \bigcup_{n=0}^{\infty} C_n$. Here $f(x)$ is as defined in remark X.1.11. A (code for a) *measurable set* E with respect to (X, μ) is defined to be a measurable characteristic function $f \in \mathsf{L}_1(X, \mu)$. We then define $\mu(E) = \mu(f)$, and the complementary set $X \setminus E$ is defined as $1 - f$. If E_1 and E_2

are measurable sets with measurable characteristic functions f_1 and f_2, then $E_1 \cup E_2$ and $E_1 \cap E_2$ are defined as $\sup(f_1, f_2)$, $\inf(f_1, f_2)$ respectively. Other set operations on measurable sets are defined similarly.

For more on measurable sets in the context of subsystems of Z_2, see Brown/Giusto/Simpson [26] and Giusto's thesis [91].

With the above notion of measurable set, we can show that $WWKL_0$ is just strong enough to prove a version of the Vitali covering theorem. We consider only Lebesgue measure μ on $[0, 1]$. Let $\mathcal{I}$ be a sequence of intervals in $[0, 1]$. We say that $\mathcal{I}$ *Vitali covers* an interval $E \subseteq [0, 1]$ if for all $x \in E$ and all $\epsilon > 0$ there exists $I \in \mathcal{I}$ such that $x \in I$ and length$(I) < \epsilon$. We say that $\mathcal{I}$ *almost Vitali covers* a Lebesgue measurable set $E \subseteq [0, 1]$ if for all $\epsilon > 0$ we have $\mu(E \setminus O_\epsilon) = 0$, where $O_\epsilon = \bigcup \{ I : I \in \mathcal{I},$ length$(I) < \epsilon \}$.

THEOREM X.1.13 (Vitali covering theorem). *The following are pairwise equivalent over* RCA_0.

1. *Weak weak König's lemma.*
2. *If $\mathcal{I}$ Vitali covers an interval E, then $\mathcal{I}$ contains a pairwise disjoint sequence of intervals I_n, $n \in \mathbb{N}$, such that $\mu(E \setminus \bigcup_{n=0}^{\infty} I_n) = 0$.*
3. *If $\mathcal{I}$ almost Vitali covers a Lebesgue measurable set E, then $\mathcal{I}$ contains a pairwise disjoint sequence of intervals I_n, $n \in \mathbb{N}$, such that $\mu(E \setminus \bigcup_{n=0}^{\infty} I_n) = 0$.*

PROOF. See Brown/Giusto/Simpson [26]. □

We now discuss the Lebesgue convergence theorems. Let μ be a Borel measure on a compact metric space X. The *monotone convergence theorem* for μ asserts that if $f, f_n \in L_1(X, \mu)$, $n \in \mathbb{N}$, and if $\langle f_n(x) : n \in \mathbb{N} \rangle$ is increasing and converges to $f(x)$ for almost all $x \in X$, then $\lim_n \| f_n - f \|_1 = 0$ and $\lim_n \int f_n \, d\mu = \int f \, d\mu$.

THEOREM X.1.14 (monotone convergence theorem). *The following are pairwise equivalent over* RCA_0.

1. *Weak weak König's lemma.*
2. *The monotone convergence theorem for Borel measures on compact metric spaces.*
3. *The monotone convergence theorem for Lebesgue measure on $[0, 1]$.*

PROOF. See Yu [278]. □

REMARK X.1.15 (dominated convergence theorem). We conjecture that the *dominated convergence theorem* is also equivalent to weak weak König's lemma over RCA_0. This is the assertion that if $f, g, f_n \in L_1(X, \mu)$, $n \in \mathbb{N}$, and if $|f_n(x)| \leq g(x)$ for all $n \in \mathbb{N}$ and $\lim_n f_n(x) = f(x)$ for almost all $x \in X$, then $\lim_n \| f_n - f \|_1 = 0$ and $\lim_n \int f_n \, d\mu = \int f \, d\mu$. For the background of this conjecture, see Yu [278].

Measure Theory in WKL_0, ACA_0, ATR_0. We end this section by noting that some set existence axioms going beyond WWKL_0 are sometimes useful in measure theory.

REMARK X.1.16 (Haar measure in WKL_0). Let G be a *separable compact group*, i.e., a compact separable metric space with continuous group operations. *Haar measure* on G is the unique invariant Borel measure on G. It is known that the existence of Haar measure for separable compact groups is equivalent over RCA_0 to weak König's lemma; this result is due to Tanaka/Yamazaki [265].

REMARK X.1.17 (measure theory in ACA_0). Clearly ACA_0 is useful in measure theory. For example, ACA_0 implies that the class of measurable sets (definition X.1.12) is closed under countable unions and intersections. Moreover, Yu [276, 277, 279] has shown that ACA_0 is equivalent over RCA_0 to several specific measure-theoretic theorems: (1) a certain form of the Radon/Nikodym theorem for Borel measures on compact metric spaces; (2) a certain form of the Riesz representation theorem for Borel measures on compact metric spaces; (3) enumerability of the set of singular points of an arbitrary Borel measure on the Cantor space.

REMARK X.1.18 (measure theory in ATR_0). Yu [277] has noted that ATR_0 suffices to prove measurability and regularity of Borel sets with respect to any Borel measure on a compact metric space. It is unclear whether ATR_0 suffices to prove measurability and regularity of analytic sets in some appropriate sense.

REMARK X.1.19. Additional results on analysis in RCA_0, WKL_0, ACA_0 and related systems are in Brown [24, 25], Giusto/Marcone [92], Giusto/Simpson [93], Hardin/Velleman [101].

Notes for §X.1. The material in this section is from Yu [275, 276, 277, 278, 279], Yu/Simpson [280], Brown/Giusto/Simpson [26], Giusto [91], Tanaka/Yamazaki [265], and Simpson [248].

X.2. Separable Banach Spaces

In this section we present some results on the theory of separable Banach spaces in subsystems of Z_2. This builds on the material that has already been presented in §§II.10 and IV.9.

Banach Separation. We begin with the so-called geometric form of the Hahn/Banach theorem. Let X be a separable Banach space. As in §IV.9, a *bounded linear functional* on X is a bounded linear operator $f : X \to \mathbb{R}$. Let A and B be convex sets in X. We say that A and B are *separated* if there exists a bounded linear functional $f : X \to \mathbb{R}$ and a real number α such that $f(x) < \alpha$ for all $x \in A$, and $f(x) \geq \alpha$ for all $x \in B$. We say that A and B are *strictly separated* if in addition $f(x) > \alpha$ for all $x \in B$.

THEOREM X.2.1 (Banach separation in WKL_0). WKL_0 *is equivalent over* RCA_0 *to the following statement. If A and B are open convex sets in a separable Banach space X, and if $A \cap B = \emptyset$, then A and B can be strictly separated.*

PROOF. This and related results are due to Humphreys/Simpson [128]. The proof of Banach separation in WKL_0 is accomplished by means of a reduction to the case of finite-dimensional Banach spaces, using a compactness argument. The reversal is obtained via the Brown/Simpson [27] reversal of the Hahn/Banach theorem; see also theorem IV.9.4.   □

REMARK X.2.2. Hatzikiriakou [111] has shown that WKL_0 is also equivalent over RCA_0 to an algebraic separation theorem for countable vector spaces over $\mathbb{Q}$. We do not see any easy way to deduce Hatzikiriakou's result from theorem X.2.1 or vice versa, but the comparison is interesting.

Dual Spaces and Alaoglu's Theorem. Next we consider dual spaces and the Banach/Alaoglu theorem. Let X be a separable Banach space. The following definitions are made in RCA_0.

DEFINITION X.2.3 (dual space, Alaoglu ball). We write $f \in X^*$ to mean that f is a bounded linear functional on X. Thus X^* is the *dual space* of X. For $0 < r < \infty$, we write $f \in B_r(X^*)$ to mean that $f \in X^*$ and $\|f\| \leq r$. Note that X^* and $B_r(X^*)$ do not formally exist as sets within RCA_0. We identify the functionals in $B_r(X^*)$ in the obvious way with the points of a certain closed set in the compact metric space $\prod_{a \in A}[-r\|a\|, r\|a\|]$, where $X = \widehat{A}$.

REMARK X.2.4 (Banach/Alaoglu theorem in WKL_0). Using definition X.2.3 and lemmas III.2.5 and IV.1.5, we see that Heine/Borel compactness of $B_r(X^*)$ is provable in WKL_0. This version of the Banach/Alaoglu theorem is very useful for the development of separable Banach space theory within WKL_0. See also Brown's [24] discussion of the Alaoglu ball.

The Weak-∗ Topology. Finally we consider the weak-∗ topology. As before, let X be a separable Banach space. We shall observe that Π_1^1 comprehension is needed to prove some basic results about weak-∗ closed subspaces of X^*.

DEFINITION X.2.5 (bounded-weak-∗ topology). A (code for a) *bounded-weak-∗-closed set* C in X^* is defined to be a sequence of (codes for) closed sets $C_n \subseteq B_n(X^*)$, $n \in \mathbb{N}$, such that

$$\forall m \, \forall n \, (m < n \to C_m = B_m(X^*) \cap C_n).$$

We write $x^* \in C$ to mean $\exists n \, (x^* \in C_n)$, or equivalently $\forall n \, (n > \|x^*\| \rightarrow x^* \in C_n)$. A *bounded-weak-*-open set* in X^* is defined to be the complement of a bounded-weak-*-closed set in X^*.

DEFINITION X.2.6 (weak-* topology). A *weak-*-open set* in X^* is defined to be a bounded-weak-*-open set U in X^* such that for all $x_0^* \in U$ there exists a finite sequence of points $x_0, \ldots, x_{n-1} \in X$ such that

$$\{x^* \in X^* : \forall k < n \, (|x^*(x_k) - x_0^*(x_k)| \leq 1)\} \subseteq U.$$

A *weak-*-closed set* in X^* is defined to be the complement of a weak-*-open set in X^*.

Clearly there is a weak-* neighborhood basis of 0 in X^* consisting of the polars of finite sets in X. Humphreys/Simpson [127, lemma 4.12] have shown that the following well known fact is provable in ACA$_0$: There is a bounded-weak-* neighborhood basis of 0 in X^* consisting of the polars of sequences converging to 0 in X. We do not know whether WKL$_0$ suffices.

THEOREM X.2.7 (Krein/Šmulian theorem in ACA$_0$). *The following is provable in ACA$_0$. Let X be a separable Banach space. Suppose that $C \subseteq X^*$ is convex and bounded-weak-*-closed. Then C is weak-*-closed.*

PROOF. This is theorem 4.14 of Humphreys/Simpson [127]. Again, we do not know whether WKL$_0$ suffices.   □

Specializing to subspaces of X^* we obtain:

COROLLARY X.2.8. *The following is provable in ACA$_0$. Let X be a separable Banach space. Let C be a closed set in $B_1(X^*)$ such that $C = B_1(X^*) \cap \mathrm{span}(C)$. Then $\mathrm{span}(C)$ is a weak-*-closed subspace of X^*.*

(Here $\mathrm{span}(C)$ denotes the linear span of C.)

The following theorem is interesting because it shows that a rather strong set existence axiom, Π_1^1 comprehension, is needed to prove a rather trivial-sounding statement about the weak-* topology: For every countable set $Y \subset X^*$, the weak-*-closed linear span of Y exists.

Recall from example II.10.2 that ℓ_1 is the separable Banach space of absolutely summable sequences of real numbers. It may be viewed as the dual of the space c_0 of sequences of real numbers which are convergent to 0, with the sup norm.

THEOREM X.2.9 (weak-* topology and Π_1^1-CA$_0$). *The following are pairwise equivalent over RCA$_0$.*

1. Π_1^1-CA$_0$.
2. *For every separable Banach space X and countable set $Y \subseteq X^*$, there exists a smallest weak-*-closed set in X^* containing Y.*
3. *For every separable Banach space X and countable set $Y \subseteq X^*$, there exists a smallest weak-*-closed convex set in X^* containing Y.*

4. *For every separable Banach space X and countable set $Y \subseteq X^*$, there exists a smallest weak-$*$-closed subspace of X^* containing Y.*
5. *Same as 2 with $X = c_0$ and $X^* = \ell_1$.*
6. *Same as 3 with $X = c_0$ and $X^* = \ell_1$.*
7. *Same as 4 with $X = c_0$ and $X^* = \ell_1$.*

PROOF. This is theorem 5.6 of Humphreys/Simpson [127]. The proof uses the notion of smooth tree (exercise VI.1.9) and is correlated to transfinite iteration of weak-$*$ sequential closure. For details, see [127].   □

Notes for §X.2. The results of this section are from Humphreys/Simpson [127, 128]. See also Humphreys [126]. For a study of the open mapping and closed graph theorems in subsystems of Z_2, see Brown/Simpson [28] and Brown [24].

X.3. Countable Combinatorics

In this section we present some results on countable combinatorics in subsystems of Z_2.

Hindman's Theorem and Dynamical Systems. We have seen in §III.7 that Ramsey's theorem for exponent 3 is equivalent over RCA_0 to ACA_0. The purpose of this subsection is to consider the status of other Ramsey-type combinatorial theorems.

DEFINITION X.3.1 (Hindman's theorem). Given $X \subseteq \mathbb{N}$, let $FS(X)$ be the set of all sums of finite nonempty subsets of X. *Hindman's theorem* says: If $\mathbb{N} = C_0 \cup \cdots \cup C_l$ then there exists an infinite set $X \subseteq \mathbb{N}$ such that $FS(X) \subseteq C_i$ for some $i \leq l$.

There has been considerable interest in the issue of whether Hindman's theorem holds constructively; see [21] for some of the history. From the standpoint of Reverse Mathematics, we conjecture that Hindman's theorem is equivalent over RCA_0 to ACA_0. We now present some partial results in this direction.

DEFINITION X.3.2 (the system ACA_0^+). Let ACA_0^+ consist of ACA_0 plus the assertion that for any $X \subseteq \mathbb{N}$ the ωth Turing jump $TJ(\omega, X)$ exists. Here ω denotes the order type of $\mathbb{N}$ under $\leq_{\mathbb{N}}$. Note that ACA_0^+ is closely related to the predicative system of Weyl [274].

We have:

THEOREM X.3.3 (Hindman's theorem and ACA_0).

1. *Hindman's theorem is provable in ACA_0^+.*
2. *Hindman's theorem implies ACA_0 over RCA_0.*

PROOF. These results are from Blass/Hirst/Simpson [21].   □

The Auslander/Ellis theorem is a well known theorem of topological dynamics. It is closely related to Hindman's theorem; see Furstenberg [84] and Graham/Rothschild/Spencer [98]. Just as in the case of Hindman's theorem, we conjecture that the Auslander/Ellis theorem is equivalent over RCA₀ to ACA₀, and we present some partial results. First we review the relevant definitions.

DEFINITIONS X.3.4 (uniform recurrence, etc.). A *dynamical system* consists of a compact metric space X and a continuous function $T: X \to X$. For $x \in X$ and $n \in \mathbb{N}$ we write

$$T^n(x) = \underbrace{TT \cdots T}_{n}(x).$$

A point $x \in X$ is called *recurrent* if for all $\epsilon > 0$ there exist infinitely many n such that $d(T^n(x), x) < \epsilon$. We say that x is *uniformly recurrent* if for all $\epsilon > 0$ there exists m such that for all n there exists $k < m$ such that $d(T^{n+k}(x), x) < \epsilon$. Two points $x, y \in X$ are said to be *proximal* if for all $\epsilon > 0$ there exist infinitely many n such that $d(T^n(x), T^n(y)) < \epsilon$. The *Auslander/Ellis theorem* says: For all $x \in X$ there exists $y \in X$ such that y is proximal to x and uniformly recurrent.

THEOREM X.3.5 (Auslander/Ellis theorem and ACA₀).
1. *The Auslander/Ellis theorem is provable in* ACA₀⁺.
2. *The existence of uniformly recurrent points is provable in* ACA₀.

PROOF. These results are from Blass/Hirst/Simpson [21]. The proof of part 1 proceeds via Hindman's theorem and uses X.3.3.1. Part 2 may be compared with Girard [90, annex 7.E]. □

REMARK X.3.6 (open problems). There are many other open problems concerning the Reverse Mathematics status of various theorems of countable combinatorics. Among these are Szemerédi's theorem (see Furstenberg [84] and Graham/Rothschild/Spencer [98]) and its generalizations due to Furstenberg/Katznelson [85, 86, 87]. There is also the Carlson/Simpson theorem [33, 34] (see also Blass/Hirst/Simpson [21] and Simpson [242]) and its generalizations due to Carlson [31, 32] (see also Hindman/Strauss [116]).

REMARK X.3.7. Another contribution to Reverse Mathematics for dynamical systems is Friedman/Simpson/Yu [80].

Matching Theory. We now turn from Ramsey theory to another branch of combinatorics known as matching theory or transversal theory. General references on this subject are Jungnickel [135], Mirsky [190], and Holz/Podewski/Steffens [124].

DEFINITION X.3.8 (matchings). A *bipartite graph* is an ordered triple $G = (X, Y, E)$ such that X and Y are sets, $X \cap Y = \emptyset$, and $E \subseteq$

$\{\{x, y\}: x \in X, y \in Y\}$. The *vertices* of G are the elements of $X \cup Y$. The *edges* of G are the elements of E. A *vertex covering* of G is a set $C \subseteq X \cup Y$ such that every edge of G has a vertex in C. A *matching* in G is a pairwise disjoint set $M \subseteq E$. Here pairwise disjointness means that no two edges in M have a common vertex.

REMARK X.3.9 (König duality theorem). For any set S we use $|S|$ to denote the cardinality of S. If G is any bipartite graph and C is any vertex covering of G and M is any matching in G, then clearly $|C| \geq |M|$. The *König duality theorem* asserts that for any finite bipartite graph G there exist a vertex covering C of G and a matching M in G such that $|C| = |M|$. In other words, $\min\{|C|: C \text{ is a vertex covering of } G\} = \max\{|M|: M \text{ is a matching in } G\}$.

DEFINITION X.3.10 (König coverings). For any bipartite graph G, a *König covering* of G is an ordered pair (C, M) such that C is a vertex covering of G, M is a matching in G, and C consists of exactly one vertex from each edge of M. (The last condition means that $C \subseteq \bigcup M$ and $|C \cap e| = 1$ for each $e \in M$.)

REMARK X.3.11. Clearly if (C, M) is a König covering of G then $|C| = |M|$. König [148] showed that every finite bipartite graph has a König covering. From this the König duality theorem follows immediately. König coverings have also been used to generalize the König duality theorem to infinite bipartite graphs. Podewski/Steffens [202] showed that every countably infinite bipartite graph has a König covering. Aharoni [5] showed that every uncountable bipartite graph has a König covering.

Consider the following instance of the Main Question: Which set existence axioms are needed to prove the *Podewski/Steffens theorem* ("every countable bipartite graph has a König covering")? The answer is arithmetical transfinite recursion, as shown by the following theorem.

THEOREM X.3.12 (Podewski/Steffens theorem in ATR_0). *The Podewski/Steffens theorem is equivalent over RCA_0 to ATR_0.*

PROOF. The reversal, i.e., the fact that the Podewski/Steffens theorem implies ATR_0 over RCA_0, is due to Aharoni/Magidor/Shore [6]. The forward direction, i.e., the fact that ATR_0 proves the Podewski/Steffens theorem, is due to Simpson [247]. The latter proof is interesting in that it employs the method of inner models, specifically countable coded ω-models of $\Sigma_1^1\text{-}\mathsf{AC}_0$. See also remark V.10.1. $\square$

We now discuss perfect matchings in countable bipartite graphs.

DEFINITIONS X.3.13 (Hall condition, perfect matchings). Let $G = (X, Y, E)$ be a bipartite graph. For $A \subseteq X \cup Y$ we write $\mathrm{N}_G(A) = \{b: \{a, b\} \in E \text{ for some } a \in A\}$. G is said to satisfy the *Hall condition* if $|\mathrm{N}_G(A)| \geq |A|$ for all finite $A \subseteq X \cup Y$. G is said to be *locally finite* if $N_G(a)$ is finite for all $a \in X \cup Y$. G is said to be *n-regular* if $|N_G(a)| = n$ for all $a \in X \cup Y$. A matching M in G is said to be *perfect* if $X \cup Y = \bigcup M$, i.e., every vertex of G is incident to an edge of M.

REMARK X.3.14. *Hall's theorem* asserts that a finite bipartite graph has a perfect matching if and only if it satisfies the Hall condition. The *marriage theorem* asserts that a finite bipartite graph which is *n*-regular for some $n \geq 1$ has a perfect matching. The marriage theorem is an easy consequence of Hall's theorem, which is an easy consequence of the König duality theorem.

THEOREM X.3.15. *The following are equivalent over* RCA$_0$.

1. ACA$_0$.
2. *If G is a countable locally finite bipartite graph, then G satisfies the Hall condition if and only if G has a perfect matching.*
3. *If $G = (X, Y, E)$ is a countable bipartite graph, and if G has matchings M_1 and M_2 such that $X \subseteq \bigcup M_1$ and $Y \subseteq \bigcup M_2$, then G has a perfect matching.*

PROOF. This is from Hirst [117, 118]. See also McAloon [182].   □

THEOREM X.3.16. *The following are pairwise equivalent over* RCA$_0$.

1. WKL$_0$.
2. *If G is a countable bipartite graph which is n-regular for some $n \geq 1$, then G has a perfect matching.*
3. *If G is a countable 2-regular bipartite graph, then G has a perfect matching.*

PROOF. This is from Hirst [117, 118]. See also Manaster/Rosenstein [168, 169].   □

WQO Theory. We now consider another branch of combinatorics: well quasiordering theory.

DEFINITION X.3.17 (well quasiordering). A *quasiordering* is a set Q together with a reflexive, transitive relation $\leq$ on Q. An *antichain* in $(Q, \leq)$ is a set of elements of Q which are pairwise incomparable under $\leq$. A *well quasiordering* (abbreviated WQO) is a quasiordering which is well founded and has no infinite antichains.

REMARK X.3.18 (equivalent characterizations). For a quasiordering $(Q, \leq)$, the following conditions are pairwise equivalent.

1. $(Q, \leq)$ is well quasiordered.
2. For every sequence $\langle a_n : n \in \mathbb{N} \rangle$ of elements of Q, there exist $m, n \in \mathbb{N}$ such that $m < n$ and $a_m \leq a_n$.
3. For every sequence $\langle a_n : n \in \mathbb{N} \rangle$ of elements of Q, there exists a subsequence $\langle a_{n_k} : k \in \mathbb{N} \rangle$, $n_0 < n_1 < \cdots < n_k < \cdots$, such that $a_{n_0} \leq a_{n_1} \leq \cdots \leq a_{n_k} \leq \cdots$.
4. Every upward closed subset of Q is finitely generated.

These equivalences are an easy consequence of Ramsey's theorem for exponent 2.

REMARK X.3.19 (WQO theory). There is a rich theory of well qua-siorderings. For instance, the Cartesian product of two well quasiorder-ings is a well quasiordering, and it follows by induction that if Q is a well quasiordering then so is the m-fold Cartesian power Q^m, for each $m \in \mathbb{N}$. One of the best known results in WQO theory is *Higman's the-orem*: If Q is a well quasiordering, then $Q^{<\mathbb{N}}$ is a well quasiordering. Here $Q^{<\mathbb{N}} = \bigcup_{m=0}^{\infty} Q^m$, the set of finite sequences of elements of Q, qua-siordered by putting $\langle a_i : i < m \rangle \leq \langle b_j : j < n \rangle$ if and only if there exist $j_0 < \cdots < j_{m-1} < n$ such that $a_0 \leq b_{j_0}, \ldots, a_{m-1} \leq b_{j_{m-1}}$. Another well known result is *Kruskal's theorem*: If Q is a well quasiordering, then the set of Q-labeled finite trees is well quasiordered under an appropriate quasiordering. See for example Simpson [239, 240].

THEOREM X.3.20 (Dickson's lemma and ω^ω). *The following are equiva-lent over* RCA$_0$.

1. ω^ω *is well ordered.*
2. *For each $m \in \mathbb{N}$, the m-fold Cartesian power $\mathbb{N}^m$ is well quasiordered.* (*This statement is sometimes known as* Dickson's lemma.)

PROOF. This follows from Simpson [245, lemma 3.6]. Note that the well orderedness of ω^ω cannot be proved in RCA$_0$, in view of theorem IX.5.4. □

REMARK X.3.21 (the Hilbert basis theorem and ω^ω). The *Hilbert basis theorem* asserts that for all countable fields K and all $m \in \mathbb{N}$, any ideal in the polynomial ring $K[x_1, \ldots, x_m]$ is finitely generated. Simpson [245] has used theorem X.3.20 to show that the Hilbert basis theorem, even for $K = \mathbb{Q}$, is equivalent over RCA$_0$ to well orderedness of ω^ω. See also Hatzikiriakou [110], who obtained a similar result in which the poly-nomial rings $K[x_1, \ldots, x_m]$ are replaced by rings of formal power series $K[[x_1, \ldots, x_m]]$. These results are of historical interest in connection with the Hilbert basis theorem's apparent lack of constructive or computational content; see Simpson [245, §1].

THEOREM X.3.22 (Higman's theorem in ACA$_0$). *The following are equiv-alent over* RCA$_0$.

1. ACA$_0$.
2. *Higman's theorem.*

PROOF. This follows by combining results of Simpson [245, lemma 4.8] (see also Schütte/Simpson [215, lemma 5.2]) and Girard [90] (see remark V.6.10). □

REMARK X.3.23 (Kruskal's theorem, etc.). There are many interesting results and open problems concerning the Reverse Mathematics status of various theorems of well quasiordering theory. Friedman (unpublished) has shown that Kruskal's theorem is not provable in ATR_0, and that a gap embedding generalization of Kruskal's theorem is not provable in $\Pi^1_1\text{-}\mathsf{CA}_0$; see Simpson [239, 240, 244]. The latter result has been used in Friedman/Robertson/Seymour [77] to show that an important theorem of graph theory is not provable in $\Pi^1_1\text{-}\mathsf{CA}_0$. This is the Robertson/Seymour *graph minor theorem*, which asserts that the class of all finite graphs is well quasiordered under minor embeddability. See also remark IX.5.11. Some generalizations of Friedman's gap embedding theorem have been proved by Kriz [153, 154, 155]; the Reverse Mathematics status of these results is unknown.

REMARK X.3.24 (minimal bad sequence lemma). An important technical lemma in WQO theory is the so-called *minimal bad sequence lemma*; see Simpson [239, 240]. Marcone and Simpson have shown that the minimal bad sequence lemma is equivalent over RCA_0 to $\Pi^1_1\text{-}\mathsf{CA}_0$; see Marcone [176, theorem 6.5].

We now consider better quasiorderings, which are useful in proving that various classes of infinite structures are well quasiordered.

DEFINITION X.3.25 (better quasiordering). A *better quasiordering* (abbreviated BQO) is a quasiordering Q with the property that for any Borel mapping $f : [\mathbb{N}]^{\mathbb{N}} \to Q$ there exists $X \in [\mathbb{N}]^{\mathbb{N}}$ such that $f(X) \leq f(X \setminus \{\min(X)\})$. This notion is originally due to Nash-Williams [195, 196]; the formulation here is due to Simpson [237]. It can be shown that any better quasiordering is a well quasiordering, and any "natural" well quasiordering is a better quasiordering.

DEFINITION X.3.26 (transfinite sequence theorem). If Q is a quasiordering, we define $\widetilde{Q}$ to be the class of countable transfinite sequences of elements of Q, quasiordered by putting $\langle a_\xi : \xi < \alpha \rangle \leq \langle b_\eta : \eta < \beta \rangle$ if and only if there exist

$$\eta_0 < \cdots < \eta_\xi < \cdots < \beta \qquad (\xi < \alpha)$$

such that $a_\xi \leq b_{\eta_\xi}$ for all $\xi < \alpha$. The *transfinite sequence theorem* says that if Q is better quasiordered then so is $\widetilde{Q}$. This result is due to Nash-Williams [195, 196].

DEFINITION X.3.27 (Laver's theorem). The class of countable linear orderings may be quasiordered by putting $X \leq Y$ if and only if X is order embeddable into Y. *Laver's theorem*, also known as *Fraïssé's conjecture*, says that the class of countable linear orderings is well quasiordered under order embeddability. This result is due to Laver [160] using BQO theory.

REMARK X.3.28. A simplified exposition of the proof of the Nash-Williams transfinite sequence theorem and Laver's theorem has been given by Simpson [237]. A further simplification has been obtained by van Engelen, Miller and Steel [271].

THEOREM X.3.29 (Nash-Williams theorem in Π^1_1-CA$_0$). *The Nash-Williams transfinite sequence theorem is provable in Π^1_1-CA$_0$ but is not equivalent to Π^1_1-CA$_0$.*

PROOF. This result is due to Marcone [176]; see also Marcone [173, 174, 175].   □

THEOREM X.3.30 (reversals). *Each of Laver's theorem and the Nash-Williams transfinite sequence theorem implies ATR$_0$ over RCA$_0$.*

PROOF. This is due to Shore [223], who actually showed that the following statement implies ATR$_0$: For all $X \subseteq \mathbb{N}$, if $\forall n \, \text{WO}((X)_n)$ then $\exists m \, \exists n \, (m \neq n \wedge (X)_m$ is order embeddable into $(X)_n)$. This is a refinement of theorem V.6.8; see also Friedman/Hirst [74, 75].   □

REMARK X.3.31 (a conjecture). We conjecture that both the Nash-Williams transfinite sequence theorem and Laver's theorem are provable in ATR$_0$. Clote [38, 39] has presented a proof of the transfinite sequence theorem in ATR$_0$, but that proof is incorrect, as Clote has acknowledged (personal communication).

REMARK X.3.32. In Downey/Lempp [48] it is shown that ACA$_0$ is equivalent over RCA$_0$ to a theorem of Dushnik and Miller: Every countably infinite linear ordering has a nontrivial self-embedding.

X.4. Reverse Mathematics for RCA$_0$

Throughout this book we have used RCA$_0$ as our base theory for Reverse Mathematics. An important research direction for the future is to weaken the base theory. We can then hope to find mathematical theorems which are equivalent over the weaker base theory to RCA$_0$, in the sense of Reverse Mathematics. There are a few results in this direction, which we now present.

DEFINITION X.4.1 (RCA$_0^*$ and WKL$_0^*$). Let L$_2$(exp) be L$_2$, the language of second order arithmetic, augmented by a binary operation symbol $\exp(m, n) = m^n$ intended to denote exponentiation. We take $\exp(t_1, t_2) = t_1^{t_2}$ as a new kind of numerical term, and for each $k < \omega$ we define the Σ^0_k and Σ^1_k formulas of L$_2$(exp) accordingly. We define RCA$_0^*$ to be the L$_2$(exp)-theory consisting of RCA$_0$ minus Σ^0_1 induction plus Σ^0_0 induction plus the exponentiation axioms: $m^0 = 1$, $m^{n+1} = m^n \cdot m$. We define WKL$_0^*$ to be RCA$_0$ plus weak König's lemma.

Thus we have

$$\mathsf{RCA}_0 \equiv \mathsf{RCA}_0^* + \Sigma_1^0 \text{ induction,}$$

and

$$\mathsf{WKL}_0 \equiv \mathsf{WKL}_0^* + \Sigma_1^0 \text{ induction.}$$

Paralleling the results of §§IX.1–IX.3, we have:

THEOREM X.4.2 (conservation theorems). *The first order part of* WKL_0^* *and of* RCA_0^* *is the* $L_1(\exp)$-*theory consisting of the basic axioms* I.2.4(i) *plus the exponentiation axioms plus* Σ_0^0 *induction plus* Σ_1^0 *bounding.* WKL_0^* *is conservative over* RCA_0^* *for* Π_1^1 *sentences.* WKL_0^* *and* RCA_0^* *have the same consistency strength as* EFA *and are conservative over* EFA *for* Π_2^0 *sentences.*

PROOF. See Simpson/Smith [250, §4].   □

REMARK X.4.3. An interesting project would be to redo all of the known results in Reverse Mathematics using RCA_0^* instead of RCA_0 as the base theory, replacing WKL_0 by WKL_0^*. The groundwork for this has been laid in Simpson/Smith [250], and much of it would be routine. Note however that bounded Σ_1^0 comprehension is not available in RCA_0^* or in WKL_0^* yet has played a key role in the proofs of several important results, including theorems III.7.2, III.7.6, IV.6.4, IV.7.9, IV.8.2, and V.6.8.

THEOREM X.4.4 (Reverse Mathematics for RCA_0). *The following are pairwise equivalent over* RCA_0^*.

1. Σ_1^0 *induction.*
2. *Bounded* Σ_1^0 *comprehension.*
3. *For every countable field* K, *every polynomial* $f(x) \in K[x]$ *has only finitely many roots in* K.
4. *For every countable field* K, *every polynomial* $f(x) \in K[x]$ *has an irreducible factor.*
5. *For every countable field* K, *every polynomial* $f(x) \in K[x]$ *can be factored into finitely many irreducible polynomials.*
6. *Every finitely generated vector space over* $\mathbb{Q}$ (*or over any countable field*) *has a basis.*
7. *Every finitely generated, torsion-free Abelian group is of the form* $\mathbb{Z}^m$, $m \in \mathbb{N}$.
8. *The structure theorem for finitely generated Abelian groups.*

PROOF. The proof of 1 ↔ 2 has been sketched in remark II.3.11. The equivalences 1 ↔ 2, 1 ↔ 3, 1 ↔ 4 and 1 ↔ 5 are from Simpson/Smith [250]. The equivalence 1 ↔ 6 is due to Friedman (unpublished). Compare theorem III.4.3. The equivalences 1 ↔ 6, 1 ↔ 7 and 1 ↔ 8 are proved in Hatzikiriakou [107, 108].   □

X.5. Conclusions

In this appendix we have mentioned a number of additional results and problems in Reverse Mathematics for RCA_0, WWKL_0, WKL_0, ACA_0, ATR_0, and $\Pi^1_1\text{-}\mathsf{CA}_0$. The mathematical statements were drawn from several branches of mathematics: measure theory, separable Banach space theory, Ramsey theory, matching theory, well quasiordering theory, and countable algebra. We have also made a start on the project of weakening the base theory in Reverse Mathematics.

BIBLIOGRAPHY

[1] STÅL AANDERAA, *Inductive definitions and their closure ordinals*, [61], 1974, pp. 207–220.

[2] OLIVER ABERTH, **Computable Analysis**, McGraw-Hill, 1980.

[3] FRED G. ABRAMSON and GERALD E. SACKS, *Uncountable Gandy ordinals*, **Journal of the London Mathematical Society**, vol. 14 (1976), pp. 387–392.

[4] JOHN W. ADDISON, *Some consequences of the axiom of constructibility*, **Fundamenta Mathematicae**, vol. 46 (1959), pp. 337–357.

[5] RON AHARONI, *König's duality theorem for infinite bipartite graphs*, **Journal of the London Mathematical Society**, vol. 29 (1984), pp. 1–12.

[6] RON AHARONI, MENACHEM MAGIDOR, and RICHARD A. SHORE, *On the strength of König's duality theorem for infinite bipartite graphs*, **Journal of Combinatorial Theory, Series B**, vol. 54 (1992), pp. 257–290.

[7] K. Ambos-Spies, G. H. Müller, and G. E. Sacks (editors), **Recursion Theory Week**, Lecture Notes in Mathematics, no. 1432, Springer-Verlag, 1990.

[8] M. M. Arslanov and S. Lempp (editors), **Recursion Theory and Complexity: Proceedings of the Kazan '97 Workshop**, W. de Gruyter, 1999.

[9] JEREMY AVIGAD, *An effective proof that open sets are Ramsey*, **Archive for Mathematical Logic**, vol. 37 (1998), pp. 235–240.

[10] Y. Bar-Hillel (editor), **Mathematical Logic and Foundations of Set Theory**, Studies in Logic and the Foundations of Mathematics, North-Holland, 1970.

[11] Y. Bar-Hillel et al.(editors), **Essays on the Foundations of Mathematics**, Magnes Press, Jerusalem, 1961.

[12] J. Barwise (editor), **Handbook of Mathematical Logic**, Studies in Logic and the Foundations of Mathematics, North-Holland, 1977.

[13] JON BARWISE, **Admissible Sets and Structures**, Perspectives in Mathematical Logic, Springer-Verlag, 1975.

[14] JON BARWISE and EDWARD FISHER, *The Shoenfield absoluteness lemma*, **Israel Journal of Mathematics**, vol. 8 (1970), pp. 329–339.

[15] JON BARWISE and JOHN SCHLIPF, *On recursively saturated models of arithmetic*, [212], 1975, pp. 42–55.

[16] E. F. Beckenbach (editor), **Applied Combinatorial Mathematics**, Wiley, New York, 1964.

[17] MICHAEL J. BEESON, **Foundations of Constructive Mathematics**, Ergebnisse der Mathematik und ihrer Grenzgebiete, Springer-Verlag, 1985.

[18] C. Berline, K. McAloon, and J.-P. Ressayre (editors), **Model Theory and Arithmetic**, Lecture Notes in Mathematics, no. 890, Springer-Verlag, 1981.

[19] GARRETT BIRKHOFF and GIAN-CARLO ROTA, **Ordinary Differential Equations**, Ginn, Boston, 1962.

[20] ERRETT BISHOP and DOUGLAS BRIDGES, **Constructive Analysis**, Grundlehren der Mathematischen Wissenschaften, no. 279, Springer-Verlag, 1985.

[21] ANDREAS R. BLASS, JEFFRY L. HIRST, and STEPHEN G. SIMPSON, *Logical analysis of some theorems of combinatorics and topological dynamics*, [227], 1987, pp. 125–156.

[22] M. Boffa, D. van Dalen, and K. McAloon (editors), **Logic Colloquium '78**, Studies in Logic and the Foundations of Mathematics, North-Holland, 1979.

[23] F. Browder (editor), **Mathematical Developments Arising from Hilbert Problems**, Proceedings of Symposia in Pure Mathematics, American Mathematical Society, 1976, Two volumes.

[24] DOUGLAS K. BROWN, **Functional Analysis in Weak Subsystems of Second Order Arithmetic**, Ph.D. thesis, The Pennsylvania State University, 1987.

[25] ———, *Notions of closed subsets of a complete separable metric space in weak subsystems of second order arithmetic*, [224], 1990, pp. 39–50.

[26] DOUGLAS K. BROWN, MARIAGNESE GIUSTO, and STEPHEN G. SIMPSON, *Vitali's theorem and WWKL*, **Archive for Mathematical Logic**, vol. 41 (2002), pp. 191–206.

[27] DOUGLAS K. BROWN and STEPHEN G. SIMPSON, *Which set existence axioms are needed to prove the separable Hahn-Banach theorem?*, **Annals of Pure and Applied Logic**, vol. 31 (1986), pp. 123–144.

[28] ———, *The Baire category theorem in weak subsystems of second order arithmetic*, **The Journal of Symbolic Logic**, vol. 58 (1993), pp. 557–578.

[29] WILFRIED BUCHHOLZ, SOLOMON FEFERMAN, WOLFRAM POHLERS, and WILFRIED SIEG, **Iterated Inductive Definitions and Subsystems of Analysis: Recent Proof-Theoretical Studies**, Lecture Notes in Mathematics, no. 897, Springer-Verlag, 1981.

[30] WILFRIED BUCHHOLZ and KURT SCHÜTTE, *Proof Theory of Impredicative Subsystems of Analysis*, Studies in Proof Theory, Bibliopolis, Naples, 1988.

[31] TIMOTHY J. CARLSON, *An infinitary version of the Graham-Leeb-Rothschild theorem*, **Journal of Combinatorial Theory, Series A**, vol. 44 (1987), pp. 22–33.

[32] ——, *Some unifying principles in Ramsey theory*, **Discrete Mathematics**, vol. 68 (1988), pp. 117–169.

[33] TIMOTHY J. CARLSON and STEPHEN G. SIMPSON, *A dual form of Ramsey's theorem*, **Advances in Mathematics**, vol. 53 (1984), pp. 265–290.

[34] ——, *Topological Ramsey theory*, [198], 1990, pp. 172–183.

[35] CHEN-CHUNG CHANG and H. JEROME KEISLER, **Model Theory**, 3rd ed., Studies in Logic and the Foundations of Mathematics, North-Holland, 1990.

[36] PETER A. CHOLAK, CARL G. JOCKUSCH, JR., and THEODORE A. SLAMAN, *On the strength of Ramsey's theorem for pairs*, **The Journal of Symbolic Logic**, vol. 66 (2001), pp. 1–55.

[37] PETER CLOTE, *A recursion theoretic analysis of the clopen Ramsey theorem*, **The Journal of Symbolic Logic**, vol. 49 (1984), pp. 376–400.

[38] ——, *The metamathematics of scattered linear orderings*, **Archive for Mathematical Logic**, vol. 29 (1989), pp. 9–20.

[39] ——, *The metamathematics of Fraïssé's order type conjecture*, [7], 1990, pp. 41–56.

[40] PAUL J. COHEN, **Set Theory and the Continuum Hypothesis**, W. A. Benjamin, New York, 1966.

[41] S. B. Cooper, T. A. Slaman, and S. S. Wainer (editors), **Computability, Enumerability, Unsolvability: Directions in Recursion Theory**, London Mathematical Society Lecture Notes, no. 224, Cambridge University Press, 1996.

[42] J. N. Crossley, J. B. Remmel, R. A. Shore, and M. E. Sweedler (editors), **Logical Methods**, Birkhäuser, 1993.

[43] NIGEL CUTLAND, **Computability, an Introduction to Recursive Function Theory**, Cambridge University Press, 1980.

[44] MARTIN DAVIS, **Computability and Unsolvability**, Dover Publications, New York, 1982.

[45] J. C. E. Dekker (editor), **Recursive Function Theory**, Proceedings of Symposia in Pure Mathematics, American Mathematical Society, 1962.

[46] OSWALD DEMUTH and ANTONÍN KUČERA, *Remarks on constructive mathematical analysis*, [22], 1979, pp. 81–129.

[47] RODNEY G. DOWNEY and STUART A. KURTZ, *Recursion theory and ordered Abelian groups*, **Annals of Pure and Applied Logic**, vol. 32 (1986), pp. 137–151.

[48] RODNEY G. DOWNEY and STEFFEN LEMPP, *The proof-theoretic strength of the Dushnik-Miller theorem for countable linear orderings*, [8], 1999, pp. 55–57.

[49] NELSON DUNFORD and JACOB T. SCHWARTZ, **Linear Operators, Part I**, Pure and Applied Mathematics, Wiley-Interscience, New York, 1958.

[50] H.-D. Ebbinghaus, G. H. Müller, and G. E. Sacks (editors), **Recursion Theory Week**, Lecture Notes in Mathematics, no. 1141, Springer-Verlag, 1985.

[51] HERBERT B. ENDERTON, **A Mathematical Introduction to Logic**, Academic Press, New York, 1972.

[52] RYSZARD ENGELKING, **General Topology**, Heldermann, Berlin, 1989.

[53] Y. L. Ershov, S. S. Goncharov, A. Nerode, and J. B. Remmel (editors), **Handbook of Recursive Mathematics**, Studies in Logic and the Foundations of Mathematics, North-Holland, 1998, Volumes 1 and 2.

[54] JURI L. ERŠOV, *Theorie der Numerierungen III*, **Zeitschrift für Mathematische Logik und Grundlagen der Mathematik**, vol. 23 (1977), pp. 289–371.

[55] SOLOMON FEFERMAN, *Arithmetization of metamathematics in a general setting*, **Fundamenta Mathematicae**, vol. 49 (1960), pp. 35–92.

[56] ———, *Systems of predicative analysis, I*, **The Journal of Symbolic Logic**, vol. 29 (1964), pp. 1–30.

[57] ———, *Systems of predicative analysis, II*, **The Journal of Symbolic Logic**, vol. 33 (1968), pp. 193–220.

[58] ———, *Impredicativity of the existence of the largest divisible subgroup of an Abelian p-group*, [212], 1975, pp. 117–130.

[59] ———, *Theories of finite type related to mathematical practice*, [12], 1977, pp. 913–971.

[60] J.-E. Fenstad, R. O. Gandy, and G. E. Sacks (editors), **Generalized Recursion Theory II**, Studies in Logic and the Foundations of Mathematics, North-Holland, 1978.

[61] J.-E. Fenstad and P. G. Hinman (editors), **Generalized Recursion Theory**, Studies in Logic and the Foundations of Mathematics, North-Holland, 1974.

[62] HARVEY FRIEDMAN, **Subsystems of Set Theory and Analysis**, Ph.D. thesis, Massachusetts Institute of Technology, 1967.

[63] ———, *Bar induction and Π_1^1-CA*, **The Journal of Symbolic Logic**, vol. 34 (1969), pp. 353–362.

[64] ———, *Iterated inductive definitions and Σ_2^1-AC*, [139], 1970, pp. 435–442.

[65] ———, *Determinateness in the low projective hierarchy*, **Fundamenta Mathematicae**, vol. 72 (1971), pp. 79–95.

[66] ———, *Higher set theory and mathematical practice*, **Annals of Mathematical Logic**, vol. 2 (1971), pp. 326–357.

[67] ———, *Countable models of set theories*, [180], 1973, pp. 539–573.

[68] ———, *Some systems of second order arithmetic and their use*, **Proceedings of the International Congress of Mathematicians, Vancouver 1974**, vol. 1, Canadian Mathematical Congress, 1975, pp. 235–242.

[69] ———, *Systems of second order arithmetic with restricted induction, I, II* (abstracts), **The Journal of Symbolic Logic**, vol. 41 (1976), pp. 557–559.

[70] ———, *Uniformly defined descending sequences of degrees*, **The Journal of Symbolic Logic**, vol. 41 (1976), pp. 363–367.

[71] ———, *On the necessary use of abstract set theory*, **Advances in Mathematics**, vol. 41 (1981), pp. 209–280.

[72] ———, *Unary Borel functions and second order arithmetic*, **Advances in Mathematics**, vol. 50 (1983), pp. 155–159.

[73] ———, *Finite functions and the necessary use of large cardinals*, **Annals of Mathematics**, vol. 148 (1998), pp. 803–893.

[74] HARVEY FRIEDMAN and JEFFRY L. HIRST, *Weak comparability of well orderings and reverse mathematics*, **Annals of Pure and Applied Logic**, vol. 47 (1990), pp. 11–29.

[75] ———, *Reverse mathematics of homeomorphic embeddings*, **Annals of Pure and Applied Logic**, vol. 54 (1991), pp. 229–253.

[76] HARVEY FRIEDMAN, KENNETH MCALOON, and STEPHEN G. SIMPSON, *A finite combinatorial principle which is equivalent to the 1-consistency of predicative analysis*, [186], 1982, pp. 197–230.

[77] HARVEY FRIEDMAN, NEIL ROBERTSON, and PAUL SEYMOUR, *The metamathematics of the graph minor theorem*, [227], 1987, pp. 229–261.

[78] HARVEY FRIEDMAN, STEPHEN G. SIMPSON, and RICK L. SMITH, *Countable algebra and set existence axioms*, **Annals of Pure and Applied Logic**, vol. 25 (1983), pp. 141–181.

[79] ———, *Addendum to "Countable algebra and set existence axioms"*, **Annals of Pure and Applied Logic**, vol. 27 (1985), pp. 319–320.

[80] HARVEY FRIEDMAN, STEPHEN G. SIMPSON, and XIAOKANG YU, *Periodic points in subsystems of second order arithmetic*, **Annals of Pure and Applied Logic**, vol. 62 (1993), pp. 51–64.

[81] SY D. FRIEDMAN, *HC of an admissible set*, **The Journal of Symbolic Logic**, vol. 44 (1979), pp. 95–102.

[82] A. FRÖHLICH and J. C. SHEPHERDSON, *Effective procedures in field theory*, **Transactions of the Royal Society of London**, vol. 248 (1956), pp. 407–432.

[83] LASZLO FUCHS, **Infinite Abelian Groups**, Pure and Applied Mathematics, Academic Press, 1970–1973, Volume I and Volume II.

[84] HILLEL FURSTENBERG, **Recurrence in Combinatorial Number Theory and Ergodic Theory**, Princeton University Press, 1981.

[85] HILLEL FURSTENBERG and YITZHAK KATZNELSON, *An ergodic Sze-merédi theorem for IP-systems and combinatorial theory*, **Journal d'Analyse Mathèmatique**, vol. 45 (1985), pp. 117–168.

[86] ———, *Idempotents in compact semigroups and Ramsey theory*, **Israel Journal of Mathematics**, vol. 68 (1989), pp. 64–119.

[87] ———, *A density version of the Hales-Jewett theorem*, **Journal d'Analyse Mathèmatique**, vol. 57 (1991), pp. 64–119.

[88] ROBIN O. GANDY, *Proof of Mostowski's conjecture*, **Bulletin de l'Academie Polonaise des Sciences, Série des Sciences Mathématiques, Astronomiques et Physiques**, vol. 8 (1960), pp. 571–575.

[89] ROBIN O. GANDY, GEORG KREISEL, and WILLIAM W. TAIT, *Set existence*, **Bulletin de l'Academie Polonaise des Sciences, Série des Sciences Mathématiques, Astronomiques et Physiques**, vol. 8 (1960), pp. 577–582.

[90] JEAN-YVES GIRARD, **Proof Theory and Logical Complexity**, Bibliopolis, 1987.

[91] MARIAGNESE GIUSTO, **Topology, Analysis and Reverse Mathematics**, Ph.D. thesis, Università di Torino, 1998.

[92] MARIAGNESE GIUSTO and ALBERTO MARCONE, *Lebesgue numbers and Atsuji spaces in subsystems of second-order arithmetic*, **Archive for Mathematical Logic**, vol. 37 (1998), pp. 343–362.

[93] MARIAGNESE GIUSTO and STEPHEN G. SIMPSON, *Located sets and reverse mathematics*, **The Journal of Symbolic Logic**, vol. 65 (2000), pp. 1451–1480.

[94] KURT GÖDEL, *Über formal unentscheidbare Sätze der Principia Mathematica und verwandter Systeme, I*, **Monatshefte für Mathematik und Physik**, vol. 38 (1931), pp. 173–198.

[95] ———, *The consistency of the axiom of choice and the generalized continuum hypothesis*, **Proceedings of the National Academy of Sciences, U.S.A.**, vol. 24 (1938), pp. 556–557.

[96] ———, *Consistency-proof for the generalized continuum hypothesis*, **Proceedings of the National Academy of Sciences, U.S.A.**, vol. 25 (1939), pp. 220–224.

[97] ———, **The Consistency of the Continuum Hypothesis**, Annals of Mathematics Studies, no. 3, Princeton University Press, 1940/1966.

[98] RONALD L. GRAHAM, BRUCE L. ROTHSCHILD, and JOEL H. SPENCER, **Ramsey Theory**, 2nd ed., Wiley, New York, 1990.

[99] A. Grzegorczyk et al.(editors), **Infinitistic Methods**, Pergamon, 1961.

[100] PETR HÁJEK and PAVEL PUDLÁK, **Metamathematics of First-Order Arithmetic**, Perspectives in Mathematical Logic, Springer-Verlag, 1993.

[101] CHRISTOPHER S. HARDIN and DANIEL J. VELLEMAN, *The mean value theorem in second order arithmetic*, **The Journal of Symbolic Logic**, vol. 66 (2001), pp. 1353–1358.

[102] L. A. Harrington, M. Morley, A. Scedrov, and S. G. Simpson (editors), *Harvey Friedman's Research on the Foundations of Mathematics*, Studies in Logic and the Foundations of Mathematics, North-Holland, 1985.

[103] LEO HARRINGTON, *A powerless proof of a theorem of Silver*, 8 pages, unpublished, November 1976.

[104] ——— , *Analytic determinacy and $0^\#$*, **The Journal of Symbolic Logic**, vol. 43 (1978), pp. 685–693.

[105] LEO HARRINGTON, DAVID MARKER, and SAHARON SHELAH, *Borel orderings*, **Transactions of the American Mathematical Society**, vol. 310 (1988), pp. 293–302.

[106] JOSEPH HARRISON, *Recursive pseudowellorderings*, **Transactions of the American Mathematical Society**, vol. 131 (1968), pp. 526–543.

[107] KOSTAS HATZIKIRIAKOU, *Algebraic disguises of Σ_1^0 induction*, **Archive for Mathematical Logic**, vol. 29 (1989), pp. 47–51.

[108] ——— , **Commutative Algebra in Subsystems of Second Order Arithmetic**, Ph.D. thesis, Pennsylvania State University, 1989.

[109] ——— , *Minimal prime ideals and arithmetical comprehension*, **The Journal of Symbolic Logic**, vol. 56 (1991), pp. 67–70.

[110] ——— , *A note on ordinal numbers and rings of formal power series*, **Archive for Mathematical Logic**, vol. 33 (1994), pp. 261–263.

[111] ——— , *WKL$_0$ and Stone's separation theorem for convex sets*, **Annals of Pure and Applied Logic**, vol. 77 (1996), pp. 245–249.

[112] KOSTAS HATZIKIRIAKOU and STEPHEN G. SIMPSON, *Countable valued fields in weak subsystems of second order arithmetic*, **Annals of Pure and Applied Logic**, vol. 41 (1989), pp. 27–32.

[113] ——— , *WKL$_0$ and orderings of countable Abelian groups*, [224], 1990, pp. 177–180.

[114] DAVID HILBERT, *On the infinite*, [272], 1967, Translated by S. Bauer-Mengelberg, pp. 367–392.

[115] DAVID HILBERT and PAUL BERNAYS, **Grundlagen der Mathematik**, 2nd ed., Grundlehren der Mathematischen Wissenschaften, Springer-Verlag, 1968–1970, Volume I and Volume II.

[116] NEIL HINDMAN and DONA STRAUSS, **Algebra in the Stone-Cech Compactification**, Walter de Gruyter, Berlin, 1998.

[117] JEFFRY L. HIRST, **Combinatorics in Subsystems of Second Order Arithmetic**, Ph.D. thesis, Pennsylvania State University, 1987.

[118] ——— , *Marriage problems and reverse mathematics*, [224], 1990, pp. 181–196.

[119] ——— , *Derived sequences and reverse mathematics*, **Mathematical Logic Quarterly**, vol. 39 (1993), pp. 447–453.

[120] ——— , *Embeddings of countable closed sets and reverse mathematics*, **Archive for Mathematical Logic**, vol. 32 (1993), pp. 443–449.

[121] ———, *Reverse mathematics and ordinal exponentiation*, **Annals of Pure and Applied Logic**, vol. 66 (1994), pp. 1–18.

[122] JEFFRY L. HIRST, *Reverse mathematics and ordinal multiplication*, **Mathematical Logic Quarterly**, vol. 44 (1998), pp. 459–464.

[123] W. Hodges, M. Hyland, C. Steinhorn, and J. Truss (editors), **Logic: From Foundations to Applications, Keele 1993**, Oxford Science Publications, Oxford University Press, 1996.

[124] MICHAEL HOLZ, KLAUS-PETER PODEWSKI, and KARSTEN STEFFENS, **Injective Choice Functions**, Lecture Notes in Mathematics, no. 1238, Springer-Verlag, 1987.

[125] TAMARA L. HUMMEL, *Effective versions of Ramsey's theorem*: *avoiding the cone above* $0'$, **The Journal of Symbolic Logic**, vol. 59 (1994), pp. 1301–1305.

[126] A. JAMES HUMPHREYS, **On the Necessary Use of Strong Set Existence Axioms in Analysis and Functional Analysis**, Ph.D. thesis, Pennsylvania State University, 1996.

[127] A. JAMES HUMPHREYS and STEPHEN G. SIMPSON, *Separable Banach space theory needs strong set existence axioms*, **Transactions of the American Mathematical Society**, vol. 348 (1996), pp. 4231–4255.

[128] ———, *Separation and weak König's lemma*, **The Journal of Symbolic Logic**, vol. 64 (1999), pp. 268–278.

[129] GERHARD JÄGER and THOMAS STRAHM, *Bar induction and ω-model reflection*, **Annals of Pure and Applied Logic**, vol. 97 (1999), pp. 221–230.

[130] THOMAS JECH, **Set Theory**, Academic Press, 1978.

[131] RONALD B. JENSEN, *The fine structure of the constructible hierarchy*, **Annals of Mathematical Logic**, vol. 4 (1972), pp. 229–308.

[132] RONALD B. JENSEN and CAROL KARP, *Primitive recursive set functions*, [216], 1971, pp. 143–167.

[133] CARL G. JOCKUSCH, JR., *Ramsey's theorem and recursion theory*, **The Journal of Symbolic Logic**, vol. 37 (1972), pp. 268–280.

[134] CARL G. JOCKUSCH, JR. and ROBERT I. SOARE, Π_1^0 *classes and degrees of theories*, **Transactions of the American Mathematical Society**, vol. 173 (1972), pp. 35–56.

[135] DIETER JUNGNICKEL, **Transversaltheorie**, Mathematik und ihre Anwendungen in Physik und Technik, Akademische Verlagsgesellschaft, Leipzig, 1982.

[136] IRVING KAPLANSKY, **Infinite Abelian Groups**, revised ed., University of Michigan Press, 1969.

[137] RICHARD KAYE, **Models of Peano Arithmetic**, Oxford University Press, 1991.

[138] ALEXANDER S. KECHRIS, **Classical Descriptive Set Theory**, Graduate Texts in Mathematics, Springer-Verlag, 1985.

[139] A. Kino, J. Myhill, and R. E. Vesley (editors), *Intuitionism and Proof Theory*, Studies in Logic and the Foundations of Mathematics, North-Holland, 1970.

[140] LAURIE A. S. KIRBY and JEFF B. PARIS, *Initial segments of models of Peano's axioms*, [159], 1977, pp. 211–226.

[141] ———, Σ_n *collection schemas in arithmetic*, [167], 1978, pp. 199–209.

[142] STEPHEN C. KLEENE, *Introduction to Metamathematics*, Van Nostrand, 1952.

[143] ———, *Arithmetical predicates and function quantifiers*, **Transactions of the American Mathematical Society**, vol. 79 (1955), pp. 312–340.

[144] ———, *Hierarchies of number-theoretic predicates*, **Bulletin of the American Mathematical Society**, vol. 61 (1955), pp. 193–213.

[145] ———, *Quantification of number-theoretic functions*, **Compositio Mathematica**, vol. 14 (1959), pp. 23–40.

[146] MOTOKITI KONDO, *Sur l'uniformisation des complémentaires analytiques et les ensembles projectifs de la seconde classe*, **Japanese Journal of Mathematics**, vol. 15 (1938), pp. 197–230.

[147] DÉNES KÖNIG, *Über eine Schlußweise aus dem Endlichen ins Unendliche*, **Acta Litterarum ac Scientarum Ser. Sci. Math. Szeged**, vol. 3 (1927), pp. 121–130.

[148] ———, *Theorie der Endlichen und Unendlichen Graphen*, Akademische Verlagsgesellschaft, Leipzig, 1936, Reprinted by Chelsea, New York, 1950.

[149] GEORG KREISEL, *Analysis of the Cantor/Bendixson theorem by means of the analytic hierarchy*, **Bulletin de L'Académie Polonaise des Sciences**, vol. 7 (1959), pp. 621–626.

[150] ———, *The axiom of choice and the class of hyperarithmetic functions*, **Indagationes Mathematicae**, vol. 24 (1962), pp. 307–319.

[151] ———, *A survey of proof theory*, **The Journal of Symbolic Logic**, vol. 33 (1968), pp. 321–388.

[152] GEORG KREISEL and J. L. KRIVINE, *Elements of Mathematical Logic*, Studies in Logic and the Foundations of Mathematics, North-Holland, 1967.

[153] IGOR KRIZ, *Proving a witness lemma in better-quasiordering theory*: *the method of extensions*, **Mathematical Proceedings of the Cambridge Philosophical Society**, vol. 106 (1989), pp. 253–262.

[154] ———, *Well-quasiordering finite trees with gap-condition*, **Annals of Mathematics**, vol. 130 (1989), pp. 215–226.

[155] ———, *The structure of infinite Friedman trees*, **Advances in Mathematics**, vol. 115 (1995), pp. 141–199.

[156] ANTONÍN KUČERA, *Measure*, Π_1^0 *classes and complete extensions of PA*, [50], 1985, pp. 245–259.

[157] ——— , *Randomness and generalizations of fixed point free functions*, [7], 1990, pp. 245–254.

[158] ——— , *On relative randomness*, **Annals of Pure and Applied Logic**, vol. 63 (1993), pp. 61–67.

[159] A. Lachlan, M. Srebrny, and A. Zarach (editors), **Set Theory and Hierarchy Theory, V**, Lecture Notes in Mathematics, no. 619, Springer-Verlag, 1977.

[160] RICHARD LAVER, *On Fraïssé's order type conjecture*, **Annals of Mathematics**, vol. 93 (1971), pp. 89–111.

[161] MANUEL LERMAN, **Degrees of Unsolvability**, Perspectives in Mathematical Logic, Springer-Verlag, 1983.

[162] AZRIEL LÉVY, **A Hierarchy of Formulas in Set Theory**, Memoirs, no. 57, American Mathematical Society, 1965.

[163] ——— , *Definability in axiomatic set theory, II*, [10], 1970, pp. 129–145.

[164] ALAIN LOUVEAU, **Ensembles Analytiques et Boréliens dans les Espaces Produits**, Astérisque, no. 78, Société Mathématique de France, 1980.

[165] ——— , *Two results on Borel orderings*, **The Journal of Symbolic Logic**, vol. 54 (1989), pp. 865–874.

[166] ALAIN LOUVEAU and JEAN SAINT-RAYMOND, *On the quasiordering of Borel linear orders under embeddability*, **The Journal of Symbolic Logic**, vol. 55 (1990), pp. 537–560.

[167] A. Macintyre, L. Pacholski, and J. Paris (editors), **Logic Colloquium '77**, Studies in Logic and the Foundations of Mathematics, North-Holland, 1978.

[168] ALFRED B. MANASTER and JOSEPH G. ROSENSTEIN, *Effective matchmaking (recursion theoretic aspects of a theorem of Philip Hall)*, **Proceedings of the London Mathematical Society**, vol. 25 (1972), pp. 615–654.

[169] ——— , *Effective matchmaking and k-chromatic graphs*, **Proceedings of the American Mathematical Society**, vol. 39 (1973), pp. 371–378.

[170] RICHARD MANSFIELD, *A footnote to a theorem of Solovay on recursive encodability*, [167], 1978, pp. 195–198.

[171] RICHARD MANSFIELD and GALEN WEITKAMP, **Recursive Aspects of Descriptive Set Theory**, Oxford Logic Guides, Oxford University Press, 1985.

[172] ALBERTO MARCONE, *Borel quasiorderings in subsystems of second-order arithmetic*, **Annals of Pure and Applied Logic**, vol. 54 (1991), pp. 265–291.

[173] ——— , **Foundations of BQO Theory and Subsystems of Second Order Arithmetic**, Ph.D. thesis, Pennsylvania State University, 1993.

[174] ——— , *Foundations of BQO theory*, **Transactions of the American Mathematical Society**, vol. 345 (1994), pp. 641–660.

[175] ———, *The set of better quasiorderings is Π_2^1-complete*, **Mathematical Logic Quarterly**, vol. 41 (1995), pp. 373–383.

[176] ———, *On the logical strength of Nash-Williams' theorem on transfinite sequences*, [123], 1996, pp. 327–351.

[177] DONALD A. MARTIN, *Borel determinacy*, **Annals of Mathematics**, vol. 102 (1975), pp. 363–371.

[178] ———, *A purely inductive proof of Borel determinacy*, [197], 1985, pp. 303–308.

[179] PER MARTIN-LØF, *The definition of random sequences*, **Information and Control**, vol. 9 (1966), pp. 602–619.

[180] A. R. D. Mathias and H. Rogers, Jr. (editors), **Cambridge Summer School in Mathematical Logic**, Lecture Notes in Mathematics, no. 337, Springer-Verlag, 1973.

[181] ADRIAN R. D. MATHIAS, *Happy families*, **Annals of Mathematical Logic**, vol. 12 (1977), pp. 59–111.

[182] KENNETH MCALOON, *Diagonal methods and strong cuts in models of arithmetic*, [167], 1978, pp. 171–181.

[183] KENNETH MCALOON and JEAN-PIERRE RESSAYRE, *Les methodes de Kirby-Paris et la théorie des ensembles*, [18], 1981, pp. 154–184.

[184] Richard McKeon (editor), **The Basic Works of Aristotle**, Random House, 1941.

[185] ELLIOTT MENDELSON, **Introduction to Mathematical Logic**, 3rd ed., Wadsworth, 1987.

[186] G. Metakides (editor), **Patras Logic Symposion**, Studies in Logic and the Foundations of Mathematics, North-Holland, 1982.

[187] GEORGE METAKIDES and ANIL NERODE, *Effective content of field theory*, **Annals of Mathematical Logic**, vol. 17 (1979), pp. 289–320.

[188] GEORGE METAKIDES, ANIL NERODE, and RICHARD A. SHORE, *On Bishop's Hahn-Banach theorem*, [209], 1985, pp. 85–91.

[189] RAY MINES, FRED RICHMAN, and WIM RUITENBURG, **A Course in Constructive Algebra**, Universitext, Springer-Verlag, 1988.

[190] LEONID MIRSKY, **Transversal Theory**, Academic Press, 1971.

[191] YIANNIS N. MOSCHOVAKIS, **Descriptive Set Theory**, Studies in Logic and the Foundations of Mathematics, North-Holland, 1980.

[192] ANDRZEJ MOSTOWSKI, *An undecidable arithmetical statement*, **Fundamenta Mathematicae**, vol. 36 (1949), pp. 143–164.

[193] ———, *Formal system of analysis based on an infinitistic rule of proof*, [99], 1961, pp. 141–166.

[194] MICHAEL E. MYTILINAIOS and THEODORE A. SLAMAN, *On a question of Brown and Simpson*, [41], 1996, pp. 205–218.

[195] CRISPIN ST. J. A. NASH-WILLIAMS, *On well-quasi-ordering infinite trees*, **Proceedings of the Cambridge Philosophical Society**, vol. 61 (1965), pp. 697–720.

[196] ———, *On better-quasi-ordering transfinite sequences*, **Proceedings of the Cambridge Philosophical Society**, vol. 64 (1968), pp. 273–290.

[197] A. Nerode and R. A. Shore (editors), **Recursion Theory**, Proceedings of Symposia in Pure Mathematics, American Mathematical Society, 1985.

[198] J. Nesetril and V. Rodl (editors), **Mathematics of Ramsey Theory**, Springer-Verlag, 1990.

[199] V. P. Orevkov, *A constructive mapping of the square onto itself displacing every constructive point*, **Soviet Math. Doklady**, vol. 4 (1963), pp. 1253–1256.

[200] Jeff B. Paris, *Independence results for Peano arithmetic*, **The Journal of Symbolic Logic**, vol. 43 (1978), pp. 725–731.

[201] Charles Parsons, *On a number-theoretic choice schema and its relation to induction*, [139], 1970, pp. 459–473.

[202] Klaus-Peter Podewski and Karsten Steffens, *Injective choice functions for countable families*, **Journal of Combinatorial Theory, Series B**, vol. 21 (1976), pp. 40–46.

[203] Marian B. Pour-El and J. Ian Richards, **Computability in Analysis and Physics**, Perspectives in Mathematical Logic, Springer-Verlag, 1988.

[204] Joseph E. Quinsey, **Applications of Kripke's Notion of Fulfilment**, Ph.D. thesis, Oxford University, 1980.

[205] Fred Richman, *The constructive theory of countable Abelian groups*, **Pacific Journal of Mathematics**, vol. 45 (1973), pp. 621–637.

[206] Wayne Richter and Peter Aczel, *Inductive definitions and reflecting properties of admissible ordinals*, [61], 1974, pp. 301–381.

[207] Raphael M. Robinson, *Recursion and double recursion*, **Bulletin of the American Mathematical Society**, vol. 54 (1948), pp. 987–993.

[208] Hartley Rogers, Jr., **Theory of Recursive Functions and Effective Computability**, McGraw-Hill, 1967.

[209] M. Rosenblatt (editor), **Errett Bishop, Reflections on Him and His Research**, Contemporary Mathematics, American Mathematical Society, 1985.

[210] Gerald E. Sacks, **Saturated Model Theory**, W. A. Benjamin, New York, 1972.

[211] ———, **Higher Recursion Theory**, Perspectives in Mathematical Logic, Springer-Verlag, 1990.

[212] D. Saracino and V. B. Weispfenning (editors), **Model Theory and Algebra**, Lecture Notes in Mathematics, no. 498, Springer-Verlag, 1975.

[213] James H. Schmerl, *Peano arithmetic and hyper-Ramsey logic*, **Transactions of the American Mathematical Society**, vol. 296 (1986), pp. 481–505.

[214] Kurt Schütte, **Proof Theory**, Grundlehren der Mathematischen Wissenschaften, no. 225, Springer-Verlag, 1977.

[215] KURT SCHÜTTE and STEPHEN G. SIMPSON, *Ein in der reinen Zahlentheorie unbeweisbarer Satz über endliche Folgen von natürlichen Zahlen*, **Archiv für Mathematische Logic und Grundlagenforschung**, vol. 25 (1985), pp. 75–89.

[216] D. S. Scott (editor), **Axiomatic Set Theory, Part 1**, Proceedings of Symposia in Pure Mathematics, vol. 13, American Mathematical Society, 1971.

[217] DANA S. SCOTT, *Algebras of sets binumerable in complete extensions of arithmetic*, [45], 1962, pp. 117–121.

[218] DANA S. SCOTT and STANLEY TENNENBAUM, *On the degrees of complete extensions of arithmetic* (abstract), **Notices of the American Mathematical Society**, vol. 7 (1960), pp. 242–243.

[219] NAOKI SHIOJI and KAZUYUKI TANAKA, *Fixed point theory in weak second-order arithmetic*, **Annals of Pure and Applied Logic**, vol. 47 (1990), pp. 167–188.

[220] JOSEPH R. SHOENFIELD, *Degrees of models*, **The Journal of Symbolic Logic**, vol. 25 (1960), pp. 233–237.

[221] ——, *The problem of predicativity*, [11], 1961, pp. 132–142.

[222] ——, **Mathematical Logic**, Addison–Wesley, 1967.

[223] RICHARD A. SHORE, *On the strength of Fraïssé's conjecture*, [42], 1993, pp. 782–813.

[224] W. Sieg (editor), **Logic and Computation**, Contemporary Mathematics, American Mathematical Society, 1990.

[225] WILFRIED SIEG, *Fragments of arithmetic*, **Annals of Pure and Applied Logic**, vol. 28 (1985), pp. 33–71.

[226] JACK H. SILVER, *Counting the number of equivalence classes of Borel and coanalytic equivalence relations*, **Annals of Mathematical Logic**, vol. 18 (1980), pp. 1–28.

[227] S. G. Simpson (editor), **Logic and Combinatorics**, Contemporary Mathematics, American Mathematical Society, 1987.

[228] S. G. Simpson (editor), **Reverse Mathematics 2001**, Lecture Notes in Logic, vol. 21, Association for Symbolic Logic, 2005.

[229] STEPHEN G. SIMPSON, *Choice schemata in second order arithmetic* (abstract), **Notices of the American Mathematical Society**, vol. 20 (1973), p. A449.

[230] ——, *Notes on subsystems of analysis*, Unpublished, typewritten, Berkeley, 38 pages, 1973.

[231] ——, *Degrees of unsolvability: a survey of results*, [12], 1977, pp. 631–652.

[232] ——, *Sets which do not have subsets of every higher degree*, **The Journal of Symbolic Logic**, vol. 43 (1978), pp. 135–138.

[233] ——, *Short course on admissible recursion theory*, [60], 1978, pp. 355–390.

[234] ——, *Set theoretic aspects of* ATR_0, [269], 1982, pp. 255–271.

[235] ——, Σ_1^1 and Π_1^1 transfinite induction, [269], 1982, pp. 239–253.

[236] ——, Which set existence axioms are needed to prove the Cauchy/Peano theorem for ordinary differential equations?, **The Journal of Symbolic Logic**, vol. 49 (1984), pp. 783–802.

[237] ——, BQO theory and Fraïssé's conjecture, [171], 1985, pp. 124–138.

[238] ——, Friedman's research on subsystems of second order arithmetic, [102], 1985, pp. 137–159.

[239] ——, Nichtbeweisbarkeit von gewissen kombinatorischen Eigenschaften endlicher Bäume, **Archiv für mathematische Logik und Grundlagenforschung**, vol. 25 (1985), pp. 45–65.

[240] ——, Nonprovability of certain combinatorial properties of finite trees, [102], 1985, pp. 87–117.

[241] ——, Reverse mathematics, [197], 1985, pp. 461–471.

[242] ——, Recursion-theoretic aspects of the dual Ramsey theorem, [50], 1986, pp. 356–371.

[243] ——, Subsystems of Z_2 and reverse mathematics, [261], 1987, pp. 432–446.

[244] ——, Unprovable theorems and fast growing functions, [227], 1987, pp. 359–394.

[245] ——, Ordinal numbers and the Hilbert basis theorem, **The Journal of Symbolic Logic**, vol. 53 (1988), pp. 961–974.

[246] ——, Partial realizations of Hilbert's program, **The Journal of Symbolic Logic**, vol. 53 (1988), pp. 349–363.

[247] ——, On the strength of König's duality theorem for countable bipartite graphs, **The Journal of Symbolic Logic**, vol. 59 (1994), pp. 113–123.

[248] ——, Finite and countable additivity, Preprint, 8 pages, November 1996.

[249] ——, **Subsystems of Second Order Arithmetic**, Perspectives in Mathematical Logic, Springer-Verlag, 1999.

[250] STEPHEN G. SIMPSON and RICK L. SMITH, Factorization of polynomials and Σ_1^0 induction, **Annals of Pure and Applied Logic**, vol. 31 (1986), pp. 289–306.

[251] DAVID REED SOLOMON, **Reverse Mathematics and Ordered Groups**, Ph.D. thesis, Cornell University, 1998.

[252] ROBERT M. SOLOVAY, Hyperarithmetically encodable sets, **Transactions of the American Mathematical Society**, vol. 239 (1988), pp. 99–122.

[253] CLIFFORD SPECTOR, Recursive well orderings, **The Journal of Symbolic Logic**, vol. 20 (1955), pp. 151–163.

[254] ——, Hyperarithmetic quantifiers, **Fundamenta Mathematicae**, vol. 48 (1959), pp. 313–320.

[255] JOHN R. STEEL, Descending sequences of degrees, **The Journal of Symbolic Logic**, vol. 40 (1975), pp. 59–61.

[256] ――――, *Determinateness and Subsystems of Analysis*, Ph.D. thesis, University of California at Berkeley, 1976.

[257] ――――, *Forcing with tagged trees*, **Annals of Mathematical Logic**, vol. 15 (1978), pp. 55–74.

[258] YOSHINDO SUZUKI, *A complete classification of the Δ_2^1 functions*, **Bulletin of the American Mathematical Society**, vol. 70 (1964), pp. 246–253.

[259] WILLIAM W. TAIT, *Finitism*, **Journal of Philosophy**, vol. 78 (1981), pp. 524–546.

[260] GAISI TAKEUTI, **Two Applications of Logic to Mathematics**, Princeton University Press, 1978.

[261] ――――, **Proof Theory**, 2nd ed., Studies in Logic and Foundations of Mathematics, Elsevier, 1987.

[262] KAZUYUKI TANAKA, *The Galvin-Prikry theorem and set existence axioms*, **Annals of Pure and Applied Logic**, vol. 42 (1989), pp. 81–104.

[263] ――――, *Weak axioms of determinacy and subsystems of analysis, I: Δ_2^0 games*, **Zeitschrift für Mathematische Logik und Grundlagen der Mathematik**, vol. 36 (1990), pp. 481–491.

[264] ――――, *Weak axioms of determinacy and subsystems of analysis, II: Σ_2^0 games*, **Annals of Pure and Applied Logic**, vol. 52 (1991), pp. 181–193.

[265] KAZUYUKI TANAKA and TAKESHI YAMAZAKI, *A non-standard construction of Haar measure and* WKL_0, **The Journal of Symbolic Logic**, vol. 65 (2000), pp. 173–186.

[266] ALFRED TARSKI, ANDRZEJ MOSTOWSKI, and RAPHAEL M. ROBINSON, **Undecidable Theories**, Studies in Logic and the Foundations of Mathematics, North-Holland, 1953.

[267] CHARLES B. TOMPKINS, *Sperner's lemma and some extensions*, [16], 1964, pp. 416–455.

[268] ANNE S. TROELSTRA and DIRK VAN DALEN, **Constructivism in Mathematics, an Introduction**, Studies in Logic and the Foundations of Mathematics, Elsevier, 1988, Volume I and Volume II.

[269] D. van Dalen, D. Lascar, and T. J. Smiley (editors), **Logic Colloquium '80**, Studies in Logic and the Foundations of Mathematics, North-Holland, 1982.

[270] B. L. VAN DER WAERDEN, **Modern Algebra**, revised English ed., Ungar, New York, 1953, Volume I and Volume II.

[271] FONS VAN ENGELEN, ARNOLD W. MILLER, and JOHN STEEL, *Rigid Borel sets and better quasi-order theory*, [227], 1987, pp. 199–222.

[272] J. van Heijenoort (editor), **From Frege to Gödel: A Source Book in Mathematical Logic, 1879–1931**, Harvard University Press, 1967.

[273] ROBERT VAN WESEP, **Subsystems of Second Order Arithmetic, and Descriptive Set Theory under the Axiom of Determinateness**, Ph.D. thesis, University of California at Berkeley, 1978.

[274] HERMANN WEYL, *Das Kontinuum: Kritische Untersuchungen über die Grundlagen der Analysis*, Veit, Leipzig, 1918, Reprinted in: H. Weyl, E. Landau, and B. Riemann, Das Kontinuum und andere Monographien, Chelsea, 1960, 1973.

[275] XIAOKANG YU, *Measure Theory in Weak Subsystems of Second Order Arithmetic*, Ph.D. thesis, Pennsylvania State University, 1987.

[276] ——, *Radon-Nikodym theorem is equivalent to arithmetical comprehension*, [224], 1990, pp. 289–297.

[277] ——, *Riesz representation theorem, Borel measures, and subsystems of second order arithmetic*, **Annals of Pure and Applied Logic**, vol. 59 (1993), pp. 65–78.

[278] ——, *Lebesgue convergence theorems and reverse mathematics*, **Mathematical Logic Quarterly**, vol. 40 (1994), pp. 1–13.

[279] ——, *A study of singular points and supports of measures in reverse mathematics*, **Annals of Pure and Applied Logic**, vol. 79 (1996), pp. 211–219.

[280] XIAOKANG YU and STEPHEN G. SIMPSON, *Measure theory and weak König's lemma*, **Archive for Mathematical Logic**, vol. 30 (1990), pp. 171–180.

[281] P. ZAHN, *Ein konstruktiver Weg zur Maßtheorie und Funktionalanalysis*, Wissenschaftliche Buchgesellschaft, 1978.

[282] OSCAR ZARISKI and PIERRE SAMUEL, *Commutative Algebra*, Van Nostrand, 1958–1960, Volumes I and II. Reprinted 1975–1976 by Springer-Verlag.

INDEX

425

For EU product safety concerns, contact us at Calle de José Abascal, 56–1°, 28003 Madrid, Spain or eugpsr@cambridge.org.

www.ingramcontent.com/pod-product-compliance
Ingram Content Group UK Ltd.
Pitfield, Milton Keynes, MK11 3LW, UK
UKHW040951090126
466816UK00019B/357

* 9 7 8 0 5 2 1 1 5 0 1 4 9 *